J.A. Schafer H.H. Ussing
P. Kristensen G.H. Giebisch (Eds.)

Membrane Transport in Biology
Volume 5

With 83 Figures and 15 Tables

Springer-Verlag
Berlin Heidelberg New York
London Paris Tokyo Hong Kong
Barcelona Budapest

Editors

James A. Schafer, Ph. D.
Departments of Physiology and Biophysics, and Medicine, University of Alabama at Birmingham, Birmingham, Alabama 35294 USA

Poul Christensen, Ph. D.
Institute of Biological Chemistry A, August Krogh Institute, 13, Universitetsparken, 2100 Copenhagen, Denmark

Hans H. Ussing, Dr. phil., med. and scient.
University of Copenhagen, Institute of Biological Chemistry A, 13, Universitetsparken, 2100 Copenhagen, Denmark

Gerhard H. Giebisch, M. D.
Department of Cellular and Molecular Physiology, Yale University School of Medicine, 333 Cedar Street, New Haven, Connecticut 06510 USA

ISBN-13:978-3-642-76985-6 e-ISBN-13:978-3-642-76983-2
DOI: 10.1007/978-3-642-76983-2

Library of Congress Cataloging-in-Publication Data (Revised for volume 5) Membrane transport in biology. Vol. 5 edited by J. A. Schafer ... [et al.]. Includes bibliographies and indexes. Contents: v. 1. Concepts and models.—v. 2. Transport across single biological membranes.—[etc.—v. 5. [without special title].
1. Biological transport. 2. Membranes (Biology) I. Giebisch, Gerhard H. II. Tosteson, D. C., 1925-. III. Ussing, H. H. (Hans H.), 1911-IV. Schafer, J. A. (James A.), 1941-. [DNLM: 1. Biological transport. 2. Cell membrane—Physiology QH509 M533] QH509. M44 574.8'75 78-17669
ISBN-13:978-3-642-76985-6 (v.1)

Softcover reprint of the hardcover 1st edition 1992

Typesetting: Macmillan India Ltd., Bangalore-25;

27/3020-5 4 3 2 1 0-Printed on acid-free paper

Preface

Well over one decade has passed since the appearance of the original four volumes of *Membrane Transport in Biology*. Since the publication of the last volume there have been spectacular advances in this field. These advances have been in part the result of the application of exciting new methodologies, and in part the result of new insights into the regulation and integration of transport processes. This volume, as well as a sixth volume, which is in preparation, are intended to cover key areas in which the development has been particularly striking.

For many years the trend in studies of membrane transport had been that of increasing specialization with regard to the transporter of interest and of the cell or tissue studied. This trend was supported by the enormous number of publications directed at understanding the cellular physiology of specific organ systems and tissues, and also by the fact that different tissues often seemed to react so differently to the same conditions that mechanisms unique to each appear to be at play. One of the happy developments in recent years has been the realization that this apparent disparity of behaviors in different tissues is based on varying combinations of a limited number of transport mechanisms, all mediated by the same or similar proteins. Some of these transport proteins have already been isolated and analyzed with respect to amino acid sequence whereas others are just entering this phase.

One of the most important transporters whose amino acid sequence is now known is the ubiquitous Na,K-ATPase. In the first chapter Dr. Jørgensen considers the subunit structure of the Na,K-ATPase and its genetic evolution. He relates this information to the organization of the transporter in the membrane and to the characteristics of the cation and other ligand binding sites. As discussed by Dr. Jørgensen, this information has led to exciting, new insights into the structural basis for the mechanism of energy transduction to cation transport.

Dr. Lever describes the identification of the Na^+/glucose symporter in intestinal and renal epithelia, and the use of monoclonal antibody approaches which has led to their purification and functional reconstitution. These approaches have provided important, new information concerning the conformational changes involved during transport. Dr. Lever continues in her chapter with a review of the new insights that have been gained by expression cloning of this symporter and the sequence homologies that exist among the Na^+/glucose symporters in different tissues, as well as with the Na^+/proline symporter. She concludes with a consideration of the regulation of gene expression for these symporters.

Over the past decade, investigators have obtained a wealth of new information about conductive ion channels, using the techniques of fluctuation or noise analysis and patch-clamping. In their chapter, Dr. Eaton and his coauthors review these experimental approaches and consider their relative advantages and limitations. They then review new findings on specific examples of these channels found in a variety of epithelia, including the amiloride-blockable Na^+ channel, K^+ channels, and Cl^- channels. They also discuss the characteristics and the possible role of nonselective cation channels, as well as calcium channels and those channels that may be activated by perturbations such as membrane stretch or osmolality changes.

Dr. Benos and his coauthors extend the discussion of amiloride-sensitive Na^+ transport by reviewing the characteristics of the Na^+/H^+ antiporter, and electrogenic Na^+/Ca^{2+} antiporters, as well as the conductive Na^+ channel. They also discuss recent information from the cloning and sequencing of the Na^+/H^+ antiporter and the challenging new avenues it provides for understanding the mechanism of operation, as well as the regulation of this important, ubiquitous molecule.

The ion exchanger associated with band 3 from erythrocytes has long been recognized to be an important transporter that allows acid-base equilibration in this cell. Dr. Gunn, in his chapter, considers how this transporter is involved in anion transport in a number of cells and tissues. He discusses the structure of this protein, as it has been determined by cloning and sequencing, as well as the heterogeneity of the gene expression for this protein in various cells and tissues.

One of the most exciting new areas of intense exploration in membrane transport has been the regulation of intracellular calcium concentration. It has become abundantly apparent that Ca^{2+} is an important intracellular signaling agent, and that the regulation of its intracellular concentration is essential not only for cation homeostasis but also for the underlying mechanisms of signal transduction. In his chapter, Dr. Bonventre considers the various calcium transport molecules that are present, both in the plasma membrane and in organelle membranes which determine the intracellular calcium concentration. He details the known transport characteristics of these mechanisms and discusses our current understanding of how they may be regulated by such factors as extracellular ion composition, transmembrane voltage, intracellular second messengers, and various autacoids.

In the final chapter, Drs. Hoffmann and Ussing consider the integration of many of the transport mechanisms discussed in the preceding chapters into the complex process of cell volume regulation. It is now known that cells can regulate their volume in spite of large changes in extracellular tonicity. In epithelia, this regulatory process is essential in maintaining a normal cell volume, in spite of drastic changes in solute entry produced by the underlying variability in the supply of substrate that is absorbed or secreted by the epithelium. Drs. Hoffman and Ussing discuss the transport mechanisms tha are involved in the response to osmotic or solute load stresses that would normally lead to either swelling or shrinkage of the cells. These mechanisms involve the fine regulation of the several interacting transport systems involved, and appears to be common among many different cells and tissues.

It is imperative that all investigators working with transport problems that may be specific to the cell or tissue of their special interest also be aware of the commonality among these various transporters that provide the basic building blocks of membrane function. It is the hope of the editors that this volume will be of assistance in unraveling the intricate combination of transport mechanisms that are present in every type of cell.

January 1992

James A. Schafer
Hans U. Ussing
Poul Kristensen
Gerhard H. Giebisch

List of Authors

D.J. Benos
Department of Physiology and Biophysics, University of Alabama at Birmingham, Birmingham, Alabama 35294, USA

J.V. Bonventre
Medical Services, Massachusetts General Hospital, Department of Medicine, Harvard Medical School, Fruit Street, Boston, Massachusetts 02114, USA

D.C. Eaton
Department of Physiology, Emory University School of Medicine, Atlanta, Georgia 30322, USA

R.B. Gunn
Department of Physiology, Emory University School of Medicine, Atlanta, Georgia 30322, USA

E.K. Hoffmann
Institute of Biological Chemistry A, August Krogh Institute, 13 Universitetsparken, 2100 Copenhagen, Denmark

P.L. Jørgensen
August Krogh Institute, Copenhagen University, 2100 Copenhagen, Denmark

J.E. Lever
Department of Biochemistry and Molecular Biology, University of Texas Medical School, P.O. Box 20708, Houston, Texas 77025, USA

B.N. Ling
Emory University School of Medicine, Renal Division, Atlanta, Georgia 30322, USA

Y. Marunaka
Division of Respiratory Research, The Hospital for Sick Children, University of Toronto, Toronto, Ontario M5G 1X8, Canada

J.B. Smith
Department of Pharmacology, University of Alabama at Birmingham, Birmingham, Alabama 35294, USA

H.U. Ussing
Institute of Biological Chemistry A, August Krogh Institute,
13 Universitetsparken, 2100 Copenhagen, Denmark

D.G. Warnock
Departments of Physiology, Biophysics and Medicine, University of Alabama at Birmingham, Birmingham, Alabama 35294, USA

Contents

Na,K-ATPase, Structure and Transport Mechanism*

P. L. Jørgensen

Introduction

The Na,K pump is found in the surface membrane of most animal cells. In coupling the hydrolysis of ATP to active transport of 3 Na^+ out and 2 K^+ into the cell, the pump is electrogenic and necessary for maintaining the membrane potential of excitable cells. Primary active Na,K pumping is a key process for the active uptake of nutrients, salts and water and the regulation of fluid and electrolyte homeostasis in mammals. The pump maintains electrochemical gradients for Na^+ ($\Delta\mu$Na) for utilization in carrier mediated secondary active transport processes in kidney, intestine and other epithelia.

Advances in understanding the structure of the Na,K pump include purification from mammalian kidney [81, 82] and characterization of the pump protein both in membrane-bound form [80] and in detergent solution where the $\alpha\beta$-unit is the minimum functional unit [83, 102]. Crystals of the Na,K pump protein in the membrane have the $\alpha\beta$-unit as the minimum asymmetric unit [68]. Low resolution models of the overall structure of the Na,K pump molecule have been constructed on the basis of diffraction analysis of p1 and p12 crystals [69, 117, 131]. A wealth of structural information about the pump proteins is available after identification of mRNA and sequencing of cloned cDNA. The amino acid sequences of α-subunit with M_r 112 500 [70, 96, 98, 133 162, 166] and β-subunit with M_r 34 500 [20, 97, 123, 163] have been deduced from cDNA of mRNA from a variety of tissues in piscine and mammalian species. Three isoforms of the α-subunit have been identified at the mRNA level [70, 157, 162] and five genes of the α-subunit are identified in the human and rat genomes [134, 165]. A similar intensity of work in adjacent fields provided the sequences of slow and fast twitch Ca pump from sarcoplasmic reticulum [18, 108], two isoforms of the Ca pump of the endoplasmic reticulum or the plasma membrane [161, 182], and the H,K pump from stomach mucosa [164]. In addition, the gene sequences are available for K pumps [71, 169] and H pumps [1, 54, 159] from micro-organisms. As paradigm for interpretation of this information, the combined X-ray crystallographic structure at high resolution and amino acid sequences are

* Work in the author's laboratory is supported by the Danish Medical Research Council, Novo's Foundation, and the Carlsberg Foundation.

available for only one membrane protein complex, the photosynthetic reaction center in bacteria [37, 113].

The intermediary steps of the Na,K exchange reactions and their relationship to cation translocation have been examined in detail. The α-subunit can reversibly assume two conformations, both in the dephosphoform and in the phosphorylated state [55, 80, 84]. The underlying mechanism involves a conformational change in the protein that alternately presents cation binding sites at the cytoplasmic and extracellular sides of the membrane, but the mechanistic details of the reaction are unknown, and the residues involved in binding and translocation of Na^+ and K^+ have not been identified.

The purpose of this article is to analyze the organization of the αβ-unit of the Na,K pump in the membrane and the molecular mechanism of active Na,K transport in light of the new information about gene structure and sequences of nucleotides and amino acids. Attempts are made to assign residues involved in formation of ligand binding sites and E_1–E_2 transition to their position in the sequence and to discuss the relationship of conformational transitions in the protein to translocation of cations across the membrane. Comparison of structure and evolution of pump proteins coupling ATP hydrolysis to active transport of H^+, K^+, Na^+, and Ca^{2+} is important for developing working hypotheses for the molecular mechanism of active cation transport. This includes the organization of sites for nucleotide and phosphorylation and other common features in the structure of the catalytic domains of the cation pumps.

Purification of Na,K-ATPase

Membrane-Bound and Soluble Na,K-ATPase

The purified membrane bound Na,K-ATPase from outer medulla of mammalian kidney has been the preparation of choice in numerous experiments, particularly in studies of ultrastructure, protein conformation and ligand binding. The purification procedure, as shown in the flow diagram in Fig. 1, is straightforward and the preparations are relatively stable [80, 81]. A recently recognized, further advantage is that the preparation from outer medulla contains only the α1-isozyme of Na,K-ATPase, while most other preparations including that from rectal gland of dogfish contains at least two α-subunit isoforms. In preparations from kidney, the purity exceeds 90% with specific activities 36–48 μmol/min mg protein, while the purity in membrane bound preparations from salt glands of *Squalus acanthias* is at best 50% in the membrane bound state.

If the membrane-bound preparation from kidney is subjected to analytical ultracentrifugation within a few hours after solubilization in nonionic detergent, the data shows that the preparation consists predominantly [80%–85%] of

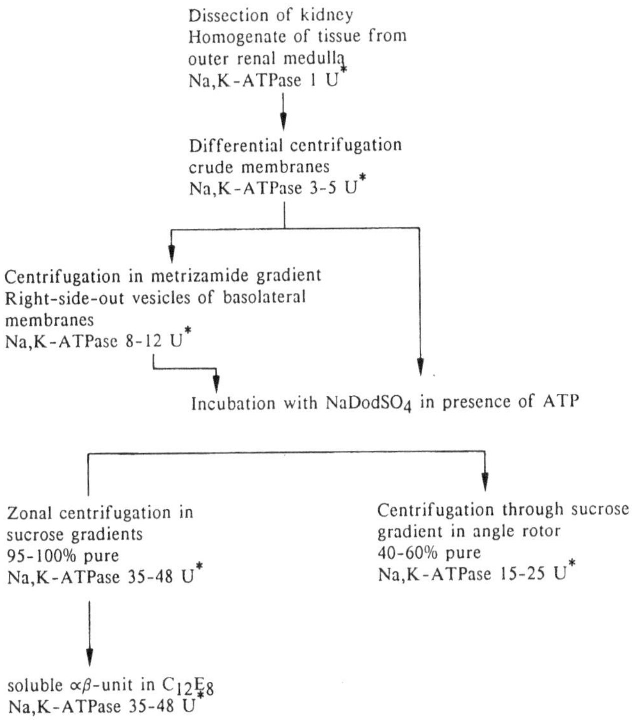

Fig. 1. Flow diagram of procedures for purification of Na,K-ATPase from outer renal medulla. The figures to the right indicate the specific activity of each preparation. *U*; μmol Pi/min mg protein at 37°C. In homoenate, crude membranes and vesicles, the activity was determined after demasking of latent activity by preincubation with sodium deoxycholate [81]

soluble $\alpha\beta$-units with M_r 143 000 [83]. After storage or prolonged chromatography the particles aggregate to $(\alpha\beta)_2$-units or higher oligomers, presumably by detergent-detergent association [80]. Soluble preparations from rectal gland of dogfish or eel electroplax must be fractionated by chromatography after solubilization in detergent. Artificial aggregation may therefore explain that they have a higher molecular weights, 280 000–380 000 and an oligomeric structure $(\alpha_2\beta_2)$ or $\alpha_2\beta_4$ (For references see [80]).

Properties of Purified Na,K-ATPase

In an ideal pure preparation of Na,K-ATPase from outer renal medulla, the α1-subunit forms 65%–70% of the total protein and the molar ratio of α to β is 1 : 1, corresponding to a mass ratio of about 3 : 1 [80]. Functionally the preparation is fully active in the sense that each $\alpha\beta$-unit binds ATP, P_i, cations and the inhibitors vanadate and ouabain. The molecular activity is close to a maximum value of 7000–8000 P_i/min. Problems related to determination of ligand binding and molecular activity have been reviewed [80, 102]. The highest reported binding capacities for ATP and phosphate are in the range of 5–6 nmol/mg protein and close to one molecule per $\alpha\beta$-unit [184], when fractions with exceptionally high specific activities of Na,K-ATPase are selected for assay.

For enzymatic or kinetic experiments, the enzyme preparation must be homogenous in the sense that all Na,K-ATPase molecules are uniform with respect to structure and function. The presence of partially damaged enzyme units and more than one isozyme of Na,K-ATPase may give rise to nonlinear behavior. The density of enzyme molecules in the membrane is another factor of potential importance for kinetic experiments. From negatively stained images similar to those shown in Fig. 2, the average density of protein in the membrane is estimated to be 12000 $\alpha\beta$-units/μm^2 [31]. This corresponds to a concentration of α-subunit in the lipid bilayer of about 7 mM or 0.5–1 g protein/ml of lipid phase [31]. These are conditions for supersaturation that allow crystallization of the protein units in the membrane. The extraordinary high concentration of sites in the membrane disks may result in apparent negative cooperativity in ligand

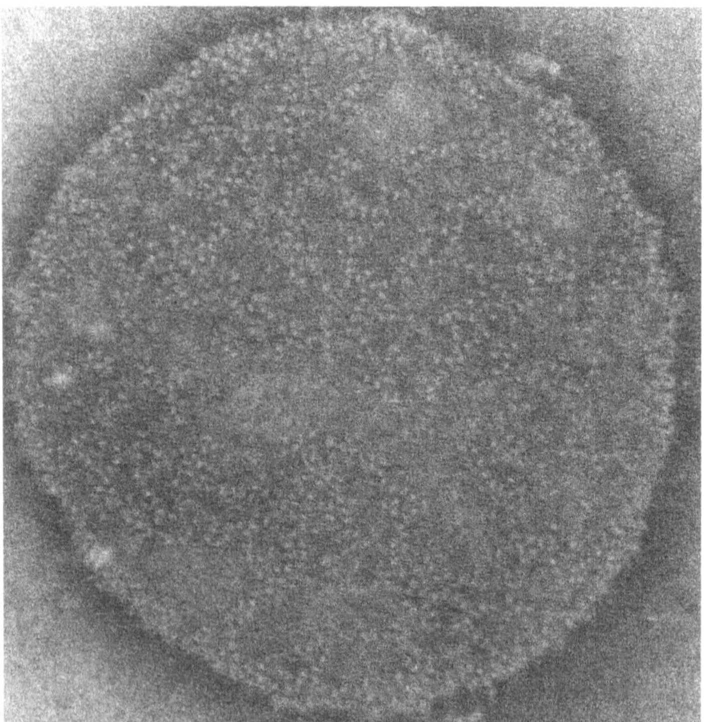

Fig. 2. Negative staining by phosphotungstic acid of Na,K-ATPase purified in membrane-bound form. The membrane surfaces are covered by particles arranged in clusters between smooth areas. (From [82]; procedure as described by Deguchi et al. [31])

binding experiments. The nonlinearity of the binding curves due to this phenomenon can be relieved upon solubilization of the $\alpha\beta$-unit in detergent solution (cf. [84]).

Subunit Structure of Na,K-ATPase

The problem of subunit structure of the Na,K pump is central to all considerations of the transport mechanism. If a single $\alpha\beta$-unit of Na,K-ATPase can carry out all catalytic reactions and bind, occlude, and translocate Na^+ and K^+, the implication is that the protein must be porous and form a cation transport pathway through the membrane. If oligomerization to $(\alpha\beta)_2$ units or higher oligomers is required, the possibility exists that the cation pathway may be formed in the interface between subunits.

The subunit structure of the pump has been studied both in membrane bound and soluble preparations of Na,K-ATPase. It is important to note the difference between the two conditions. In the tightly packed fragments of the membrane-bound Na,K-ATPase the difference in free energy required for self association of the $\alpha\beta$-units is several fold less than that required for association of protein molecules in solution. Even minor changes of intermolecular forces may therefore precipitate association of the membrane embedded particles that are not necessarily of functional importance for Na,K pumping.

Evidence for Quaternary Structure from Electron Microscopy and Crystal Analysis

The purified membrane bound Na,K-ATPase consists of disc shaped membrane fragments, 1000–3000 Å in diameter, with no tendency for vesicle formation. The densely packed protein particles with diameters of 30–50 Å represent $\alpha\beta$-units that can be visualized by negative staining with phosphotungstic acid (Fig. 2) or uranyl acetate. They are arranged in irregular clusters or stands and appear to be free to move in the plane of the membrane without formation of well defined oligomeric structures [31]. Freeze fracture analysis shows that the particles are asymmetrically arranged like in the basolateral membranes of the tubules cells. The particles of the fracture faces are larger with diameters near 90 Å and their density is lower (4000–6000 μ^2) than after negative staining, but these images are difficult to interprete in terms of the selfassociation state of the protein.

Crystal formation of the densely packed protein particles of the membrane-bound Na,K-ATPase is induced in the presence of vanadate or other ligands that stabilize the protein in the E_2 conformation [167]. The predominant crystal form, shown in the upper part of Fig. 3, has the two-sided plane group

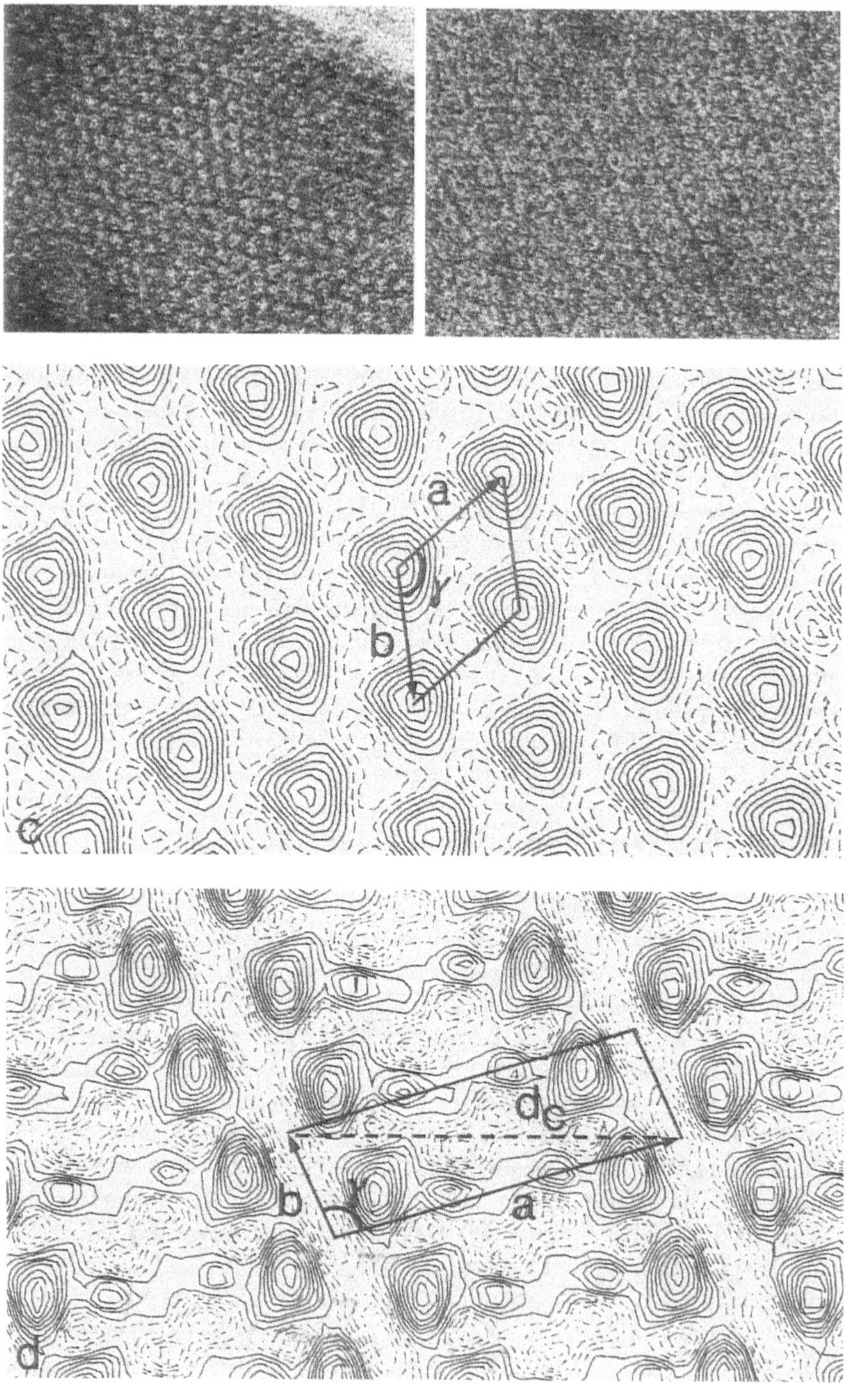
a
b
γ
c
d
d_c

symmetry, p1, and contains one protomeric $\alpha\beta$-unit per unit cell. Crystals with two sided plane group symmetry, p21, with two $\alpha\beta$-units occupying one unit cell, are transient and less frequent [68, 88], lower part of Fig. 3. The appearance of two crystal forms shows that the protein in the membrane exists in equilibrium between the protomeric $\alpha\beta$-unit and oligomeric $(\alpha\beta)_2$-forms. The high rate of crystal formation of the protein in vanadate solution shows that transition to the E_2 form reduces the difference in free energy required for self-association of the protein. These methods for crystallization have been very reproducible [117, 131, 193], and the procedure for incubation in vanadate also induced formation of crystalline arrays of Ca-ATPase in sarcoplasmic reticulum [41, 178] and in preparations of H,K-ATPase from stomach mucosa [144].

Three-Dimensional Models

Low-resolution models (20–30 Å) based on diffraction analysis of membrane crystals of Na,K-ATPase [69, 117, 131] and Ca-ATPase [22, 178] show that the cytoplasmic protrusions of the proteins are remarkably similar. A notable difference is a 10–20 Å protrusion on the extracellular surface of the model for Na,K-ATPase while the Ca-ATPase model has a smooth extracytoplasmic surface. Superposition of a model of β-subunit crystals formed in Mg solution shows that the β-subunit forms the extracellular protrusion [84]. This agrees with estimates of distances in fluorescence studies [105]. As illustrated in Fig. 4, the mass of the intramembranous portion [69] estimated from this $\alpha\beta$-unit model is 30%–40% of the total mass of the three-dimensional model of the $\alpha\beta$-unit. Similar estimates of the intramembrane portion of Ca-ATPase are in the range of 30%–50% [178].

The relatively low resolution (25 Å) of these models does not allow further assignment of structural detail. Assignment of the coordinates of individual residues requires preparation of crystals with diffractions to < 3 Å that are suitable for X-ray crystallography. Methods developed for the photosynthetic reaction center [37, 113] have not been directly applicable to other membrane proteins. Three-dimensional crystals consisting of sheets of protein arrays separated by lipid layers, were prepared from soluble Ca-ATPase in nonionic

Fig. 3. Crystalline arrays of Na,K-ATPase in the membrane with (*a*, *c*) a protomeric $\alpha\beta$-unit as minimum asymmetric unit in a p1 crystal or (*b*, *d*) with an oligomeric $(\alpha\beta)_2$-unit in the unit cell of a p21 crystal. The p1 crystal was formed after incubation of purified membrane-bound Na,K-ATPase in 0.25 m*M* sodium monovanadate, 1 m*M* $MgCl_2$ at 4°. For formation of the p21 crystal the purified membrane-bound Na,K-ATPase was incubated in 12.5 m*M* phosphate, 1 m*M* $MgCl_2$, and 10 m*M* Tris-HCl, pH 7.5 at 4°C. The membranes were negatively stained with uranyl acetate and micrographs were obtained at 235000 × magnification. For *c* and *d* images suitable for further analysis were densitometered at 20-μm intervals. Projection maps were calculated using the Fourier transform amplitudes and phases collected at the reciprocal lattice points. *Unbroken contour lines*, protein rich regions are drawn with *dashed lines*, negative stain regions. In the reconstructed images 1 mm corresponds to 2.8 Å. The unit cell dimensions are in *c*: a = 53 Å, b = 51 Å, γ = 120°; and in *d*: a = 135 Å, b = 44 Å, γ = 101°. (From [88])

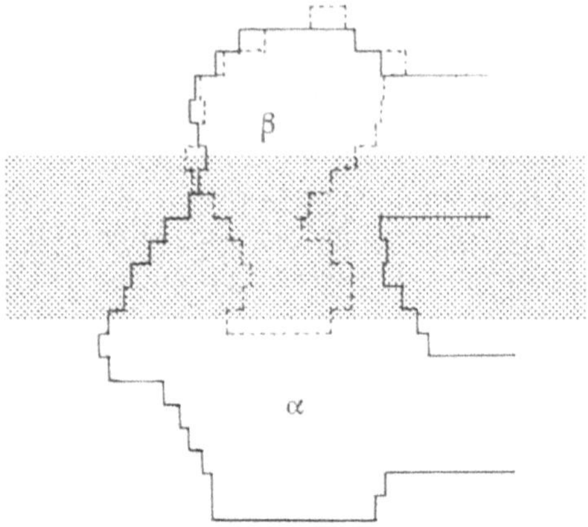

Fig. 4. Model of distribution of protein mass of $\alpha\beta$-unit between extra and intramembranous domains. *Shaded area*, position of the bilayer. (Redrawn from models of the $\alpha\beta$-unit in [69] and a model of a crystal of the β-subunit [131])

detergent [42]. A critical factor for formation of these crystals is stabilization of the soluble protein in an enzymatically active form for several months. They diffract to < 10 Å, but the crystals are fragile and so far unsuitable for X-ray analysis.

Molecular Weight Determination on Soluble Na,K-ATPase

After gradual refinement of techniques for solubilization and chromatography, reliable data defining the mass and subunit structure of cation pump proteins have become available [3, 11, 80, 110]. Molecular weights of the soluble enzymes are determined by sedimentation equilibrium (Fig. 5) and sedimentation velocity studies [5, 83] agree with those calculated from amino acid composition within $\pm$ 10% (147000 for $\alpha\beta$-unit of Na,K-ATPase). Sedimentation coefficients are 6–7 S for the protomeric $\alpha\beta$-unit of Na,K-ATPase. High-resolution gel chromatography combined with low angle laser light scattering gives molecular weights for soluble Na,K-ATPase [66, 67] that are in agreement with those from sedimentation equilibrium analysis [83].

Soluble Na,K-ATPase undergoes time dependent denaturation and aggregation as a function of temperature and cation composition of the medium [83]. High-resolution chromatography in TSK columns rapidly inactivates soluble Na,K-ATPase, but the addition of phosphatidylserine [67] allows assay of Na,K-ATPase activity during passage over the TSK column. This experiment confirms that protomeric $\alpha\beta$-units have Na,K-ATPase activity.

The decision whether the $\alpha\beta$-unit can catalyze ATP-driven Na,K transport requires demonstration that the protein remains monomeric during the transport reaction. In the absence of phase borders, studies on the soluble Na,K-ATPase are limited to ligand binding, intermediary reactions and conformational transitions in the protein. The dephosphoforms of soluble $\alpha\beta$-unit can undergo E_1–E_2 transitions and the two phosphoenzyme forms, E_1P and E_2P,

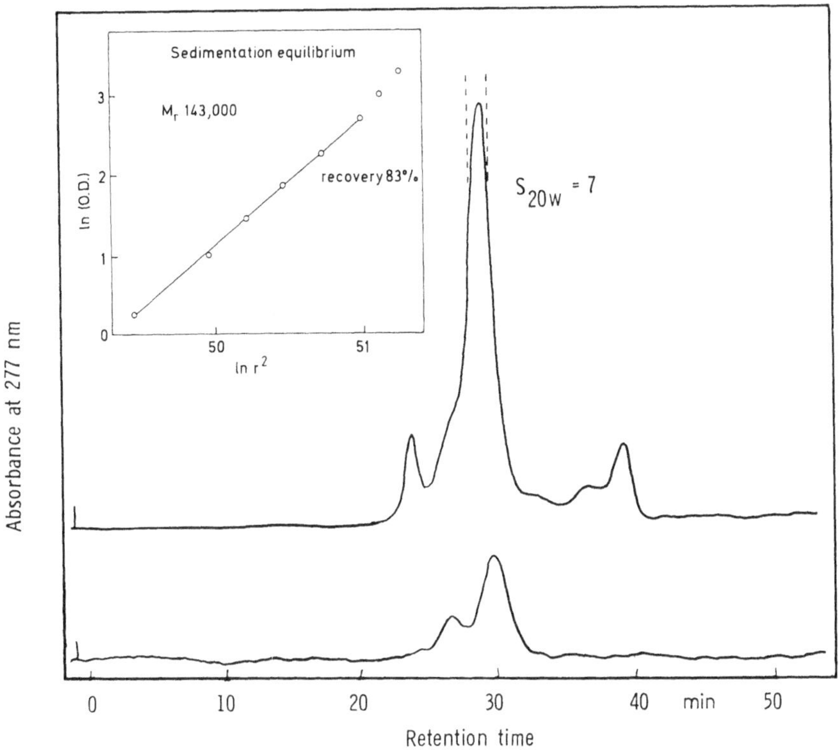

Fig. 5. Molecular weight of soluble Na,K-ATPase in $C_{12}E_8$. HPLC chromatography on TSK gel G 3000 SW column in solution containing $C_{12}E_8$. The protein in the large peak eluting at 28–29 min after charging soluble Na,K-ATPase on the column has a molecular weight 143 000 as determined by sedimentation equilibrium analysis in the analytical ultracentrifuge (*insert*). Analytical ultracentrifugation of soluble Na,K-ATPase in $C_{12}E_8$ after elution from a 7.5 × 600 mm TSK gel G 3000 SW column equilibrated with 5 mg/ml $C_{12}E_8$ in 150 m*M* potassium acetate, 25 m*M* Tris acetate, pH 7.0 and operated at 0.5 ml/min. The membrane bound Na,K-ATPase was solubilized at 5 mg protein/ml in 12.5 mg/ml $C_{12}E_8$ in the buffer above and a 100-μl aliquot was injected into the column. The fractions containing the peak of protein eluting at 28–29 min were divided in two aliquots. One was used as a sample for sedimentation equilibrium in a Beckman analytical ultracentrifuge. The other was stored for the at 20°C and rechromatographed after 16 h to produce *lower tracing*. *Inset* shows the sedimentation equilibrium data. Molecular weight was calculated from the linear part of the plot after correction for carbohydrate (0.15 mg/mg protein), and detergent (0.70 mg/mg protein). (From [83])

can also be demonstrated. The only reasonable possibility for attributing the Na,K-ATPase activity to oligomer forms of the protein is that ATP binding or formation of the E_1P form induces oligomerization of the $\alpha\beta$-unit [83]. This possibility has been examined in occlusion experiments demonstrating that the soluble $\alpha\beta$-unit can occlude 2 Rb^+ ions, and that it forms the phosphorylated E_1P with 2 or 3 Na^+ ions occluded [184]. The soluble $\alpha\beta$-unit can therefore assume all of the four major conformations that Na,K-ATPase must pass to carry out ion translocation in the E_1–E_2 reaction cycle [184]. These data support the notion that the $\alpha\beta$-unit can form a cation transport pathway and that this is the minimum functional unit of Na,K-ATPase for active Na,K transport. In the occluded state, ions are bound in a presumptive cavity in the protein and rendered unable to exchange freely with ions in the medium. This cavity is part of the transport pathway, and it is formed within the structure of the $\alpha\beta$-unit and not in the interface between subunits in an oligomeric structure.

Evolution of Cation Pump Proteins and Structure of Na, K Pump Genes

Evolution of Cation Pump Proteins

A comparison of Na,K, H,K, Ca, H, and K pumps shows that overall sequence homology is moderate, 17%–24% except for 63% between H,K-ATPase and Na,K-ATPase, while tertiary structure is remarkably similar with respect to localization of transmembrane segments relative to domains involved in ATP binding and phosphorylation. The homologies are most pronounced in segments connecting transmembrane helices (M4 and M5) with the phosphorylation site and the nucleotide binding area, respectively, while there is little or no homology between the amino acid sequences of N-termini and transmembrane helices in Na,K-ATPase, Ca-ATPase, H-ATPase, and K-ATPase.

Homologies between cation pumps and other ATP binding proteins (see "Conformations of the Nucleotide Binding Area") suggest that they have a common origin. The molecular basis for evolution of the family of cation pumps from a common ancestor may be gene duplication and drift or fusion and splicing of the gene sequences [39]. Comparison of sequences show that divergence of cation pump genes may have proceeded by recombination of exons involved in recognition, binding of cation and formation of transcellular pathways, in accordance with the needs for cation specificity. Exons coding for domains involved in nucleotide binding, phosphorylation and transduction of energy to the cation sites are ancient and the best conserved.

From the comparison of nucleotide and deduced amino acid sequences it appears that the K-ATPase of *Streptococcus faecalis* [169] is the most simple and primitive form of cation pump that may represent an ancestral form, but the properties of the protoorganism existing 3–4 billion years ago [112] are not known. The time span of evolution of isoforms of the α-subunit of Na,K-ATPase is estimated from mutation rates obtained from comparison of sequences from various species and the time of divergence of phyletic lines [39, 134]. The mutation rates vary considerably among exons within each gene. Average mutation rates of α-subunit are relatively low, 0.4 per residue per 10^9 years, as compared to rates of 1 or 2 per residue per 10^9 years for other proteins [134]. The divergence of the α1-, α2- and α3-isozymes of Na,K-ATPase may thus have occurred about 300 million years ago at the time of separation of mammalian from piscine lines. It is of interest to note that the amino acid sequences of plasma membrane Ca-ATPase shows more homology to the α-subunit of Na,K-ATPase than to the Ca-ATPase of sarcoplasmic reticulum. Both the cation pumps and ($\Delta\mu$Na)-driven secondary active cotransport systems in eukaryotic cells as alternative to ($\Delta\mu H^+$)-driven systems may have evolved before division of prokaryotes and eukaryotes occurred [168]. Development of Na and Ca pumps and secondary active transport systems for nutrients may form the basis for adaptation to changes in food supply and appearance of a variety of new animal species during transition from precambrian to cambrian period, 700–500 million years ago [112].

Genes of Na,K-ATPase

The genes of the Na,K pump are complex and diverse with at least five genes of the α-subunit isoforms with sizes near 20–25 kb. They have multiple introns, comprising more than 80% of known gene sequences [134, 165]. At least three mRNA isoforms of the α-subunit are expressed in rat brain and muscle [70, 129, 162]. Clarification of DNA structure and mechanisms behind expression of isozymes is important for understanding the organization of the protein in the membrane and physiological pump functions. The positions of the introns may correspond to functional subdivision of the protein [134], in particular to the position of presumed membrane spanning helices [101].

Screening of human genomic libraries with α-subunit ^{32}P-labeled cDNA so far revealed five genes. Two were identified in a library from human placenta [134], one related to the α1 isoform, the other to the α3-isoform. In a genomic library from human leukocytes, four genes (αA, αB, αC, and αD) were identified by restriction mapping. Using isoform-specific DNA probes it was shown that αA and αB encode α1- and α2-isoforms, respectively. The two remaining genes (αC and αD) do not correspond to previously identified isoforms and it is not known if they are expressed [165].

Chromosome mapping shows that the α1-subunit gene is located on human chromosome 1p and on chromosome 3 in mouse. Hybridization of α1 subunit ^{32}P-labeled cDNA probe from human placenta to chromosome spot blots shows that the α1 gene is localized to region 1p11–1p13 in chromosome 1, in the proximal short arm close to the centromere. A polymorphic ECoRI band has been detected in some humans [24] and in genomic DNA from a HeLa cell line that is resistant to ouabain, while the band is absent from the ouabain sensitive cell line [70]. The genetic linkage map of the human genome [38] shows a relatively high density of inherited polymorphic loci or RFLPs (restriction fragment length polymorphisms) in that region of the α1-gene.

Hybridization of specific human genomic DNA fragments to Southern filters containing panels of rodent × human somatic cell hybrid lines allows more precise assignment of loci for the α- and β-subunit genes [190]. The αA gene of the α1-subunit is located to human chromosome 1, region 1p21, the α2-isoform αB gene to chromosome 1 region q32, and the α3 gene to human chromosome 19. The αD gene with unknown expression product has been mapped to chromosome 13.

The α3 gene has been sequenced [134]. It is a large, 20–25 kb, and complex structure with 22 introns separating the exons. In contrast, only 4 introns are found in the gene of H-ATPase in *Neurospora Crassa* [1], and the gene of H-ATPase in yeast has no introns [159]. Examination of intron-exon boundaries in the sequence of α3 gene, shows that most introns start with GT and ends with a stretch of pyrimidines (CT) followed by AG, and assignment of intron is in agreement with the AGGT consensus sequence criterion [136].

These and other intron positions correspond to boundaries between functional domains of the α-subunit [84]. Thus, several transmembrane segments (M1, M2, M3, M6, M7, and M8) are bordered by introns, like in the murine gene of capnophorin [101]. Similarly, introns separate exons corresponding to domains contributing to the different components of the area for nucleotide binding (exons 11, 15, and 16) or phosphorylation (exon 9) that are among the best conserved. Thus, exon 13 and exon 16 code for segments with close to 100% amino acid homology between the isoforms and higher than average homology with H,K-ATPase and Ca-ATPase. Other exons (nos. 1–3, 10, 19) relate to domains in which homologies between α-isoforms are well below overall homology (89%). Calculation of relative homologies within exons is of interest for identifying specific functional regions. For example, in exons 7 and 17, α-isoforms show homology mutually and with H,K-ATPase, but not with Ca-ATPase suggesting that segments contributing to K sites may be located here [84]. Exons 20 and 21 may correspond to domains that are specific for the Na,K pump, such as Na sites, the ouabain site, or areas for interaction with β-subunit, since there is high mutual homology between α1-isoforms, while homology with H,K-ATPase and Ca-ATPase is well below the average value. This illustrates that localization of intron positions in genomic DNA will be important for understanding functional subdivision of α-subunit.

Expression of α-Subunit Isoforms

The discovery of a family of α-subunit genes, mRNAs and expressed proteins with different primary structures is a probable basis for functional diversity of Na,K pumps in various tissues and cells. This may in particular apply to the differential expression of the isozymes in specific tissues with different mechanisms for hormonal and cellular regulation of Na,K pump activity. The background for these presumptive physiological functions of the isozymes is not yet clear, as the regulation of isozyme expression at the genomic level and the precise cellular localization of the isozymes remain to be clarified. Equally important, it has not yet been possible to isolate the α2- and α3-isozymes of Na,K-ATPase for comparison of their enzymatic properties, regulation and transport behavior with those of the well known α1-isozyme of Na,K-ATPase from the outer renal medulla.

During ontogenesis, the α-isoforms appear to be regulated in coordination with expression of the β-subunit. Maximum expression of both subunit mRNAs occur between 15 and 25 days of age [129]. Examples of the different quantitative distributions of mRNA of the three α-isoforms and the β-subunit in brain, lung, kidney, and muscle are shown in Fig. 6. Overall homology is 97%–98% between amino acid sequences of α1-isoforms of rat [162], sheep [166], or porcine [133] kidney, rat brain [162], and human HeLa cells [98], while it is 85%–86% between α-, α2-, and α3- isoforms. The tertiary structural organization of the three isoforms is identical and close to 100% homology is found in ATP binding and phosphorylation segments and in the hydrophobic domains (M2, M5, and M7). Substantial differences in amino acid sequence occur in the N-terminal region and in the region between the phosphorylation site (Asp 369) and Lys 501 that binds fluorescein-isothiocyanate (FITC) [84]. The N-terminal regions show considerable variation in number of charged residues and this may be important for catalytic differences between isozymes that can be related to different rate constants for the E_1–E_2 transitions [84, 174]. Phosphorylation experiments using brain and kidney Na,K-ATPase suggest that the poise of the E_1P-E_2P equilibrium is shifted towards the E_1P form for the α2-isozyme. For example, at 150 m*M* NaCl the fraction of E_1P in kidney Na,K-ATPase (α1) is 14% of total EP, while E_1P forms 55% of total EP in preparations from brain (α2 or α3) [86].

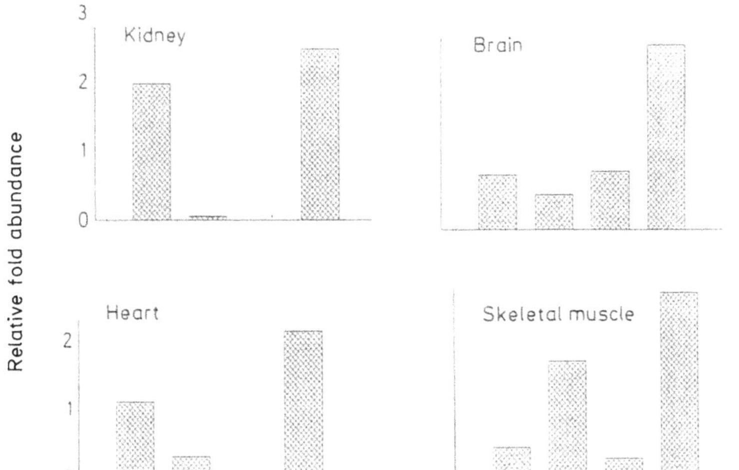

Fig. 6. Relative abundance of α1-, α2-, α3-isoforms and β-subunit mRNA in tissues from 25-day-old rats. Total cellular RNA was hybridized with specific ^{32}P-labeled cDNA probes of α1 (nt 89–421), α2 (nt 121–502), α3 (nt 51–331), and β-subunit (nt 459–759). Staining intensities were obtained by densitometry of autoradiography films and expressed relative to values of α1 mRNA at 14 days of gestation (−8 days). (Data from Orlowsky and Lingrel [129])

The lower electrophoretic mobility of the α2-isoform under certain conditions is not due to a higher molecular weight of α2 since it has fewer residues (1015) and lower M_r (111 736) than α1 (1018 residues and M_r 112 573) [162]. The catalytic properties of the α1-isoform are well known from studies on purified Na,K-ATPase from kidney [80]. The overall Na,K pump reaction appears to be similar for the two other isozymes, but the actual properties of the α2- and α3-isozymes are uncertain, since both forms are abundant in brain where the α2-isozyme have been examined [170]. Recently the α3-isozyme has been identified in microsomes from rat brain and the α3-isozyme of Na,K-ATPase has been identified through N-terminal sequencing as an ouabain sensitive form in axolemma [74]. In rats, the α2-isozyme is more sensitive to digitalis than the α1-isozyme and the α3-isozyme of Na,K-ATPase has been expressed with even higher ouabain sensitivity in cell culture [64]. The properties and occurrence of the isozymes of Na,K-ATPase in different tissues have recently been systematically reviewed [174].

Titration of the abundance of mRNA transcripts using cDNA probes specific for each α-subunit isoform, suggests that the α-isoforms and β-subunit are coordinately regulated throughout development in rats. The highest levels of α1- and β-subunit mRNA are exhibited by the kidney at all stages of development. Posttranslational regulation may thus serve to maintain equimolar amounts of α-subunit and β-subunit in the tissue. In fetal muscle the α1 mRNA is the predominant α-isoform. After 4 days and until adulthood, muscle expresses predominantly the α2 mRNA and in coordination with β-subunit mRNA, but unlike other tissues the β-subunit mRNA decreases at 15–55 days of age by 50% without concomitant decreases in the levels of α-isoform mRNA [129]. In general, the β-subunit mRNA levels exceed those of the total α-subunit mRNA levels, except in lung and muscle where opposite ratios are observed.

In whole kidney the ratio of α1- to α2-isoform is $> 280:1$ [129] and only the α1-isoform seems to be present in thick ascending limb of Henle. It is likely that the small amounts of α2 mRNA in whole kidney is localized to cortical collecting ducts where Na,K-ATPase is more sensitive to digitalis than in other segments and may be subject to hormone regulation [40].

In brain the α1- and α2-isoforms have a characteristic cellular distribution that may be of importance for function. The α1-isoform occurs in astrocytes and unmyelinated sympathetic neurones while α2 predominates in myelinated neurons [47, 170, 174]. The three known isoforms are all expressed in rat brain with α3 as the predominant isoform. As shown in Fig. 6, the amount of α2 mRNA is high in skeletal muscle with a ratio of α2 to α1 mRNA of 30:1. The mRNA of the α2-isoform is also found in heart muscle, adipose tissue, stomach and lung, whereas the mRNA of the α3-isoform is detected in brain, stomach, and lung [129]. It is interesting to note the distribution of isoforms in heart muscle, where the α2-isoform is found in much higher concentration than α1-isoform in ventricles, while the converse is true for atria where the α1-isoform predominates [129]. This may form the basis for the inotropic effect of digitalis on the hypertrophied heart since the α2-isoform in the enlarged ventricles is more sensitive to cardiac glycosides than the α1-isoform.

Two noninteracting pools of isozymes with high and low affinity for ouabain and different sensitivity to alkylation of sulfhydryl groups are found in the salt gland of *Squalus achantias* [111], but the structural relationship of these isoforms to those in mammalian tissues has not been clarified.

Gene Expression of β-Subunit

In the human genome two different loci hybridize with specific probes for the β-subunit, one on the long arm of chromosome 1 and the other on chromosome 4. Previous evidence had suggested that only a single gene for the β-subunit [190] would translate multiple mRNA for the β-subunit. Two isoforms of the β-subunit have also been observed in rats.

Coordinate synthesis of α-subunit and β-subunit is required for expression of Na,K-ATPase in the surface plasma membrane. Synthesis of α-subunit and β-subunit is concurrent and assembly of the αβ-unit of Na,K-ATPase occurs during or immediately after polypeptide synthesis in neuronal cells [176]. Coordinate synthesis is also observed in myogenic cells [188] and epithelial cells of toad bladder [53]. Expression of Na,K-ATPase in *Xenopus laevis* oocytes thus requires microinjection of both α and β mRNAs [124]. Experiments with glycosylation inhibitors show that the β-subunit plays a role in efficient accumulation of α-subunit in the endoplasmic reticulum [192]. Multiple mRNAs have been identified in rat kidney and brain, but they appeared only to encode a single β-subunit protein [191], and β-subunit isoforms have not been identified. Endonuclease mapping and DNA sequencing show that the multiple β-subunit mRNAs can be grouped into classes based on the lengths of their 5' and 3' untranslated segments. They are transcribed from two sets of initiation sites that are separated by more than 300 nucleotides. These mRNA forms are expressed at different levels in kidney and brain suggesting that the different mRNA forms are transcribed from two promotors within a single gene. The process may have different tissue specific or cell-specific patterns of expression [191].

Transfection of β-subunit from avian cells with high ouabain affinity and expression in the cell surface of mouse cells with relatively low affinity receptors for ouabain shows that ouabain sensitivity does not follow the β-subunit [175]. The transfected mouse cell line expresses the avian β-subunit of Na,K-ATPase from an uncontrolled promotor. This cell line may provide an interesting tool for examining control of gene expression in situations where the cells regulate their content of Na,K-ATPase up or down in response to altered concentrations of Na^+ in the cell cytoplasm.

Gene Regulation and Biosynthesis

The amount of Na,K-ATPase in cell membranes is modulated according to the demand for transport capacity through effects of Na^+ concentrations in cytoplasm on transcription or biosynthesis rates. This feedback regulation of the amount of Na,K-ATPase protein in the cells was proposed as a hypothesis for explaining adaptive changes in amount of Na,K-ATPase following changes in supply of Na to adrenalectomized rats [77]. In this model a sustained change in Na/K ratio in cytoplasm following an increased influx of Na^+ results in adaptive increases in amount of Na,K-ATPase that appear slowly after 16–24 h in vivo. As illustrated in Fig. 7, the feedback cycle is completed when the capacity for active Na,K transport has been increased to a level (from $V_m = 1$ to $V_m = 2$) allowing the increased flux rates to be maintained at the original levels of cytoplasmic Na/K activities. A similar induction of Na,K-ATPase is elicited after incubation of cultured cells in ouabain or low-K media [14]. Also in myocytes [188] and in suspensions of kidney tubule cells [146] it can be demonstrated that an increase in Na/K ratio after 2–4 h triggers a marked increase in rate of synthesis of Na,K-ATPase protein. The mechanism for the effect of the change in Na/K ratio on synthesis of the $\alpha\beta$-unit of Na,K-ATPase is not clarified, but recent quantification of mRNA by hybridization with [^{32}P]-labeled cDNA shows an early rise in amount of specific mRNA followed by a late increase in rate of synthesis of the α-subunit and β-subunit and in abundance of $\alpha\beta$-unit in the cells [15, 76, 142]. Northern blot analysis of the low-K induced increase in a kidney cell line showed that the mRNA of the α1-isozyme of Na,K-ATPase was increased by up to 200% in response to the cation stimulus, while the α2- or α3-isozymes are indetectable in these cells [76]. On the basis of minor changes in activity during vitro incubation, a defective homeostatic response of Na,K-ATPase to salt induction has been related to the pathogenesis of hypertension [146]. The time required for affecting a change in

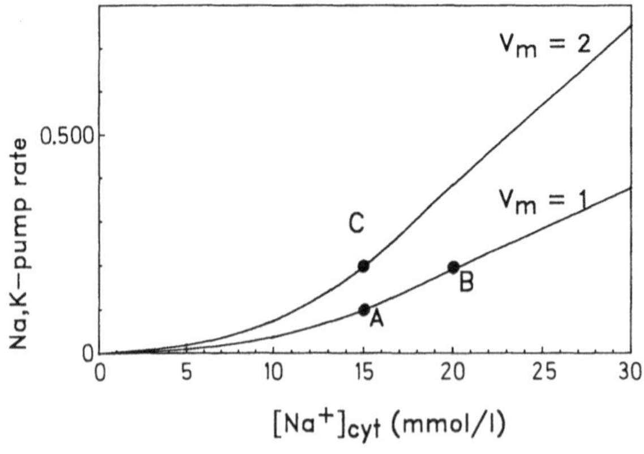

Fig. 7. Na,K pump rate as a function of $[Na]^+$ activity in the cell cytoplasm before ($V_m = 1$) and after ($V_m = 2$) adaptation of the amount of Na,K-ATPase to a twofold higher level. The acute adaptation following an increased rate of Na^+ influx and a rise of $[Na^+]$cyt from 15 mmol/l (*A*) to 20 mmol/l (*B*) causes an increase in the turnover rate of existing pumps from 0.1 to 0.2. After chronic adaptation involving a doubling of the amount of Na,K-ATPase corresponding to an increase from $V_m = 1$ to $V_m = 2$, the increased turnover rate (0.2) of Na,K pumping (*C*) can be sustained at the original level of $[Na]^+$cyt. (Adopted from [77])

amount of Na,K-ATPase [16–24 h] agrees with the rate of change in response to in vivo changes in Na^+ load. Rates of degradation of $\alpha\beta$-unit are slow with $T_{1/2} = 20–40$ h. In outer medulla, a 60%increase in amount of Na,K-ATPase activity can be elicited by incubation with ouabain [147]. The time course of this response is similar to that previously observed in the intact kidney [77]. Measurements of rates of synthesis and degradation over the time course shows that a transient increase in rate of synthesis is observed after 12 h. After 24 h the rate of synthesis returns to control levels, while the rate of degradation is reduced to maintain the higher level of Na,K-ATPase activity [147]. It remains to be examined whether the reduced degradation rate in tubule suspensions also can be observed in the intact kidney.

Mechanisms for direct hormone effects on biosynthesis rates have not been resolved. In principle hormones may act on transcription or translation of mRNA, on recruitment of an intracellular pool or via changes in cytoplasmic cation concentrations. Aldosterone effects on the rate of mRNA synthesis have been analyzed in relation to stimulation of Na,K-ATPase synthesis [183]. In cortical collecting ducts, aldosterone in presence of triiodothyronine increases the amount of Na,K-ATPase by de novo synthesis of pump units independent of changes in cytoplasmic Na^+ [10], while changes in Na^+ concentration may stimulate recruitment of a latent pool of Na,K-ATPase independent of protein synthesis. An intracellular pool amounting to 30%–70% of total Na,K-ATPase has been demonstrated in muscle cells [188] and cultures [142, 176].

Organization of the Na,K-Pump Proteins in the Membrane

Presumptive membrane-spanning segments can be identified in the amino acid sequences, but the availability of the amino acid sequences of the cation pumps has not led to solution of the overall topology of the lipid-associated domains. Biochemical studies, such as selective proteolytic or chemical cleavage in combination with specific chemical labeling from the membrane surfaces or the hydrophobic interior, are required to determine the orientation of intramembrane segments and the sidedness of N- and C-terminal segments. It is important to evaluate whether the topology of the membrane protein as determined from these data correlates with the distribution of mass between extra- and intramembranous domains as estimated in models derived from physical studies of membrane crystals.

Insertion in Endoplasmic Reticulum Membrane

Information about the mechanism of biosynthesis may help to establish the orientation of N- and C-termini of the pump proteins. Orientation of the membrane protein in endoplasmic reticulum is conserved during transport to the surface cell membrane and in maturation of sarcoplasmic reticulum so that domains exposed on the luminal side in ER are eventually exposed on the noncytoplasmic side. The subunits of Na,K-ATPase [176] and Ca-ATPase [108, 118] are synthesized without a cleavable hydrophobic N-terminal signal sequences, yet they are cotranslationally inserted into the endoplasmic reticulum membrane due to the presence of insertion signals further down the sequence. In vitro expression of deletion mutants of cDNA of human α-subunit and β-subunit shows that membrane insertion signals are contained in M1–M4 in α-subunit [73]. The presence of one of these segments is required for insertion of α-subunit in the membranes of an in vitro biosynthesis system. Insertion of the β-subunit is possible as long as at least 16 residues remain in the transmembrane segment (residues 34–53). The hydrophobic signal sequence of the nascent protein emerging from the ribosome is recognized by a part of the signal recognition particle (SRP). The complex attaches to the SRP receptor or docking protein. After release of the signal sequence from the docking protein, it interacts with a signal sequence receptor (SSR), an integral protein of the rough endoplasmic reticulum membrane [187]. This SSR protein may mediate the insertion of the first hydrophobic segment in the membrane in an orientation determined by the net charge of neighboring residues such that the end of the hydrophobic segment which is attached to a hydrophilic segment with a net negative charge will be oriented towards the extracellular membrane surface. This model is in agreement with the orientation of the $\alpha\beta$-unit, since the first 10 residues at the extracellular end of the first transmembrane segment of both subunits have more negative charges than the first 10 residues at the N-terminal end at the cytoplasmic surface.

Cotranslational insertion signals span the membrane from cytoplasm to noncytoplasmic surface in the N-C direction. The cation pump proteins belong to a group of transmembrane proteins with the N-terminus exposed on the cytoplasmic surface [72] all having a net negative charge at the outside of the first transmembrane segment. Multiple crossings can be generated by a sequence of alternating cotranslational insertion and halt transfer signals [151]. The hydrophobic parts of a halt transfer signal would span the membrane with opposite direction. The location of the C-terminus during insertion of the membrane protein in the ER membrane will thus depend on whether the last transmembrane segment serves as part of an insertion or halt transfer signal. In agreement with this scheme studies of the structure of Na,K-ATPase [80] and sulfhydryl modification of Ca-ATPase [148] shows that their N-termini are cytoplasmic, while localization of their C-termini may vary.

Mass of Intramembrane Protein

Calculation of mass distribution of the model in Figs. 8 with 9 transmembrane segments in the α-subunit and in the β-subunit shows that 27% of $\alpha\beta$-unit mass

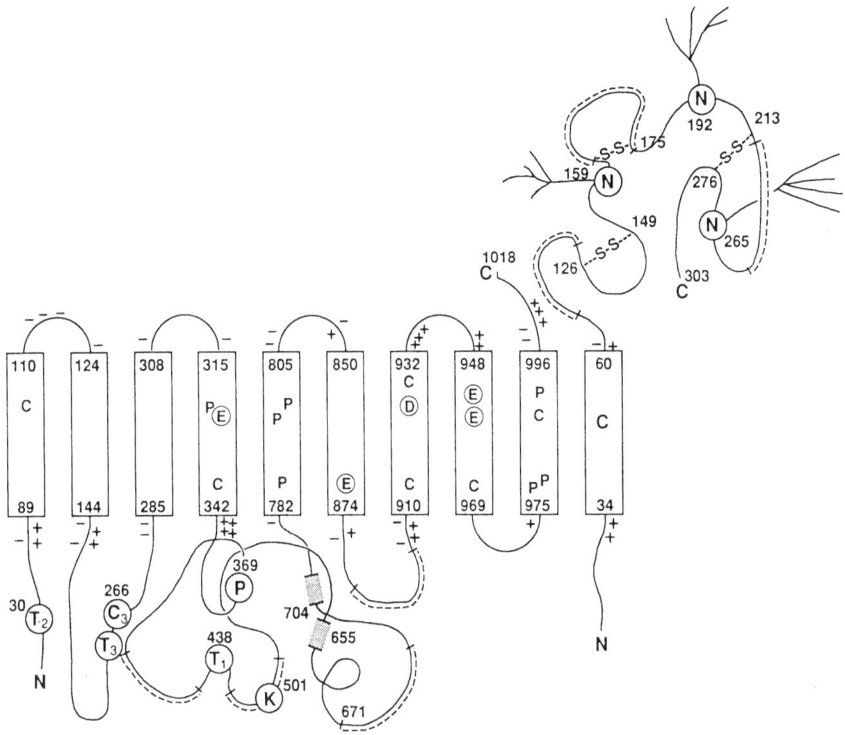

Fig. 8. Disposition of $\alpha\beta$-unit in the membrane, based on sequence information [96, 97], selective proteolytic digestion of the α-subunit [86], identification of C-terminal antibody epitopes at the extracellular surface [132], and labeling with FSBA [29, 127]. The model for the β-subunit is based on sequencing of surface peptides [126] and identification of S-S bridges [100, 126]. T_1, T_2, T_3, and C_3 show location of proteolytic splits. N are glycosylated asparagines in the β-subunit. (+) and (−) indicate all extracellular charged residues and charges within the first 10 residues from the transmembrane segments at the cytoplasmic face. (shaded box) shows FSBA labeling and (dashed line) segments located at the protein surface

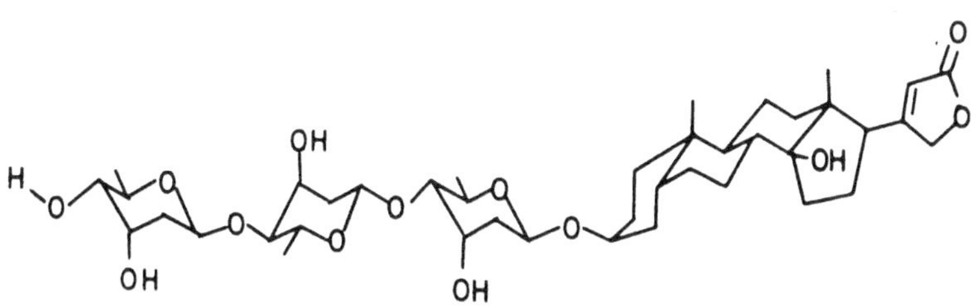

Digitoxin

Fig. 9. The structure of digitoxin with (*from the right*) the lactone moiety, the steroid moiety and the three sugar molecules

is extracellular, 56% is exposed on the cytoplasmic surface, and only 17% of $\alpha\beta$-unit mass is inside the membrane.

An alternative estimate of mass distribution is obtained by proteolysis. Extensive digestion with excess trypsin or unspecific proteases shows that 40%–50% of the protein of Na,K-ATPase [91] or Ca-ATPase [189] is protected by lipid or secondary structures that are resistant to proteolysis.

These data suggest that the mass of 9 + 1 transmembrane segments in Na,K-ATPase may be less than the mass of intramembrane protein determined by other methods. With this in mind it is appropriate to consider the possibility that the cation pumps possess intramembrane protein structures in addition to those of predicted transmembrane helices.

Paradigms for Secondary Structure of Intramembrane Protein

Combined knowledge of sequence and structure at high resolution is available only for the photosynthetic reaction center of bacteria [37, 113]. The amino acid sequences of three subunits fit well into the electron density map based on X-ray crystallography with identification of individual residues. A total of 11 transmembrane structures is found typically consisting of α-helices of 19–23 amino acid residues without basic or acidic side chains. Using a window of 19, the hydroplot analysis [103] shows that four of the helices have a hydropathic index peak above 2.0 and another four have a peak above 1.5, while three have index peaks between 1 and 1.5.

Formation of β-sheet in integral membrane proteins has been predicted in porin and gap junction protein [137, 138] and X-ray diffraction measurements reveal a transmembrane domain with a high proportion of β-sheet running parallel to the membrane surfaces in gap junction [109]. In a β-sheet structure an extensive network of hydrogen bonds may allow polar residues to exist in environments of low dielectric constants. Only 10 residues may be required to traverse the hydrophobic membrane core in an extended β-configuration of a polypeptide. Porin spans the outer membrane lipid bilayer of *E. coli* to form voltage dependent channels in a compact structure with little material protruding into the aqueous phase [138], but hydroplots of its protein shows at most one sizeable hydrophobic domain with a hydropathic index above 1.5.

Another possibility for explaining the presence of both hydrophobic and polar or ionizable residues within the membrane is the packing in amphipathic helices possessing hydrophobic nonpolar faces and relatively narrow charged polar faces in helical wheel plots (cf. [17]). Critical evaluation of algorithms for predicting secondary structure shows that they are inadequate for membrane proteins, particularly when used on hydrophobic segments [186]. As an example, secondary structure prediction according to Chou and Fassmann [27] of the photosynthetic reaction center [37] shows higher propensity for β-sheet than for α-helix for hydrophobic segments that are indeed organized in α-helical structures in the high resolution model from X-ray crystallography.

Intramembrane Structure of Na,K-ATPase

In the model of the $\alpha 1$ subunit of Na,K-ATPase in Fig. 8, the transmembrane segments M1, M2, M3, M4, and M5 are predicted by a hydropathic index of > 2.0 and three segments M6, M7, and M8 have a peak index of > 1.5 using a window of 19 residues for the hydroplot [103]. These segments consist of 21–25 amino acid residues with overrepresentation of the hydrophobic residues Phe, Ile, Leu, Val, Trp, Tyr, but also of Pro and Cys. This may suggest that S-S bridge formation is part of stabilizing intramembrane structures. Prolines or glycines break the continuity of membrane helices and the excess of proline is interesting in view of a recent survey demonstrating membrane-buried proline residues in transport proteins, while they are excluded from the membranous domains in nontransport proteins [17]. *Cis-trans* isomerizations of peptide bonds or mobility around bonds involving prolines and glycines may thus be part of the conformational transitions associated with the transport process. Few basic or acidic side chains are found in the transmembrane segments in Fig. 8, but the segments carry charged residues close to their cytoplasmic ends. These charges may react with headgroups of lipids to stabilize the structure in the membrane [19, 44]. During biosynthesis the charged residues may have served as stop signals [151] preventing transfer across the membrane.

The presence of four transmembrane segments in the N-terminal half of the α-subunit of Na,K-ATPase was predicted from the results of controlled tryptic cleavage combined with selective chemical labeling with photosensitive ouabain, phosphorylation [80], and insertion of small hydrophobic probes, [^{125}I]iodonaphtylazide (INA) [13], [^{125}I]trifluoromethyl-iodo-phenyldiazirine (TID) [21, 83], or [^{3}H]adamantane diazirine [122]. In contrast, neither chemical labeling experiments nor hydroplot analysis lead to a decision as to whether 2, 4, or 6 transmembrane segments are formed by the C-terminal part of the α-subunit (residues 779–1016) and orientation of the C-terminus remains uncertain. Recent immunological studies suggest the presence of an extracellular epitope near the C-terminus of α-subunit [116]. This led to a suggestion of only 7 transmembrane segments in the α-subunit [132], but the model in Fig. 8 with 9 transmembrane segments is also compatible with these results.

Possible candidates for additional intramembrane structures in the α-subunit are relatively hydrophobic segments with high β-sheet propensity in the cytoplasmic domains, e.g., segments 175–202, 242–260, 410–430, and 560–590 in α-subunit, Fig. 8. These segments alternate with α-helices and may form flexible structures that contribute to cation binding and E_1–E_2 transition.

Structure of the β-Subunit of Na,K-ATPase

In the family of cation pumps, only the Na,K-ATPase and H,K-ATPase [161a] posess a β-subunit glycoprotein. A glycoprotein is found in the sarcoplasmic reticulum [106], but its relationship to the Ca-ATPase is uncertain. Sequences of the β-subunit (302 residues, M_r 34 528) of Na,K-ATPase are now available from human HeLa cells [97], sheep [163], pig [133], and dog [20] kidney and *Torpedo electroplax* [123]. The initiating methionine is removed in posttranslational processing, but otherwise the NH_2-terminal sequence from cDNA is identical to that determined by gas phase sequencing [85].

The bulk of hydrophilic residues of β-subunit are exposed on the extracellular surface [43]. The β-subunit sequence has one hydrophobic segment (residues 34–53) with hydropathic index above 2.0 that may form a transmembrane helix. This is basis for the model in Fig. 8. The β-subunit is labeled from the bilayer by INA [91] and TID [80]. As alternative to this model three transmembrane segments are proposed on basis of papain digestion [26] and immunological studies [194], but there is no evidence for additional hydrophobic segments in the amino acid sequence.

The β-subunit is not as well conserved as the α-subunit, with 91% overall homology between β-subunit of sheep, pig, and human and 61% between β-subunit of human and *Torpedo*. This might suggest a dissociation in development between α-subunit and β-subunit, but closer inspection reveals a wide variation in conservation. Some segments are invariant among mammalian, avian and piscine β-subunits like the best conserved segments of α-subunit. This applies to the N-terminus (residues 1–94) and C-terminus (residues 235–302) and the location in the sequence of 3 glycosylation sites, 7 cysteins and 4 tryptophans. In other domains, sequence homology is low, e.g., the sequences around the 1st (N-157) and 2nd (N-192) glycosylation sites show less than 50% homology in alignment among the mammalian species. In agreement with this the oligosaccharide chains seems to be without importance for the catalytic activity of Na,K-ATPase. Oligosaccharide deficient αβ-units expressed in *Xenopus* oocytes in presence of tunicamycin, exhibits the same ATPase activity, ouabain binding, and Rb^+ transport activity as fully glycosylated Na,K-ATPase [174]. Three disulfide bonds are formed by residues Cys^{125}–Cys^{148}, Cys^{158}–Cys^{174}, and Cys^{212}–Cys^{275} [100, 126], and their reduction is accompanied by loss of Na,K-ATPase Na,K-ATPase activity [45, 99, 100]. These observations agree with the hypothesis [192] that segments of the β-subunit may be important for Na,K-ATPase function, e.g., for proper insertion of the α-subunit in the membrane during biosynthesis, while the less conserved parts are of purely structural importance. Otherwise the appearance of the amino acid sequence data have not contributed to our understanding of the function of the β-subunit.

Extracellular Cardiac Glycoside Binding Site

The pharmacologically important action of digitalis compounds is that of a positive inotropic effect, i.e., an increased force of contraction of the failing or insufficient heart. The only known specific binding site for cardiac glycosides is found on the α-subunit of Na,K-ATPase and high-affinity binding to compounds other than Na,K-ATPase has never been demonstrated. Schatzman in

1953 [155] observed that cardiac glycosides inhibits the active coupled Na,K transport and ascribed this to formation of a complex of glycoside with a pump molecule. More recent studies have shown that this is a saturable, reversible binding with a stoichiometry of one cardiac glycoside molecule per $\alpha\beta$-unit [63, 130]. There is a direct relationship between binding to the Na,K pump protein, inhibition of Na,K transport and the increase of contractility of the failing heart. A probable explanation for this is the increase of cytoplasmic $[Na^+]$ and $[Ca^{2+}]$ activities following inhibition of the active Na,K exchange (for references. see [158]).

The three isozymes of the Na,K pump show a pronounced difference in sensitivity to ouabain. Na,K pumps with α1-subunits are relatively resistant while α2- and α3-isozymes are ouabain sensitive [64, 181]. In rats and mice the binding to the α1-isozyme is so weak that only Na,K pump sites containing α2- or α3-subunits contribute to the number of pump sites determined in $[^3H]$ouabain binding experiments. Sensitivity and resistance to ouabain may thus become a key to understanding the physiological significance of the difference in properties among the isozymes of the Na,K pump.

Ouabain Sensitivity and Resistance

The ouabain binding area has at least two functionally distinct regions (Fig. 9). The primary region binds the lactone ring and steroid portion and this interaction determines the association rate and inhibitory potency. Binding of the carbohydrate moieties of the cardiac glycosides to the secondary sugar sites predominantly influences the dissociation rate of the complex (for references see [80, 158]). Covalent labeling with light sensitive compounds shows that the α-subunit forms the binding area for the lactone, the steroid and the first two carbohydrate moieties of the cardiac glycoside molecule. Only if a light sensitive group is inserted into the 4′ position of the third sugar residue of digitoxin has it been possible to label the β-subunit from the extracellular surface [62]. These data suggest that the α-subunit provides the area for binding of the digitalis glycoside molecule, while the β-subunit only contributes to the binding site if a third carbohydrate moiety is attached to the cardiac glycoside (cf. [80]).

The α1- and α2- isozymes in brain, cardiac muscle, striated muscle and adipocytes are distinguished by the high ouabain sensitivity of the Na,K pumps containing the α2-isoform. A high ouabain sensitivity of the α3-isozyme in rat brain is suggested on basis of tryptic digestion experiments [181]. The α3-isozyme was identified on the basis of cDNA cloning in rat brain [162], and the enzymatic activity of the α3 mRNA product has now been assayed in rat brain [128] and axolemma [74]. Expression of the rat brain α3-subunit in a functional Na,K pump was achieved by cotransfection into a mouse 3T3 cell culture [64]. The ouabain sensitivity of the Rb^+ uptake in these cells shows that the α3-isozyme of Na,K-ATPase has a high sensitivity to ouabain with K_i around 8×10^{-8} M, relative to a K_i of the wild type Na,K pump in the 3T3 cells of 4

$\times 10^{-5}$ *M*. In the transfected cells a hybrid Na,K-ATPase containing rat brain α3- and β-subunit of the mouse 3T3 cells was formed. This confirms that ouabain sensitivity is exclusively determined by the properties of the α-subunit.

Ouabain Sensitivity of α1-Subunit Chimeras

Human, porcine, and sheep Na,K-ATPase containing the α1-isoform are sensitive to ouabain while rat α1-Na,K-ATPase is ouabain resistant. Construction of chimeric α-subunits using ouabain-sensitive α1-subunits from *Torpedo* [124] or sheep [143] and ouabain resistant α-subunits from rat, shows that the determinants of ouabain sensitivity are located to the N-terminal half of the α-subunit. Here only two peptide segments are exposed to the extracellular surface where the area for ouabain binding is located. The sequences of the second extracellular segment (residues 306–314; Fig. 8) are identical in all isozymes of rats and other mammals. This segment may be involved in binding, but it can not be responsible for the ouabain resistance of the rat Na,K-ATPase.

Site-Directed Mutagenesis of Ouabain Sensitive α1-Subunits

The only extracellular sequence in the segment (Met^{1} to Val^{165}) conferring ouabain resistance is the junction between the first and second transmembrane segment, i.e., residues 110–122 (Fig. 8). The amino acid sequences of this extracellular domain in α1-subunits from rat, *Torpedo*, and sheep are shown in Table 1 in comparison with the sequences of ouabain-sensitive α2- and α3-subunits.

A mutated sheep α1 in which the sequence of the first extracellular domain is changed to that of the rat α1, is able to confer ouabain resistance to HeLa cells [143]. As seen from Table 1, the sheep α1 sequence differs from the rat α1 in four positions (Gln^{111}, Ala^{112}, Gln^{119}, and Asn^{122}), but mutation of only two of these residues ($Gln^{111} \rightarrow$ Arg and $Asn^{122} \rightarrow$ Asp) confers the full complement of ouabain resistance. Mutation of either of the two charged residues results in pumps with ouabain sensitivity intermediate between the rat α1 and the wild type HeLa cell Na,K-ATPase. Comparison of the rat isozymes shows that the ouabain sensitive α2 and α3 forms have uncharged residues at the borders of the segment in Table 1. It is therefore the presense of a positive (Arg^{114}) and a negative (Asp^{124}) charge in rat α1 that explains the generation of ouabain-resistant Na,K-ATPase.

The negatively charged residues in the first extracellular segment (Table 1) may engage in binding of the unsaturated lactone attached to the C-17 position in the steroid moiety of ouabain. It is proposed that the charged Arg and Asp in the ouabain resistant α1-isozyme of the rat may prevent a conformational change in the protein that allow stabilization of the complex of ouabain with Na,K-ATPase. The charges may thus interfere with an adaptive fit of the

Table 1. Charged residues in the amino acid sequences of the N-terminal extracellular domain of the α-subunit of ouabain resistant α1-isoform from rat [162], ouabain-sensitive α1-isoforms *Torpedo* [96] and sheep [166], and ouabain-sensitive α2- and α3-isoforms from rat [162]

					+				−	−	−				−	−	
α1 Rat	110	Y	G	I	R	S	A	T	E	E	E	P	P	N	D	D	L
									−	−	−				−		
α1 Sheep	108	Y	G	I	Q	A	A	T	E	E	E	P	Q	N	D	N	L
									−	−	−				−		
α2 Rat	108	Y	G	I	L	A	A	M	E	D	E	P	S	N	D	N	L
									−	−	−				−		
α3 Rat	105	Y	G	I	Q	A	G	T	E	D	D	P	S	G	D	N	L
										−					−		
α1 Torpedo	115	Y	G	I	Q	V	A	T	V	D	N	P	A	N	D	N	L

binding area that prevents ouabain dissociation from the sensitive α2- or α3-Na,K-ATPase isozyme. The segment in Table 1 forms only a part of the ouabain binding area. The C-terminal half of the α-subunit also contributes to the binding area [80], and binding to the extracellular region between the third and fourth transmembrane segment is suggested from photoaffinity labeling [58]. Association with the steroid moiety is probable because a segment in this region (EYTWLE) shows resemblance to estrogen (ECAWLE) and glucocorticoid (QYSWMF) receptors [9].

Cytoplasmic Domains for Nucleotide Binding and Phosphorylation

A characteristic structural feature of the Na,K pump proteins is a cytoplasmic protrusion with approximate dimensions 45 × 65 Å in the plane of the membrane and a length of 50–60 Å in the plane perpendicular to the membrane. As illustrated in Fig. 8, the transmembrane segments separate the protrusion into subdomains. Selective cleavage and chemical labeling established structure-function correlates for some of these domains. The bulk of the protrusion is formed by the large central domain (residues 340–780 in α-subunit) that forms sites for ATP binding and phosphorylation. The second cytoplasmic domain (142–284 in α-subunit) between M-2 and M-3 contains peptide bonds susceptible to proteolytic cleavage in E_1 forms, and we suggested that this domain is involved in energy transduction [88]. The N-terminus is attached to M1 and is important for control of E_1–E_2 transition in Na,K-ATPase and of the rate of Na,K pumping.

Conformations of the Nucleotide Binding Area in Na,K-ATPase

The segments contributing to nucleotide binding and phosphorylation domains undergo structural changes accompanying E_1–E_2 transition as the ATP binding region adapts for tight binding in the E_1 form with K_D 0.1 μM, while binding to the E_2 form is weak requiring millimolar concentrations of ATP for saturation. In addition, structural evidence [88] suggests that motion of phosphorylated residue may be part of the transition from E_1 to E_2. Studies of other ATP binding proteins show that functional residues are contributed by regions of different protein sequences and secondary structure [144, 185]. An indication of the functional groups required to form a nucleotide site may be obtained from examining sites in dehydrogenases [150], phosphofructokinase (cf. [185]) and in particular adenylate kinase [51, 114].

Determination of the conformation of ATP bound in the active site is an interesting approach at the structure of the nucleotide site [172]. Transferred nuclear Overhauser effect (TRNOE) measurements have been used to determine the three dimensional conformation of an ATP analogue, $Co(NH_3)_4ATP$ at the active site of renal Na,K-ATPase (Figs. 10, 11). Data of nine unique proton-proton distances are consistent with a single nucleotide conformation at the active site. The bound ATP adopts an anticonformation ($\chi = 0 \pm 90°$) with a glycosidic torsion angle (χ) of 35°. A striking feature of the TRNOE spectrum is the strong interaction between the H_8 resonance of the adenine and the $H_{2'}$ resonance of the ribose. This is consistent with an anticonformation for the adenine ring with respect to the ribose. As shown in Fig. 11, the conformation of the ribose ring is N type ($C_{2'}$-exo, $E_{3'}$-endo) with a torsional angle $\gamma = 100°$. The orientation of O_5 relative to the ribose is determined by the torsional angle, $\delta = 178°$, a typical value for protein bound MgATP [172].

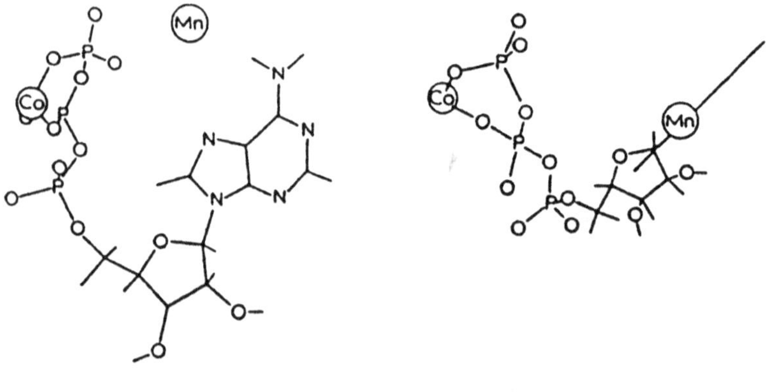

Fig. 10. Conformation of $Co(NH_3)_4ATP$ and Mn^{2+} bound at the active sites of purified Na,K-ATPase from kidney. The $Mn^{2+} - Co^{3+}$ distance is obtained by Mn^{2+} electron paramagnetic resonance. The other distances were determined from the two-dimensionally transferred nuclear Overhauser effect measurements and paramagnetic relaxation data [172]. The nucleotide adopts a bent conformation with the triphosphate chain lying nearly parallel to the adenine moiety

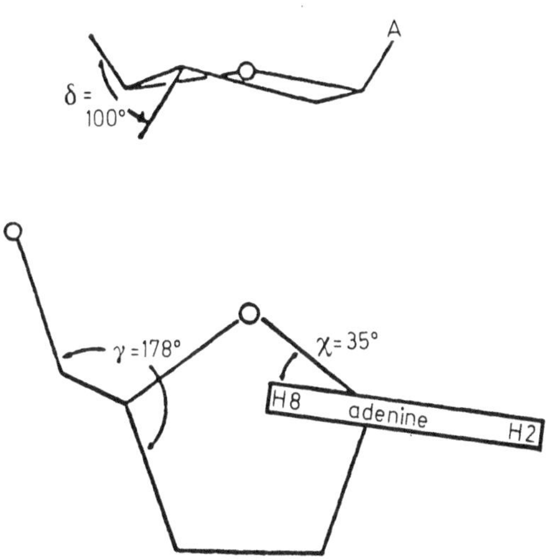

Fig. 11. Torsion angles of ATP bound to Na,K-ATPase. Two views of the ribose ring and the adenine moiety with the torsional angle $\chi = 35°$ about the $C_{1'}-N_9$ bond that links base to sugar. The torsional angles, $\delta = 100°$ about the $C_{5'}$ $-C_{4'}$, $-C_{3'}$ $-O_{3'}$ is indicative of the puckering in the ribose, and $\gamma = 178°$ about $O_{5'}$ $-C_{5'}$ $-C_{4'}$ $C_{3'}$ determining the orientation of O_5 relative to the ribose ring. (From [172])

Identification of Peptide Segments Involved in Nucleotide Binding

The comparison with adenylate kinase is particularly interesting in view of the information available on the structure of the nucleotide binding area in that protein and the homologies demonstrated in Table 2. An important feature in the model of the ATP site in adenylate kinase is a hydrophobic pocket for accomodation of the adenine and ribose moieties which is formed by Ile, Val, His, and Leu residues. The triphosphate moiety is flanked by a hydrophobic strand of parallel β-pleated sheet terminated by Asp [51]. As seen from Table 2, such a segment in adenylate kinase and in the β-subunit of F_1-ATPase shows some homology with respect to charges and hydrophobic residues to segments in α-subunit of Na,K-ATPase (543–561). The N-terminal part of this segment is rich in glycines and may form a flexible loop that undergoes conformational transitions related to altered affinity of the site or to reallocation of catalytic groups. In adenylate kinase, other segments contribute an α-helix with hydrophobic residues separating two lysines thay may interact with phosphate moieties. The sequence data complement the TRNOE measurements in the sense that part of the ATP moiety is organized in a manner similar to that in adenylate kinase while there is a significant difference in torsion angle about the bond between adenine and ribose. This agrees with the homology between amino acid sequences belonging to the backbone of the nucleotide binding sites, while other parts of the domain, probably involved in more specific cation dependent reactions and ion translocation are specific for the cation pump proteins.

A sequence of 8 amino acids around the phosphorylation site of Na,K-ATPase (Asp^{369}) also shows homology with the β-subunit of F_1-ATPase of mitochondria and chloroplasts (cf. [84]). The Asp^{369} of the α-subunit in Na,K-ATPase corresponds to a Thr in the β-subunit and in agreement with this a

Table 2. Sequence homologies of a nucleotide binding domain: Na,K-ATPase, α-subunit [166], H, K-ATPase [164], Ca-ATPase (SR) [18], Ca-ATPase (PM) [161], H-ATPase, yeast [159], KdpB [71], K-ATPase, *Streptococcus faecalis* [169], F1-ATPase, β, *E. coli* and bovine, ADP-ATP exchange, adenylate kinase, and phosphofructokinase [185]

Na,K-, α	543	L	G	E	R	V	-	L	G	F	C	H	L	F	L	P	D	E	Q	F	P
H,K-	613	L	K	C	R	T	-	A	G	I	R	V	I	M	V	T	G	D	H	P	I
Ca-(SR)	611	Q	L	C	R	D	-	A	G	I	R	V	I	M	I	T	G	D	N	K	G
Ca-(PM)	673	K	C	Q	R	-	-	A	G	I	T	V	R	M	V	T	G	D	N	I	N
H-, yeast	545	S	E	A	R	H	-	L	G	L	R	V	K	M	L	T	G	D	A	V	G
F1-β bovine	243	E	Y	F	R	D	Q	E	G	Q	D	V	L	L	F	I	D	N	I	F	R
F1-α E. coli	267	E	Y	F	R	D	-	R	G	E	D	A	L	I	I	Y	D	D	L	S	K
ATP-ADP exchange	277	V	L	-	R	G	N	G	G	A	F	V	L	V	L	Y	D	E	I	K	K
Adenylate kinase	104	E	F	E	R	K	-	I	G	Q	P	T	L	L	L	Y	V	D	A	G	P
Phosphofructokinase	87	E	Q	L	K	K	-	H	G	I	Q	G	L	V	V	I	G	G	D	G	S

phosphorylated intermediate has not been demonstrated in the catalytic cycle of F_1-ATPase. Replacement of Thr^{285} of the β-subunit of F_1-ATPase with Asp by in vitro mutagenesis led to reduction in the rate of ATP binding and hydrolysis [125]. The mutant complex could not be phosphorylated, suggesting that other residues than Asp^{369} may be important for formation of the phosphorylated intermediates in Na,K-ATPase.

In addition to the segment around Asp^{369}, labeling with ATP analogues provides evidence for contribution from three additional segments in the α-subunit of Na,K-ATPase. The first indication came from ATP sensitive covalent insertion of FITC into Lys^{501} in α-subunit [90]. FITC also inserts covalently in Ca-ATPase of SR [108], in Ca-ATPase of plasma membranes [161, 182], and in H,K-ATPase [164]. The strong fluorescence signal provides a convenient probe for monitoring conformational transitions in the proteins. Fluorescein is a relatively large molecule and the specific insertion at Lys^{501} does not necessarily indicate that this residue also forms part of the nucleotide binding area.

The C-terminal part of the central domain contains two peptide segments (655–664 and 704–722 that are labeled by 5-(*p*-fluorosulfonyl)-benzoyl-adenosine (FSBA) [127], an ATP analogue that labels the Na,K-ATPase with concomitant inhibition of ATP hydrolysis, while K-phosphate and ouabain binding remain intact. The peptides are located at residues 655–664 and 704–722 of the α-subunit. Another ATP analogue, γ-[4-(*N*-2-chloroethyl-*N*-methylamino)]benzoylamide-ATP (CIR-ATP) inserts covalently in Asp^{710} in the

704–722 segment [135]. Labeling by these ATP analogues may relate to charged residues coordinating phosphate moieties in ATP.

Peptides released from the surface of the central domain of the α-subunit by tryptic cleavage have also been sequenced [126], and a provisional arrangement of the ATP binding domain in the cytoplasmic protrusion has been drawn as in Fig. 8 without attempts to illustrate molecular detail of the binding site. From the considerations above, the α-subunit provides the necessary segments for formation of a nucleotide binding area, but the question remains open whether there are more than one nucleotide site per $\alpha\beta$-unit. Two separate ATP sites with high and low affinity have been proposed on the basis of kinetic studies or inactivation experiments [156]. The different affinities for ATP have been explained [115] by the alternating E_1 and E_2 conformations with high and low ATP affinities, respectively. Other possibilities are complex interactions between subunits in oligomeric Na,K-ATPase, as suggested from negative cooperativity between binding sites in KCl medium, or the actual presence of two ATP sites per enzyme molecule. Since the maximum capacity of ATP binding in equilibrium experiments never exceeds one ATP bound per $\alpha\beta$-unit, one may assume that a presumptive additional site in the $\alpha\beta$-unit has low affinity and fewer coordinating groups than a site for high-affinity binding.

In search for additional nucleotide binding segments in the cation pump proteins attention has been drawn to the second cytoplasmic domain (residues 142–284 in Fig. 8). In this domain, bonds T_3 (Arg^{262}) and C_3 (Leu^{266}) are exposed to proteolytic cleavage in the E_1 form of the α-subunit [85]. The rate of cleavage is reduced by high affinity binding of ATP or ADP suggesting that the nucleotides protect the bonds by simple shielding [87]. Covalent insertion of FITC at Lys^{501} also reduces the rate of chymotryptic cleavage of C_3. The cleavage of C_3 has drastic consequences for the catalytic properties of Na,K-ATPase, since the cleaved protein remains in the E_1 conformation and binds ATP with high affinity without potassium antagonism, while vanadate binding is abolished [87]. Recently Goffeau and coworkers [54] found that in a mutant H pump in *Schizosaccharomyces pompe*, Gly^{268} is replaced by Asp. The mutant pump has reduced H-ATPase activity which is vanadate resistant. As shown in Table 3, this Gly^{268} is conserved in the α-subunit of Na,K-ATPase and in other cation transport pumps. The sequence TGD is present in the FSBA binding segments [84], and it is also found in the segment near T_3 and C_3 (Table 3). This together with the vanadate resistance of the mutant *S. pombe* led to the conjecture that this domain engages in low-affinity ATP binding or in formation of a leaving site for phosphate [161]. In the α-subunit of Na,K-ATPase, Gly^{261} is located next to the C_3 cleavage site (Leu^{266}). This cut with chymotrypsin does not interfere with ATP binding but interrupts the conformational transition and stabilizes the protein in E_1 conformation which does not bind vanadate. The most likely explanation for the vanadate resistance of the mutant *S. pombe* H-ATPase is therefore that the replacement of Gly^{268} with the negatively charged Asp interferes with the structural change of the protein in the same manner as the chymotryptic cleavage in the α-subunit of Na,K-ATPase.

Table 3. Amino acid sequences around the T_3 and C_3 cleavage sites in the second cytoplasmic loop of α-subunit of Na,K-ATPase [85, 166] and homologous segments in H,K-ATPase [164], H-ATPase from *Schizosaccharomyces pombe* [54], Ca-ATPase of sarcoplasmic reticulum [108], and Ca-ATPase of plasma membranes [161]

	Residue number	Sequence (T3 and C3 cleavage sites marked above)
Na,K-ATPase α-subunit	244	E G T A R G I V V Y T G D R T V M G R I A T L A S G
H,K-ATPase	260	E G T A Q G L V V S T G D R T I I G R I A S L A S G
H-ATPase (SP)	250	R G E G L M V V T A T G D S T F V G R I A T L A S G
H-ATPase (Y)	250	R G E G L M V V T A T G D N T F V G R A A A L V N K
Ca-ATPase (S)	216	A G K A L G I V - A T G V S T E I G K I R D Q M A A
Ca-ATPase (P)	268	E G S G R M V V T A V G V N S Q T G I I F T L L G A

Sites for Cation Binding and Occlusion

It is necessary to consider the question whether the pump has independent Na^+ and K^+ sites or if the cations bind alternately to the same set of sites. With a stoichiometry of 3Na/2K per ATP split this would mean that two sites bind Na and K alternately, while one site only binds Na^+. The possibility that the stoichiometry of the pump is variable and dependent on the concentrations of $[Na^+]_i$ or $[K^+]_o$, or membrane potential have been discussed and reviewed repeatedly. The consensus now appears to be that the stoichiometry is fixed at 3Na/2K per ATP split, independent of the conditions for transport [35, 36].

Experimentally, there are two approaches at identification of cation binding residues. One is specific chemical labeling, the other is oligonucleotide directed site-specific mutagenesis, but as yet there are only indirect inferences of the nature and identity of sites for Na^+ and K^+ in the Na,K pump protein.

Capacity for Binding and Occlusion of Na^+ or $K^+(Rb^+)$

For determination of the absolute capacity for cation binding it is important to work with preparations of high purity, since they allow absolute estimates of cation binding and occlusion per αβ-unit as well as relative estimates of cation binding capacity per site for phosphorylation or binding of ouabain and vanadate. Using CrATP for stabilizing the Na-occluded $[Na_n^+]E_1P$ conformation, a stoichiometry 2.6 Na^+ ions/EP was obtained. In similar preparations of Na,K-ATPase, the capacity for occlusion of $^{86}Rb^+$ was 8–10 nmol/mg protein

[184] and close to 2 Rb^+ ions per α-subunit. With extrapolation to saturating concentrations of Rb^+, this is close to 2 Rb^+ per phosphorylation site for both membrane bound and soluble Na,K-ATPase. Preparations of less purity were capable of occluding 3 Rb^+ ions per phosphorylation site for membrane-bound enzyme, but in this case the phosphorylation capacity was low relative to the specific activity [12, 57]. Using oligomycin to stabilize occlusion of Na^+ ions a stoichiometry of 3 Na^+ and 2 Rb^+ per phosphorylation sites were obtained in preparations with purity of 30%–40% [160].

These data correspond to binding of 1 mol ouabain and vanadate per mole of soluble or membrane bound $\alpha\beta$-unit. The cation binding data above therefore correspond to $2Rb^+$ and 3 Na^+ occluded per $\alpha\beta$-unit both in the soluble and in the membrane bound state.

Transport Stoichiometry and Net Charge of Na^+ and K^+ Complexes with Na,K-ATPase

The membrane potential can be used as a tool for characterization of the properties and net charge of the major conformations of Na,K-ATPase and their complexes with the cations. The potential difference across the membrane represents large electric fields of up to 500 kV/cm. If the conformational states of the Na,K pump have different electrical properties with respect to net charge or movable dipoles, the equilibria between conformations should interact with the field. The field may also affect binding of Na^+ or K^+ by altering the concentration of cation near the pump protein or by altering the position of charged groups in the cation sites.

Looking at overall Na,K pump rates, i.e., K^+ uptake or Na^+ extrusion, only the effects of the electric field on rate limiting reactions can be observed. Similarly, the field can only affect intrinsic Na^+ or K^+ binding constants if the cation concentration is below saturation. In red blood cells, the electrical field has been altered by the pulsed field method to induce small transmembrane potentials of 12 mV [179]. This stimulates ^{86}Rb uptake at low temperature (3°C) where the conformational change is expected to be rate limiting, but Na efflux was not activated. The significance of the observation is therefore uncertain.

In phospholipid vesicles reconstituted with renal Na,K pumps, the membrane potential can be controlled using gradients of K^+ or Li^+ and appropriate ionophores. Working with high or low ATP concentrations, Karlish and coworkers [59, 160] found that the conformational change E_1P3Na–E_2P3Na is accelerated by voltage, whereas the $E_2[2K]$–$E_1[2K]$ step is independent of potential. At limiting low Na^+ concentrations the imposed potential stimulates Na transport showing that the electric field has an effect on the intrinsic binding affinity for Na^+. This may mean the Na^+ has to cross part of the electric field in order to reach its sites, thus becoming concentrated by the field. Alternatively, the voltage may cause charges to move or dipoles to reorient in the cation binding area of the protein. Studies of the conformational transitions using fluorescein attached covalently to Lys^{501} (149) or to Cys^{457} (89) also shows that the E_1P–E_2P isomerization is voltage dependent while the E_2K–E_1K step is electrically silent.

The purified membrane bound Na,K-ATPase from kidney adsorbs to planar lipid bilayers and transient currents can be elicited upon release of caged ATP [6, 120]. The transient pump currents depend on Mg^{2+} + Na^+, but not K^+, and they are abolished by chymotryptic cleavage [6]. This agrees with the inference that Na^+ translocation and charge movements precede the K^+ transport step. Similar conclusions are reached from studies of presteady state fluxes of Na^+ ions in plasma membrane or reconstituted vesicles [48]. With the assumption that 3 Na^+ ions are transported into and 2 K^+-ions out of the vesicles in a ping pong reaction, these results mean that the net charge is + 1 with 3 Na^+ ions bound and zero with 2 K^+ ions bound [149].

In intact cells, studies of voltage dependence of the Na,K pump have given more controversial results. Gadsby and coworkers [8, 52, 121] demonstrated voltage dependence of the ouabain sensitive current in myocytes and observed that K-translocation by the pump is voltage insensitive. In squid axons [34] and *Xenopus* oocytes [104], interpretation of current voltage relationships has been more difficult. Regions of negative slope were demonstrated in the current voltage curves, and the data were interpreted to provide evidence for more than one voltage dependent step in the overall pump cycle [34, 104]. Rakowski and Paxon [145] reexamined the problem and found significant hysteresis in repeated measurements. These slow drifts of the currents with time suggest contaminating ionic currents, either through channels or through the open ends of the squid axons. Such data can be excluded, since the Na,K pump current reaches its steady state within few milliseconds following voltage steps [145]. With this scrutiny, the current voltage curves do not provide evidence for more than one voltage dependent step in the Na,K pump cycle.

DCCD Reaction with Carboxyls Engaged in Cation Binding

Hydrophobic *N*,*N*'-dicyclo-hexyl-carbodiimide (DCCD) reacts with carboxyl groups in relatively hydrophobic environments. In F_0 of ATP synthetase, proton pumps, and Ca-ATPase, inhibition of ATP hydrolysis is due to reaction with residues that are involved in binding H^+ or Ca^{2+} (for references see [60]). Inhibition of Na,K-ATPase by DCCD is also sensitive to the presence of Na^+, K^+, or ATP [60, 160], and the rate of crosslinking of the α-subunit appears to be lower than the rate of inactivation of the Na,K-ATPase activity. Gorga suggested that Na,K-ATPase in addition to Asp^{369}, contains two carboxyls that are important for the conformational change in the protein [60]. At high DCCD/protein ratios intramolecular crosslinking may cause inhibition of phosphorylation with less inhibition of the binding of Rb^+ to Na,K-ATPase [139].

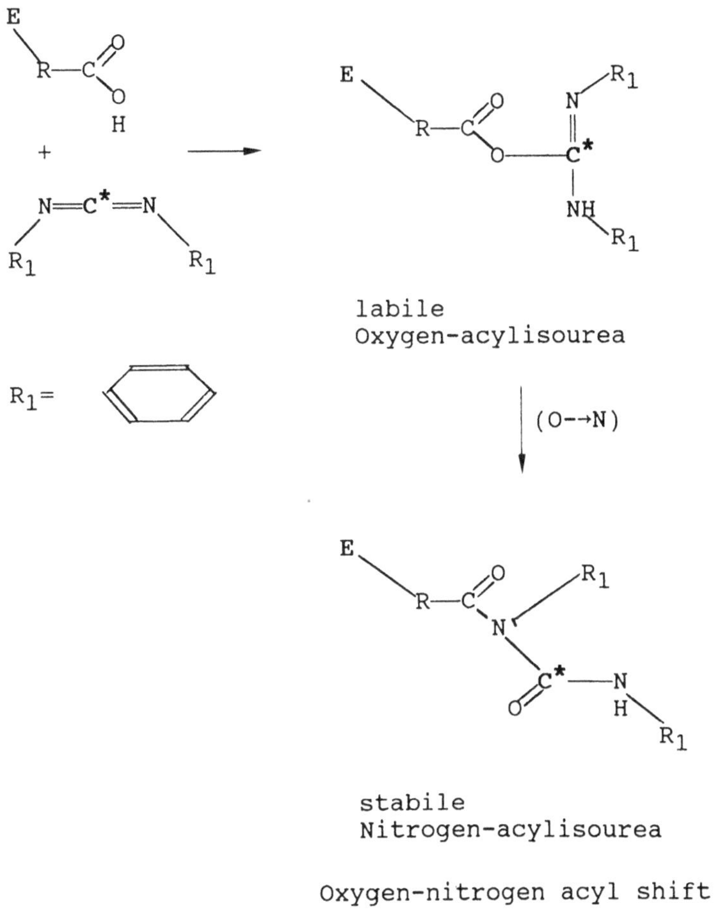

Fig. 12. Reaction of dicyclohexyl carbodiimide with a protonated carboxyl group in the enzyme protein (*E*). The carbodiimide becomes protonated and activated and reacts with the carboxylate to form a labile *O*-acylisourea adduct that can undergo an intramolecular *O*-to-*N* acyl shift to form a stable *N*-acylurea with incorporation of ^{14}C into the protein. Alternatively the labile *O*-acylisourea can be substituted by an intrinsic nucleophile to form an intramolecular crosslink or by an exogenous nucleophile (e.g., ethylenediamine) to form an irreversibly modified carboxyl group with the amino group of the nucleophile bound to the carbonyl group in an amide linkage. Both substitutions lead to release of [^{14}C] dicyclohexylurea and leave the enzyme protein devoid of radioactive label

As shown in Fig. 12, reaction of DCCD with a carboxyl group in the protein (E) leads to formation of an unstable O-acylisourea. Rearrangement in a oxygen-nitrogen acyl shift produces a stable *N*-acylisourea with retention of the ^{14}C label in the protein. The labile oxygen-acylisourea can undergo three alternative reactions: (a) hydrolysis with formation of [^{14}C]dicyclohexylurea; (b) substitution by an exogenous nucleophile (e.g., ethylenediamine or glycinethylester) with block of the residue R and loss of ^{14}C label from the protein; or (c) crosslinking by nucleophilic attack of a intrinsic nearby residue ($-$ R′) yielding

[^{14}C]dicyclohexylurea (for references see [2]). In all of these alternative reactions, the ^{14}C labeling is not retained in the protein, but the carboxyl residue may become blocked or crosslinked.

The complex nature of the DCCD reactions may obscure the causes of inactivation of a protein like Na,K-ATPase with multiple ligand binding sites. During reaction with [^{14}C] DCCD, the loss of catalytic activity may be due to blocking of charged groups or to internal crosslinking without retention of ^{14}C label in the protein. Simultaneously, another charged group may incorporate ^{14}C in a reaction which does not affect catalysis. However displacement of the labeling from a given peptide by Na^+ or K^+ and correlation of labeling to reduction of cation occlusion, could indicate a specific labeling of a cation binding residue. In addition to altering the capacity for occlusion, the blocking of carboxyls in the cation site should also increase the pK value determined from the pH dependence of cation occlusion.

In both Ca-ATPase [2] and the α-subunit of Na,K-ATPase [160], conditions have been specified for specific cation dependent incorporation of ^{14}C from DCCD into the proteins. In Ca-ATPase, the ^{14}C labeling from DCCD inserts into the N-terminal A_2 fragment (residues 1–198), but Ca^{2+} also protects against internal crosslinking in the A_1 fragment (residues 198–505) suggesting that conformational changes in this segment accompanies Ca^{2+} binding and occlusion [2].

In Na,K-ATPase, DCCD inactivates Na^+ or Rb^+ occlusion at identical rates in a first-order process [160]. Na^+ or Rb^+ occlusion are equally well protected by $Rb^+(K^+)$ or Na^+ against inactivation by DCCD, and the loss of occlusion capacity is correlated to loss of Na,K-ATPase activity. These results suggest that Na^+ and $K^+(Rb^+)$ bind to the same residues in the protein, and that the ions are transported sequentially in a ping-pong reaction. This inactivation by DCCD is accelerated about two fold by hydrophobic amines suggesting that the carboxyl residues reside in a hydrophobic environment. After labeling with [^{14}C] DCCD, it is seen from Fig. 13 that all of the K^+-sensitive incorporation is located to the α-subunit, but a large fraction (80%–90%) of the radioactivity is found in lipid. A major fraction of the K^+-sensitive label is located to hydrophobic peptide fragments of the α-subunit after cleavage. Isolation of the hydrophobic peptides by HPLC suffers from low recoveries (10%–30%) and alternative methods for isolation of the intramembrane fragments are therefore required.

Isolation of the Cation Occlusion and Transport Path After Tryptic Digestion

As shown by Karlish et al. [95], K^+ protects the intramembrane part of Na,K-ATPase against tryptic digestion. Figure 14 shows inactivation of ^{86}Rb occlusion or Na,K-ATPase activity in the presence of excess trypsin (20 μg trypsin per 100 μg membrane protein). The exchange of tris-Cl for RbCl protects against

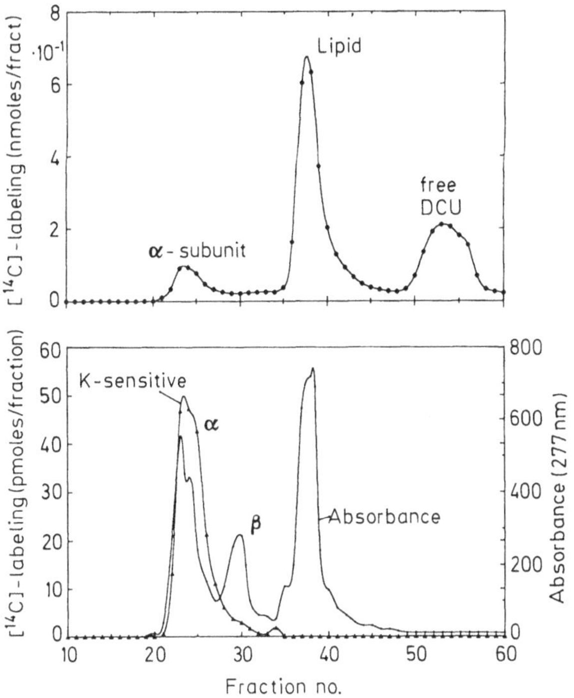

Fig. 13. Potassium-sensitive [^{14}C] DCCD labeling of the α-subunit of Na,K-ATPase. Membrane-bound Na,K-ATPase was incubated at 37°C for 10 min with [^{14}C] DCCD 0.4 μmol/mg protein in MES-Tris 10 m*M*, pH 5.5 and choline chloride 95 or 100 m*M*, KCl 0 or 5 m*M*, with the sum of the salt concentrations being constant at 100 m*M*. The reaction was stopped by adding defatted bovine serum albumin to a final concentration of 10 mg protein per ml. The membranes were collected by centrifugation at 110000 × *g* for 20 min in the Beckman Airfuge and washed once in 20 m*M* MES-Tris, pH, 4.2–7 and solubilized in 2% (w/v) $NaDodSO_4$, 100 m*M* sodium acetate, pH 4.5 at 1.5 mg protein per ml and injected into a TSK G 3000 SW column equilibrated with 1% (w/v) $NaDodSO_4$, 100 m*M* sodium acetate, pH 4.5 at 20°C and operated at 0.5 ml/min. About 5% of the total ^{14}C label is incorporated in the α-subunit (*upper frame*) and 34% of this labelling is K sensitive (*lower frame*), while the β-subunit remains unlabeled

inactivation of occlusion sites for both Rb^+ and Na^+ and NaCl in the presence of oligomycin protects against inactivation of occlusion sites for Rb^+ or K^+. Na,K-ATPase activity, ATP binding, and phosphorylation are rapidly abolished as the entire cytoplasmic protrusion is digested away. The fragments remaining in the membrane are the same whether Rb^+ or Na^+ ions are protecting the sites against inactivation. This demonstrates that Na^+ and Rb^+ share the same coordinating groups, and that they all are located in the intramembrane part of the enzyme. The sites seem to remain accessible from both membrane surfaces after the extensive cleavage.

The intramembrane cation occluding structure comprises a 19-kDa fragment plus several minor fragments containing the remaining hydrophobic transmembrane segments. The N-terminal sequence of the 19-kDa fragments

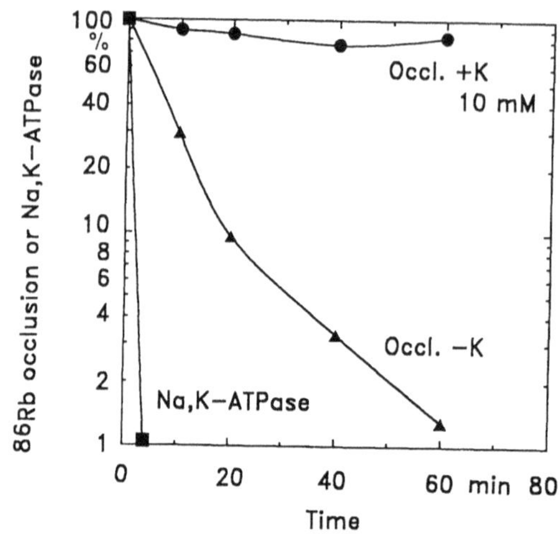

Fig. 14. Tryptic inactivation of ^{86}Rb occlusion in KCl 100 m*M* (●) or Tris-HCl 12 m*M* (▲) or of Na,K-ATPase (with Tris-HCl 12 m*M* and absence of Kcl) during digestion and 37°C with 20 μg trypsin per 100 μg Na,K-ATPase protein in Tes-Tris 10 m*M* pH 8.0, EDTA 1 m*M*. At the indicated times, aliquots are taken out amd mixed with trypsin inhibitor and assayed for ^{86}Rb occlusion and Na,K-ATPase activity as previously described [87, 95]

starts at 830-Asn-Pro-Lys-Thr-Asp-Lys-Leu-Val [95] in the α1-subunit of pig kidney Na,K-ATPase [133] probably extending Val^{996}–Arg^{997} yielding a segment of M_r 19447. The presence of residues coordinating cations is demonstrated by [^{14}C]DCCD binding to the fragments. Thus K^+-protected labeling was demonstrated in the 19-kDa fragment [95], suggesting that Glu^{954} and Glu^{955} are candidates for K^+-binding carboxyls.

The excellent separation in the tricine gel system (Shagger & Jagow) of minor hydrohobic peptides in the range 3–20 kDa opens possibilities for higher resolution in studies of chemical labeling of cation sites and the transmembrane segments. The approach of tryptic digestion thus opens an alternative method for isolating the cation sites and the pathway for cations across the membrane as a structure quite separate from the ATP binding and phosphorylation sites. Interaction between these domains in the cytoplasmic protrusion and the cation sites inside the membrane may thus require relatively long-range conformational transitions in the protein.

Site-Specific Mutagenesis and Models of Cation Binding Sites

An example of involvement of carboxyls in formation of cation sites is found in Ca^{2+} binding sites of the EF hand type characterized by a helix-loop-helix configuration. They are arranged pairwise and permit cooperative Ca^{2+} binding. Protein engineering of the cation site in Calbindin shows that clusters of negative surface residues (Glu and Asp) surround the inner sphere of cation coordination groups which consists of two Glu and two Asp residues [107]. The surface charges do not coordinate the cation, but they contribute about 7 kcal/mol to the free energy of binding of cation, and they affect the cooperativity of binding.

In the Ca-ATPase from sarcoplasmic reticulum, oligonucleotide-directed, site-specific mutagenesis has been applied to identify amino acids involved in

Ca^{2+} binding. Mutation of 30 glutamate and aspartate residues, singly or in groups, in a stalk sector near the transmembrane domain has little effect on Ca^{2+} transport. In contrast mutations to Glu^{309}, Glu^{771}, Asn^{796}, Thr^{799}, Asp^{800} or Glu^{908} resulted in loss of Ca^{2+} transport [29]. As an indication that these mutations abolished Ca^{2+} binding, phosphorylation from inorganic phosphate was observed even in the presence of Ca^{2+}. These data suggest that the carboxylate or carboxamide side chains or hydroxyl groups form coordinating groups in high affinity binding sites for Ca^{2+} near the center of the transmembrane domain. The homology around these residues between Ca-ATPase and α-subunit of Na,K-ATPase and other pump proteins suggest that their cation sites may have a similar location and overall structure (Tables 3 and 4).

Two types of compounds can be relevant models for cation binding and occlusion, ionophores and crown ethers. In models for ionophores, cation binding sites consist of electrophilic carbonyl groups that are located in a fixed cavity to accommodate the size of the selected cations inside the ionophore. Transition from an open to a closed or occluded configuration of the cation binding sites in the ionophore involves only limited changes in conformation of carbonyl residues and a substitution of solvent molecules from the inner coordination sphere of the cation [61]. This arrangement allow for excellent cation specificities, e.g., the affinity of valinomycin for K^+ is about six orders of magnitude higher than for Na^+. The complex of a crown ether (cyclohexyl-18-crown-6) with Na^+ has five nearly coplanar oxygens surrounding the Na^+ with the ion coordination completed by the remaining oxygen of the hexaether and a water molecule, and with Na-O distances of 2.5–2.6 Å. [16, 75]. The corresponding complex with K^+ can be formed with minimal distortion of the cyclic ether and with bond lengths of 2.8 Å. Such arrangements may provide the necessary equipment for the type of conformational transitions described below.

Table 4. Homology among cation pump proteins around E-309 of Ca-ATPase from sarcoplasmic reticulum: Ca-ATPase SR [18, 29], Na,K-ATPase [98, 162, 166], H,K-ATPase [164], Ca-ATPase PM [161, 182], H-ATPase yeast [159], K-ATPase, *E. Coli* [71], K-ATPase, *Streptococcus faecalis*[169]

											*												
Ca-ATPase SR	300	V	A	L	A	V	A	A	I	P	E	G	L	P	A	V	I	T	T	C	L	A	L
Na,K-ATPase α1-3	323	I	G	I	I	V	A	N	V	P	E	G	L	L	A	T	V	T	V	C	L	T	L
H,K-ATPase rat	334	M	A	I	V	V	A	Y	V	P	E	G	L	L	A	T	V	T	V	C	L	S	L
Ca-ATPase PM iso-1	424	V	T	V	L	V	V	A	V	P	E	G	L	P	L	A	V	T	I	S	L	A	Y
H-ATPase yeast	327	L	G	I	T	I	I	G	V	P	V	G	L	P	A	V	V	T	T	T	M	A	V
K-ATPase E.coli B	256	V	A	L	L	V	C	L	I	P	T	T	I	G	G	L	L	S	A	-	S	A	V
K-ATPase Str. faecalis	227	V	T	V	F	I	I	A	C	P	H	A	L	G	L	A	I	P	L	V	V	A	R

Structural Transitions in the Protein Related to Energy Transformation and Na,K-Transport

Exposure and protection of bonds on the surface of the cytoplasmic protrusion provides unequivocal evidence for structural changes in α-subunit accompanying E_1–E_2 transition in Na,K-ATPase [78]. Localization of the proteolytic splits provided a shortcut to identification of residues involved in E_1–E_2 transition [85, 88] and to detection of structure-function correlations [88]. The conformational changes involve residues in ATP binding and phosphorylation domains, and they are transmitted to the extracytoplasmic surface with changes in binding affinity for cations and ouabain. In addition to bonds exposed to proteolysis, the transition involves tryptophans, sulfhydryl groups, protonizable groups and residues binding FITC [90] and iodoacetamide fluorescein [89] (see previous reviews for details [55, 80]). This treatise will focus on attempts to identify residues in the amino acid sequences that undergo spatial rearrangements as part of E_1–E_2 transition and cation translocation.

Conformation Dependent Proteolytic Cleavage of Na,K-ATPase

Definition of E_1 and E_2 conformations of α-subunit of Na,K-ATPase involves identification of cleavage points in the protein as well as association of cleavage with different rates of inactivation of Na,K-ATPase and K-phosphatase activities [78, 80]. In the E_1-form of Na,K-ATPase the cleavage patterns of the two serine proteases are clearly distinct. Chymotrypsin cleaves at Leu^{266} (C_3) (Fig. 8), and both Na,K-ATPase and K-phosphatase are inactivated in a monoexponential pattern [87, 88]. Trypsin cleaves the E_1-form rapidly at Lys^{30} (T_2) and more slowly at Arg^{262} (T_3) (Fig. 8) to produce the characteristic biphasic pattern of inactivation. Localization of these splits was determined by sequencing N-termini of fragments after isolation on high resolution gel filtration columns [85].

The E_2 form is not cleaved by chymotrypsin, but trypsin cleaves at Arg^{438} (T_1) and subsequently at Lys^{30} (T_2) and tryptic inactivation of E_2K or E_2P forms is linear and associated with cleavage at Arg^{438} (T_1) [78, 85]. Inactivation of K-phosphatase is delayed because cleavage of T_1 and T_2 in sequence is required for inactivation of K-phosphatase activity [79]. A bond in the C-terminal half of α-subunit may be exposed in ouabain bound complexes [23].

Thus, transition from E_1 to E_2 consists of an integrated structural change involving protection of bond C_3 or T_3 in the second cytoplasmic domain and exposure of T_1 in the central domain, while position of T_2 in the N-terminus is altered relative to the central domain (T_1) so that cleavage of T_2 becomes secondary to cleavage of T_1 within the same α-subunit in the E_2 form.

Cleaved Derivatives; Cleavage of Bond 2 and Regulatory Function of N-Terminus

Selective cleavage of bonds in α-subunit of Na,K-ATPase is important for examining structure-function relationships for the protein. The N-terminus of the α-subunit is strongly hydrophilic with clusters of alternating positive and negative residues between residues 15 and 60. It is a flexible structure with a strong propensity for α-helix formation and several predicted turns, notably at residues 14–16, 35–36, 49–50, and 70–72. The removal of residues 1–30 by selective tryptic cleavage at Lys^{30} (T_2) is possible because the rate of cleavage of this bond is up to 60-fold higher than the rate of tryptic cleavage at Arg^{262} (T_3) in the second slow phase of inactivation. T_2 cleavage reduces Na,K-ATPase activity by 50%–60% [78] with a parallel loss of Na,K-transport [93]. The loss of activity is explained by a poise of equilibria between cation bound (E_1Na–E_2K) and phosphoforms (E_1P–E_2P) in direction of E_1 forms [88]. These properties of the selectively cleaved derivative suggested that charged residues in the N-terminus engage in salt bridge formation as part of E_1–E_2 transition [79, 87, 90], and that this is important for control of the rate of active Na,K-transport by pumps containing the α1-isoform.

These data suggest that the substantial differences between the amino acid sequences of the N-terminal region in the α1-, α2-, and α3-isoforms [162] reflect different regulatory functions of these segments. The first 30 residues of the α1 isoform has a high frequency of charges with 8 lysines and 2 arginines. In the α2-isoform there are 2 negative and 3 positive charges fewer in this segment than in the α1-isoform suggesting that the strength of salt bridges formed with other segments may be weaker. The first 11 residues of α3-isoform are not homologous with the other isoforms, but all isoforms of the α-subunit have a stretch of lysines around residue 30 [162]. Also the N-terminus of H,K-ATPase [164] possesses a lysine-rich sequence with strong homology with the α1-subunit around T_2. Biphasic cleavage patterns resembling those in Na,K-ATPase have also been demonstrated in H,K-ATPase [152], but so far without identification of cleavage points.

Effect of C_3 Cleavage on E_1P–E_2P Transition and Cation Exchange

The alternating exposure of T_1 (Arg^{438}) in the E_2 form and C_3 (Leu^{266}) or T_3 (Arg^{262}) in the E_1 form reflects that motion within the segment (M_r 18 170) between these bonds including the phosphorylated residue (Asp^{369}) is an important element in E_1–E_2 transition. This is illustrated by the widely different consequences of selective cleavage of C_3 and T_1 for E_1–E_2 transition and cation exchange.

C_3 cleavage is a selective and particularly efficient tool for examining structure-function relationships of the second cytoplasmic domain. Binding affinities for ADP and ATP are reduced four- to five fold, while TNP-ATP binds with the same affinity as in native Na,K-ATPase [87, 88]. Nucleotide binding is not affected by K^+ or Rb^+ although cation sites are undamaged. Conversely the cleaved enzyme also binds ^{86}Rb with high affinity [87], and it occludes the cations, but cation binding and occlusion are unaffected by nucleotide.

Combination of C_3 and T_1 cleavages allowed comparison of the properties of four phosphopeptides, the intact α-subunit, and the 83-kDa, 47-kDa, and 18-kDa fragments [88]. Transfer of phosphate from ATP to the fragments remains sensitive to Na^+ (Fig. 15), but the 83-kDa and 18-kDa fragments are stabilized in the E_1P form. This is a specific effect of C_3 cleavage because comparison with the properties of the 47-kDa fragment shows that after cleavage of T_1 alone, the 47-kDa fragment forms both E_1P and E_2P and the Na-Na and K-K exchange reactions are preserved. Stabilization of the protein in the E_1P form after C_3 cleavage allows demonstration of ^{22}Na occlusion [55], and the split abolishes the transient current associated with the translocation of Na^+ in the first turnover of the pump [6].

Transport studies in reconstituted vesicles show that C_3 cleavage blocks the relatively fast Na-Na or K-K exchange (20–40 s^{-1}) and Na-K exchange (500 s^{-1}; Fig. 16), but the slow passive ouabain-sensitive Rb-Rb exchange (1 s^{-1}) and occlusion of K^+ or Na^+ are only partially affected. This is basis for the conclusion that C_3 cleavage interferes with coordination between structural changes of the phosphorylated segment and cation sites that may be formed by the segment between C_3 and the phosphorylated residue Asp^{369} [84, 88].

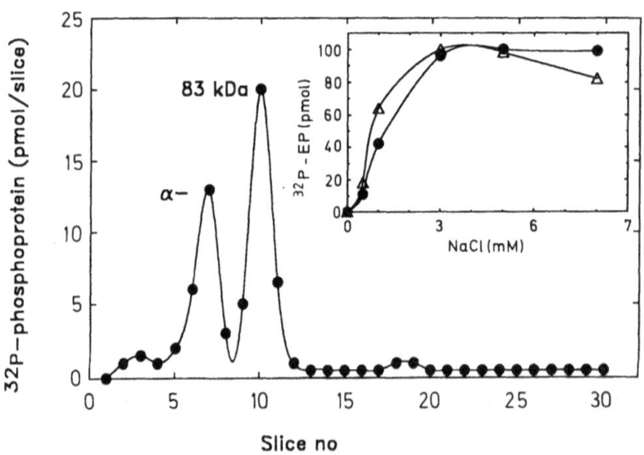

Fig. 15. Na^+-dependent phosphorylation of the intact α1-subunit and the 83-kDa fragment after cleavage purified membrane bound Na,K-ATPase with 5 μg α-chymotrypsin per 100 μg protein in 1 ml 15 m*M* tris, 10 m*M* NaCl, pH 7.5 for 30 min. Phosphorylation for 5 s with 25 μ*M* [g^{32}P]ATP in 3 m*M* $MgCl_2$, 25 m*M* Tris-HCl, pH 7.5, and 3 m*M* NaCl. After precipitation and wash in 4% perchloric acid, the phosphoproteins were dissolved in SDS and fractionated by electrophoresis at pH 2.4. The gels were sliced (3 mm) and counted. *Inset* shows the amount of ^{32}P in the α-subunit (●) and the 83-kDa fragment (△) as a function of NaCl concentration. (Redrawn from [88])

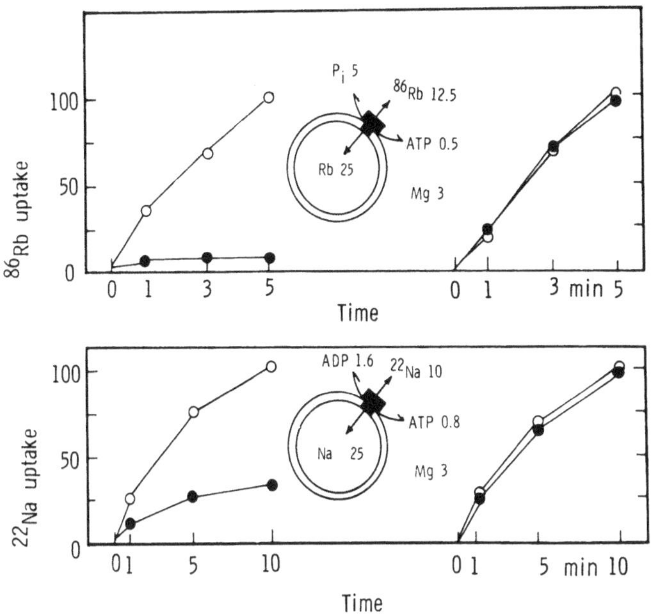

Fig. 16. Effect of cleavage of bond C_3 (*left*) or bond T_1 (*right*) on the rate of Rb-Rb exchange or Na-Na exchange in reconstituted vesicles. Purified membrane-bound Na,K-ATPase was inactivated with chymotrypsin in NaCl to 23% or with trypsin in KCl to 29% Na,K-ATPase activity remaining. Controls were incubated without protease. After solubilization in cholate, Na,K-ATPase was reconstituted into phospholipid vesicles by the freeze-thaw technique with composition of intra-vesicular and external media as indicated. Na-Na exchange at 24°C was initiated by adding ^{22}Na, ADP, and ATP. Intravesicular ^{22}Na was measured after passage over an ion-exchange column and Na-Na exchange was calculated as the difference in rate with and without nucleotides in the medium. Rb-Rb exchange was calculated as the difference in rate of ^{86}Rb accumulation with and without 1 m*M* vanadate in the external medium. (From [88])

The notion that the segment containing C_3 and T_3 is important for conformational adaptability of the protein is supported by mutations in yeast. Serrano et al. [28] induced mutations in plasmids containing the genes H-ATPase of *Saccharomyces cerevisiae* and found a thermosensitive mutant after mutation of Gly 254–Ser (Table 3). Goffeau and coworkers [54] showed that the Gly-268-Asp substitution is responsible for a mutant phenotype in *Schizosaccharomyces pombe*. The mutant H-ATPase has reduced ATPase and proton pumping activity and it is vanadate resistant. The mutation may thus produce a higher concentration of E_1 forms and less of the vanadate binding E_2 form during steady state ATP hydrolysis. These observations agree with the proposal that this conserved region of the cation pumps (Table 3) is involved in the structural transitions accompanying isomerization of E_1P–E_2P phospho-forms that are important for transduction of the energy from ATP hydrolysis to movement of cations.

Tryptophan Fluorescence and Secondary Structure Changes

Movement of peptide segments from cytoplasmic to hydrophobic environments accompanying E_1–E_2 transitions in Na,K-ATPase may be related to changes in intensity of intrinsic fluorescence in the proteins. Quenching analysis of Na,K-ATPase suggests that the tryptophan fluorescence changes are directly related to change of overall protein conformation (180). Tryptophan fluorescence is increased 2%–3% by transition from E_1 to E_2 in Na,K-ATPase (cf. [90, 94]). This quantitative difference in fluorescence of E_1 and E_2 forms may be understood in terms of different locations of the tryptophans in the protein structures. In Na,K-ATPase, only 2 of a total of 12 tryptophans in the α-subunit are located in predicted transmembrane segments.

Circular dichroism spectroscopy may detect changes in ratio between secondary structure elements, but even large shifts in position of α-helices and β-sheets relative to each other do not affect CD spectra. The results of experiments addressing the question whether E_1–E_2 transitions are accompanied by changes in ratio among secondary structure elements raised some controversy. CD spectroscopy shows that Na,K-ATPase contains a roughly equal mixture of α-helical, β-sheet, and random coil structures [62], and changes in CD spectra accompanying exchange of Na^+ for K^+ are interpreted to involve α-helix to β-sheet transition of about 7%. A conversion in the opposite direction of 100 residues from β- to α-helical conformation is suggested by Raman spectroscopy [119]. In apparent contrast to this, another group found that addition of K^+ to a Tris-HCl medium did not cause changes in CD spectra [65]. Infrared spectroscopy in the amide I region shows that Na^+ and K^+ bound forms of renal Na,K-ATPase have almost the same structure [25], but it is not known if the enzyme is active in D_2O. It is not clear to what extent these contradictory results concerning secondary structure transitions are related to the special structure of the membrane bound Na,K-ATPase.

Transposition of Mass Accompanying Translocation

Altered degrees of immersion of protein into the hydrophobic environment of the membrane may be monitored by photoactivable hydrophobic reagents [91]. Labeling of the α-subunit of Na,K-ATPase [80, 91] and Ca-ATPase [4] provide evidence that movement of peptide segments between hydrophilic and hydrophobic environments may accompany E_1–E_2 transitions. The hydrophobic photoactivable label [^{125}I]iodonaphtylazide [13] labels E_2 forms 10%–25% more than E_1 forms of Na,K-ATPase [91]. A more detailed analysis shows that the hydrophobic photolabel TID [4, 80] may be used to determine relative accessibility of the peptide chain from the lipid phase. Thus in Ca-ATPase from sarcoplasmic reticulum we were able to demonstrate a 10%–20% higher degree of hydrophobic labeling in E_2 forms than in E_1 forms. The preferential hydrophobic labeling of E_2 forms was located to the phosphorylated A1 fragments (residues 199–505). This difference between hydrophobic labeling of E_1 and E_2 forms corresponds to movement of a minimum of 38 residues into the bilayer [4].

A monoclonal antibody binds to extracellular aspects of the α-subunit and stabilizes the protein in the E_2K or E_2P forms [154]. Antibody binding enhances K-phosphate activity and the rate of ouabain binding. It will be of interest to learn the exact site of binding since the antibody epitope may be engaged in the E_1–E_2 conformational transition.

Effect of Chymotryptic Cleavage on Conformation of Occluded Complexes of Na,K-ATPase

Occluded cations are bound within the structure of the pump protein so that they are prevented from exchange with medium cations [141]. The presumptive occlusion cavity may be part of the transport pathway, because occlusion and release of cations from the cavity is governed by physiological ligands [92]. Recently, it was shown that the $\alpha\beta$-unit is the minimum functional unit for occlusion of ^{22}Na or ^{86}Rb in Na,K-ATPase [184]. This observation shows that the presumptive occlusion cavity is formed within the structure of an $\alpha\beta$-unit and that interaction between α-subunits is not required for occlusion.

Understanding the relationship between occlusion-deocclusion and E_1–E_2 transitions in the protein may be a key to solving the cation transport mechanism. It is therefore important to determine the protein conformations of the occluded complexes of Na,K-ATPase with Na^+ or Rb^+. ATP and ADP act with low affinity (K_d 0.3–1 mM) to induce transition from $E_2(K2)_{occl}$ to E_1(K2)ATP and rapid release of both ions [55, 92]. These are normal steps in the reaction cycle (Fig. 17) and represent a significant rate limitation in the pump cycle at low ATP. Phosphorylation from P_i releases occluded Rb^+ or K^+ ions at the extracellular surface of the pump with rates of 5–15 s^{-1} at 20°C. In presence of Na^+ both ions are rapidly released, but with K^+ in the medium there is slow release of one occluded ion. This lends support to a flickering gate hypothesis for deocclusion to the extracellular surface [50, 56].

After cleavage of bond C_3 (Leu^{266}) with chymotrypsin [87, 88] the E_1P form occludes Na^+ ions with a stoichiometry of 3Na/EP [55]. Lower capacities are seen after treatment with N-ethylmaleimide or oligomycin [46]. Using CrATP as phosphorylating substrate values of ^{22}Na occlusion as high as 11 nmol/mg protein were reached in selected purified preparations of Na,K-ATPase with 6 nmol EP/mg protein [184]. This yields a stoichiometry of 1.8 Na/EP, but appropriate extrapolation for unsaturation of binding sites corrects this value to 2.7 Na/EP. Phosphorylation from CrATP stabilizes a Na-occluded conformation, CrADP-E_1P[Na]. Examination of tryptic cleavage patterns of the CrADP-E_1P[Na] form, confirms that the α-subunit is cleaved according to the pattern that is characteristic for the E_1 form. Occlusion of Na^+ in CrADP-E_1P[Na] has been compared with occlusion of Na^+ in complexes with ouabain

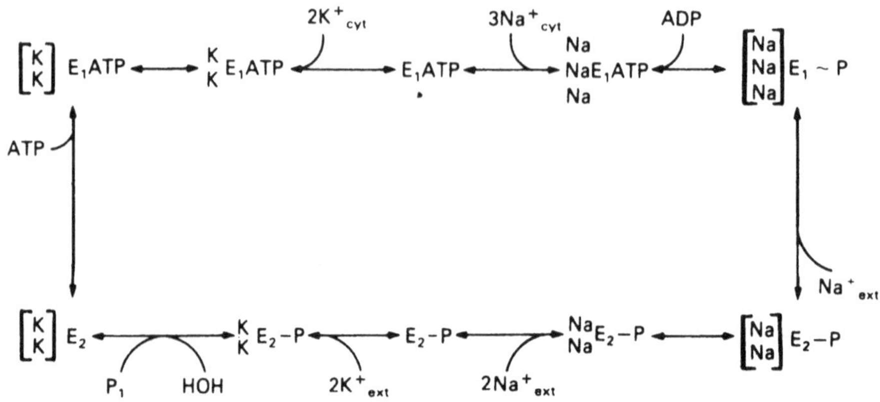

Fig. 17. $E_1 - E_2$ reaction cycle of the Na,K pump with four major occluded conformations and ping-pong sequential cation translocation. The phosphoforms can occlude Na^+ and dephosphoforms can occlude K^+ or Rb^+. Na^+ and K^+ *without brackets* are cations bound to an open form such that they can exchange with medium cation; [Na^+] or [K^+] *within brackets* are occluded and prevented from exchanging with medium cations. It is proposed that release of Na_{ext} accompanies transition from $E_1P[Na_3]$ to $E_2P[Na_2]$, since the capacity for occlusion of Na^+ in the ouabain-stabilized E_2P-form is lower than in the E_1P-form prepared by incubation with CrATP [84]

(cf. [49]). The capacity for occlusion of Na^+ is low in presence of Mg^{2+} and ouabain, but it is increased to 5–6 nmol/mg protein in presence of 1 nmol/l phosphate without changing the apparent affinity for Na^+. Phosphorylation of the aspartyl residue may thus be necessary for occlusion or directly involved in coordinating the cation in the E_2 form. The observation that the capacity for Na^+ occlusion by the Ouab-E_2P[Na] complex is lower than that of the CrADP-E_1P[Na] complex suggests that the transition from E_1P to E_2P involves a change in number of cation coordinating groups. The change in orientation accompanying the transition therefore results in the release of one of the occluded Na^+ ions to the extracellular medium. The change in number of cation coordinating groups and in orientation of the sites may be mediated by a twist of transmembrane helices carrying cation coordinating groups (Tables 4, 5) as previously proposed by Tanford [177].

Rb^+(or K^+) ions are occluded in the E_2[Rb] form but there is also evidence for occlusion of Rb^+ in E_1 forms [87]. Formation of a ternary complex of Na,K-ATPase, K^+, and ATP is clearly demonstrated after cleavage by chymotrypsin or after blocking sulfhydryl groups. The chymotrypsin cleaved by enzyme occludes Rb^+(or K^+) with high affinity in a process that is unaffected by ATP and without changes in fluorescence from intrinsic or extrinsic probes. Na^+ and Rb^+(or K^+) compete for occlusion in the chymotrypsin cleaved E_1ATP form and in complexes of Na,K-ATPase with ouabain and P_i. This is in agreement with the notion that a single set of cation sites may bind and occlude either Na^+ or Rb^+(or K^+). The evidence for formation of occluded complexes of the $\alpha\beta$-unit of Na,K-ATPase with Na^+ or Rb^+(or K^+) in both E_1 and E_2 forms is basis for the reaction cycle (Fig. 17).

Table 5. Homologies among cation pump proteins around N-796, T-799, and D-800 of Ca-ATPase from sarcoplasmic reticulum (for references see Table 3)

														*			*	*							
Ca-ATPase SR	783	P	E	A	L	I	P	V	Q	L	L	W	V	N	L	V	T	D	G	L	P	A	T	A	L
Na,K-ATPase α1	797	P	L	P	L	G	T	V	T	I	L	C	I	D	L	G	T	D	M	V	P	A	I	S	L
H,K-ATPase rat	808	P	L	P	L	G	C	I	T	I	L	F	I	E	L	C	T	D	I	F	P	S	V	S	L
Ca-ATPase PM	886	D	S	P	L	K	A	V	Q	M	L	W	N	V	L	I	M	D	T	L	A	S	L	A	L
H-ATPase yeast	714	D	N	S	L	D	I	D	L	I	V	F	I	A	I	F	A	D	-	V	A	T	L	A	I

High- and Low-Energy Phosphate Forms, E_1P–E_2P, and Ion Translocation

A fundamental property of E_1 and E_2 forms of Na,K-ATPase is their specific chemical reactivities. E_1 forms of both enzymes accept γ-phosphate from ATP, whereas E_2 forms react with P_i. E_1P forms are ADP sensitive, whereas E_2P forms do not transfer phosphate to ADP [55] (Fig. 17). The molecular basis for these distinctions has not been elucidated in detail, but proteolytic cleavage and hydrophobic labeling studies suggest that rearrangement of peptide structures formed from the segment connecting the second cytoplasmic domain with the phosphorylated aspartic acid residue may play an important role [88]. A separation of phosphate from the nucleotide site in E_2 forms was proposed as an interpretation of the effects of proteolytic cleavage of bond 1 and bond 3 in the α-subunit of Na,K-ATPase on phosphorylation, dephosphorylation and cation exchange [80, 88].

Based on a series of studies of the effect of organic solvent on the reaction of Ca-ATPase with P_i and ATP synthesis, DeMeis proposed that a different solvent structure in the phosphate microenvironment in E_1 and E_2 forms the basis for existence of high- and low-energy forms of the aspartyl phosphate [32, 33, 140]. Acyl phosphates have relatively low free energy of hydrolysis when the activity of water is reduced, due to the change of solvation energy. The covalently bound phosphate may also reside in a hydrophobic environment in E_2P of Na,K-ATPase since increased partition of P_i into the site is observed in presence of organic solvent [84]. It is seen that DMSO increases the affinity of Na,K-ATPase for inorganic phosphate in the same manner as in Ca-ATPase.

In the reaction scheme for Na,K pumping in Fig. 17, the E_1–E_2 transitions constitute the cation translocation steps, Na extrusion involving the E_1P–E_2P transitions and K uptake the E_2K–E_1ATP transition. Na^+ and K^+ are transported in sequence, Na^+ first and then K^+ in a ping-pong, sequential reaction, but this model is challenged by some authors. In a review, Glynn [55] evaluates problems related to interpretation of kinetic studies with [γ^{32}P]-ATP and examines in detail the arguments for and against the competence of E_1P and E_2P in this transport cycle.

Recent developments has allowed more direct studies of the correlation between E_1–E_2 transitions and cation translocation. New techniques have been introduced for studying effects of membrane potential is reconstituted vesicles [149], pre-steady-state cation fluxes in vesicles [48], measurement of transient Na transport currents in planar lipid bilayers reconstituted with purified Na,K-ATPase [6, 120] and for voltage clamping in intact cells [121, 145] (see "Transport Stoichiometry and Net Charge of Na^+ and K^+ Complexes with Na,K-ATPase"). Data from this work provide convincing evidence for correlation of ping-pong transport of Na^+ and K^+ with E_1–E_2 transitions in the protein.

An important problem in examining the relevance of E_1–E_2 transitions for active cation transport was to determine the sidedness of ligand effects on the conformation of the pump and the correlation between conformational changes and cation translocation. Tryptic digestion of Na,K pumps reconstituted with their cytoplasmic aspects facing outward gives biphasic or linear patterns of inactivation of Na,K transport [93] that are similar to those for inactivation of purified membrane bound Na,K-ATPase. These experiments show that K^+ and Na^+ ions stabilize E_2K or E_1Na, respectively, after combination with sites on the cytoplasmic surface.

In a new reconstitution technique [120], membrane fragments containing purified Na,K-ATPase from kidney [80] are adsorbed to planar lipid bilayers. Light-induced conversion of caged ATP is used to study transient currents coupled to Na^+ ion transport in absence and presence of K^+ [6, 120]. The method allows differentiation between a K^+-independent, transient current and a ($Na^+ + K^+$)-dependent, stationary current reflecting the continuous operation of the Na,K pump. Selection of the charge translocating step is possible since the transient current is selectively blocked by chymotryptic cleavage [6]. In view of the earlier demonstration that the cleavage of bond C_3 (Leu^{266}) of the α-subunit stabilizes Na,K-ATPase in the Na-occluded $E_1P[Na]$ form [87, 88], the data show that reactions preceding $E_1P[Na3]$ are electrically silent. This identifies the E_1P–E_2P transition as the charge translocating step. In the lipid bilayer, the Na^+ transport step coupled to E_1P-E_2P transititon is electrogenic both in absence and presence of K^+ since Na^+ stimulates dephosphorylation at a low rate in absence of K^+ [30, 120]. Thus, measurement of transient Na^+ currents from transport by Na,K-ATPase in black lipid membranes shows that Na^+ is released as an early electrogenic event before K^+ is bound. The transient Na^+ current elicited upon lysis of caged ATP is in agreement with ping-pong transport of Na^+ first and then K^+, as shown in Fig. 17. These conclusions from work on the purified Na,K-ATPase in artificial bilayers are in perfect agreement with the detailed kinetic analysis of Na,K fluxes through the pump in red blood cells [153, 171].

The basis for using the $\alpha\beta$-unit as a functional unit in the transport cycle in Fig. 17 is the evidence for the structure and function of cation pumps that was discussed in the preceding sections of this article. The functional unit is an $\alpha\beta$-unit (see "Subunit Structure of Na,K-ATPase"), with cytoplasmic (see "Extracellular Cardiac Glycoside Binding Site") and extracellular protrusions as revealed in three-dimensional models of crystals of Na,K-ATPase (see "Subunit Structure of Na,K-ATPase"). The demonstration of occluded E_1 and E_2 forms in combination with a single set of sites for binding of Na^+ or Rb^+ (see "Structural Transitions in the Protein-Related Energy Transformation and Na,K Transport"), led to the proposal of a model in which Na^+ ions are occluded in both the $E_1P[Na3]$ and $E_2P[Na2]$ forms with the $E_1P[Na3]$–$E_2P[Na2]$ transition coupled to charge translocation. In the model in Fig. 17, dehydration of the cation and coordination in an occluded conformation are therefore considered to be principal conditions for the conformational changes in the protein leading to translocation. The cation binding sites are assumed to consist of electrophilic carbonyl groups located in a cavity to accommodate the cations inside the protein. Dehydration of Na^+ takes place on the cytoplasmic surface and after coordination of the cation, the carbonyl groups rearrange from open to occluded form. The rearrangement of the inward (E_1) or outward (E_2) facing configurations of the cation sites involves only limited changes in conformation of carbonyl residues and a substitution of solvent molecules from the inner coordination sphere of the cation [87]. There are at least two aspects of the subsequent E_1–E_2 conformational transitions of the protein. One is a change in capacity for Na^+ binding and in apparent cation affinity. The conformational transition accompanying E_1P–E_2P isomerization may alter the position of one or more of the coordinating groups resulting in release of one Na^+ ion as suggested in Fig. 17. The other is a change in orientation of the residues possessing the coordinating carbonyl groups. In the E_1 form they are exposed to the bottom of the cytoplasmic mouth, while in the E_2 form they become accessible from the bottom of the extracellular mouth of the cation pathway.

At the extracellular mouth of the cation pathway the cation site of the E_2 form shifts from closed to open configuration and Na^+ is rehydrated and released. The same set of coordinating groups, except phosphate, bind K^+ from the outside. The mouth of the cation pathway is similar to the mouth of the K channel which has the function of dehydrating K^+ before the ions can pass the narrow part of the channel (cf. [171]).

References

1. Addison R (1986) Primary structure of neorospora plasma membrane H-ATPase. J Biol Chem 261:14896–14901

2. Ancos JC, Inesi G (1988) Patterns of proteolytic cleavage and carbodiimide derivatization in sarcoplasmic reticulum adenosinetriphosphatase. Biochemistry 27:1793–1803
3. Andersen JP (1989) Monomer oligomer equilibria of sarcoplasmic reticulum Ca-ATPase and the role of subunit interaction in the Ca-pump mechanism. Biochim Biophys Acta 988:47–72
4. Andersen JP, Vilsen B, Collins JH, Jørgensen PL (1986) Localization of $E_1 - E_2$ conformational transitions of sarcoplasmic reticulum Ca-ATPase by tryptic cleavage and hydrophobic labeling. J Membr Biol 93:85–92
5. Andersen JP, Vilsen B, Nielsen H, Møller JV (1986) Characterization of detergent solubilized sarcoplasmic reticulum Ca-ATPase by HPLC. Biochemistry 25:6439–6447
6. Apell H-J, Borlinghaus R, Lauger P (1987) Fast charge translocations associated with partial reactions of the Na,K-pump. II. Microscopic analysis of transient currents. J Membr Biol 97:179–191
7. Aquila H, Link TA, Klingenberg M (1987) Solute carriers involved in energy transfer of mitochondria form a homologous protein family. FEBS Lett 212:1–5
8. Bahinski A, Nakao M, Gadsby DC (1986) Potassium translocation by the Na,K-pump is voltage insensitive. Proc Natl Acad Sci USA 85:3412–3416
9. Baker ME (1986) Computer based search for steriod and DNA binding sites on estrogen and glucocorticoid receptors. Biochem Biophys Res Commun 139:281
10. Barlet-Bas C, Khadouri C, Marsy S, Doucet A (1988) Sodium independent invitro induction of Na,K-ATPase by aldosterone in renal target cells: permissive effect of triiodothyronine. Proc Natl Acad Sci USA 85:1707–1711
11. Barrabin H, Scofano HM, Inesi G (1984) Adenosinetriphosphatase site stoichiometry in sarcoplasmic reticulum vesicles and purified enzyme. Biochemistry 23:1542–1548
12. Beauge JA, Glynn IM (1979) Occlusion of K ions in the unphosphorylated sodium pump. Nature 280:510–512
13. Bercovici T, Gitler C (1978) Synthesis of $[^{125}I]$iodonaphtylazide. Biochemistry 17:1484–1489
14. Boardman LJ, Huett M, Lamb JF, Newton JP, Polson JM (1974) Evidence for the genetic control of the sodium pump density in HeLa cells. J Physiol 241:771–794
15. Bowen WJ, McDonough A (1987) Pretranslational regulation of Na,K-ATPase in cultured canine kidney cells by low K. Am J Physiol 52:C179–C189
16. Boyer PD (1988) Bioenergetic coupling to protonmotive force. Trends Biochem Sci 13:5–7
17. Brandl CJ, Deber CM (1986) Hypothesis about the function of membrane-buried proline residues in transport proteins. Proc Natl Acad Sci USA 83:917–921
18. Brandl CJ, Green NM, Korczak B, MacLennan DH (1986) Two Ca-ATPase genes; homologies and mechanistic implications. Cell 44:597–607
19. Brotherus J, Jost, PC, Griffith OH, Keana JFW, Hokin LE (1980) Charge selectivity of the lipid-protein interface of membranous Na,K-ATPase. Proc Natl Acad Sci USA 77:272–276
20. Brown TA, Horowitz B, Miller RP, McDonough AA, Farley RA (1987) Molecular cloning of Na,K-ATPase β-subunit from dog kidney. Biochim Biophys Acta 912:244–253
21. Brunner J, Franzusoff AJ, Lüscher B, Zugliani C, Semenza G (1985) Membrane protein topology: aminoacid residues in a putative transmembrane α-helix of bacteriorhodopsin labeled with the hydrophobic carbene-generating reagent 3-(trifluoromethyl)-3-(m$[^{125}I]$iodophenyl)diazirine. Biochemistry 24:5422–5430
22. Castellani L, Hardwicke PMD, Vibert P (1985) Dimer ribbons in the three dimensional structure of sarcopoasmic reticulum. J Mol Biol 185:579–594
23. Castro J, Farley RA (1979) Proteolytic fragmentation of the catalytic subunit of Na,K-ATPase. J Biol Chem 154:2221–2228
24. Chehab FF, Kan WY, Law ML, Hartz J, Kao FT, Blostein R (1987) Human placental Na,K-ATPase α-subunit: cDNA cloning, tissue expression, DNA polymorphism and chromosomal localization. Proc Natl Acad Sci USA 84:7901–7905
25. Chetverin AB, Brazhnikov EV (1985) Do Na and K-forms of Na,ATPase differ in their secondary structure? J Biol Chem 260:7817–7819
26. Chin GJ (1985) Papain fragmentation of Na,K-ATPase β-subunit reveals multiple membrane-bound domains. Biochemistry 24:5943–5947
27. Chou PY, Fassman GD (1974) Predictions of protein conformation. Biochemistry 13:222–245
28. Cid A, Serrano R (1988) Mutations of the yeast plasma membrane H-ATPase which cause thermosensitivity and altered regulation of the enzyme. J Biol Chem 263:14134–14139
29. Cooper JB, Winter CG (1980) 5′-p-fluorosulfonyl benzoyladenosine as an ATP site affinity probe for Na,K-ATPase. J Supramol Struct 3:165–174

30. Cornelius F, Skou JC (1988) Non-equivalent cytoplasmic Na^+ sites and their susceptibility to transmembrane interaction from extracellular Na^+. Prog Clin Biol Res 268A:485–492
31. Deguchi N, Jørgensen PL, Maunsbach AB (1977) Ultrastructure of the sodium pump: comparison of thin secitoning, negative staining and freeze fracture of purified, membrane-bound Na,K-ATPase. J Cell Biol 74:619–634
32. De Meis L (1981) The sarcoplasmic reticulum. Wiley, New York
33. De Meis L, Martins OB, Alves EW (1980) Role of water hydrogen ions, and temperature on the synthesis of ATP by the sarcoplasmic reticulum adenosine triphosphatase in the absence of a calcium ion gradient. Biochemistry 19:4252–4261
34. De Weer P, Rakowski RF (1984) Current generated by backward running electrogenic Na pump in squid giant axons. Nature 309:450–452
35. De Weer P, Gadsby DC, Rabowsky RF (1988) Voltage dependence of the Na pump. Annu Rev Physiol 50:225–241
36. De Weer P, Gadsby DC, Rabowsky RF (1988) Stoichiometry and voltage dependence of the Na pump. Prog Clin Biol Res 268A:421–434
37. Deisenhofer J, Epp O, Miki K, Huber R, Michel H (1985) Structure of protein subunits in photosynthetic reaction centre of Rhodopseudomonas viridis at 3 Å resolution. Nature 318:618–624
38. Donis-Keller H et al. (1987) A genetic linkage map of the human genome. Cell 51:319–337
39. Doolittle RF (1985) The genealology of some recently evolved vertebrate proteins. Trends Biochem Sci 10:233–237
40. Doucet A, Barlet C (1986) Evidence for differences in the sensitivity to ouabain of Na,K-ATPase along the nephron of rabbit kidney. J Biol Chem 261:993–995
41. Dux L, Taylor KA, Ting-Beall HP, Martonosi A (1985) Crystallization of the Ca-ATPase of sarcoplasmic reticulum by calcium and lanthanide ions. J Biol Chem 260:11730–11743
42. Dux L, Pikula S, Mullner N, Martonosi A (1987) Crystallization of Ca-ATPase in detergent solubilized sarcoplasmic reticulum. J Biol Chem 262:6439–6442
43. Dzhandzhugazyan KN, Jørgensen PL (1985) Asymmetric orientation of amino groups in α-subunit and β-subunit of Na,K-ATPase in tight right side our vesicles of basolateral membranes from outer medulla. Biochim Biophys Acta 817:165–173
44. Esmann M, Marsh D (1985) Spin-label studies on the origin of the specificity of lipid-protein interactions in β-subunit membranes from Squalus acanthias. Biochemistry 24:3572–3578
45. Esmann M (1982) Sulphydryl groups of Na,K-ATPase from rectal glands of Squalus acanthias, detection of ligand induced conformational changes. Biochim Biophys Acta 688:251–259
46. Esmann M, Skou JC (1985) Occlusion of Na by the Na,K-APTase in the presence of oligomycin. Biochem Biophys Res Commun 127:857–863
47. Felsenfeld DP, Sweadner KJ (1988) Fine specificity mapping and topography of an isozyme specific epitope of the Na,K-ATPase catalytic subunit. J Biol Chem 263:10932–10942
48. Forbush B III (1984) An apparatus for rapid kinetic analysis of isotopic effelux from membrane vesicles and of ligand dissociation from membrane proteins. Anal Biochem 140:159–163
49. Forbush B III (1988) Occluded ions and Na,K-ATPase. Prog Clin Biol Res 268A:229–248
50. Forbush B III (1987) Rapid release of ^{42}K or ^{86}Rb from two distinct transport sites on the Na,K-pump in the presence of Pi or VO4. J Biol Chem 262:11116–11127
51. Fry DC, Kuby SA, Mildvan AS (1986) ATP binding site of adenylate kinase: mechanistic implications of its homology with ras-encoded p 21, F_1-ATPase, and other nucleotide binding proteins. Proc Natl Acad Sci USA 83:907–911
52. Gadsby DC, Kimura J, Noma A (1985) Voltage dependence of Na,K-pump current in isolated heart cells. Nature 315:63–65
53. Geering K, Meyer DI, Paccolat M-P, Kraehenbühl J-P, Rossier BC (1985) Membrane insertion of α- and β-subunits of Na,K-ATPase. J Biol Chem 260:5154–5160
54. Ghislain M, Schlesser A, Goffeau A (1987) Mutation of a conserved glycine residue modifies the vanadate sensitivity of the plasma membrane H-ATPase from Schizosaccharomyces pombe. J Biol Chem 262:17549–17555
55. Glynn IM (1985) The Na,K-transporting adenosine triphosphatase. In: Martonosi AM (ed) The enzymes of biological membranes, vol 3. Plenum, New York, pp 35–114
56. Glynn IM, Howland JL, Richards DE (1985) Evidence for the ordered release of rubidium ions occluded within the Na,K-ATPase of mammalian kidney. J Physiol 368:453–459
57. Glynn IM, Richards DE (1982) Occlusion of rubidium ions by the sodium-potassium pump: its implication for the mechanism of potassium transport. J Physiol 330:17–43

58. Goeldner MP, Hirth GG, Rossi B, Ponzio G, Lazdunski M. (1983) Specific photoaffinity labeling of the digitalis binding site of Na,K-ATPase by energy transfer. Biochemistry 22:4685–4690
59. Goldshlegger R, Karlish SJD, Rephaeli A, Stein WD (1987) The effect of membrane potential on the mammalian Na,K-pump reconstituted into phospholipid vesicles. J Physiol 387:331–355
60. Gorga FR (1985) Inhibition of Na,K-ATPase by dicyclohexylcarbidiimide. Evidence for two carboxyl groups that are essential for enzymatic activity. Biochemistry 24:6783–6788
61. Grell E, Oberbaumer I (1977) Dynamic aspects of carrier mediated cation transport through membranes. Mol Biol Biochem Biophys 24:371–380
62a. Gresalfi TV, Wallace BA (1984) Secondary structural composition of the Na,K-ATPase E1 and E2 conformers. J Biol Chem 259:2622–2628
62b. Hall C, Rouho A (1980) Quabain binding site photoaffinity probes that label both subunits of Na,K-ATPase. Proc Natl Acad Sci USA 77:4529–4533
63. Hansen O (1984) Interaction of cardiac glycosides with Na,K-ATPase. A biochemical link to digitalis-induced inotropy. Pharmacol Rev 36:143–163
64. Hara Y, Nikamoto A, Kojima T, Matsumoto A, Nakao M (1988) Expression of sodiumpump activities in BALB/c3T3 cells transfected with cDNA encoding α3-subunits of rat brain Na,K-ATPase. FEBS Lett 238:27–30
65. Hastings DF, Reynolds JA, Tanford C (1986) Circular dichroism of the two major conformational states of mammalian Na,K-ATPase. Biochim Biophys Acta 860:566–569
66. Hayashi Y, Mimura K, Matsui H, Takagi T (1987) Determination of minimum functional unit of solubilized Na,K-ATPase by estimation of molecular weight during ATPase reaction. Prog Clin Biol Res 268A:205–210
67. Hayashi Y, Takagi T, Kafzawa S, Matsui H (1983) Molecular weights of αβ protomeric and oligomeric units of soluble Na,K-ATPase determined by low-angle laser light scattering after HPLC. Biochim Biophys Acta 748:153–167
68. Hebert H, Jørgensen PL, Skriver E, Maunsbach AB (1982) Crystallization patterns of membrane bound Na,K-ATPase. Biochim Biophys Acta 689:571–574
69. Hebert H, Skriver E, Maunsbach AB (1985) Three-dimensional structure of renal Na,K-ATPase from membrane crystals. FEBS Lett 187:182–187
70. Herrera VLM, Emanuel JR, Rouz-Opazo N, Levenson R, Nadal-Ginard B (1987) Three differentially expressed Na,K-ATPase α-subunit isoforms: structural and functional implications. J Cell Biol 105:1855–1865
71. Hesse JE, Wieczorek L, Altendorf K, Reicin AS, Dorus E, Epstein W (1984) Sequence homology between two membrane transport ATPase, the kdp ATPase of E. coli and the Ca-ATPase of sarcoplasmic reticulum. Proc Natl Acad Sci USA 81:4746–4750
72. Holland EC, Drickamer K (1986) Signal recognition particle mediates the insertion of a transmembrane protein which has a cytoplasmic NH_2 terminus. J Biol Chem 261:1286–1292
73. Homareda H, Kawakami K, Nagano K, Matsui H (1988) Membrane insertion of α and β subunits of human Na,K-ATPase. Prog Clin Biol Res 268B:77–84
74. Hsu YM, Guidotti G (1989) Rat brain has the α3 form of the Na,K-ATPase. Biochemistry 28:569–573
75. Izatt RM, Bradshaw JS, Nielsen SA, Lamb JD, Christensen JJ (1985) Synthesis and modification of crown ethers. Chem Rev 85:271
76. Ismail-Beigi F, Pressley TA, Haber RS, Gick GG, Loeb JN, Edelman IS (1988) Kinetic analysis of Na,K-ATPase induced by low external K^+ in a rat liver cell line. J Biol Chem 263:8162–8167
77. Jørgensen PL (1972) The role of aldosterone in the regulation of Na,K-ATPase in rat kidney. J Steroid Biochem 3:181–191
78. Jørgensen PL (1975) Purification and characterization of Na,K-ATPase. V. Conformational changes in the enzyme. Transitions between the Na-form and the K-form studied with tryptic digestion as a tool. Biochim Biophys Acta 401:399–415
79. Jørgensen PL (1977) Purification and characterization of Na,K-ATPase. VI. Differential tryptic modification of catalytic functions of the purified enzyme in presence of NaCl and KCl. Biochim Biophys Acta 466:97–108
80. Jørgensen PL (1982) Mechanism of the Na,K-ion pump. Protein structure and conformations of the pure Na,K-ATPase. BBA reviews on Biomembranes. Biochim Biophys Acta 694:27–68
81. Jørgensen PL (1974 and 1988) Isolation of Na,K-dependent ATPase. Methods Enzymol 32:277–294; 156:29–143

82. Jørgensen PL (1975) Isolation and characterization of the components of the sodium pump. Q Rev Biophys 7:239–274
83. Jørgensen PL, Andersen JP (1986) Thermoinactivation and aggregation of $\alpha\beta$-unit in soluble and membrane-bound (Na,K)-ATPase. Biochemistry 25:2889–2897
84. Jørgensen PL, Andersen JP (1988) Structural basis for $E_1 - E_2$ conformational transition in Na,K-ATPase and Ca-ATPase. J Membr Biol 103:95–120
85. Jørgensen PL, Brunner J (1983) Labeling of intramembrane segments of the α-subunit and β-subunit of pure membrane bound Na,K-ATPase with 3-trifluoromethyl-3-(m-[^{125}I]iodophenyl)diazirine. Biochim Biophys Acta 735:291–296
86. Jørgensen PL, Collins JH (1986) Tryptic and chymotryptic cleavage sites in the sequence of α-subunit of Na,K-ATPase from outer medulla of mammalian kidney. Biochim Biophys Acta 860:570–576
87. Jørgensen PL, Petersen J (1985) Chymotryptic cleavage of α-subunit in E_1-forms of renal Na,K-ATPase: effects on enzymatic properties, ligand binding and cation exchange. Biochim Biophys Acta 821:319–333
88. Jørgensen PL, Skriver E, Hebert H, Maunsbach AB (1982) Structure of the Na,K-pump: crystallization of pure membrane-bound Na,K-ATPase and identification of functional domains of the α-subunit. Ann NY Acad Sci 402:203–219
89. Kapakos JG, Steinberg M (1986) 5-iodoacetamidofluorescein-labeled Na,K-ATPase; steady state fluorescence during turnover. J Biol Chem 261:2084–2089
90. Karlish SJD (1980) Characterization of conformational changes in Na,K-ATPase labeled with fluorescein at the active site. J Bioenerg Biomembr 12:111–136
91. Karlish SJD, Jørgensen PL, Gitler C (1977) Identification of a membrane embedded segment of α-subunit of Na,K-ATPase. Nature 269:715–717
92. Karlish SJD, Lieb WR, Stein WD (1982) Combined effects of ATP and phosphate on rubidium exchange mediated by Na,K-ATPase reconstituted into phospholipid vesicles. J Physiol 328:333–350
93. Karlish SJD, Pick U (1981) Sideness of the effects of sodium and potassium ions on the conformational state of the Na,K-pump. J Physiol 312:505–529
94. Karlish SJD, Yates DW, Glynn IM (1978) Conformational transitions between Na-bound and K-bound forms of Na,K-ATPase studied with formycin nucleotides. Biochim Biophys Acta 525:252–264
95. Karlish SJD, Goldshleger R, Stein WD (1990) A 19-kDa C-terminal tryptic fragment of the α-chain of Na,K-ATPase is essential for occlusion and transport of cations. Proc Natl Acad Sci USA 87:4566–4570
96. Kawakami K, Noguchi S, Noda M, Takahashi H, Ohta T, Kawamura M, Nojima H, Nagano K, Hitose T, Inayama S, Hayashida H, Miyata T, Numa S (1985) Primary structure of α-subunit of Torpedo californica Na,K-ATPase. Nature 316:733–736
97. Kawakami K, Nojima H, Ohta T, Nagano K (1986) Molecular cloning and sequence analysis of human Na,K-ATPase β-subunit. Nucleic Acids Res 14:2833–2844
98. Kawakami K, Ohta T, Nojima H, Nagano K (1986) Primary structure of α-subunit of human Na,K-ATPase. J Biochem 100:389–397
99. Kawamura M, Nagano K (1984) Evidence for essential disulfide bonds in the β-subunit of Na,K-ATPase. Biochim Biophys Acta 744:188–192
100. Kirley TL (1989) Determination of three disulfide bonds and one free sulfhydryl in the β-subunit or Na,K-ATPase. J Biol Chem 264:7185–7192
101. Kopito RR, Anderson M, Lodish HF (1987) Structure and organization of the murine band 3 gene. J Biol Chem 262:8035–8040
102. Kyte J (1981) Molecular considerations relevant to the mechanism of active transport. Nature 292:201–204
103. Kyte J, Doolitle RD (1984) A simple method for displaying the hydrophatic character of a protein. J Mol Biol 157:105–132
104. Lafaire AV, Schwarz W (1986) The voltage dependence of the rheogenic sodium pump in the membrane of oocytes of Xenopus Laevis. J Membr Biol 91:43–51
105. Lee JA, Fortes PAG (1986) Spatial relationship and conformational changes between the cardiac glycoside site and β-subunit oligosaccharides in Na,K-ATPase. Biochemistry 25:8133–8144
106. Leonards KS, Kutchai H (1985) Coupling of Ca transport to ATP hydrolysis by the Ca-ATPase of sarcoplasmic reticulum: potential role of the 53-kilodalton glycoprotein. Biochemistry 24:4876–4884

107. Linse S, Brodin P, Johansson C, Thulin E, Grundstrom T, Forsen S (1988) The role of protein surface charges in ion binding. Nature 335:651–652
108. MacLennan DH, Brandl CJ, Korczak B, Green NM (1985) Aminoacid sequence of a Ca-ATPase from rabbit muscle sracoplasmic reticulum, deduced from its complementary DNA sequence. Nature 316:696–700
109. Makowsky LD, Caspar LD, Goodenough DA, Philips WC (1982) Gap junction structures. III. The effect of variations in the isolation procedure. Biophys J 37:189–191
110. Martin DW (1983) Active unit of solubilized sarcoplasmic reticulum Ca-ATPase: an active enzyme centrifugation analysis. Biochemistry 22:2276–2282
111. Marver D, Lear S, Marver LT, Silva P, Epstein FH (1986) Cyclic AMP-dependent stimulation of Na,K-ATPase in shark rectal gland. J Membr Biol 94:205–215
112. McMenamin MAS (1987) The emergence of animals. Sci Am 256:84–92
113. Michel H, Weyer KA, Gruenberg H, Dunger I, Oesterhelt D, Lottspeich F (1986) The heavy subunit of the photosynthetic reaction centre from Rhodopseudomonas viridis: isolation of the gene, nucleotide and aminoacid sequence. EMBO J 5:1149–1158
114. Mildvan AS (1989) NMR studies of the interactions of substrates with enzymes and their peptide fragments. FASEB J 3:1705–1714
115. Moczydlowski EG, Fortes PAG (1981) Inhibition of Na,K-ATPase by TNP-adenine nucleotides. Implications for the structure and mechanism of the Na,K-pump. J Biol Chem 256:2357–2366
116. Modyanov NN, Arzamazova NM, Arystarkhova EA, Gevondyan NM (1985) Ovchinnikov, Yu. A. Sequence analysis of Na,K-ATPase revealed common structural elements of ion-transporting ATPases. Biol Membr (USSR) 2:844–855
117. Mohraz M, Simpson MV, Smith PR (1987) The three-dimensional structure of the Na,K-ATPase from electron microscopy. J Cell Biol 105:1–8
118. Mostov KE, DeFoor P, Fleischer S, Blobel G (1981) Cotranslational membrane integration of calcium pump protein without signal sequence cleavage. Nature 292:87–88
119. Nabiev IR, Dzandzhugazyan KN, Efrenov RG, Modyanov NN (1988) Binding of monovalent cations induces large changes in the secondary structure of Na,K-ATPase as probed by Raman Spectroscopy. FEBS Lett 236:235–239
120. Nagel G, Fendler F, Grell E, Bamberg E (1987) Na^+ currents generated by the purified Na,K-ATPase on planar lipid membranes. Biochim Biophys Acta 901:239–249
121. Nakao M, Gadsby DC (1986) Voltage dependence of the Na translocation by the Na,K-pump. Nature 323:628–630
122. Nicholas RA (1984) Purification of the membrane-spanning tryptic peptides of the α-polypeptide from Na,K-ATPase labeled with adamantane diazirine. Biochemistry 23:888–898
123. Noguchi S, Noda M, Takahashi H, Kawakami K, Ohta T, Nagano K, Hirose T, Inayama S, Kawamura M, Numa S (1986) Primary structure of the β-subunit of Torpedo californica Na,K-ATPase deduced from the cDNA sequence. FEBS Lett 196:315–319
124. Noguchi S, Ohta T, Takeda K, Ohtsubo M, Kawamura M (1988) Ouabain sensitivety of a chimeric α-subunit (torpedo/rat) of the Na,K-ATPase expressed in Xenopus oocyte. Biochem Biophys Res Commun 155:1237–1243
125. Noumi T, Maeda M, Futai M (1988) A homologous sequence between H-ATPase F_0F_1) and cation transporting ATPases. J Biol Chem 263:8765–8770
126. Ohta T, Yoshida M, Nagano K, Hirano H, Kawamura M (1986) Structure of the extra-membranous domain of the β-subunit of Na,K-ATPase revealed by the sequences of its tryptic peptides. FEBS Lett 204:297–301
127. Ohta T, Nagano K, Yoshida M (1986) Active site structure of Na,K-ATPase: location of 5-p-fluorosulfonyl-benzoyl-adenosine binding site. Proc Natl Acad Sci USA 83:2071–2075
128. Orayama O, Sweadner KJ (1988) Ouabain sensitivity of the α-3-isozyme of rat Na,K-ATPase. Biochem Biophys Res Commun 156:796–800
129. Orlowski J, Lingrel JB (1988) Tissue-specific and developmental regulation of rat Na,K-ATPase catalytic α-isoform and β-subunit mRNA's. J Biol Chem 263:10436–10442
130. Ottolenghi P, Nørby JG, Jensen J (1986) Solubilization and further chromatographic purification of highly purified, membrane bound Na,K-ATPase. Biochem Biophys Res Comm 135:1008–1014
131. Ovchinnikov YA, Demin VV, Barnakov AN, Kuzin AP, Lunev AV, Modyanov NN, Dzhandzhugazyan KN (1985) Three dimensional structure of Na,K-ATPase revealed by electron microscopy of two-dimensional crystals. FEBS Lett 190:73–76

132. Ovchinnikov YA, Luneva NM, Arystarkhova EA, Gevondyan NM, Arzamazova NM, Kozhich AT, Nesmeyanov VA, Modyanov NN (1988) Topology of Na,K-ATPase. Identification of the extra- and intracellular hydrophilic loops of the catalytic subunit by specific antibodies. FEBS Lett 227:230–234
133. Ovchinnikov YA, Modyanov NN, Broude NE, Petrukhin KE, Grishin AV, Arzamazova NM, Aldanova NA, Monastyrskaya GS, Sverdlov ED (1986) Pig kidney Na,K-ATPase, primary structure and spatial organization. FEBS Lett 201:237–245
134. Ovchinnikov YA, Monastyrskaya GS, Broude NE, Ushkaryov YA, Melkov AM, Smirnov YV, Malyshev IV, Allikmets RL, Kostina MB, Dulubova IE, Kiyatkin NI, Grishin AV, Modyanov NN, Sverdlov ED (1988) Family of human Na,K-ATPase genes. Structure of the gene for the catalytic subunit (α-III-form) and its relationship with structural features of the protein. FEBS Lett 233:87–93
135. Ovchinnikov YA, Dzhandzugazyan KN, Lutsenko SV, Mustayev AA, Modyanov NN (1987) Affinity modification of E_1-form of Na,K-ATPase revealed Asp-710 in the catalytic site. FEBS Lett 217:111–116
136. Padgett RA, Grabowski PJ, Konarska MM, Seiler S, Sharp PA (1986) Splicing of messenger RNA precursors. Ann Rev Biochem 55:1119–1150
137. Paul DL (1986) Molecular cloning of cDNA for at liver gap junction protein. J Cell Biol 103:123–134
138. Paul C, Rosenbusch JP (1985) Folding patterns of porin gnabacteriorhodopsin. EMBO J 4:1593–1597
139. Pedemonte CH, Kaplan JH (1986) Carbodiimide inactivation of Na,K-ATPase, via intramolecular cross-link formation, is due to inhibition of phosphorylation. J Biol Chem 261:3632–3639
140. Pedersen PL, Carafoli E (1987) Ion motive ATPases, II. Energy coupling and work output. TIBS 12:186–189
141. Post RL, Hegyvary C, Kume S (1972) Activation by adenosine triphosphate in the phosphorylation kinetics of Na,K-ATPase. J Biol Chem 247:6530–6540
142. Pressley TA, Edelman IS (1986) Reduced ouabain inhibition of Na,K-ATPase in cultured cell recipients of the ouabain resistance gene. J Biol Chem 261:9779–9786
143. Price EM, Lingrel JB (1988) Structure-function relationships in the Na,K-ATPase α-subunit: Site directed mutagenesis of Glu-111 to Arg and Asn-122 to Asp generates a ouabain-resistant enzyme. Biochemistry 27:8400–8408
144. Rabon E, Wilke M, Sachs G, Zampighi G (1986) Crystallization of the gastric H, K-ATPase. J Biol Chem 261:1434–1439
145. Rakowski RF, Paxson CL (1988) Voltage dependence of Na,K-pump current in Xenopus oocytes. J Member Biol 106:173–182
146. Rayson BM (1985) Na,K-ATPase regulation in Dahl sensitive and Dahl resistant rats. J Biol Chem 260:12740–12743
147. Rayson BM (1989) Rates of synthesis and degradation of Na,K-ATPase during chronic ouabain treatment. Am J Physiol 256:C75–C80
148. Reithmaier RAF, MacLennan DH (1981) The NH_2-terminus of the Ca-ATPase is located on the cytoplasmic surface of the sarcoplasmic reticulum membrane. J. Biol Chem 256:5957–5960
149. Rephaeli A, Richards DE, Karlish SJD (1986) Electrical potential acceleerates the $E_1P(Na)$-E_2P conformational transition of Na,K-ATPase in reconstituted vesicles. J Biol Chem 261:6248–6254
150. Rossmann MG, Liljas A, Branden CI, Babaszak LJ (1975) The protein. In: Boyer PD (ed) vol 11a. Academic, New York, pp 62–103
151. Sabatini DD, Kreibich G, Morimoto T, Adesnik M (1982) Mechanisms for the incorporation of proteins into the plasma membrane. J Cell Biol 92:1–22
152. Saccomani G, Dailey DW, Sachs G (1979) The action of trypsin on the gastric H,K-ATPase. J Biol Chem 254:2821–2827
153. Sachs JR (1986) The order of addition of sodium and release of potassium at the inside of the sodium pump of the human red cell. J Physiol 381:149–168
154. Satoh K, Nakao T, Nagai F, Kano I, Nakagawa A, Ushiyama K, Yrayama O, Hara Y, Nakao M (1989) A monoclonal antibody against horse kidney Na,K-ATPase inhibits sodium pump and E_2K to E_1 conversion of Na,K-ATPase from outside of the cell membrane. Biochim Biophys Acta 994:104–113

155. Schatzmann HJ (1953) Herzglykoside als hemmstoffe des aktiven kalium- und natriumtransportes durch die erythrocytenmembran. Helv Physiol Acta 111:346–354
156. Scheiner-Bobis G, Buxbaum E, Schoner W (1988) The catalytic subunits cooperate in Na,K-ATPase, demonstration with $MgPO_4$ and MgATP complex analogues. Prog Clin Biol Res 268A:219–226
157. Schneider JW, Mercer RW, Benz EJ (1988) Molecular cloning and characterization of α-subunit isoforms of the Na,K-ATPase. Prog Clin Biol Res 268A:119–126
158. Schwartz A, Grupp G, Wallick E, Grupp IL, Ball WJ (1988) Role of the Na,K-ATPase in the cardiotonic action of cardiac glycosides. Prog Clin Biol Res 268B:321–338
159. Serrano R, Kielland-Brandt MC, Fink GR (1986) Yeast plasma membrane ATPase. Nature 316:689–693
160. Shani-Sekler M, Goldscleger R, Tal DM, Karlish SJD (1988) Inactivation of Rb and Na occlusion on Na,K-ATPase by modification of carboxyl groups. J Biol Chem 263:19331–19341
161. Shull GE, Greeb J (1988) Molecular cloning of two isoforms of the plasma membrane Ca-ATPase from rat brain. J Biol Chem 263:8646–8657
162. Shull GE, Greeb J, Lingrel JB (1986) Molecular cloning of three distinct forms of Na,K-ATPase α-subunit from rat brain. Biochemistry 25:8125–8132
163. Shull GE, Lane LK, Lingrel JB (1986) Aminoacid sequence of the β-subunit of Na,K-ATPase. Nature 321:429–431
164. Shull GE, Lingrel JB (1986) Molecular cloning of rat stomach H,K-ATPase. J Biol Chem 261:16788–16791
165. Shull MM, Lingrel JB (1987) Multiple genes encode the human Na,K-ATPase α-subunit. Proc Natl Acad Sci USA 84:4039–4043
166. Shull GE, Schwartz A, Lingrel JB (1985) Aminoacid sequence of the catalytic subunit of Na,K-ATPase. Nature 316:691–698
167. Skriver E, Maunsbach AB, Jørgensen PL (1981) Formation of two-dimensional crystals in pure membrane-bound Na,K-ATPase. FEBS Lett 131:219–222
168. Skulachev VP (1985) Membrane-linked energy transductions. Bioenergetic functions of sodium: H is not unique as a complexing ion. Eur J Biochem 151:199–208
169. Solioz M, Mathews S, Furst P (1987) Cloning of K-ATPase of streptococcus faecalis J Biol Chem 262:7458–7362
170. Specht S, Sweadner K (1984) Two different Na,K-ATPase in the optic nerve cells or origin and axonal transport. Proc Natl Acad Sci USA 81:1234–1238
171. Stein WD (1986) Transport and diffusion across cell membranes. Academic, New York
172. Stewart JMM, Jørgensen PL, Grisham CM (1989) Nuclear Overhauser effect studies of the conformation of $Co(NH_3)_4ATP$ bound to kidney Na,K-ATPase. Biochemistry 28:4695–4701
173. Takeda K, Noguchi S, Sugino A, Kawamura M (1988) Functional activity of oligosaccharide deficient Na,K-ATPase expressed in Xenopus oocytes. FEBS Lett 238:201–204
174. Sweadner KJ (1989) Isozymes of the Na,K-ATPase. Biochim Biophys Acta 988:185–220
175. Takeyasu I, Tamkun MM, Siegel NR, Fambrough DM (1988) Expression of hybrid Na,K-ATPase molecules after transfection of mouse Ltk cells with DNA encoding the β-subunit of an avian brain sodium pump. J Biol Chem 263:4347–4354
176. Tamkun MM, Fambrough DM (1986) The Na,K-ATPase of chick sensory neurons. Studies on biosynthesis and intracellular transport. J Biol Chem 261:1009–1019
177. Taylor KA, Dux L, Martonosi A (1986) Structure of vanadate-induced crystals of sarcoplasmic reticulum Ca-ATPase. J Mol Biol 187:417–427
178. Tonomura Y (1986) Energy transducing ATPases-Structure and Kinetics. Cambridge University Press London
179. Tsong TY, Astumian RD (1988) Electroconformational coupling: how membrane-bound ATPase transduces energy from dynamic electric fields. Ann Rev Physiol 50:273–290
180. Tyson P, Steinberg M (1987) Accessibility of tryptophan residues in Na,K-ATPase. J Biol Chem 262:4644–4648
181. Urayama O, Sweadner KJ (1988) Ouabain sensitivity of the α-3-isozyme of rat Na,K-ATPase. Biochem Biophys Res Commun 156:796–800
182. Verma AK, Filoteo AG, Stanford DR, Wieben ED, Penniston JT, Strehler EE, Fischer R, Heim R, Vogel G, Mathews S (1988) Complete primary structure of a human plasma membrane Ca^{2+} pump. J Biol Chem 263:14152–14159
183. Verrey F, Schaerer E, Moerkler P, Paccolat MP, Gehring K, Kraehenbuhl JP, Rossier B (1987) Regulation by aldosterone of Na,K-ATPase mRNAs, protein synthesis and sodium transport in cultured kidney cells. J Cell Biol 104:1231–1237

184. Vilsen B, Andersen JP, Petersen J, Jørgensen PL (1987) Occlusion of ^{22}Na and ^{86}Rb in membrane bound and soluble protomeric $\alpha\beta$-units of Na,K-ATPase. J Biol Chem 262:10511–10517
185. Walker JE, Saraste M, Ronswick MJ, Gay NJ (1982) Sequences of ATP-requiring enzymes and a common nucleotide binding fold. EMBO J 1:945–951
186. Wallace B, Cascio M, Mielke DL (1986) Evaluation of prediction methods for membrane proteinsecondary structures. Proc Natl Acad Sci USA 83:9423–9428
187. Wiedmann M, Kurzchalia TV, Hartmann E, Rapoport TA (1987) A signal sequence receptor in the endoplasmic reticulum membrane. Nature 328:830–833
188. Wolitzky BA, Fambrough DM (1986) Regulation of the Na,K-ATPase inclutured chick skeletal muscle. J Biol Chem 261:9990–9999
189. Yamanaka N, Deamer W (1976) Protease digestion of membranes, Ultrastructural and biochemical effects. Biochim Biophys Acta 426:132–147
190. Yang-Feng TL, Schneider JW, Lindgren V, Shull MM, Benz EJ, Lingrel JB, Francke U (1988) Chromosomal localization of human Na,K-ATPase α-subunit and β-subunit genes. Genomics 2:128–138
191. Young RM, Shull GE, Lingrell JB (1987) Multiple mRNAs from rat kidney and brain encode a single Na,K-ATPase β protein. J Biol Chem 262:4905–4910
192. Zamofing D, Rossier BC, Geering K (1988) Role of Na,K-ATPase β-subunit in the cellular accumulation and maturation of the enzyme as assessed by glycosylation inhibitors. J Membr Biol 104:69–77
193. Zamphigi G, Kyte J, Freytag W (1984) Structural organization of Na,K-ATPase in purified membranes. J Cell Biol 98:1851–1864
194. Zibirre R, Hippler-Feldtmann G, Kühne J, Poronnik P, Warnecke G, Koch G (1987) Detection and localization of a cytoplasmic domain on the β-subunit of Na,K-ATPase; a monoclonal antibody study. J Biol Chem 262:4349–4354

Cell and Molecular Biology of Na^+/Glucose Symport

J.E. Lever

Introduction

Renal proximal tubule and intestinal epithelial cells contain apical membrane transporters (symporters) which catalyze the coupled translocation of glucose and Na^+ in a symport (cotransport) mechanism [55]. The Na^+/glucose symporter is restricted to the apical membrane in these epithelia, where it mediates the uphill movement of glucose across the apical membrane, against its concentration gradient. The energy source for this process, termed secondary active transport, is the transmembrane Na^+ gradient generated by the ATP-driven Na^+ pump (the Na^+, K^+, ATPase) localized on the basolateral membrane of these polarized epithelia. A Na^+-independent passive glucose transporter localized in the basolateral membrane mediates the downhill exit of glucose at the basolateral cell surface. Working together, these two unrelated and mechanistically distinct glucose transporters mediate the transepithelial transport of glucose from the apical (luminal) side to the basolateral side of the epithelium (Fig. 1). While nonepithelial cell types lack Na^+/glucose symporters, a similar Na^+/symport mechanism is used to drive active uptake of certain amino acids and phosphate ion [7].

In each case, the symporter mediates uphill movement of the organic solute by directly and obligatorily coupling it to the downhill flux of Na^+ such that the overall process has a negative free energy. Na^+ diffuses down its concentration gradient since intracellular Na^+ concentration is maintained lower than external by the activity of the Na^+, K^+, ATPase.

The molecular basis of the coupling between Na^+ and organic solute flux resides in the structure of the symporter itself. Binding of Na^+ increases the affinity of the symporter for the cotransported solute (e.g., glucose). This implies that coupling is mediated by a conformational change in the symporter subunits. Direct evidence for such a model has been obtained [45, 47, 49, 66].

The Na^+/glucose symporter was the first eukaryotic Na^+/cotransporter to be purified to homogeneity [67] and cloned [16]. The Na^+/glucose symporter has no sequence homology to the Na^+-independent facilitative glucose transporters of epithelial basolateral membranes and nonepithelial tissues, but does bear a significant homology to the *Escherichia coli* Na^+/proline cotransporter [19]. This conservation of structure during evolution, taken together with the several well-characterized mutants which affect Na^+ coupling in the bacterial

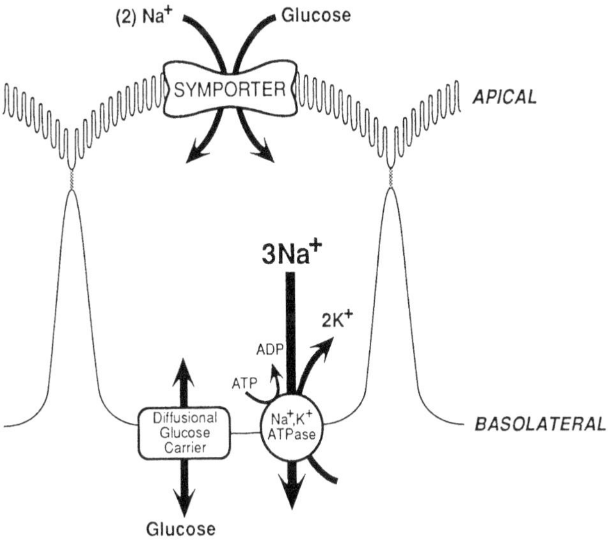

Fig. 1. Transepithelial glucose transport

proline transporter, promise important new insights into Na^+ cotransporter structure and function. The cDNA sequences of other Na^+-coupled symporters such as amino acid transporters or the plasma membrane phosphate transporter have not yet been reported. Therefore it is not yet known if additional homologies exist.

This chapter will review recent progress in the molecular identification and characterization of the Na^+/glucose symporter subunit and its cDNA. Kinetic and earlier biochemical studies have been reviewed elsewhere [7, 29, 55].

Biochemical Identification

The very low abundance of the Na^+/glucose symporter in the apical membrane presents a considerable obstacle in achieving its identification and purification to homogeneity. Early attempts to identify protein subunits of the Na^+/glucose symporter included solubilization and reconstitution [8, 9, 24, 27, 31, 38, 56] and photoaffinity labeling [12, 21, 43]. These approaches did not permit unequivocal identification of the symporter. Candidates for identification as the renal Na^+/glucose symporter included species of 52 and 94 kDa identified after partial purification and reconstitution [31], 47, 64 and 82 kDa identified by affinity labeling [43], and 110 kDa estimated from radiation inactivation measurements [61]. In the case of the intestinal Na^+/glucose symporter, a candidate polypeptide of 72 kDa was identified by negative purification and thiol labeling [30], photoaffinity labeling [27], and substrate-protectable FITC labeling [47]. Thus, at least in the intestine, a strong consensus favored the identification of a 72-kDa polypeptide as a subunit of the symporter, but direct evidence, such as reconstitution of activity in a preparation purified to homogeneity, was lacking.

In the presence of Na^+, Na^+/glucose symporters bind the transport inhibitor phlorizin with high affinity and specificity [55]. Binding of phlorizin to the symporter requires Na^+ and is competitively inhibited by glucose. This ligand inhibits the glucose transport function of the symporter but is not itself transported across the membrane. Thus Na^+-dependent [^{3}H] phlorizin binding activity is frequently used as a sensitive and specific ligand-binding assay of the number of Na^+/glucose symporters.

A monoclonal antibody was isolated after screening for inhibition of phlorizin binding that recognized a 72-kDa polypeptide in intestinal brush border membranes [54]. This strategy was based on the knowledge that high-affinity, Na^+-dependent phlorizin binding is specific for the Na^+/glucose symporter and the expectation that one subset of monoclonal antibodies which bind the symporter will interact with the phlorizin binding site to inhibit binding. The 72-kDa polypeptide was eluted from an immunoaffinity column by the substrates Na^+ and glucose but this was not sufficient evidence for its identification as a Na^+/glucose symporter. In the 7 years since this initial report appeared, no additional characterization of the immunoaffinity-purified intestinal protein or the monoclonal antibody has been reported.

A more recent study from the same laboratory demonstrated the isolation of monoclonal antibodies that stimulated intestinal glucose transport [20]. Antibodies were screened after immunization of mice with intestinal brush border membranes. Proteins of apparent molecular weights 175, 78, and 65 kDa were immunoprecipitated by antibodies which influenced transport.

A two-pronged approach was used to identify a 75-kDa protein as a subunit of the renal Na^+/glucose symporter in order to avoid the pitfall of circular reasoning [65, 67]. The first part of this approach employed a similar strategy to that described above except that renal brush border membranes were used as antigen. Monoclonal antibodies were obtained after immunization of mice with purified renal brush border membranes. A screening strategy based on inhibition of ligand binding offers increased sensitivity and decreased interference from nonspecific inhibition than a screening strategy based on transport inhibition. Furthermore, a higher frequency of specific monoclonal antibodies are expected to be identified based on recognition of a specific binding site (e.g., epitope) than by inhibition of transport function. This is because in other systems where such information is available, only a small fraction of antibodies that bind a transporter inhibit its function. For example, only 1 in 60 monoclonal antibodies that bound the purified *E. coli. lac* transporter inhibited transport [6].

Monoclonal antibodies were screened for inhibition of Na^+-dependent phlorizin binding to membranes [65]. All antibodies which influenced specific phlorizin binding also recognized a 75-kDa antigen after Western blot analysis of renal brush border membranes. Na^+-dependent sugar uptake in vesicles was not affected after incubation with these antibodies. However, these antibodies significantly prevented transport inhibition by bound phlorizin [65].

In the second part of this strategy, conventional purification and reconstitution methods were employed. Solubilized renal brush border membranes were resolved using high pressure liquid chromatography and reconstituted into phosphatidyl choline vesicles [67]. Na^+-dependent phlorizin binding was assayed after reconstitution as a measure of Na^+/glucose symporter function. The peak of phlorizin binding activity was associated with fractions greatly enriched in a 75-kDa protein detectable by silver stain. The 75-kDa protein was purified to homogeneity by preparative SDS-PAGE. Western blot analysis of the purified protein indicated that it was specifically recognized by monoclonal antibodies shown to functionally interact with phlorizin binding to the Na^+/glucose symporter [67]. This observation indicated that the epitope recognized by a monoclonal antibody shown to functionally interact with the symporter was associated with a protein which copurified with phlorizin binding activity. Since the antibody was not involved in the purification of the protein, this provides independent confirmation of the identification of the 75-kDa protein as a symporter subunit. Also, this represented the first purification to homogeneity of a mammalian Na^+-coupled transporter component.

The 75-kDa renal symporter subunit is a glycoprotein as shown by the glucosamine content of the purified protein and contained a blocked N-terminus [67].

The role, if any, of the carbohydrate moiety in symporter function is unclear. Our preliminary data suggests that glycosylation of the symporter is not required for transport function but may enhance the stability of this protein to degradation within the cell. These studies were carried out by monitoring Na^+/glucose symport in LLC-PK_1 cultures after treatment with tunicamycin, an inhibitor of N-linked glycosylation. By contrast, tunicamycin prevented the expression of the symporter after transfection of the cloned cDNA into COS cells [3]. It is possible that these results may be explained by decreased stability of the unglycosylated symporter in COS cells.

Peerce and Clarke [46] have recently reported the isolation and reconstitution of the intestinal Na^+/glucose cotransporter. The final preparation consisted of a 75-kDa polypeptide and demonstrated Na^+-dependent, phlorizin-sensitive, osmotic strength-sensitive glucose uptake after reconstitution in phosphatidyl choline:cholesterol liposomes. This was the first demonstration of transport activity in a Na^+/glucose symporter preparation purified to homogeneity.

Conformational Changes

Conformational changes in the intestinal Na^+/glucose symporter were investigated using chemical modification and fluorescence studies [45–49]. Binding of

Na^+ induces a conformational change in the symporter which could be monitored at a glucose binding site separated from the ion-binding site by 35 Å [48a]. Binding of glucose to the symporter is not required to observe this conformational change since it can be observed even when access of glucose to its binding site is blocked by covalent modification with isothiocyanates [47, 66]. Also transport activity is not required [47, 66]. The Na^+-induced conformational change results in increased exposure of lysine residues near the glucose binding site and a redistribution of tryptophan residues at the surface of the protein [49].

Conformational changes in the symporter were investigated using monoclonal antibodies to the 75-kDa subunit [66]. Exposure of membranes to specific monoclonal antibodies mimicked the conformation-dependent FITC labeling of the symporter triggered by Na^+ and glucose. A synergystic additional stimulation was observed when antibodies and substrates were added together. These observations were interpreted in terms of a model in which a monoclonal antibody may bind and stabilize a Na^+-activated conformation of the symporter. In this hypothesis, the symporter exists in an equilibrium between an inactive conformation and one or more activated conformations. The accumulation of active conformers is favored in the presence of Na^+. Activation of the symporter upon binding of Na^+ is accompanied by exposure of previously buried residues which can be labeled by FITC. Specific monoclonal antibodies would stabilize an active conformation of the symporter. This could be accomplished by preferential binding to an epitope exposed in the active conformation, forming a tight complex and pulling the equilibrium in favor of accumulation of the active conformer. This interpretation was favored by the observation that binding of ^3H-labeled anti-symporter monoclonal antibody to brush border membranes was stimulated in the presence of Na^+, while binding of control antibodies specific for other brush border membrane proteins was not influenced by Na^+. While monoclonal antibodies to the symporter could mimic the conformational effects of Na^+ and glucose on carrier activation monitored by exposure of residues to isothiocyanates, antibodies could not replace the Na^+ requirement for activation of the transport activity. This suggests that the obligatory role of Na^+ in the symport mechanism as cosubstrate and determinant of the net charge of the translocating complex cannot be replaced by antibody stabilization of an activated conformation. The lack of antibody effects on transport rates suggests that symport does not occur by a rotating carrier mechanism. Transport models based on carrier rotation are also eliminated by the extremely large energy barrier posed by the necessary rotation of hydrophilic residues across the membrane. The Na^+-induced conformational change of the symporter is apparently a partial reaction which serves to increase the affinity of the symporter for glucose. In the absence of Na^+, the K_m for glucose is 10 mM or higher whereas maximal Na^+ activation can decrease this value to 100 μM [55].

While the above studies were carried out with intact membranes, Peerce [45] has now examined conformational changes in a purified intestinal Na^+/glucose symporter preparation. Tryptophan quenching was used to provide evidence for a second conformational change. It is postulated that this second conformational change upon binding of D-glucose shifts the protein into a transport-competant conformation.

Expression Cloning of the Intestinal Symporter

The intestinal Na$^+$/glucose symporter cDNA was cloned using a strategy which did not rely on the use of antibody or oligonucleotide probes [16]. This approach, based on expression cloning in oocytes, can be used for cloning of proteins that are not endogenously expressed in oocytes and for which a highly sensitive assay is available. An advantage of this approach is that it may be used to clone sequences that are low abundance even if the gene product has not been biochemically identified. Expression of Na$^+$/glucose symporter function in *Xenopus* oocytes was observed after microinjection of total rabbit intestinal polyA-RNA [17]. mRNA was size-selected and used to construct a cDNA plasmid library in an in vitro expression vector [16]. Rather than individually testing mRNA's from thousands of clones, a shortcut was used. Pooled groups of clones were transcribed in vitro and the resultant mRNA mixtures tested for symporter expression after injection into oocytes. Negative pools were rejected and positive pools were subdivided and the test repeated. After repeated subdivision of positive pools, a single 2.23-kb Na$^+$/glucose cotransporter cDNA clone was identified [16]. Injection into oocytes of mRNA transcribed from this clone elicited expression of phlorizin-sensitive Na$^+$-dependent α-methyl-D-glucoside uptake.

The cDNA sequence of this rabbit intestinal cotransporter clone contained an open reading frame coding for 662 amino acids. The calculated relative molecular mass of 73 080 for the polypeptide is consistent with the M_r of 75 000 observed for renal and intestinal Na$^+$/glucose symporters. Eleven membrane spanning regions are predicted by hydrophobicity analysis of the predicted amino acid sequence (Fig. 2). Two long hydrophilic regions near the C-terminus were noted.

Two potential N-glycosylation sites were predicted from the sequence, one on either side of the membrane as analyzed by the hydropathy plot. A potential N-linked glycosylation sequence is associated with the amino acid sequence Asn-Xaa-Thr/Ser. In vitro translation studies suggest that only the external Asn 248 is glycosylated [40].

No DNA sequence homology was noted between the Na$^+$/glucose cotransporter and the various mammalian facilitated glucose carriers or bacterial H$^+$/sugar cotransporters [16]. At present, five tissue-specific isomeric forms of the facilitative glucose transporter have been sequenced: GLUT1, the erythrocyte, HepG2, brain form; GLUT2, the liver form; GLUT3, the fetal muscle form;

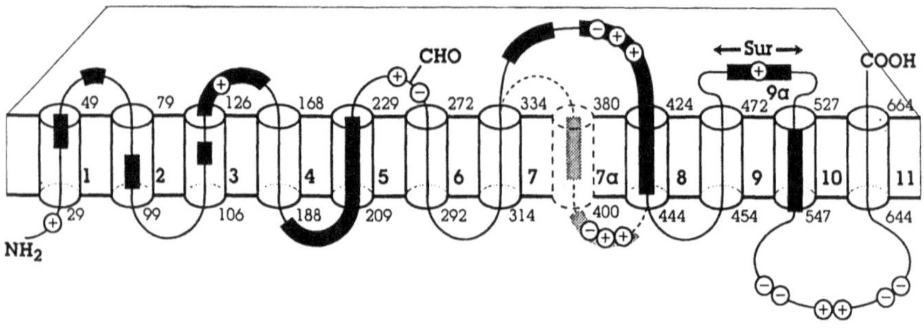

Fig. 2. Model of the membrane orientation of the human intestinal Na^+/glucose symporter. Postulated membrane-spanning segments are numbered 1–11. CHO indicates a potential N-linked glycosylation site. Thick lines indicate regions of particular sequence homology with the *E. coli* Na^+/proline transporter. (Based on data from [19])

GLUT4, the insulin-regulatable muscle/adipocyte form; and GLUT5, the small intestine basolateral isomer (reviewed in [13]). There are also extensive sequence homologies between these facilitative glucose transporters and the *E. coli.* arabinose and xylose H^+ cotransporters [37].

The detailed transport characteristics of the cloned rabbit intestinal symporter expressed in oocytes [23] and COS-7 cells [3] were closely similar to the native brush border membrane symporter. The cloned protein expressed in oocytes catalyzed Na^+-dependent α-methyl-D-glucopyranoside uptake (Hill coefficient 1.5). Uptake was inhibited by phlorizin ($K_i = 6$ μM) and saturable ($K_m = 110$ μM). The rank order of substrate specificity was D-glucose > α-methyl-D-glucopyranoside > D-galactose > 3-O-methylglucoside ⋙ D-mannose = L-glucose [23]. These observations indicate that the classical symporter is encoded by this full-length cDNA.

The Na^+/glucose symporter cDNA was transfected into COS-7 cells using the vector pEUK-C1 [3]. Transfectants exhibited Na^+-dependent α-methyl-D-glucopyranoside uptake with the expected characteristics of the intestinal symport system. Treatment of the cells with tunicamycin resulted in loss of the transport activity suggesting that N-linked glycosylation is required for symporter function. Alternately, this result may be explained by decreased stability of the unglycosylated symporter in these host cells.

The rabbit intestinal cotransporter cDNA clone was used as a probe to screen a human intestinal cDNA lambda gt10 library [19]. A 2.6-kb clone was obtained which coded for an RNA transcript capable of expressing Na^+/glucose cotransport after injection into ooctytes. The probe hybridized with 2.2-, 2.6-, and 4.8-kb species of human mRNA, which differed only in the lengths of the 3′ noncoding region [19]. DNA sequence analysis revealed an 82% similarity between the rabbit and human Na^+/glucose cotransporter clones. At the amino acid level, there is 84% identity between the human and rabbit Na^+/glucose cotransporter sequences, with an additional 10% similarity based on conservative amino acid substitutions. The human symporter was two amino acid residues longer than the rabbit. Two potential N-linked glycosylation sites were noted, as in the rabbit. The predicted secondary structure of the human intestinal Na^+/glucose cotransporter is similar to that predicted for the rabbit (Fig. 2), with 11 potential membrane-spanning regions and two major C-terminal hydrophilic regions.

The human intestinal Na^+/glucose cotransporter gene (SGLT1) is located on the q11.1–qter region of chromosome 22 [15]. This map position differs from those deduced for the facilitative glucose transporters. GLUT1 and GLUT5 map on chromosome 1, GLUT2 on chromosome 3, GLUT3 on chromosome 12, and GLUT4 on chromosome 17 [13].

Sequence Homologies with The Renal Na^+/Glucose Symporter

In preliminary studies [18], a cDNA clone with partial sequence homology to the intestinal symporter was isolated from a human renal cortical cDNA library. Also, we have sequenced a Na^+/glucose symporter cDNA from the LLC-PK_1 pig renal epithelial cell line (S.F. Yet, C.T. Kong, J.E. Lever, unpublished results). The amino acid sequence deduced for the LLC-PK_1 clone is more than 85% homologous to that of the human intestinal symporter. These observations indicate that there is a strong sequence homology between the renal and intestinal glucose symporters. It is not known whether the regions of non-homology represent species-specific or tissue-specific differences, since renal and intestinal sequences of the same species were not compared. Our anti-symporter monoclonal antibody also recognized an antigen in rat kidney and rat intestine as analyzed by Western blot, indicating that there is conservation of this epitope among these tissues and species.

We have demonstrated (S.F. Yet, C.T. Kong, J.E. Lever, unpublished results) that the protein encoded by symporter cDNA after expression in bacteria is specifically recognized by our monoclonal antibody described above which binds the 75-kDa symporter subunit. This antibody binds the fusion protein expressed by our LLC-PK_1 renal symporter cDNA. The levels of this protein in bacteria are regulated by induction of the *lac* operon as expected for expression of a protein genetically fused to β-galactosidase. This significant finding proves that the 75-kDa symporter subunit purified to homogeneity is closely similar to the protein expressed by the symporter cDNA since both are specifically recognized by the same monoclonal antibody.

It has been proposed that kidney [62] and intestine [4] each contain more than one form of the Na^+/glucose symporter. A low-affinity system was found in renal outer cortex (early proximal tubule) which exhibited a stoichiometry of 1:1 Na^+:glucose [62]. By contrast, a high-affinity system with a 2:1 Na^+:glucose stoichiometry was found in renal outer medulla (late proximal tubule). In the intestine evidence for two systems with 1:1 and 2:1 Na:glucose stoichiometries was also presented [4]. It was speculated that in the kidney most of the glucose was absorbed in the early proximal tubule with the energetic cost of 1 Na^+ per molecule of glucose absorbed. A higher energetic cost ($2Na^+$ per glucose molecule) was involved in the higher affinity system required to scavenge the lower glucose concentrations remaining in the late proximal tubule. The system with the 2:1 stoichiometry would be capable of maintaining a glucose concentration ratio an order of magnitude higher than that maintained by the system with a 1:1 stoichiometry. The energetic cost is paid in the form of ATP hydrolyzed by the Na^+, K^+, ATPase in pumping out the additional Na^+ ions. Appart from this kinetic evidence for the two systems, no convincing biochemical evidence has been obtained for the existence of the two systems. Thus far, extensive screening of rabbit intestinal libraries with the cloned rabbit intestinal symporter probe has failed to reveal any additional homologous sequences in the same tissue which encode a Na^+/glucose symporter. Since the gene maps on an autosomal chromosome (chromosome 22 in the human) [15], two gene copies are expected per cell. Even if two sequence variants were identified in the genome of one individual, it would have to be established that they coded for gene products differing in proximal tubule location (early or late) rather than allelic variants of the same gene coding for products with the same microlocalization.

Inherited disorders of Na^+/glucose symporters have been described in the human. Intestinal glucose/galactose malabsorption is associated with a defect in intestinal glucose symport and renal glycosuria involves a defect in the renal proximal tubule symporter. The inheritance patterns of these disorders is unlinked suggesting that different genes encode the renal and intestinal symporters. It is not yet known whether these disorders represent structural gene mutations resulting in an altered amino acid sequence or premature chain termination (nonsense mutation). Alternately, a separate regulatory gene may account for altered symporter expression in certain cases. These possibilities may now be directly investigated with the availability of cloned gene probes.

Sequence Homologies with the *E. coli* Proline Cotransporter

A 44% DNA sequence homology was observed between the human intestinal Na^+/glucose cotransporter and the *E. coli* proline cotransporter (*putP*) [19]. Amino acid sequence identity between these transporters was 28%, with an additional 25% similarity based on conservative amino acid substitutions. The proline cotransporter is 24% shorter than the glucose symporter, and lacks the hydrophilic C-terminal region [42]. Particular regions of homology are shown by the thick lines in Fig. 2.

A tyrosine residue had been postulated to be located at or near the Na^+-binding site in the intestinal Na^+/glucose symporter [48]. It is not clear which of the 27 tyrosine residues contained in the protein represent this site. Analysis of the much smaller conserved region common to the human and rabbit intestinal Na^+/glucose symporters and the bacterial proline cotransporter potentially narrows the search for the tyrosine implicated in Na^+ binding to only 4 conserved tyrosine residues at amino acid residues 131, 191, 410, and 528, respectively, in the human.

Chemical modification studies [47, 63] had suggested that multiple lysine residues may reside in a cleft at or near the glucose binding site. It is reasonable to expect that multiple charged residues are involved in hydrogen bonding to glucose during transport. Only three lysine residues, at positions 415, 420, and 549, are conserved in the rabbit, human and bacterial transporters.

Analysis of bacterial Na^+/proline cotransporter mutants has implicated Arg 257 in Na^+ binding and energetic coupling to the Na^+ gradient [44]. It is interesting to note that this residue is conserved in the mammalian Na^+/glucose cotransporter. Other proline carrier mutants which cause a reduction in affinity for Na^+ for proline transport and binding involve substitutions at Gly 22 and Cys 141 [69]. Gly 22 is also conserved in the intestinal Na^+/glucose cotransporter. A model is proposed for the Na^+ binding site of the proline cotransporter in which Gly 22 of helix I, Cys 141 in helix III, Arg 257 between helix VI and VII, Cys 281 in helix VII, and Cys 344 in helix VIII are in close proximity [69].

A weak homology has also been detected between the primary amino acid sequences of the *E. coli* Na^+/glutamate cotransporter and the rabbit intestinal Na^+/glucose symporter [69]. A distinct consensus sequence, Gly-Ala-Leu-Gly-Arg, is observed in the primary amino acid sequences of these three Na^+ cotransporters [69]. This sequence may be relevent to the Na^+ symport process.

The striking kinetic and structural homologies between these bacterial and mammalian cotransporters suggests a common molecular mechanism with respect to coupling to a Na^+ gradient. Analysis of the conserved regions between the rabbit and human glucose symporter and bacterial proline cotransporter is potentially very informative about structure/function relationships. Assignment of functional domains will await X-ray crystallography analysis and site-directed mutagenesis studies. As yet, no membrane transporter has been crystallized. Site-directed mutagenesis of the symporter will require expression of the full-length cDNA in a suitable host cell system. Currently, this is feasible by microinjection of the mRNA in the oocyte system, but the approach is limited by the small number of cells that can be processed. Also, the symporter can be expressed by transfection into COS-7 cells [3]. As an alternative approach, the facilitative glucose transporter has been expressed in *E. coli* [53, 60]. This tactic was feasible because glycosylation of the facilitative transporter is not required for function or insertion in the membrane. Also unlike most mammalian integral membrane proteins, the facilitative glucose transporter can insert into the membrane after its biosynthesis, e.g., posttranslational insertion [60]. Assuming the same holds true for the Na^+/glucose symporter, functional expression in bacteria is theoretically feasible. However, it is debatable whether this is a valid system for site-directed mutagenesis studies since the stability and folding of the unglycosylated protein in a bacterial cell membrane lipid and ionic environment may differ from that of the glycosylated protein in a mammalian cell. In fact glycosylation has been shown to influence the insertion of the facilitative glucose transporter [14] and appears to be a requirement for symporter expression in COS cells [3]. Stable transfection of a suitable mammalian host cell appears to offer the most promising approach where large numbers of cells are required.

Kaback and coworkers [5, 25, 26, 32, 36, 50–52] have carried out site-directed mutagenesis studies of the *lac* permease of *E. coli*. This transporter is currently the most extensively characterized symporter in terms of structure/-function analysis of specific sequence alterations. Specific residues required for coupling between H^+ and lactose in a symport mechanism have been identified. This permease is proposed to function as a monomer. Although the *lac* permease (H^+/lactose symporter) has no sequence homology to the Na^+/glucose symporter, the number of membrane-spanning regions predicted by hydropathy analysis is similar. These elegant studies serve as a general model in the mutational analysis of ion-linked symport.

Oligomeric Structure

The observation that oocytes injected with a transcript encoded by the 75-kDa symporter subunit cDNA express the full range of the transport and ligand binding properties associated with this transporter [23] strongly suggests that only one type of subunit is required for function.

The radiation inactivation method provides a measurement of the size of functional complexes in their native membrane state, without the need for isolation or even biochemical identification of the protein. The main requirement is that the activity be stable to freezing in cryoprotectant buffer at liquid Nitrogen temperatures. Suitable controls have established that nonspecific vesicle leakiness does not occur even at the highest radiation dose. This technique assigns the size of the minimum functional complex under the assay conditions.

While the oligomeric state of the symporter is not well understood, recent radiation inactivation analysis of the renal Na^+/glucose symporter suggests a homotetrameric structure of 288–298 kDa [2, 35, 58, 61]. Similarly, the target size to radiation inactivation of the rabbit intestinal Na^+/glucose symporter (290 kDa) can be interpreted as a homotetramer of 70-kDa subunits [57]. J_{max} rather than K_m varied as a function of radiation dose, consistent with complete rather than partial destruction of each unit by a radiation hit [57]. In the same study [57], the target size of the 70-kDa intestinal symporter monomer on

Western blots after SDS-PAGE was estimated as 66-kDa. Detection in irradiated membranes was accomplished by anti-peptide antibodies based on hydrophilic portions of the cloned intestinal symporter sequence and densitometry of the immunostained blot. The close correspondence between these two values provides independent confirmation of the accuracy of the radiation inactivation method and an in situ estimate of the monomeric size of the symporter subunit.

Regulation of Gene Expression

The long-term renal epithelial cell line LLC-PK_1 exhibits cell-density-dependent regulation of symporter activity, together with other markers characteristics of renal proximal tubule [34]. Na^+/glucose symport activity is almost undetectable in sparse, actively dividing cultures, but is expressed following the development of a confluent cell density [1]. Studies using membrane vesicles isolated from differentiated LLC-PK_1 cultures indicated a Na^+:glucose stoichiometry of 2:1 for symport [33]. Immunofluorescence studies using a monoclonal antibody to the renal Na^+/glucose symporter indicated that the symporter antigen became apparent on the apical membrane after development of a confluent monolayer, in correlation with the expression of transport activity [68].

Na^+/glucose symport activity and other differentiation markers were dramatically increased after treatment of confluent cultures with the differentiation inducer hexamethylene bisacetamide (HMBA) [1, 70]. Increased expression of the 75-kDa symporter antigen after inducer treatment was analyzed quantitatively by Western blot and qualitatively by immunofluorescence staining [68]. These studies indicated that the number of symporter-positive cells was increased after inducer treatment, as predicted for differentiation-regulated expression [68]. A similar conclusion was reached when symporter activity was compared in populations of differentiated and undifferentiated cells separated by transport-dependent cell density shifts [64].

The mechanism of HMBA effects on glucose symporter expression is not understood. By contrast, the effects of this inducer on differentiated gene expression have been extensively characterized in murine erythroleukemia cells [39]. The availability of this inducer permits the initiation at a defined time of a relatively synchronous differentiation event in a large proportion of the cell population as compared with the slow asynchronous differentiation observed spontaneously after the development of confluence. HMBA is relatively nontoxic and is one of the most potent inducers identified thus far. The identification of intracellular mediators of HMBA action should provide insight into the physiological signals which regulate the committment event and subsequent symporter expression in renal epithelial cells. Addition of HMBA to renal epithelial cell cultures caused increased intracellular cyclic AMP levels and increased transepithelial resistance [59]. In a recent review, a working hypothesis to explain intracellular mediators of HMBA is discussed [34]. It appears likely that HMBA does not act directly, but via one of its intracellular metabolites, *N*-acetyl-1,6-diaminohexane. Changes in Na^+ levels, protein kinase C and protein kinase A activities, polyamines and DNA methylation have been implicated in mediating HMBA effects on cell differentiation [34].

Our preliminary results indicate that HMBA addition to confluent LLC-PK_1 cultures stimulates increased accumulation of Na^+/glucose symporter mRNA (S.F. Yet, C.T. Kong, J.E. Lever, unpublished results). These studies were carried out using a symporter cDNA probe cloned from a LLC-PK_1 cell library. The probe hybridized with 2.4-kb and 4.0 kb species in LLC-PK_1 cells but did not hybridize with mRNA from MDCK kidney epithelial cells which do not express Na^+/glucose symporter activity. It remains to be established whether symporter mRNA accumulation is due to increased transcription or decreased turnover of symporter mRNA.

Symporter activity is down-regulated by glucose in LLC-PK_1 cultures [41] in common with glucose effects on facilitative glucose transport. The mechanism of regulation of symporter expression by glucose is not understood.

Kempson et al. [28] demonstrated using LLC-PK_1 cultures that perturbations in the rate of endocytosis are accompanied by V_{max} changes in Na^+/solute cotransport. This could provide a mechanism for rapid up- or down-regulation of apical Na^+/glucose symport activity independent of changes in the rate of symporter protein synthesis.

Summary and Conclusions

It is now conclusively established that the 75-kDa protein purified from kidney brush border membranes is a subunit of the Na^+/glucose symporter. A monoclonal antibody which influences phlorizin binding to the symporter in brush border membranes specifically recognized a 75-kDa protein which was purified to homogeneity from these membranes based on its phlorizin binding activity after solubilization and reconstitution. The same monoclonal antibody also specifically recognized the fusion protein obtained after expression of the cloned pig kidney Na^+/glucose symporter cDNA in *E. coli*. Symporter cDNA contains all information necessary for expression of the classical Na^+/glucose symporter in frog oocytes after microinjection of the mRNA. Thus previous theories that more than one type of subunit (e.g., a separate Na^+-binding subunit) are required for symport have not been substantiated. Radiation inactivation analysis indicates that a tetramer of 75-kDa subunits may form the active complex involved in transport function.

Questions which remain concern the secondary structure and detailed conformation of the symporter in the membrane and identification of the specific amino acid residues which participate in the movement of Na$^+$ and glucose across the membrane and the obligatory coupling of these fluxes. Conformational changes appear to be intimately related to the symport mechanism. Important clues to identify the active residues may be derived by analysis of sequence homologies between various mammalian Na$^+$/glucose symporters and the bacterial Na$^+$/proline symporter. Significant progress towards answering these questions will await the results of mutagenesis studies and X-ray crystallography. Chemical modification studies have been of limited utility due to the relative lack of specificity of the approach to target one specific residue and due to effects of introducing bulky groups into the protein. A recent demonstration [11] that the facilitative glucose transporter serves as a water channel poses the question of whether the Na$^+$/glucose channel performs a similar function.

Other questions concern the cellular expression of the symporter. The renal and intestinal forms of the symporter are very homologous as shown by nucleic acid hybridization studies, functional and biochemical similarities and immunological cross-reactivity. It remains to be established whether kidney and intestine each express different but homologous Na$^+$/glucose symporter genes, or whether the same gene is expressed in both tissues. Alternately, splice variants or regulatory genes may play a role. A direct test of these possibilities is now feasible. Similarly, it is now possible to determine the precise molecular defect in human inherited renal glycosuria and intestinal glucose/galactose malabsorption.

Another area for future investigation is the possible existence of distinct fetal isoforms of the symporter in early embryos. Epitopes shared with the pig renal Na$^+$/glucose symporter were detected on the apical membranes of precompaction eight-cell mouse embryos [64a].

The signal necessary for targeting the symporter to the apical membrane during its biosynthesis has not yet been identified. A number of apical membrane hydrolases have been shown to be anchored to the apical membrane by a glycosyl-phosphatidylinositol structure [10]. However, the symporter is clearly not linked to the membrane by this type of mechanism. Possible signal sequences specifying apical membrane localization of the symporter have not yet been identified. Glycosylation of the symporter does not appear to be a requirement for apical membrane insertion but the role of carbohydrate residues in symporter function versus stability requires more critical investigation.

LLC-PK$_1$ cultures exhibit differentiation-specific expression of the symporter. This cell line will provide a valuable model system for studying transcriptional and posttranslational regulation of symporter expression, hormonal regulation and identification of biosynthetic precursors and pathways of recycling and turnover. Additional questions which can be explored using this cell line are the effects of transfected oncogenes on symporter expression and the possibility of co-regulation of expression of isoforms of the facilitative glucose transporter with expression of the symporter.

References

1. Amsler K$^+$ Cook JS (1982) Development of Na$^+$-dependent hexose transport in a cultured line of porcine kidney cells. Am J Physiol 242 (Cell Physiol 11):C94–C101
2. Beliveau R, Demeule M, Ibnoul-Khatib H, Bergeron M, Beauregard G, Potier M (1988) Radiation-inactivation studies on brush-border membrane vesicles. General considerations and application to the glucose and phosphate carriers. Biochem J 252:807–813
3. Birnir B, Lee HS, Hediger MA, Wright EM (1990) Expression and characterization of the intestinal Na$^+$/glucose symporter in COS-7 cells. Biochim Biophys Acta 1048:100–104
4. Brot-Laroche E, Serrano MA, Delhomme B, Alvarado F (1986) Temperature sensitivity and substrate specificity of two distinct Na$^+$-activated D-glucose transport systems in guinea pig jejunal brush border membrane vesicles. J Biol Chem 261:6168–6176
5. Carrasco N, Puttner IB, Antes LM, Lee JA, Larigan JD, Lolkema JS, Roepe PD, Kaback HR (1989) Characterization of site-directed mutants in the lac permease of *Escherichia coli.* 2. Glutamate-325 replacements. Biochemistry 28:2533–2539
6. Carrasco N, Viitanen P, Herlinger D, Kaback HR (1984) Monoclonal antibodies against the *lac* carrier protein from *Eschericia coli.* I. Functional studies. Biochemistry 23:3681–3687
7. Crane RK (1977) The gradient hypothesis and other models of carrier-mediated active transport. Rev Physiol Biochem Pharmacol 78:99–159
8. Crane RK, Malathi P, Prieser, H (1976) Reconstitution of specific Na$^+$-dependent D-glucose transport in liposomes by Triton X-100-extracted proteins from purified brush border membranes of rabbit kidney cortex. FEBS Lett 67:214–216
9. Ferguson DR, Burton KA (1977) Reconstitution in phospholipid vesicles of a glucose transport system from pig small intestine. Nature 265:639–642
10. Ferguson MA, Williams AF (1988) Cell-surface anchoring of proteins via glycosyl-phosphatidylinositol structures. Ann Rev Biochem 57:285–320
11. Fischbarg J, Kuang KY, Vera JC, Arant S, Silverstein SC, Loike J, Rosen OM (1990) Glucose transporters serve as water channels. Proc Natl Acad Sci USA 87:3244–3247
12. Gibbs EM, Hosang M, Reber BFX, Semenza G, Diedrich DF (1982) 4-Azidophlorizin, a high-affinity probe and photoaffinity label for the glucose transporter in brush border membranes. Biochim Biophys Acta 688:547–556
13. Gould GW, Bell GI (1990) Facilitative glucose transporters: an expanding family. TIBS 15:18–23
14. Haspel HC, Revillame J, Rosen OM (1988) Structure, biosynthesis and function of the hexose transporter in Chinese hamster ovary cells deficient in N-acetylglucosaminyltransferase 1 activity. J Cell Physiol 136:361–366
15. Hediger MA, Budarf ML, Emanuel BS, Mohandas TK, Wright EM (1989) Assignment of the human intestinal Na$^+$/glucose cotransporter gene (SGLT1) to the q11.2–qter region of chromosome 22. Genomics 4:297–300
16. Hediger MA, Coady MJ, Ikeda TS, Wright EM (1987) Expression cloning and cDNA sequencing of the Na$^+$/glucose co-transporter. Nature 330:379–381
17. Hediger MA, Ikeda T, Coady M, Gunderson CB, Wright EM (1987) Expression of size-selected mRNA encoding the intestinal Na$^+$/glucose co-transporter in *Xenopus laevis* oocytes. Proc Natl Acad Sci USA 84:2634–2637
18. Hediger MA, Turk E, Pajor AM Mohandas TK, Wright EM (1989) The human renal Na$^+$/glucose co-transporters. Fed Proc 3:A561
19. Hediger MA, Turk E, Wright EM (1989) Homology of the human intestinal Na$^+$/glucose and *Escherichia coli* Na$^+$/proline co-transporters. Proc Natl Acad Sci USA 86:5748–5752
20. Honold K, Ludeke B, Hengartner H, Semenza G (1988) Stimulation of intestinal Na$^+$/D-glucose cotransport by monoclonal antibodies. J Membr Biol 105:165–175
21. Hosang M, Gibbs EM, Diedrich DF, Semenza G (1981) Photoaffinity labeling and identification of (a component of) the small intestinal Na$^+$, D-glucose transporter using 4-azidophlorizin. FEBS Lett 130:244–247
22. Hosang M, Vasalla A, Semenza G (1981) Specific photoaffinity inactivation of the D-glucose transporter in small intestinal .brush border membranes using new phlorizin analogues. Biochemistry 20:5844–5854

23. Ikeda TS, Hwang ES, Coady MJ, Hirayama BA, Hediger MA, Wright EM (1989) Characterization of a Na^+/glucose cotransporter cloned from rabbit small intestine. J Membr Biol 110:87–95
24. Im WB, Ling KY, Faust RG (1982) Partial purification of the Na^+-dependent D-glucose transport system from renal brush border membranes. J Membr Biol 65:131–137
25. Kaback HR (1987) Use of site-directed mutagenesis to study the mechanism of a membrane transport protein. Biochemistry 26:2071–2076
26. Kaback HR (1988) Site-directed mutagenesis and ion-gradient driven active transport: on the path of the proton. Annu Rev Physiol 50:243–256
27. Kano-Kameyama A, Hoshi T (1983) Purification and reconstitution of Na^+/D-glucose cotransport carriers from guinea pig small intestine. Jpn J Physiol 33:955–970
28. Kempson SA, Ying AL, McAteer JA, Murer H (1989) Endocytosis and Na^+/solute cotransport in renal epithelial cells. J Biol Chem 264:18451–18456
29. Kessler M, Semenza G (1983) The small intestinal Na^+, D-glucose cotransporter: an asymmetric gated channel (or pore) responsive to delta psi. J Membr Biol 76:27–56
30. Klip A, Grinstein S, Semenza G (1979) Partial purification of the sugar carrier of intestinal brush border membranes. Enrichment of the phlorizin-binding component by selective extractions. J Membr Biol 51:47–73.
31. Koepsell H, Menuhr H, Ducis I, Wissmuller TF (1983) Partial purification and reconstitution of the Na^+-D-glucose cotransport protein from pig renal proximal tubules. J Biol Chem 258:1888–1894
32. Lee JA, Puttner IB, Kaback HR (1989) Effect of distance and orientation between arginine-302, histidine-322 and glutamate-325 on the activity of lac permease from *Escherichia coli.* Biochemistry 28:2540–2544
33. Lever JE (1982) Expression of a differentiated transport function in apical membrane vesicles from an established epithelial cell line. Sodium electrochemical potential-mediated active sugar transport. J Biol Chem 257:8680–8686
34. Lever JE (1989) Chemical inducers of differentiation in a long-term renal cell line. Environ Health Perspect 80:173–180
35. Lin JT, Szwarc K, Kinne R, Jung CY (1984) Structural state of the Na^+/D-glucose cotransporter in calf kidney brush border membranes. Target size analysis of Na^+-dependent phlorizin binding and Na^+-dependent D-glucose transport. Biochim Biophys Acta 777:201–208
36. Lolkema JS, Puttner IB, Kaback HR (1988) Site-directed mutagenesis of pro-327 in the lac permease of *Escherichia coli.* Biochemistry 27:8307–8310
37. Maiden MCJ, Davis, EO, Baldwin SA, Moore DCM, Henderson PJF (1987) Mammalian and bacterial sugar transport proteins are homologous. Nature 325:641–643
38. Malathi P, Preiser H (1983) Isolation of the Na^+-dependent D-glucose transport protein from brush border membranes. Biochim Biophys Acta 735:314–324
39. Marks PA, Rifkind RA (1978) Erythroleukemic differentiation. Annu Rev Biochem 47:419–448
40. Mendlein J, Hediger M, Coady M, Wright EM (1988) In vitro translation and glycosylation of the intestinal Na^+-glucose transporter. FASEB J2:A1021
41. Moran A, Turner RJ, Handler JS (1983) Regulation of sodium-coupled glucose transport by glucose in a cultured epithelium. J Biol Chem 258:15087–15090
42. Nakao T, Yamato I, Anraku Y (1987) Nucleotide sequence of *PutP* the proline carrier gene of *Escherhichia coli* K-12. Mol Gen Genet 208:70–75
43. Neeb M, Kunz U, Koepsell H (1987) Identification of D-glucose-binding polypeptides which are components of the renal Na^+-D-glucose cotransporter. J Biol Chem 262:10718–10727
44. Oshawa M, Mogi T, Yamamoto H, Yamato I, Anraku Y (1988) Proline carrier mutant of *Escherichia coli* K-12 with altered cation sensitivity of substrate binding activity: Cloning, biochemical characterization and identification of the mutation. J Bacteriol 170:5185–5191
45. Peerce BE (1990) Examination of substrate-induced conformational changes in the Na^+/glucose cotransporter. J Biol Chem 265:1737–1741
46. Peerce BE, Clarke RD (1990) Isolation and reconstitution of the intestinal Na^+/glucose cotransporter. J Biol Chem 265:1731–1736
47. Peerce BC, Wright EM (1984) Sodium-induced conformational changes in the glucose transporter of intestinal brush borders. J Biol Chem 259:14105–14112
48. Peerce BE, Wright EM (1985) Evidence for tyrosyl residues at the Na^+ site on the intestinal Na^+/glucose cotransporter. J Biol Chem 260:6026–6031

48a. Peerce BE, Wright EM (1986) Distance between substrate sites on the Na-glucose cotransporter by fluorescence energy transfer. Proc Natl Acad Sci USA 83:8092–8096

49. Peerce BE, Wright EM (1987) Examination of the Na^+-induced conformational change of the intestinal brush border sodium/glucose symporter using fluorescent probes. Biochemistry 26:4272–4279

50. Puttner IB, Sarkar HK+, Padan E, Lolkema JS, Kaback HR (1989) Characterization of site-directed mutants in the lac permease in *Escherichia coli*. 1. Replacement of histidine residues. Biochemistry 28:2525–2533

51. Roepe PD, Kaback HR (1989) Site-directed mutagenesis of tyrosine residues in the lac permease of *Escherichia coli*. Biochemistry 28:6127–6132.

52. Roepe PD, Zbar RI, Sarkar HK, Kaback HR (1989) A five-residue sequence near the carboxyl terminus of the polytopic membrane protein lac permease is required for stability within the membrane. Proc Natl Acad Sci USA 86:3992–3996

53. Sarkar HK, Thorens B, Lodish HF, Kaback HR (1988) Expression of the human erythrocyte glucose transporter in *Escherichia coli*. Proc Natl Acad Sci USA 85:5463–5467

54. Schmidt UM, Eddy B, Fraser CM, Venter JC, Semenza G (1983) Isolation of (a subunit of) the Na^+/D-glucose cotransporter(s) of rabbit intestinal brush border membranes using monoclonal antibodies. FEBS Lett 161:279–283

55. Semenza G, Kessler M, Hosang M, Weber J, Schmidt U (1984) Biochemistry of the Na^+, D-glucose cotransporter of the small intestinal brush-border membrane. The state of the art in 1984. Biochim Biophys Acta 779:343–379

56. Silverman M, Speight P (1986) Isolation and partial purification of a Na^+-dependent phlorizin receptor from dog kidney proximal tubule. J Biol Chem 261:13820–13826

57. Stevens BR, Fernandez A, Hirayama B, Wright EM, Kempner ES (1990) Intestinal brush border membrane Na^+/glucose cotransporter functions in situ as a homotetramer. Proc Natl Acad Sci USA 87:1456–1460

58. Takahashi M, Malathi P, Preiser H, Jung CY (1985) Radiation inactivation studies on the rabbit kidney sodium-dependent glucose transporter. J Biol Chem 260:10551–10556

59. Thomas SR, Schultz SG, Lever JE (1982) Stimulation of dome formation in MDCK kidney epithelial cell cultures by inducers of differentiation: dissociation from effects on transepithelial resistance and cyclic AMP levels. J Cell Physiol 113:427–432

60. Thorens B, Sarkar HK, Kaback HR, Lodish HF (1988) Cloning and functional expression in bacteria of a novel glucose transporter present in liver, intestine, kidney and beta pancreatic islet cells. Cell 55:281–290

61. Turner RJ, Kempner ES (1982) Radiation inactivation studies of the renal brush-border membrane phlorizin binding protein. J Biol Chem 257:10794–10797

62. Turner RJ, Moran A (1982) Further studies of proximal tubule brush border membrane D-glucose transport heterogeneity. J Membr Biol 70:37–45

63. Weber J, Semenza G (1983) Chemical modification of the small intestinal Na/glucose cotransporter by amino groups. Evidence for a role of amino groups in binding of the sugar. Biochim Biophys Acta 731:437–447

64. Weiss ER, Cook JS (1986) Separation of hexose-transporting from nontransporting $LLC\text{-}PK_1$ cells on density gradients. Am J Physiol 250 (Cell Physiol 19): C199-C206

64a. Wiley LM, Lever JE, Pape C, Kidder GM (1991) Antibodies to a renal Na^+/glucose contransport system localize to the apical plasma membrane domain of polar mouse embryo blastomeres. Dev Biol 143: 149–161

65. Wu J-SR, Lever JE (1987) Monoclonal antibodies which bind the renal Na^+/glucose symport system. I. Identification. Biochemistry 26:5783–5790

66. Wu J-SR, Lever JE (1987) Monoclonal antibodies which bind the renal Na^+/glucose symport system. II. Stabilization of the active conformation. Biochemistry 26:5790–5796

67. Wu J-SR, Lever JE (1987) Purification and reconstitution of a 75 kD protein identified as a component of the renal Na^+/glucose symporter. Biochemistry 26:5958–5962

68. Wu J-SR, Lever JE (1989) A developmentally regulated 75 kD protein expressed in $LLC\text{-}PK_1$ cultures is a component of the renal Na^+/glucose cotransport system. J Cell Biochem 40:83–89

69. Yamato I, Oshawa M, Anraku Y (1990) Defective cation-coupling mutants of *E. coli* Na^+/proline symport carrier. Characterization and localization of mutations. J Biol Chem 265:2450–2455

70. Yoneyama Y, Lever JE (1984) Modulation of a differentiated phenotype in cell culture: induction of microvillar hydrolase activities by cell density and exogenous differentiation inducers in an established kidney epithelial cell line ($LLC\text{-}PK_1$). J Cell Physiol 121:64–73

Ion Channels in Epithelial Tissue: Single-Channel Properties

D. C. Eaton, Y. Marunaka, and B. N. Ling

Introduction

A major function of a cell membrane is to form a barrier between the internal environment of the cell and the external world. However, a complete barrier to the outside world is unacceptable since to survive cells require some ability to communicate with their external environment. Communication consists of a variety of different processes mediated by membrane proteins which serve as signal transduction pathways or as mediators to maintain the intracellular milieu. One subset of these proteins allows movement of ions across the cell membrane. An even more restrictive class of ion transport proteins are those which involve transmembrane movement of ions through "channels." Ion channels form an interesting class of membrane proteins which have specifically evolved to provide an essentially aqueous pathway through the otherwise highly hydrophobic barrier of the cell membrane. As such, they are capable of carrying relatively large numbers of ions per unit time. In general, to regulate the number of ions which cross the cell membrane, a channel protein undergoes conformational changes which produce states of the molecule which either contain an aqueous pathway across the protein or which have no such pathway for ions. Typically, there may be several "open" states which allow ion movement as well as several "closed" states in which there is no ionic current across the protein. Since the primary function of channel proteins is the transport of ions across the membrane, the conformational changes which lead to alterations in current flow have been the subject of significant investigation. The examination of the transitions between open and closed states of channel proteins is also motivated by the fact that regulation of ion transport is often mediated by controlling the rates of transitions between open and closed states and, thus, controlling the relative amount of time the channel is open and conducting ions compared to the amount of time closed.

Ion channels are in general characterized by having relatively high rates of ion flux through pores which act as if they are water filled. In addition, ion flux through channels is, by definition, in response to passive electrochemical gradients; that is, there is no direct expenditure of energy. This implies that transport through channels does not involve any obligatory coupling between different molecular species and that the transport is "conductive;" that is, there can be net charge movement. The uses to which cells put channel proteins are quite varied, but channels are often involved in cellular signaling because of the relatively large numbers of ions which can be translocated per unit time. Thus, sodium channels in excitable tissues generate voltage responses which signal excitation; ligand-gated channels signal the presence of high concentrations of specific molecules in the cell's bathing solution; while influx of calcium through voltage-gated calcium channels can signal the release of a variety of cellular secretory products including transmitters and hormones.

In an evolutionary context, epithelial tissues have used channels in many of the same ways as other cell types; however, they have also adapted them to more tissue specific purposes. Because of the polarized insertion of membrane proteins in the apical versus the basolateral surfaces of epithelial tissues, ion channel proteins can contribute to the vectorial transport of ions from one body compartment to another. The classical example of such vectorial transport is the uptake of sodium across frog skin [362].

The purpose of this article is to consider the different channel types which are present in epithelial tissue and, to the extent possible, to describe the physiological role that the ion channels play in the normal function of epithelial tissues. We have generally divided the subject into four major areas primarily based on the volume of literature available on the separate subjects: (a) sodium conductive channels, (b) potassium conductive channels, (c) anion channels, and (d) other channels including calcium permeable channels, nonselective channels, and stretch-activated channels. Epithelial sodium channels and to a lesser extent chloride channels are the only channels in epithelial tissues which have received much biochemical attention as proteins per se. Because of this attention the chemical and molecular biological properties of sodium channels will be dealt with in a separate chapter of this volume (Benos et al.) in the context of all amiloride-blockable sodium transport systems. Therefore, this chapter will concentrate on the cellular physiology of the four major classes of channels enumerated above. In an attempt to survey the various ion channels of epithelial tissues, we will, when possible, tend to emphasize reports which describe the properties of single epithelial channels when available and use references to the macroscopic epithelial conductances primarily in a historic or supportive sense or in those cases when no single channel data is yet available. Our rationale for such an approach is threefold. First, a survey of the enormous literature of the past 30 years which describes all the various macroscopic conductances in epithelial cells would not be manageable; second, different aspects of the literature on macroscopic properties of epithelial tissue have been reviewed recently (for other recent reviews, see [79, 89, 379, 407]); and, third, descriptions of single channel properties provides a description of membrane events much closer to the properties of the individual membrane proteins than the lumped macroscopic properties of whole tissues or whole membranes.

Measurement Methods

The properties of single channels in epithelial tissues have generally been obtained from one or the other of two different types of measurements: power spectral or "noise" analysis and single channel analysis using patch clamp methods. We will present here the fundamentals of these two techniques and compare their strengths and weaknesses with particular emphasis on their application to epithelial tissues. Epithelial tissues present a special situation which make either analysis method very attractive, but which also poses special problems for interpretation of both approaches.

Power Spectral or "Noise" Analysis

Historically, power spectral analysis, which is also known as noise or fluctuation analysis, antedates single channel measurements. The initial description of its use in epithelial tissues dates from 1977 [216].

Determination of Channel Properties from Fluctuation Measurements: The Power Spectral Density Function

A much more detailed treatment of methods and the theoretical aspects of noise analysis has been given elsewhere; however, a simple description of the method is appropriate here to provide a basis for comparison of single channel and fluctuation measurements.

While the chemical kinetic events associated with opening and closing of single channels and the current fluctuations produced by these kinetic transitions are temporal phenomenon, it is often easier to consider temporal events in the frequency domain rather than the time domain. This is particularly true when examining chemical kinetic systems since chemical kinetic rates are directly related to the frequency of transitions. While in the time domain, rates must be extracted from the characteristic times obtained from autocovariance functions [56]. But more importantly as pointed out by Fishman [68], handling data in the time domain is significantly more difficult than in the frequency domain; calculation of time domain representation of fluctuations (the autocovariance function) is time consuming compared to the generation of power spectra using any of several fast Fourier transform methods. In general for a channel which contains at least one transition from an closed to an open state as in the following scheme:

$$\text{-CLOSED} \underset{k_{-1}}{\overset{k_1}{\rightleftharpoons}} \text{OPEN-} \tag{1}$$

where k_1 is the rate of transitions from closed to open conformations and k_{-1} is the rate in the reverse direction, then the generalized power spectral density function which describes these transitions is:

$$S(f) = \sum_{i=1}^{n} S_i(0)/(1 + (2\pi f/(k_1 + k_{-1}))^2) \tag{2}$$

If Eq. 2 represents the power spectrum of a single channel, the power spectrum of M independent channels, $S_M(f)$, is $M \cdot S(f)$. The power spectral density function contains one component, a lorentzian, for each separate set of transition rates leading from a closed to an open state of the channel. Moreover the half-power point or corner frequency, f_c, is directly related to the transition rates by the relationship:

$$2\pi f_c = k_1 + k_{-1} \tag{3}$$

Generalized power spectra provide estimates for the plateau levels, S(0), and the corner frequencies, f_c of state transitions. For a simple two-state model (or other simple kinetic schemes) additional information can be obtained from the power spectrum. The integral of the power spectral density function is equal to the variance of the fluctuations. That is:

$$\sigma^2 = S(0)\pi f_c/2 \tag{4}$$

and if the total macroscopic current $\langle I \rangle$, passing through the channel is known and the reverse rate constant, k_{-1}, is also known or can be derived from other experimental measurements, then, unit current, i, can be derived from the following expression:

$$i = S(0)^2 f_c^2/k_{-1}\langle I \rangle \tag{5}$$

Once the unit current is calculated, the total number of channels, M, contributing to the power spectrum can also be estimated since:

$$M = \frac{2\pi f_c \langle I \rangle}{k_1 i} \tag{6}$$

similar methods may be used to derive corresponding expressions for more complicated kinetic schemes [15, 56, 72, 189, 267].

Advantages and Disadvantages of Power Spectral Analysis

The advantage of current fluctuation analysis lies in its ability to provide information about the properties of entire populations of channels. Unfortunately, the price for obtaining information about large numbers of channels is the necessity for assuming the properties of individual channels. In particular, it requires the assumption that all the channels contributing to the fluctuations are identical and behave independently. Moreover, to estimate the unit current of a channel (from Eq. 5) requires specific knowledge about the rate constants of the proposed model. Neither of these assumptions need be excessively restrictive. A more serious problem arises when fluctuation measurements are used as a basis to provide kinetic information. The form of the fluctuation functions can give information about the minimum number of closed and open states if the frequency resolution is adequate; however, even with appropriate frequency resolution, fluctuation analysis cannot provide information about how the states communicate and can provide only limited information about closed states which do not communicate directly with an open state. Therefore, to interpret the fluctuation data some kinetic model which contains the minimum number of states must be assumed. In addition, the corner frequencies of power spectra do not give information about individual rate constants, but, rather, provide the sum of at least two rates. Unless one of these rates can be experimentally manipulated, determination of individual rate constants is usually not possible. In the absence of a frequency resolution which includes the time constants for all of the channel transitions, the situation becomes more problematic since unresolved transitions may contaminate the otherwise well-resolved fluctuations. Thus, the ideal situation in which to use fluctuation analysis would be one in which the true kinetic scheme has been independently verified by some other experimental method, but one in which large populations of channels must be monitored. A close to ideal situation may be provided by the combination of fluctuation measurements and single channel measurements.

Single-Channel Analysis

Recently, the ability to measure the currents associated with single membrane channels has dramatically changed the manner in which current fluctuations can be analyzed. Since single channel analysis methods have been dealt with extensively by others, we will only briefly describe the methods that are directly relevant to characterizing epithelial ion channels.

One result of observing single channels is an absolutely clear picture of the mechanisms which govern the magnitude of membrane ion permeability. In general, the ionic current, I_X, which flows through a single class of channels which are permeable to ion X is given by the expression:

$$I_x = \gamma_x N P_o (E_m - E_x) \tag{7}$$

where N is the total number of channels present per unit area of membrane, γ_x is the unit conductance of a single channel, P_o is the open probability of single channel, E_m is the membrane potential and E_x is the equilibrium potential for ion X. This equation distinguishes the conduction properties of channels from their gating properties. The conduction property of a channel is described by the unit conductance which is a measure of the rate at which ionic current flows through a single open channel in the presence of a voltage driving force $(E_m - E_x)$. The gating property of a channel is described by P_o, the open probability, which is the fraction of time that a single channel remains in an open or conducting state. The product, NP_o, is the mean number of open channels. In theory, any of the parameters on the right side of Eq. 7 could be altered to change the magnitude of current through a group of channels. In fact, the most frequent mechanism for altering ion flow is an alteration of N or P_0. Therefore, many of the single-channel analysis methods concentrate on counting the density of channels or examining the determinants of open probability, i.e., the mean duration of open intervals and the mean duration of closed intervals for a single channel.

Although all of the formulations which were developed in conjunction with power spectral density functions are applicable to the fluctuations of a single channel, there are other methods which can only be applied to the fluctuations of a single channel. These methods are particularly useful in providing estimates of minimum kinetic schemes which represent the conformational transitions of channel proteins. There are two types of information available from a single-channel record which cannot be obtained from the fluctuations produced by multiple channels. First, the duration of individual open and closed intervals can be directly measured; and, second, the sequence of intervals can be determined. In the first case, histograms of the duration of open and closed intervals can be used to estimate the number of different classes of intervals. In the second, once classes of intervals are established, then the temporal relationship between classes can be determined; i.e., does an interval of one identified class always follow an interval of another identified class.

Distributions of Intervals

The distribution of durations for any state of a channel can be described by an exponential function [35]. Experimentally the underlying distribution is estimated by generating a histogram of all open intervals and a similar histogram of all closed intervals. The form of these discrete distributions would in the limiting case of a very large number of measured intervals and a large number of histogram bins each representing a very narrow range of durations simply be an exponential distribution. The number of exponential components in the distribution of open intervals represents the minimum number of open states of the channel while the number of exponential components in the closed interval histogram gives an estimate of the minimum number of closed states. The mean of each exponential component is just the mean lifetime of one of the states of the channel and the reciprocal of the mean duration is the sum of all the rate constants for leaving that state. As mentioned above, noise analysis can only give information about the minimum number of states present in a kinetic model. Without additional experimental manipulation, the rate constants associated with these states are often difficult to determine individually and the relationship between states, i.e, which states communicate with one another, is often impossible to determine unequivocally. A better determination of a correct kinetic model can often be made by analysis of a single-channel record. There are two reasons for this. First, the mean duration of measurable states is determined only by the rate constants leaving the state, i.e., half as many rate constants contribute to the mean durations compared to corner frequencies. Therefore, experimental manipulations can more easily isolate individual rate constants. In addition to simpler expressions to determine rate constants, there is also specific information about communication between states in the single-channel record. By observing which classes of events occur temporally together, it is possible to determine which states are "adjacent" or communicate in a kinetic model.

Other Measurement Methods

There are several other methods for gaining information about ion channels in epithelial tissues. Probably the most used is measurement of so-called short-circuit current. This technique pioneered by Ussing and Zerhan [363] recognizes that if there is no potential gradient and no concentration gradients across an epithelial tissue, then any transepithelial current must be associated with energy requiring processes, i.e, active transport. In some tissues, short-circuit current measurements are particularly useful since it can be demonstrated experimentally in these tissues that all of the short-circuit current is associated with the active transport of one ion. Examples include frog skin (and most other sodium-transporting tight epithelia) where the current can all be attributed to active, apical-to-basal sodium transport or turtle bladder under appropriate conditions when the short-circuit current is due to active chloride flux. There does not appear to be under ordinary conditions any chordate epithelial tissue in which the current can be attributed exclusively to active potassium transport; however, some insect epithelial tissue seem to display a potassium-sensitive short-circuit current. Despite these generalizations, it is still necessary in every tissue exposed to new conditions to test for the equivalence of the short-circuit current and the flux of a specific ion. Even when such experimental tests are done, however, short-circuit current is a relatively indirect measure of the properties of single ion channels in epithelial cell membranes. This is primarily because it reflects the properties of the two membranes, apical and basolateral, in series with quite different transport processes taking place at the two membranes and because there is often little or no access to the intracellular compartment, thus, making it difficult to know the ion composition on the inner face of ion channels or the absolute potential gradient driving ions through the channels. Since the ion composition and the potential may play a critical role in regulating the properties of individual channels more information than short-circuit current alone is usually necessary to understand specific ion channels. One major aid to gaining additional information about the intracellular environment is the use of microelectrodes. Conventional microelectrodes can provide specific information about individual membrane potentials while ion-specific microelectrodes can provide information about the content of intracellular ion composition. Nonetheless, even with microelectrodes, it may not be possible to associate a membrane conductance with a specific class of ion channel proteins. One alternative method which does isolate individual conducting channel molecules is bilayer reconstitution in which membrane isolates of epithelial cells are recombined with artificial lipid bilayers and the current through single, reincorporated channels can be measured. This method has intrinsic merit, but has been used infrequently to examine epithelial tissues. We will not discuss the method in this chapter, but examples of reconstitution methods have recently been reviewed [22].

Amiloride-Blockable Channels Including Sodium Channels

In "tight" epithelia (i.e., epithelia with very high junctional resistance), Na^+ transport is a two-step process: first, passive entry of sodium from the lumen into the cell through apical sodium channels followed by ATP-dependent extrusion from the cell to the serosal compartment via the basolateral Na^+,K^+-ATPase. Specific sodium-transporting epithelial tissues, such as the mammalian distal nephron, are the major sites for discriminatory control of total body Na^+ balance. Because of the importance of sodium transport to homeostatic balance, a primary emphasis of transport physiology focuses on modulatory mechanisms which can provide the tight feedback control necessary for the maintenance of this sodium homeostatic balance. While some regulation of sodium transport may take place at the basolateral exit step by controlling Na^+,K^+-ATPase activity, almost all regulation of Na^+ transport is, apparently, via control of the apical step, namely the amiloride-blockable Na^+ channel. Modulation of the channel involves multiple factors which constitute a threefold hierarchy: first, regulation at an organismic level by circulating hormones; hormonal interaction with basolateral membrane receptors trigger the second level of modulation which is regulation at a cellular level by intracellular second messenger cascades which transduce basolateral receptor signals to the apical membrane; and finally, the third level of regulation occurs at a molecular level within the apical membrane and involves second messenger interaction with apical membrane-associated molecules including the Na^+ channel protein itself.

Unfortunately, the complex geometry and morphology of many sodium transporting epithelia makes the interpretation of a data describing Na^+ channel regulation difficult. In addition, lack of access to the cytoplasmic compartment precludes examination of the intracellular regulatory factors controlling apical membrane sodium transport. Fortunately, fluctuation analysis and patch voltage clamp methodology permits resolution of the picoampere currents which correspond to the individual openings and closings of single Na^+ channels, and thus these methods allow a detailed examination of Na^+ transport regulation.

Macroscopic Properties and Fluctuation Measurements

Some of the first fluctuation measurements on any epithelial tissue were performed to investigate the properties of sodium channels in frog skin [216]. The hope was that, as with other types of channels in excitable tissue, it would be possible to measure the spontaneous openings and closings of the epithelial sodium channel. However, the conclusion from these original measurements was that there was no obvious spontaneous channel noise in the range 0.2–100 Hz. These first fluctuation measurements and every subsequent similar measurement all observed power spectra from intact tissue that were characterized by a uniform rolloff of power at a rate proportional to a value somewhere between $1/f$ and $1/f^2$ with no observable corner frequency above 0.1 Hz. Theoretically, fluctuations of channels in an intact membrane would be expected to produce a power spectrum consisting of the sum of $1/f$ noise (of unknown origin) and at least one lorentzian component ($1/f^2$ noise) with a distinct plateau at low frequencies and an inflection or "corner" at the frequency at which the power reaches half the power of the plateau value [56]. This corner frequency, as pointed out above (see "Measurement Methods"), is simply related to the sum of the rates for leaving the open and closed states of the channel. Since the observed spectra are somewhere between the background $1/f$ noise and the $1/f^2$ noise of channel fluctuations with no obvious corner, the implication appears to be that there are spontaneous fluctuations, but their corner frequency is below that easily measurable by fluctuation methods. Only after application of channel blockers like amiloride is it possible to observe a lorentzian component of the power spectra. Because of the lack of any observable spontaneous lorentzian, Lindemann and his colleagues [214, 215] argued that, at least as a good approximation, the sodium channel block by amiloride could be modeled as a two-state process; i.e.:

$$\text{OPEN} \underset{}{\overset{\text{amiloride}}{\rightleftarrows}} \text{OPEN BLOCKED} \tag{8}$$

This notion effectively side-stepped the issue of the underlying kinetic scheme for the sodium channel in the absence of blocking agents. Lindemann [215] was always careful to point out that there might be spontaneous fluctuations of the sodium channel, but that he had no experimental basis for concluding whether they existed. Unfortunately, this presentation also tended to cloud the fact that there was also no a priori basis for assuming that the kinetic scheme above was a completely correct description of amiloride's interaction with the sodium channel. It is true that the corner in the power spectra induced by amiloride and other similar blockers does imply that one of the mechanisms of the amiloride block is an interaction with open channels, but such an interaction does not preclude other types of interaction as well.

Nonetheless, the simple model of amiloride block was useful for a variety of studies involving the nature of the amiloride binding site at the external surface of the channel [168, 169, 202–204] and parameters obtained from fluctuation analysis have been used to provide a qualitative estimate of the number of amiloride-blockable sodium channels present in tight epithelia [1, 131, 223, 224, 423, 425]. However, when more quantitative estimates of the sodium channel density were required, the two-state model shown above was not generally adequate [15, 132]. The problems developed because of a variety of interesting features of the epithelial sodium channel. As might be expected considering the channels role in maintaining homeostatic balance, the sodium flux through the apical membrane via these channels is strongly modulated by a variety of physiological factors. One of the mechanisms by which this modulation takes place is to alter the channel density in the apical membrane; the hormones aldosterone and antidiuretic hormone (ADH) appear to regulate the apical sodium entry in this manner. In addition, there appears to be at least one mechanism (which has been termed "homocellular" [327]) to maintain a balance between apical sodium entry and total sodium transport. This modulation of sodium entry has also been termed "sodium self-inhibition" [73, 83, 205, 373]. In one explanation of "self-inhibition," sodium ions are capable of directly or indirectly closing open channels. Of course, this concept of "self-inhibition" requires that there must be a closed state of the channel and, therefore, at least a three-state kinetic scheme is necessary (if amiloride block is included). This scheme is usually written as:

$$\text{CLOSED} \rightleftharpoons \text{OPEN} \underset{}{\overset{\text{amiloride}}{\rightleftharpoons}} \text{OPEN BLOCKED} \tag{9}$$

Such a three-state model has been used to describe the effect of a variety of blocking agents [1, 204], several hormones [131], and a few drugs [14, 15]. The properties of this model have been reviewed elsewhere [132]. Nonetheless, despite the apparent necessity for a closed state of the channel in the absence of blocking agents, there was no direct evidence for such a state.

To summarize these points: examination of the microscopic properties of individual sodium channels using power spectrum analysis methods has been complicated by the fact that there is no discernible corner to the spectra from intact tissues at least down to frequencies of about 0.5 Hz [1, 131, 132, 198, 215]. This implies that either channels are open all of the time [215] or that their rate of transitions between open and closed states is less than 1 s^{-1} [1].

Single-Channel Measurements: Why No Corner is Measurable; Difficulties with Measuring Sodium Channel Fluctuations

Since these fluctuation measurements were made, single-channel measurements [55, 78, 230, 282, 283] have demonstrated in several epithelial preparations that sodium channels do have both an open and closed state in the absence of blockers and that, as expected, the spontaneous transition rates between the closed and open states are in the range 0.4–0.7 s^{-1}. This corresponds to a corner frequency of about 0.1 Hz. To resolve a spontaneous corner would require a frequency resolution down to approximately 0.01 Hz; a frequency not generally accessible to investigators measuring fluctuations from whole tissues. Thus, only the blocker-induced fluctuations of the channel are normally accessible for fluctuation measurements.

The sodium channels which have been observed in freshly dissected native tissue [78, 283] and in tissue culture cells grown under appropriate conditions [55, 218, 230] have low unit conductance (about 4.5 pS) with mean open and closed times of several seconds.

Single-Channel Measurements: A Different View of Amiloride Block

As mentioned above, single-channel measurements demonstrate that sodium channels in the absence of blockers clearly have regular transitions which occur on a time scale of the order of seconds.[1] Blockers of the channels produce effects on single channel currents which are consistent with the block of open sodium channels [55, 230]. Weak blockers, such as 6-chloro-3,5-diamino-pyrazine-2-carboxamide (CDPC), are particularly striking in adding many fast closures to what would otherwise be a normally open channel. The effect of the strong blocker, amiloride, is more interesting since it does not apparently produce the fast, flickery closures characteristic of the CDPC block. Interval histograms of the open and closed intervals consist of one class of open intervals and one class of closed intervals in the absence of blockers. In the presence of blockers there is still only one class of open intervals, but with a significantly shorter mean open time. Closed interval histograms in the presence of blockers have two components, one of relatively short duration and one of much longer duration.

Because there are two exponential components in the closed interval histogram and one exponential contribution to the open interval histogram in the presence of blocking agents, the single-channel interval data appears superficially consistent with the three-state kinetic scheme suggested by fluctuation analysis. However, the interaction of blockers with open channels does not preclude the possibility that the blockers might also interact with closed channels to produce an occluded form of the closed channel. Additionally, if such an occluded form is possible, then open occluded channels may be capable of closing with the blocking molecule in place thus entering the closed occluded

[1] Marunaka and Eaton [55, 230] have reported channel kinetics which involve at least two open and two closed states for the normal sodium channel. Besides the long open and long closed states which would be anticipated from the fluctuation and macroscopic measurements, they also report a short duration open and closed state. While these short duration states may be important for a complete understanding of sodium channel conformation, the short duration events are relatively infrequent and the frequency varies with the culture conditions. This low frequency of occurrence coupled with their short duration mean that these states contribute little to the overall open probability and are, thus, not very important in a physiological context. Because of small effect on the open probability, the short duration states can, for all practical purposes, be excluded from consideration when examining channel properties.

conformation. That is, rather than the three-state model above, the channel might best be described by the following kinetic model in which both long closed and long open states of the channel can interact with blocking molecules and occluded channels can make transitions between the OPEN OCCLUDED state and the CLOSED OCCLUDED state:

$$\begin{array}{ccc} \text{CLOSED} & \underset{k_{-1}}{\overset{k_1}{\rightleftharpoons}} & \text{OPEN} \\ k_3 \uparrow\downarrow k_{-3} & & k_2 \uparrow\downarrow k_{-2} \\ \text{CLOSED OCCLUDED} & \underset{k_{-4}}{\overset{k_4}{\rightleftharpoons}} & \text{OPEN OCCLUDED} \end{array} \tag{10}$$

There are a number of quantitative characteristics of amiloride block which are only consistent with this model and not the three-state model [55, 230]. Specifically, if there is only one pathway from the OPEN to the CLOSED state, then the total time from when the channel enters the OPEN state from the CLOSED state until it leaves the OPEN state to enter the CLOSED state should increase with increasing blocker concentration [5, 256]. If an open occluded channel can become a closed occluded channel, this time should not increase as much or may even decrease. Since the duration between leaving and reentering the CLOSED state does decrease in the presence of blockers, there must exist a new path from the OPEN to the CLOSED state. If there were no such alternative pathway, the burst duration should increase [256]. Thus the simplest explanation of the properties of sodium channels (including the effects of blockers) can be represented by the kinetic model given by Eq. 10.

There are several implications of such a cyclic, four-state model for fluctuation measurements. The consequences of the model depends upon the nature of the experiment and also on the blocking agent used to induce fluctuations. In particular, the action of CDPC can be very well approximated by a three-state model in which CDPC blocks only open channels. Because of the very rapid off-rate from the channel, a CDPC-blocked channel rarely closes with the blocking molecule in place. Thus, the transitions reflected in the power spectra do not reflect a significant contribution of states other than the OPEN and OPEN OCCLUDED states. Therefore, as pointed out by Helman and Kizer [132], CDPC is an easier blocker to use experimentally for the induction of noise. Amiloride, on the other hand, is more complicated. At high concentrations of amiloride, a significant fraction of the channel molecules reside in a closed configuration with an amiloride molecule bound to the channel. The off-rate constant for amiloride is so slow that there is a significant reduction in the number of channels which are available to open and, therefore, are available for open channel block. In the absence of large numbers of open channel blocking events, the total power produced by amiloride block is less than expected for a true open channel blocker of similar affinity.

Another potential problem involves the use of fluctuation measurements to "count" channels. As pointed out above (see "Measurement Methods"), if the appropriate kinetic scheme is known, as well as the macroscopic current and some estimate of the open probability of the channel can be made, then the unit current and total number of channels contributing to the total current can be calculated from the values of the plateau and the corner frequency of the power spectrum. Choosing an inappropriate kinetic scheme implies applying the wrong formulae to make the calculations of channel number. If rate constants from the two-state model are used to calculate open (or closed) probability even though the correct kinetic model is the four-state model, then the unit current will be underestimated and the calculated number of channels will be overestimated particularly at high blocker concentration. Therefore, it is essential to use the correct kinetic scheme if a detailed description of the sodium channel is to be obtained from fluctuation measurements.

Single-Channel Measurements: A Family of Amiloride-Blockable Channels

On the basis of the macroscopic, fluctuation, and single-channel measurements described above, one would predict that single, amiloride-blockable channels on the apical surface of sodium-transporting epithelia should be highly selective for sodium over potassium and should open and close infrequently. Indeed, as described above, channels with such properties can be observed in several epithelial preparations [55, 230, 282, 283], but in addition, other amiloride-blockable channels have also been described. Previous patch clamp studies have identified different channels with either high selectivity, low selectivity, or no selectivity for Na^+ over K^+ in the apical membranes of a variety of cultured renal cells (A6, LLC-PK_1, and cultured inner medullary collecting duct (IMCD) cells [25, 53, 119–121, 211]). Cell culture conditions are apparently the primary determinant of the frequency with which these different channel types are found in the apical membrane [53]. Properly defined incubation media with cells grown on very permeable supports always leads to numerous, highly selective Na^+ channels in the apical membrane with few, if any, other types of amiloride-blockable channels. But multiple types of amiloride-blockable channels are also found in other epithelial tissues as well. Primary cultures of rabbit cortical collecting tubule (CCT) cells also have three channels of differing sodium selectivity, two of which are blockable by submicromolar concentrations of amiloride [218]. There is also a well-characterized nonselective cation channel in both primary cultures and isolated rat IMCD cells [211].

There is some experimental evidence to suggest a relationship between amiloride-blockable, nonselective cation channels and amiloride-blockable, highly sodium-selective, low-conductance, apical Na^+ channel [53]. Lewis and his colleagues [194, 196, 198, 427] postulate that an intramembranous pool (presumably cytoplasmic vesicles) containing highly Na^+-selective channels is inserted into the apical membrane of rabbit urinary bladder, and that the insertion is mediated through interaction of the vesicle with cytoplasmic microfilaments. The newly inserted Na^+-selective channels then undergo hydrolytic degradation by renal tubular urokinase or kallikrein resulting in a channel with progressive loss of amiloride sensitivity and Na^+ selectivity until the only conductance remaining is the so-called "leak" pathway. This non-selective, partially degraded channel finally becomes unstable in the apical membrane and is released into the luminal fluid and possibly the cytosol.

This hypothesis is supported by the experimental observation that, after stimulation of vesicle fusion by bladder stretching [196], fluctuation analysis revealed that the newly inserted, amiloride-blockable conductance starts with a ratio of sodium permeability to potassium permeability of about 30. As the channels degrade they lose this selectivity with partially degraded channels having a Na to K permeability ratio of between 2.6 and 9.0. Finally, the fully degraded channel has little, if any, Na to K selectivity and is not blockable by amiloride [198]. Zweifach and Lewis [427] have successfully inserted a non-selective, 6-pS "leak" channel from rabbit urinary bladder washings into lipid bilayers. Patch clamp measurements of this latter channel yield a Na to K permeability ratio of 0.5, very close to the selectivity estimates obtained from noise analysis of the urokinase/kallikrein-degraded, "leak" channel described above. Besides these functional relationships which may imply some specific relationship between the selective Na^+ channel and some cation nonselective channels, both the Na^+-selective and nonselective cation channel are inhibited by the indolamine, serotonin, with similar half-maximal inhibitory concentrations. Consistent with this similarity in properties and possible common evolutionary ancestry is the identification using patch clamp methods of an amiloride-blockable, Na^+-selective channel and an amiloride-insensitive, nonselective cation channel possessing the same conductance (5 pS) and gating kinetics in the apical membrane of toad urinary bladder [78]. In cultured sweat gland secretory epithelium from normal and cystic fibrosis patients, an amiloride-blockable 5-pS non-selective cation channel has also been described [176]. Amiloride-blockable Na^+ channels from the *Xenopus* distal nephron cell line, A6, reconstituted into lipid bilayers have consistently low selectivities (P_{Na}/P_K = 2–3) with a wide range of unit conductances (4–80 pS) [279, 320]. High-conductance (40 pS) nonselective cation channels have also been described in rabbit cortical collecting tubule primary cultures [398]. That these channels may represent a continuum of differentiation or degradation states may explain the identification of both high-selectivity, low-conductance and low-selectivity (P_{Na}/P_K 3–4), high conductance (7–10 pS) Na^+ channels and 21 pS nonselective cation channels in A6 cells depending on their transporting state or permeability of the cells growth supports [119, 120]. Evidence for the existence of more than one set of distinguishable channel characteristics (either due to different channel proteins or different forms of the same protein) depending on exposure to aldosterone has also been demonstrated in both A6 cells [162] and rabbit urinary bladder [200]. The P_{Na}/P_K in urinary bladder, calculated from changes in apical membrane potential was 2.6 for animals maintained on a normal diet (low circulating levels of aldosterone) and 9 for animals on a low-Na^+ diet (high levels of circulating aldosterone). Eaton and Hamilton [53] postulated a scheme for the regulation of apical Na^+ permeability in which exposure to aldosterone switches existing sodium channels from a non-conducting or "cryptic" state to a functional, conducting state with no change in the number of apical membrane channel proteins. Recent experimental work by Kleyman et al. [170] on A6 cells in which sodium channels are labelled with channel specific antibodies shows that there is no alteration in the number of marked channels in the presence and absence of aldosterone even though the total cellular conductance increases after aldosterone treatment. On the other hand, in this scheme, exposure to vasopressin does not alter the properties of preexistent channels, but rather promotes the insertion of new channels from a vesicular pool. In addition, besides the functional appearance and disappearance of channels in response to changes in exposure to aldosterone, the scheme also suggested that, associated with aldosterone removal, there was a change in the selectivity of apical channels from being highly selective for sodium to being much less selective for sodium over potassium. There is some evidence for aldosterone exposure changing the functional conductance of preexistent channels ([88, 90, 286] and see also Benos et al., this volume); however, aldosterone-induced transitions from sodium-selective to cation nonselective channels are still hypothetical since there have been no reported single channel recordings documenting the spontaneous conversion of a highly sodium-selective channel to a poorly selective channel.

Another link between amiloride-blockable, poorly selective channels and highly Na^+-selective channels is a common response to several hormones and second messenger agents known to regulate sodium reabsorption. Amiloride-blockable sodium channels and amiloride-blockable nonselective cation channels are both activated by vasopressin, adenosine 3′, 5′-cyclic monophosphate (cAMP) analogues, phosphodiesterase inhibitors, and protein kinase A (PKA) in toad urinary bladder [2, 55, 78, 230].

There are, however, some differences in regulation among the various amiloride-blockable cation channels. G-protein activation appears to activate poorly selective sodium channels [25] and nonselective cation channels [208, 209, 211, 212] but inhibit highly selective sodium channels (Ohara, personal communication).

Nonetheless, there are enough similarities between the various amiloride-blockable channels that there is a possibility that they may be related. Only isolation of the genes for the individual channels will provide the final answer to this question. Since both selective and nonselective sodium channels have been expressed in oocytes [92, 138, 281] and monoclonal antibodies against at least one of the channel proteins are available (Benos et al., this volume), the answer may be near.

Cellular Regulation of Amiloride-Blockable Channels

As mentioned above, one of the main reasons for examining sodium channels in the first place is to understand the control and regulation of total body sodium homeostasis at a cellular level. In addition, mechanisms for regulation also seem to be one method for categorizing and differentiating different types of amiloride-blockable channels. In this regard, patch clamp methods are particularly suited to an examination of channel regulation at a molecular level. When apical membrane patches are "excised" there is direct access to the intracellular surface of the channel protein, the presumptive site of much of the channel regulation. Such access is not possible with other methods of studying epithelial tissue. Thus, it is possible to study the modulation of individual Na^+ channels at the proximal site of regulation; that is, localized apical membrane events. This is important since the mechanisms associated with regulation of Na^+ transport form at least a threefold hierarchy. The first level involves hormonal regulation at the level of the entire organism. At a cellular level, hormonal agents interact with specific cellular receptors at the basolateral (serosal) surface of epithelial cells to activate the second level of regulatory events. Intracellular second messenger cascades transduce the hormonal signal from the basolateral membrane to the apical membrane where the primary control of Na^+ transport occurs. The third level involves molecular events confined to the local microenvironment of the apical membrane itself. At this level membrane-bound regulatory elements are activated by cytosolic second messengers. Activated membrane-bound factors directly modulate the activity of single, Na^+ channel molecules. This last level of modulation is difficult to unequivocally examine in any preparation except an excised patch of apical membrane. The difficulty in examining regulation at a molecular level is refected in the fact that only recently has their been significant progress using single-channel methods to examine single amiloride-blockable channels in epithelial tissues. As mentioned above (see "Single-Channel Measurements: A Family of Amiloride-Blockable Channels"), based on their properties and mechanisms for regulation, there appear to be three major classes of amiloride-blockable channels: those with high selectivity for sodium over potassium, those with low selectivity, and those with little or no selectivity among univalent cations. We will describe their properties separately, but the characteristics of the different amiloride-blockable channel types are summarized in Table 1; non-specific channels are described in Table 2.

Regulation of Poorly Selective Amiloride-Blockable Channels

We define poorly selective channels to be those channels with a low selectivity for sodium over potassium (P_{Na}/P_K about 4–5). Single poorly selective, amiloride-blockable channels were first described in the amphibian renal cell line, A6 [120], and have subsequently also been described in the porcine kidney cell line, LLC-PK_1 [25]. The channel has a unit conductance of 9 pS and P_{Na}/P_K ratio of 4. It is slightly voltage dependent with the open probability increasing with depolarization. The increase in open probability is due to a voltage-dependent increase in the opening rate and a decrease in the closing rate of about tenfold for a 180 mV depolarization of the membrane potential. The channels are not blocked by Ba^{2+} or tetraethylammonium (TEA) [122]. The mean open time is 30–40 ms, and the mean closed time is approximately 100–120 ms. The mean

Table 1. Comparison of Amiloride-Blockable Channels

	High-selectivity channel	Low-selectivity channel	Non-selective channel
Predominant occurrence	1. Mineralocorticoid-replete CCT cells 2. Aldosterone-treated A6 cells grown on permeable supports 3. DOC-treated rabbit distal colon	1. Untreated CCT cells in primary culture 2. Aldosterone-depleted A6 cells grown on permeable supports 3. A6 cells grown on impermeable supports 4. Medullary CT cells	1. IMCD cells in primary culture 2. A6 cells grown on glass
Na: K selectivity	Greater than 20:1	less than 5:1	Close to 1:1
Unit conductance	4–5 pS	8–9 pS	24–28 pS
Kinetics	Long open time in mineralocorticoid-treated tissues (about 2 s); much shorter in depleted tissues (11 ms); closed time about 3–4 s	Similar under all conditions (open time about 30–50 ms; closed time 100–120 ms)	Similar under all conditions (open time about 80 ms; closed atime 200 ms)
Voltage dependence	Open probability increases with hyperpolarization.	Open probability increases with depolarization.	No effect of voltage
Amiloride 1/2 block	47 n*M*	200 n*M*	50 n*M*
PK-C phosphorylation	Decreases open probability	?	Decreases open probability
Apical G-protein activation	Decreases open probability	Increases open probability	Increases open probability
Increased Intra-cellular calcium[a]	Decreases open probability	No effect	Increases open probability
Increased Intra-cellular acidity[b]	Decreases open probability	No effect	No effect
Aldosterone	Increased mean open time	Little or no effect	No effect
ADH	Increased number of channels	No effect	No effect

[a] Cell-attached patches; intracellular calcium elevated with calcium ionophore.
[b] Excised, inside-out patches.

Table 2. Amiloride-Blockable Channels

Epithelium	Method[a]	Selectivity or P_{Na}/P_K[b]	Conductance (pS)	Voltage Activated[c]	Amiloride 1/2 Block	Inhibition/Stimulation[d]	References
Frog skin							
Rana pipiens	Noise	High	–	–	Inhibits	Quinine, quinidine, indomethacin-inhibits, ADH, forskolin stimulate	[1, 73, 130, 139]
Urinary bladder							
Bufo marinus	Noise	High	–	–	–	–	[375]
Bufo marinus	PC	High	4.8	–	Inhibits	–	[78]
Rabbit	Noise	High	–	–	Inhibits	Pressure stimulates	[197]
Descending colon							
Rabbit	Noise	High	–	–	Inhibits	–	[426]
Human	Noise	High	–	–	Inhibits	–	[402]
Kidney							
Rabbit							
CCT	PC	> 40	4–5	Hyperpol	≈50	Aldosterone stimulates	[217]
PT (S3)	PC	High	12	–	Inhibits	–	[97, 99]
Rat							
CCT	PC	> 40	4–5	Hyperpol	≈50	Aldosterone stimulates, Ca^{2+}, low pH, phorbol, ANF, cGMP, PTX inhibit, GTP-S, α_{i-3} stimulate	[282, 284]
IMCD culture	PC	≈1	27	None	≈50		[208–211]

Mouse							
PT (S3)	PC	High	12	–	Inhibits	–	[97]
A6 cell line	PC	> 20	4.5	Hyperpol	48	ADH, aldosterone, PTX, GDP-β-S stimulate, GTP-S, phorbol, DAG inhibit	[54, 57, 160, 216, 161, 219]
	PC	4–5	8.9	Depol	124	Ca^{2+}, Ba^{2+}, no effect	[120]
	PC	≈1	28	None	Inhibits	Ca^{2+}, Ba^{2+}, no effect	[219]
LLC-PK_1 cell line	PC	3–4	9	Depol	Inhibits	ANF, cGMP, PTX inhibit, GTP-S, α_{i-3}, LT_4 stimulate	[25, 26]
Lens (frog, rat, mouse)	PC	2.0–4.0	25–30	Depol	Inhibited	Stretch stimulates; Ca^{2+} inhibits	[36, 151, 293, 296]

CCT, Cortical collecting tubule; PT, proximal tubule; IMCD, inner medullary collecting duct; TEA, tetraethytlammonium; PTX, pertussis toxin; ADH, antidiuretic hormone; ANF, atrial natriuretic factor.

[a] PC, Patch clamp; noise, spontaneous or blocker-induced fluctuation analysis.

[b] Where P_{Na}/P_K ratios are not available, relative permeability of sodium over potassium is given.

[c] Depol, open probability increases with depolarization; hyperpol, open probability increases with hyperpolarization.

[d] Ca^{2+} inhibition, inhibition by increases in intracellular calcium in cell attached patches.

open time is decreased and mean closed time increased when the apical surface of the channel is exposed to submicromolar concentrations of amiloride. The amiloride block is also voltage dependent with increasing hyperpolarization increasing the block. The amiloride binding site is approximately 35% of the way through the membrane field. With no applied potential, 200 n*M* amiloride reduces the mean open time by close to 50%. The channel has similar kinetics in cell-attached and excised patches.

G-protein activation by a variety of methods stimulates poorly selective, amiloride-blockable channels in A6 cells grown on nonpermeable supports, LLC-PK_1 cells and toad urinary bladder cells [25, 54, 91]. Recently, Ausiello and associates [9] have shown that A6 cells contain both 41-kDa and 95-kDa proteins that are ADP-ribosylated by pertussis toxin (PTX) and immunoprecipitated by antibodies against an amiloride-binding protein that has been putatively identified as the poorly selective channel. Thus it appears that the poorly selective channel is directly modulated by membrane-associated G-proteins in a manner similar to several types of channels in excitable tissues [23]. This idea is supported by several recent reports demonstrating localization of G-proteins in the apical membranes of renal epithelial cells using either ADP-ribosylation stimulated by cholera toxin (CTX)/PTX or a specific antibody against a G-protein subunit [255, 308, 366].

While most evidence suggests that there are apical membrane-associated G-proteins, experiments do not clearly indicate that G-proteins directly interact with the channel proteins themselves. In fact, there is some evidence to the contrary. Application of lipooxygenase metabolites to poorly selective channels in A6 cells mimics the effects of activation of G-proteins by GTPγS or direct application of the activated $G_{\alpha_{i-3}}$ subunit [26].

Regulation of Nonselective Amiloride-Blockable Channels

The amiloride-blockable, nonselective cation channel was originally described in inner medullary collecting duct cells obtained either from the native tissue or primary cultures of IMCD cells [211]. These channels have no significant selectivity of sodium over potassium (P_{Na}/P_K about 1). The unit conductance is 27.5 pS, and the channel does not discriminate between Na^+, K^+ and NH_4^+. It is not voltage or pH dependent and is not blocked by potassium channel blockers, Ba^{2+} or TEA or the chloride channel blocker, diphenylamine-2-carboxylic acid (DPC). The mean open and closed times are relatively short (mean open time approximately 35 ms and mean closed time approximately 80 ms). However, like all of the amiloride-blockable channels in tight epithelia, the mean open time is decreased and mean closed time increased when the apical surface of the channel is exposed to submicromolar concentrations of amiloride. With 500 n*M* amiloride the mean open time is reduced to 20% of control, implying a half-blocking concentration of about 50 n*M*. Increasing cytoplasmic Ca^{2+} does not affect the channel. The channel is usually closed in cell-attached patches, but opens when patches are excised suggesting the presence of some intracellular inhibitory substance. Atrial naturietic peptide (ANP) and its second messenger cGMP inactivate the channel [209] consistent with the ANP induced inhibition of macroscopic Na^+ reabsorption seen in IMCD cells [365, 417]. Also, a PTX-sensitive, guanosine-nucleotide dependent protein (G-protein) in the IMCD apical membrane directly activates this amiloride sensitive non-selective cation channel [208, 209, 212]. In excised patches, PTX or GDPβS decreases, and GTPγS or the activated α_{i-3} subunit of G_i increases the open probability of the channel. The role of this nonselective cation channel in the IMCD is not completely clear. However, besides Na^+ reabsorption, the IMCD is capable of either K^+ reabsorption or secretion in response to changes in total body potassium load. The nonselective cation channel described above could provide an ion pathway for these macroscopically observed potassium movements. Alternatively, the channel might play a role in renal ammonia handling. Renal medullary accumulation of ammonia increases during metabolic acidosis and is a major determinant of inner collecting duct ammonia secretion [123]. Since the nonselective cation channel described above is quite permeable to NH_4, transport of medullary interstitial ammonium through this nonselective, cation channel in IMCD cells could play a role in acid/base homeostasis.

Regulation of Highly Selective Amiloride-Blockable Channels

As mentioned above, fluctuation analysis and the properties of sodium transport in whole tissues suggests that the major pathway for transepithelial sodium movement should be highly selective and that individual channels should remain open or closed for relatively long periods of time. The properties of single highly selective channels correspond to these expectations. The sodium to potassium selectivity of these channels is very high (P_{Na}/P_K greater than 40) with a unit conductance of 4–5 pS. Single highly selective, amiloride-blockable channels were first described in the apical membranes of principal cells from isolated rat cortical collecting tubules [283]. They have subsequently been described in the amphibian renal cell line, A6 [55, 230], and may also be present in the porcine kidney cell line, LLC-PK_1 (Ausiello, personal communication), and IMCD cells (Stanton, personal communication). It is slightly voltage dependent with the open probability decreasing with depolarization. The decrease in open probability is due to a voltage dependent decrease in the opening rate and a increase in the closing rate of about tenfold for a 230 mV depolarization of the membrane potential. The channels are not blocked by Ba^{2+} [230]. The mean open is typically 1–2 s and the mean closed time is approximately 2–3 s. The mean open time is decreased and mean closed time increased when the apical surface of the channel is exposed to submicromolar concentrations of amiloride. The amiloride block is voltage dependent with increasing hyperpolarization increasing the block. With no applied potential, 52 n*M* amiloride reduces the mean open time by 50%. The channel usually disappears rapidly in excised patches. This makes examination of channel regulation significantly more difficult.

Regulation by Hormonal Agents. The steroid hormone, aldosterone, is the major hormonal mechanism for maintaining Na^+ homeostasis at the organismic level [53, 88, 215, 319]. Increasing the circulating concentration of aldosterone markedly increases apical membrane Na^+ permeability in tight epithelial tissues such as the distal nephron and descending colon. This increase in the tissue sodium transport occurs over the span of hours [88] and obligatorily involves gene expression and protein synthesis [33, 181, 307, 309]. The traditional view of aldosterone action is that the newly synthesized proteins were sodium channels which were inserted into the apical membrane to increase sodium permeability by increasing the number of transporting units. Fluctuation analysis [286] and measurements of whole cell currents from rat cortical collecting tubule cells [77] seemed to be consistent with this view since the apparent number of channels or the whole cell conductance after aldosterone treatment appeared to increase. However, a number of reports have questioned this view. The observations which suggest an alternative view of aldosterone action have all relied on the ability to mark individual sodium channels to allow counting of marked and unmarked channels before and after aldosterone treatment. The general observation was that aldosterone increased sodium transport, but that associated with increased transport, there was no increase in the number of sodium channels in the apical membrane ([88, 90, 286] and see also chapter by Benos et al.). This contrasted with the effect of ADH which dramatically increases the number of markable apical channels [90, 131, 286, 384, 385, 407]. Patch clamp experiments [393] provided an interesting explanation of the differences. As expected from noise fluctuation experiments on tight epithelia, exposure of A6 cells to aldosterone increases the apparent density of functional channels detectable in the apical membrane with a time course similar to the increases in sodium transport observable in whole tissues. In addition, aldosterone also alters the kinetics of single sodium channels: exposure to aldosterone increased the mean open time and open probability by about 100 times (mean open time typically increased from about 20–30 ms to 2–3 s; open probability increased from values near zero to values near 0.4). In the absence of aldosterone, the mean open time is so short that there is a significant probability of underestimating the number of channels per unit area of membrane, i.e., the *apparent* number of functional channels is smaller. These results imply that the traditional view of aldosterone action (i.e., that the aldosterone-induced proteins are sodium channels) cannot be completely correct. That the kinetics of preexisting channels is dramatically changed in the presence and absence of aldosterone implies that at least one of the induced proteins must modulate the transport of channels and that the major action of aldosterone is to cause activation of preexisting "cryptic" channels by increasing their mean open time via a post-translational modification of membrane proteins or lipids. Possible covalent modifications might involve membrane methylation or phosphorylation ([53, 217, 319] see also Benos et al., this volume).

As alluded to above, the mechanism by which vasopressin increases sodium transport in sodium-transporting, tight epithelia, primarily involves the insertion of new channels. Numerous studies support the notion of addition of new channel protein [90, 190, 299] and the addition to or turnover of the apical membrane [48, 127, 193, 341, 343, 384]. Patch clamp studies also support this conclusion [57, 78]. In one study [57], single amiloride-blockable Na channels were examined in membrane patches from cultured distal nephron cells (A6) either before or after treatment with arginine vasopressin. Treatment of cells with vasopressin increases the number of conductive Na channels with little or no change in the open probability of individual Na channels. Treatment of cells with N^6,2′-*O*-dibutyryladenosine 3′, 5′-cyclic monophosphate (db-cAMP), which is thought to be a second messenger for vasopressin, or cholera toxin which should increase the intracellular levels of cAMP through activation of a stimulatory guanine-nucleotide binding protein produce similar results. The distribution of channels within patches suggest that pretreatment of cells with vasopressin, db-cAMP or CTX may induce membrane "hot spots" containing many more conductive Na channels than the surrounding membrane whose channel density is similar to that of untreated cells which is consistent with the idea that vasopressin may act by promoting insertion of clusters of new sodium channels.

Sodium "Self-Inhibition:" Regulation by Calcium and Protein Kinase C. Because steroid hormone action occurs via gene expression, the time course of aldosterone action on Na^+ transport in target tissue is usually hours. The time course of action in the kidney is even longer since aldosterone production is increased only after an increase in renin secretion [217]. Given the large flux of Na^+ through sodium-transporting epithelial cells, more rapidly acting control mechanisms are necessary to control the minute-to-minute flux of Na^+. Cellular processes which modulate Na^+ transport within minutes can change Na^+ reabsorption in response to changes in Na^+ delivery to the apical surface of the transporting cells [327]. These processes are characterized by an inverse relationship between intracellular Na^+ activity and apical Na^+ permeability [217]. While some investigators suggest that self-inhibition is a direct effect of luminal Na^+ ions on a "modifier" site exposed to intraluminal fluid, this type of negative-feedback inhibition appears to involve more complex mechanisms than a direct action of intracellular Na^+ on the inner surface of the Na^+ channels since alterations in intracellular sodium do not affect the open probability of Na^+ channels in excised patches at concentrations comparable to that which induces "self-inhibition" [53, 282]. The mechanism probably requires secondary changes in cytosolic Ca^{2+} which, in turn, are due to changes in the Na^+ gradient driving the basolateral $Na^+ - Ca^{2+}$ exchanger. It is well established that increased cytosolic Ca^{2+} reduces the total apical Na^+ conductance [354] and reduces the open probability of single sodium channels [285]. However, in excised, inside-out patches from isolated rat CCT cells, high cytosolic Ca^{2+} concentrations applied directly to the inner surface of the channels does not affect the open probability of single apical Na^+ channels [285]. These results imply that calcium activates a cytosolic factor which then inhibits apical Na^+ channels indirectly.

One mechanism for Na^+ channel self-inhibition is mediated by an apical membrane delimited process involving protein kinase C (PKC) [217, 284]. This is interesting since activation of PKC by diacylglycerols and phorbol esters is potentiated by increases in intracellular Ca^{2+} [233] and since many other ion channels are modulated by PKC-mediated phosphorylation [27]. Whole tissue and biochemical data also support the role of PKC in down-regulating sodium channels. Yanase and Handler [413] have found that activators of protein kinase C inhibit I_{sc} in A6 cell cultures. Isotopic Na^+ uptake in the LLC-PK_1 porcine kidney cell line is inhibited by phorbol esters [258]. Rabbit CCT cell cultures possess PKC that translocates from a soluble to a particulate fraction (an indication of PKC activation) with application of phorbol ester, diacylglycerol or diacylglycerol kinase inhibitor. In addition, net Na^+ reabsorption is inhibited by phorbol esters in microperfused rabbit cortical collecting tubules [128, 243].

Patch clamp experiments on A6 cells show that the primary effect of PKC activation by phorbol esters and synthetic diacylglycerols is to reduce open probability with little change in the number of channels [217, 220] although there may be some reduction in the number of channels in rat CCT cells treated with phorbol ester [284]. Increasing intracellular calcium by application of ionophore also seems to activate PKC [217]. The effect of PKC is interesting since at least one amiloride-inhibitable apical channel in A6 cells (consisting of six nonidentical subunits with molecular masses ranging from 35 to 330 kDa) contains two subunits (55 and 130 kDa) which are phosporylated by PKC ([17]; Benos, ASN Annual Meeting, 1989; Benos et al., this volume).

Besides its effect on PKC, intracellular Ca^{2+} may activate other enzymes which modulate Na^+ channels. Several neutral proteases are Ca^{2+} dependent. This is interesting because of the hypothesized regulation of Na^+ transport in mammalian urinary bladder by proteolytic degradation of the apical Na^+ channel protein [427]. The Ca^{2+}-dependent, neutral protease, calpain I is activated by micromolar changes in cytoplasmic free Ca^{2+} levels; and may play a role in the proteolytic processing of membrane-associated PKC to an active fragment (protein kinase M) [167]. Limited proteolysis of the membrane-bound PKC molecule by calpain also more completely proteolyses PKC leading to cellular down regulation. Unfortunately, the effect of these proteases on single epithelial sodium channels has not been reported.

G-Protein Regulation of Sodium Channels. Like other amiloride-blockable channels, another modulator of highly selective, apical Na^+ channels is the family of G-proteins. As mentioned above, in A6 cells, activation of basolateral G-proteins associated with V_2 receptors, increases protein kinase A activity and leads to an increase in the number of apical sodium channels probably by promoting insertion of new channels [57]. However, there are additional G-protein effects. CTX increases apical insertion of sodium channels, but before there is a significant insertion of new channels, it also produces a transient increase in open probability which returns to baseline levels 30 min after the peak increase [54, 220]. PTX produces a similar increase in P_o, the effect is persistent, and there is no significant increase in the number of sodium channels. The unit conductance and sodium selectivity of the channel were unaffected by either toxin [220].

The issue with these experiments is determining the location and mechanism of action of the G-proteins. They could either act indirectly to promote production of second messenger agents (cyclic nucleotides or inositol polyphosphates) or G-proteins could directly gate channels as they do for K^+ channels and Ca^{2+} channels in excitable cells and for amiloride-sensitive, low-selectivity Na^+ channels in renal epithelial cells (see above). In excised patches, the activated PTX produces a increase in the mean open time of single sodium channels from 1–2 s in untreated patches to 15–40 s in treated patches [220]. A nonhydrolysable GTP analogue (GTPγS) applied to the cytoplasmic surface of excised, inside-out patches decreases the open probability of single Na^+ channels while application of GDPβS increases open probability (Ohara, personal communication). These effects are consistent if PTX is acting in a manner comparable to its action on other G-proteins; that is, to uncouple a G-protein from its receptor and, thus, deactivate the G-protein. If this mechanism is correct, the G-protein activation (by GTPγS) inhibits sodium channels (by reducing P_o) and GDPβS and PTX inhibit G-protein activation and, thereby, relieve a persistent inhibition. The problem with this interpretation is in identifying an appropriate "receptor" present in the apical membrane of A6 cells. There are no clear candidates; however, several investigators have sug-

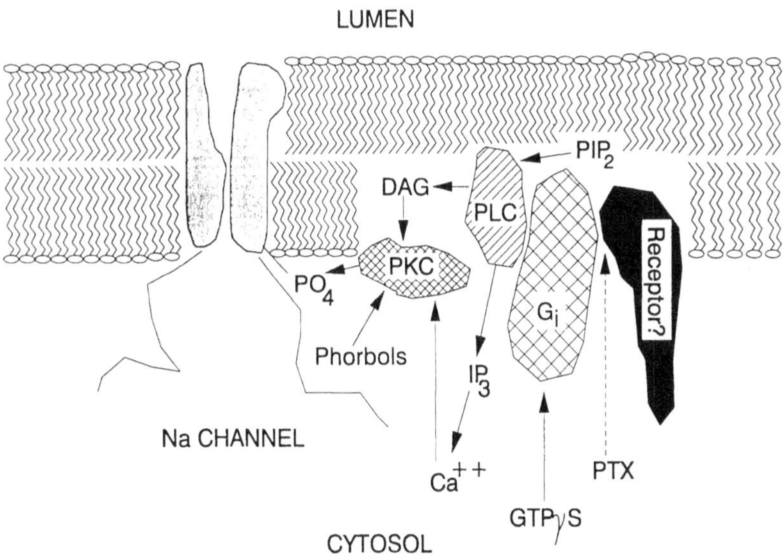

Fig. 1. One possible scheme for the regulation of highly selective sodium channels. All of the components of this scheme are membrane associated and can be activated in excised patches. Such a membrane associated regulatory cascade provides for second to second modulation of sodium transport

gested that locally produced autocoids (e.g., ANP or bradykinin) may be present in the luminal solution and inhibit sodium transport [21, 107]. Alternatively, in rat proximal tubule, at least one G-protein is modulated by intracellular Na^+ concentration [18]. Such "intracellular sodium" receptor could be the basis for the sodium self-inhibition associated with changes in cytoplasmic Ca^{2+} and protein kinase C activation. In this context, fluoroaluminate ion, an activator of G-proteins, applied to the apical surface of A6 cells produces an increase in the production of inositol polyphosphates. The actions of fluoroaluminate are mimicked by levels of luminal sodium large enough to activate sodium self-inhibition (Hinton and Eaton, personal communication). This implies that there may be a cascade of events which involves sensing intracellular sodium, activation of a apical G-protein, activation of phospholipase C, release of diacylglycerols and inositol polyphosphates, increases in intracellular calcium, activation of protein kinase C with final inhibition of the sodium channel. One hypothetical scheme for such a cascade is shown in Fig. 1.

Regulation by Other Factors. One class of possible candidates for the autocoids and agonists which might interact with apical, G-protein-linked receptors are lipoprotein metabolites produced at or within the apical membrane. Phospholipase A_2 (PLA_2) directly cleaves side-chain fatty acids, including arachidonic acid, from membrane phospholipids and is activated by G-proteins in some cells. PLA_2 activity is apparently stimulated by release of $\beta\gamma$-subunits from activated G-protein [163, 164, 265]. Alternatively, diglyceride lipase can release arachidonic acid from the diacylglycerols produced by phospholipase C [246]. Arachidonic acid can activate certain subspecies of protein kinase C [332]; the primary importance of arachidonic acid is that the release of arachidonic acid is the rate-limiting step for the biosynthesis of eicosanoids.

Cyclooxygenase metabolites derived from arachidonic acid in the kidney have been reported to decrease NaCl reabsorption in the thick ascending loop of Henle and the CCT [20]. Marver [243] has also demonstrated that PGE_2 inhibits Na^+ permeability of the rabbit CCT apical membrane by stimulating PKC; basolateral Na^+, K^+-ATPase activity falls secondary to decreased luminal Na^+ entry. PGE_2- and phorbol ester-induced decrease in transepithelial Na^+ transport was reversed by staurosporine (a PKC inhibitor) or monensin (a Na^+ ionophore). In A6 cells, the phorbol ester-induced decrease in distal nephron single Na^+ channel activity is reversed by another PKC inhibitor, sphingosine [217]. PGE_2 also inhibits Na^+ transport in isolated rabbit CCT by inducing an increase in cytosolic Ca^{2+} [130].

Summary of Amiloride-Blockable Channels

Amiloride-blockable channels form a family of functionally related channels present in the apical membranes of most sodium transporting epithelia. Whether they are genetically related must await more detailed information on the properties of the channel genes themselves. Nonetheless, single-channel and fluctuation measurements have provided a rich picture of the regulation of these channels. Virtually every modulatory cascade known seems to affect the activity of these channels either by altering their number or kinetics. This may be appropriate since the channels are the control point for one of the most tightly controlled homeostatic variables, systemic sodium concentration. As with any similar complicated regulatory scheme, the potential for "cross-talk" between secondary messenger systems (involving PLA_2, PLC, and G-proteins, among others) modulating membrane Na^+ permeability underscores the need for an examination of the influence of these signaling pathways on a single Na^+ channels.

Potassium Channels

In this review, we apply the label potassium channel to any channel which is significantly more selective for potassium over sodium. With the possible exception of nonselective channels, potassium channels appear, in an evolutionary context, to be among the oldest and most diverse of all ion channel proteins. There are documented examples of potassium channels as phylogenetically early as the coelenterates [135]. As might be expected because of their evolutionary age, potassium channels are widely distributed among a variety of tissues including most, if not all, epithelial tissues. It may be that this diversity is the single most important factor when considering the physiological role of potassium channels. It is not uncommon to find a single cell which contains several different types of potassium channel, with each type subject to differential regulation [43]. In excitable tissues, a primary role of potassium channels appears to be in the control of membrane potential [187]. In some epithelial tissues, the control of membrane potential also appears to be a function of potassium channels. Particularly, in tight epithelia the presence of potassium channels maintains the membrane potential of both the apical and basolateral membrane at relatively hyperpolarized potentials which aids in the reabsorption of sodium [379]. However, as with several other types of channels in epithelial tissues, potassium channels are also used in ways that contribute uniquely to the epithelial function of vectorial transport between different body compartments. In this sense, some potassium channels appear only in either the apical or the basolateral membrane and thereby promote K^+ secretion or reabsorption [23, 24, 31, 146, 148, 149, 199, 201, 376, 399, 400, 402–405, 407]. In this section, we will survey the biophysical and pharmacological characteristics of different potassium channels found in epithelial tissues and then describe in greater detail the properties of one family of potassium channels, the calcium-activated potassium channels, which have received an significant amount of attention particularly from investigators patch clamping epithelial cells.

Occurrence of Potassium Channels in Epithelial Tissues

Potassium channels are very widely distributed in epithelial tissues. Some idea of the diversity is given by the size of Table 3 (adapted and expanded from a similar table by Wills and Zweifach [407]) which presents only those reports of potassium channels in vertebrates. Examination of potassium channels regardless of their origin has been significantly hampered by the enormous diversity in different potassium channel types even within the same tissue and a lack of specific high-affinity blocking agents for individual channels. As an exmaple of the confusion of different potassium channels, Rae et al. [298] have identified no less than 11 different K^+ channels in occular epithelia. What the role of so many types of potassium channels could be is unclear. As for blocking agents, those which are generally used (e.g., Ba^{2+}) block with relatively low affinity and often block many different types of potassium channels. Only charybdotoxin, a specific blocker of calcium-activated potassium channels, may have the requisite specificity and affinity to allow its use in the isolation and characterization of this class of K^+ channels. Coupled with the difficulty in finding high-specificity, high-affinity blocking agents, the study of potassium channels have been hampered by the fact that classical methods of membrane protein chemistry which have been successful in allowing isolation of the acetycholine receptor and the sodium channel from excitable tissue and at least one type of amiloride-blockable sodium channel from epithelial tissue do not appear very successful for potassium channels. Only recently has significant progress been made in isolation of the gene for one type of potassium channel from neural cells [317, 355, 356] using molecular genetics methods. Complementary DNA for this channel appears to bind with low stringency to several other membrane proteins from several tissue types including some epithelial tissues. In addition to this work, Nakanishi [347] has recently described a gene for a potassium channel which appears to be unique to renal tissue. This identification also implies the possibility of identifying families of renal potassium channels.

Until that time when extensive molecular biological information is available concerning epithelial potassium channels, it will be necessary to rely on the currently available biophysical characteristics of the potassium channels.

Apical Potassium Channels in Epithelia

As indicated in Table 3, apical membrane channels have been reported in numerous different epithelial tissues including ventral skin from Rana temporaria, rabbit gall bladder, toad gallbladder, Necturus gall bladder, frog gastric mucosa, rabbit renal cortical collecting duct, proximal tubule and medullary thick ascending limb, Amphiuma diluting segment, cultured chick renal cells, Xenopus distal nephron cells, rabbit descending and human proximal and descending colon, and Necturus choroid plexus. These potassium channels fall into several general categories that make differing contributions to the normal

Table 3. Potassium channels

Epithelium	Method[a]	Channel Blockers[b]	Unit Current (pA)	Conductance (pS)	Density (channels/μ^2)	Voltage dependence	Activation[c]	References
Apical membrane								
Frog skin								
Rana temporaria	Noise	B, C, E	1	20	0.2	Yes	–	[368–370, 378, 409, 421]
Gallbladder								
Bufo marinus	Noise	A	–	–	–	–	–	[375]
Necturus	Noise	A, B, E	1.3	6–40	0.1–1.9	–	–	[104–106]
Triturus	PC		–	140		Yes	Ca^{2+}	[233]
Rabbit	Noise	–	–	–	–	–	–	[380]
Gastric mucosa								
Rana esculenta, *Rana temporaria*	Noise	A, B	–	–	–	–	–	[419]
Descending colon								
Rabbit	Noise	A, C	–	–	–	–	–	[407]
Human	Noise	A	–	–	–	–	–	[402]
Kidney								
Rabbit								
CCT[d]	PC	B	–	90	–	Yes	Ca^{2+}	[147]
PT	PC	B	–	–	–	–	–	[98]
PCT	PC	–	–	30–35	–	–	–	[287]
MTAL culture	PC	B	–	133	–	Yes	Ca^{2+} ADH	[37, 112, 113]
Rat								
CCT	PC	B	–	9	–	No	–	[76]
CCT	PC	A, B	–	135	–	Yes	Ca^{2+}	[75]
MDCK cell line	PC	A, F	–	221	145/cell	Yes	Ca^{2+}	[19]
JTC-12.P3	PC	–	–	200	–	Yes	Ca^{2+}	[174]
Amphiuma DT	PC	B, H	–	120	–	Yes	Ca^{2+}	[146, 155]

Table 3. Contd.

Epithelium	Method[a]	Channel Blockers[b]	Unit Current (pA)	Conductance (pS)	Density (channels/μ^2)	Voltage dependence	Activation[c]	References
Chick cell culture	PC	B	–	97	–	–	Forskolin	[112, 113]
Lens								
Chick	PC	–	–	Several (10–230)	–	–	–	[293, 294, 298]
Frog	PC	–	–	45; 60	–	–	–	[293, 296, 298]
Rat	PC	–	–	45	–	–	–	[293, 298]
Corneal endothelium	PC	–	–	30; 35; 100	–	–	–	[293, 295, 298]
Choroid plexus; *Necturus*	PC	A, B, H		150; 220	–	Yes	Ca^{2+}	[24, 30]
Mammary epithelial cells	PC	A, B		12	–	Yes	Ca^{2+}	[85]
Pigment epithelium	PC	–	–	25, 45, 100	–	No	–	[70]
Basolateral membrane								
Isolated oxyntic cells	PC	–	–	67	–	Yes	Ca^{2+}	[362]
				33	–	No	cAMP	[362]
Descending Colon								
Rabbit	Noise, SM	B, C	0.1	1.4	12	–	–	[377, 400, 403]
Rabbit	PC	A, B, H		131		Yes	cAMP, Ca^{2+}	[221]
Turtle	Noise, SM, PC	F, G	–	8–20	–	–	Cell swelling	[44, 46, 117, 301]
Jejunum (rabbit)	PC	–	–	35; 90	–	–	–	[332]
Frog Skin								
Rana pipiens	Noise, SM	B, C	–	–	–	–	–	[376, 377]
Rana catesbeiana (larval skin)	Noise, SM	B, F, G, H	0.2–0.3	–	7–28	–	–	[376, 377]
Locust hindgut	Noise, SM	B	1–2	77	1.0–1.5	–	–	[126]
Moth *Manduca sexta* (larval)	Noise	B	–	–	–	–	–	[425]
Urinary bladder (rabbit)	PC	B	–	200	0.07	Yes	–	[124, 196]

Kidney								
Rabbit PT	PC	A	–	30–40	–	–	–	[98]
Rabbit PCT	PC	–	–	30–35	–	–	–	[286, 287]
Necturus PT	PC	B	–	26–47	–	Weak	Stretch	[313]
Necturus PT	PC	B	–	21	–	–	Stretch	[157]
Pancreatic acinar cells (pig)	PC	–	–	200	0.3	Yes	Ca^{2+}, bombesin, ACH, CCK	[237, 238, 240]
Lacrimal gland								
rat	PC	I	–	250	0.2–0.7	Yes	Ca^{2+}, CCH	[228]
Mouse	PC	–	–	200–400	–	Yes	Ca^{2+}, ACH	[66]
Salivary acinii	PC							[86, 232, 235]
Trachea (dog)	PC	–	–	19	–	–	Ca^{2+}, epinephrine	[396]
Lens (frog)	PC	–	–	45	–	–	–	[293]
Lens (frog)	PC	–	–	40	–	–	–	[150]

CCT, Cortical collecting tubule; PT, proximal tubule; PCT, proximal convoluted tubule; MTAL, medullary thick ascending limb; DT, distal tubule.

[a] Noise, current fluctuation analysis; PC, patch clamp; RC, reconstituted membranes in planer bilayer membrane; SM, nystatin-permeabilized apical membrane or the "single-membrane" preparation.

[b] A, tetraethylammonium; B, Ba^{2+}; C, Cs^{+}; D, divalent cations; E, H^{+}; F, quinidine; G, lidocaine; H, pH; I, high Ca^{2+} concentration; J, quinine

[c] ACH, Acetylcholine; CCK, cholecystokinin; CCH, carbamylcholine

function of the tissues. It is interesting that despite the presence of apical potassium channels many of these epithelial tissues do not secrete or reabsorb significant amounts of potassium. Many of the tissues use the potassium channels to facilitate the transport of other ions [e.g., K^+ channels in gastric mucosa which seems to promote recycling of potassium to produce maximal activity of the (H^+, K^+)-ATPase].

As might be expected from the wide distribution of potassium channels across many cell types, many of the potassium channels appear to be similar to potassium channels in excitable tissues where the properties of the channels have been more thoroughly described (because the preparations are usually easier to study). On the basis of this comparison, epithelial potassium channels which have been well characterized can be general divided into four groups: (a) voltage-dependent, outwardly rectifying channels ("delayed rectifiers"); (b) voltage-dependent, inwardly rectifying channels ("inward" or "anomolous rectifiers"); (c) Ca^{2+}-activated channels ("maxi-K"); and (d) other types, some of which may be unique to epithelial tissues. Outwardly rectifying channels are mostly found in basolateral membranes while Ca^{2+}-activated channels are found almost exclusively in apical membranes (except for some secretory epithelial cells). The other two classes of channels are found in one membrane or the other, but usually not both membranes in the same cell type.

Large-Conductance Calcium-Activated Potassium Channels

Large-conductance, calcium-activated potassium channels (sometimes known as BK channels or maxi-K channels) are present in a large variety of cell types including epithelial cells, nerve and muscle cells, and numerous other cell types. Single-channel studies have shown that, regardless of their origin, the properties of these channels are generally similar, particularly with respect to their response to voltage and internal Ca^{2+} concentration [12, 147, 382]. In epithelial cells, BK or presumptive BK channels have been identified in the apical membranes of a variety of epithelia (for a review, see [147]) including medullary thick ascending limb cells in culture, rabbit and rat CCT cell [75, 148, 149, 218], *Amphiuma* distal tubule [145], *Triturus* gall bladder [234], isolated colonic cells [302, 303], and *Necturus* proximal tubule [158], and gall bladder [379]. In addition, contrary to the rule that BK channels are found in the apical membranes of epithelial tissues, a few secretory epithelia appear to have basolateral BK channels including salivary acinar cells [236], pancreatic acinar cells [238, 240], and possibly tracheal epithelial cells [395]. BK channels have also been found in the basolateral membranes of rat enterocytes [263], avian salt glands [304], rabbit distal colon crypt cells [225], and human respiratory epithelial cells [179].

Identifying Characteristics of BK Channels. As mentioned above, in epithelial tissues, BK channels seem to be confined to apical membranes of nonsecretory epithelia while other types of potassium channels are more typically (but not exclusively) found in the basolateral membrane. The specific identification of a channel as a BK channel is usually based on several criteria. First, either depolarization or an increase in intracellular Ca^{2+} concentration can increase channel open probability; when membrane potential and intracellular calcium are increased together, the effect appears synergistic (but see below, "Inwardly Rectifying Potassium Channels"). Second, the unit conductance of the channels (particularly in symmetric potassium solutions) in large, typically about 100–250 pS. Third, the kinetic characteristics of BK channels are relatively distinctive. The channels are usually characterized by having several distinct classes of open and closed intervals, the easiest of which to see are the long openings interrupted by rapid flicker closures and groups of openings or "bursts" separated by long closures [257, 382]. In addition, the channel is unusually selective for K^+ over other univalent ions with the exception of Rb^+ to which they are almost as permeable as to K^+. In particular, BK channels are typically 40 times more selective for K^+ than Na^+. Pharmacologically, some but not all BK channels are blocked with relatively low affinity by quarternary ammonium ions and Cs^+ and very low affinity by Na^+. Quinidine and *N*-phenylanthranilic acid (also known as phenylamine 2-carboxylate) have been used to block BK channels in epithelial cells [28, 302, 303]. Ba^{2+} is also a moderately potent blocker of the channel [252], but the most effective blocker is charybdotoxin, a component of the venom of certain scorpions, which blocks with nanomolar affinity [253]. Block by charybdotoxin has come to be an additional signature of BK channels and may provide a specific method for the isolation of enough the BK channel proteins to allow production of antibodies to the protein. Antibodies would allow specific localization of the channels within tissues and should also allow isolation of the BK channel gene by expression screening of cDNA libraries prepared from cells expressing the channel.

Interaction of Intracellular Calcium and Depolarization. The more than additive increase in the open probability of BK channels when membrane potential and intracellular calcium are increased together has been assumed to imply some effect of voltage on the intracellular calcium binding site [257]. Recently, however, experiments in which only a few individual amino acid residues on the intracellular face of the BK channel protein have been modified by group specific reagents casts significant doubt on this assumption. One reagent, *N*-bromoacetamide (NBA), which cleaves peptides at histidine, tryptophan, and tyrosine residues eliminates the Ca^{2+} dependence of channel opening yet leaves the voltage dependence of channel opening unaffected. In addition, the channel can be protected from the effects of NBA by continuous exposure of the channel to high calcium solutions [37, 280]. The modification by NBA suggests the presence of an essential tyrosine, tryptophan, or histidine residue at or near the Ca^{2+} binding site of the BK channel. In contrast, when amino reactive reagents are applied to the inner surface of the BK channel protein, the channel behaves similarly to the calcium activated state regardless of the intracellular concentration of calcium [71]. This modification also occurs without affecting the voltage dependence of the channel. The amino group modification can only occur in the presence of calcium. Thus there are two separate modifications, one of which locks the channel out of the calcium activated state and one which locks it into the state. The information implied by these studies will be useful in identifying the calcium-binding site at that time when the primary amino acid structure of the channel is known.

Physiological Role of BK Channels in Epithelia. In many secretory cells, the importance of BK channels in modulating the secretory response induced by calcium influx is well established [37, 183, 187, 238, 240, 254, 280, 305, 306]. By definition, BK channels are modulated by intracellular levels of calcium, but the physiological role of calcium in modulating potassium flux in epithelia has been most convincingly demonstrated in secretory exocrine cells. The basolateral membranes of salivary [233, 237], lacrimal [66, 229], and pig and human pancreatic [242, 291] acinar cells as well as salt gland cells [304] contain BK channels. Muscarinic agonists or cholecystokinin, agents which stimulate secretion increase the open probability of the BK channels in specific target cells [87, 242, 304, 348]. Channel activation presumably occurs in response to elevated intracellular Ca levels induced by these secretagogues [262, 339] and provides a pathway for recycling of K across the basolateral membrane.

The role of BK channels in other cell types including epithelia is much less clear. The primary reason is the combination of conditions which are necessary to activate BK channels. To achieve any significant open probability for the channel requires the combination of a relatively depolarized membrane potential coupled with a very high intracellular calcium activity. The half-activating concentration of calcium necessary to activate BK channels is typically greater than 2 μM calcium and can range as high as 10 μM [147, 148, 257]. In secretory cells, secretion is initiated by strong depolarization and influx of calcium leading to exactly the conditions necessary to activate BK channels (and repolarize the secretory cells). Such conditions do not typically exist under normal physiologial conditions in nonsecretory epithelial tissues. As such BK channels in most epithelial tissues would seem to serve only an "emergency" function; that is, under conditions when there was abnormal depolarization of the tissue and abnormally high intracellular calcium concentrations, the BK channel would tend to restore the normally hyperpolarized state of the tissue. One could imagine that if there were damage to some cells in an epithelial synctium, that BK channels could prevent normal cells which were adjacent to the damaged cells from depolarizing and also being damaged. Such a mechanism might be important for normal turnover of surface cells particularly in such tissues as the intestinal, airway and urinary bladder epithelia.

One interesting alternative possibility has been suggested for some epithelial tissues. Guggino et al. [113] observed that addition of ADH or forskolin (an activator of adenylate cyclase) caused an increase in the open probability of BK channels in cultured chick renal cells. Since both these agents increase intracellular concentrations of cAMP, a second messenger which is likely to activate

PKA, they suggest that phosphorylation of the BK channels might shift the calcium activation curve toward lower intracellular calcium concentrations. If this were so, then it might be possible that BK channels could contribute a significant potassium conductance at normal or only slightly elevated intracellular calcium concentrations. This hypothesis is consistent with the effect of cAMP and protein kinase A on an invertebrate calcium-activated potassium channel (with properties somewhat different than the vertebrate channel) [61, 191, 192], with the effect of cAMP on rat smooth muscle [314], and on crypt cells from rabbit distal colon, where cAMP increases the open probability of BK channels in the basolateral membrane [225]. A particular case where such a change in calcium sensitivity might be important would be in the mammalian thick ascending limb cells. Reabsorption of sodium and chloride in this segment is via a Na/K/2Cl cotransporter. Normal function of this transporter requires extensive apical recycling of K^+ from the intracellular pool to the luminal solution [108, 109, 323]. Phosphorylation of BK channels could regulate the extent of recycling and consequently the magnitude of sodium chloride reabsorption.

Besides cAMP, other nucleotides also regulate the activity of BK channels. ATP inhibits a BK channel in human respiratory epithelial cells [180] possibly due to a local change in ATP level or in the ATP/ADP ratio altering the open probability of these channels. The authors hypothesize that such a channel sensitivity could be one mechanism to couple the basolateral potassium conductance to changes in the rate of the (Na, K)-ATPase. Hunter and Giebisch [145] have also reported that somewhat higher concentrations of ATP block BK channels in the apical membranes of Amphiuma distal tubule, possibly by reducing the sensitivity of the channel to cytoplasmic Ca.

Other Apical Potassium Channels

Categorizing apical potassium channels other than calcium-activated potassium channels is more difficult because there seems to be enormous diversity of channel types coupled with the fact that many reports do not provide information which allows easy comparison between different channels. Nonetheless, there do seem to be several very broad groups of potassium channels, other than BK channels, which can be found in the apical membrane: outwardly rectifying potassium conductances, inwardly rectifying conductances, and other channels which cannot readily be classed in either of the first two groups. The names of the first two groups tend to hide a fundamental difference in the properties of these two different classes of channels. The inward rectification of the second class is an intrinsic property of the single channel, i.e., the single channel conductance, itself, increases as the potential across single-channel patches is hyperpolarized. On the other hand, for the outwardly rectifying class of potassium channels, the unit conductance is relatively constant for both hyperpolarizing and depolarizing stimuli. The rectification is a property of the macroscopic conductance (a population of channels) and is due to an increase in the open probability of single channels at depolarized potentials. This increase in open probability increases total outward current even though the amplitude and unit conductance of individual channels does not change.

Unfortunately, the current-voltage characteristics of potassium channels cannot be or are not always examined. In particular, experiments which involve fluctuation measurements generally do not provide information about the I-V characteristics of channels although they may provide some information about the open probability. In the absence of the ability to categorize potassium channels on the basis of their current-voltage relationships, it is possible to examine the pharmacology of the channels and infer possible similarities to other potassium channels for which more information is available. Of course, this approach is fraught with potential problems since, as mentioned above, potassium channels are notorious for the paucity of specific, high-affinity blockers.

Outwardly Rectifying Potassium Channels. Rae and Levis [296, 297] were among the first, if not the first, to report the properties of single channels in epithelial tissues using patch clamp methods. As summarized in later work [293, 294], they found no fewer than 11 distinguishable types of apical potassium channels in the different ocular epithelia they examined (see Table 3). In lens epithelium, they found eight different potassium channels all but one of which had an increased open probability when the patch membrane was depolarized. Other than the characteristic increase in open probability, each channel appeared to have different kinetic properties or unit conductance. Such a varied mix of channels poses the serious problem of explaining the physiological significance and contribution of the various channels. As yet there seems to be no straightforward explanation for such channel diversity.

Inwardly Rectifying Potassium Channels. There are fewer well-documented examples of inwardly rectifying potassium channels than other types of potassium channels in epithelial tissues. As with outwardly rectifying channels, inwardly rectifying potassium channels were first thoroughly described in excitable tissues [154, 315, 316]. A particularly good example of the characteristics of such channels in epithelial tissues can be found in chick lens epithelium [294]. The unit conductance of this channel strongly rectifies so that there is little or no measurable single channel current for potentials more positive than the potassium equilibrium potential even though the open probability of the channel increases as the channel is depolarized. In addition, this channel like other inwardly rectifying channels is blocked in a characteristic voltage-dependent manner by cesium ion. Cesium reduces the mean open time of the channel by inducing many short closures with the number of closures per open interval increasing as the channel is hyperpolarized. Although it was not demonstrated directly on this channel, inwardly rectifying potassium channels are also characteristically blocked by barium ion in a manner similar to cesium and by hydrogen ion in a non-voltage-dependent fashion. Based on this characteristic pharmacology, Van Driessche and coworkers have suggested that an apical potassium conductance which they have identified using fluctuation measurements [377] is likely to be of the inward rectifier type [380, 381, 408, 420]. There is a characteristic voltage-dependent block by both cesium [419] and barium [377, 408, 419, 421]. They suggest that the role of these channels, at least in frog skin, is to serve as an activatable pathway for potassium secretion across the apical membrane [369, 378]. While such a role is possible, inwardly rectifying channels do not form a very efficient mechanism for potassium secretion since there tends to be so little outward potassium current through the channel (indeed, in patch clamp experiments, there is no measurable outward current).

One other possible role for such channels might relate to truly "tight" epithelial tissues which reabsorb very few cations besides sodium. Such tissues have two problems. If the sodium conductance of the apical membrane becomes high enough, then the apical membrane will depolarize until there is very little driving force for the entry of sodium. One way to increase the driving force for sodium would be to incorporate potassium channels in the apical membrane, but this would provide a significant leak pathway for potassium out of the cells which is usually not appropriate for many tight epithelia. Most tight epithelia can maintain an adequate driving force and yet not have a leaky apical membrane by hyperpolarizing the basolateral membrane with some current flowing from the basolateral battery through the paracellular junctional resistance and hyperpolarizing the apical membrane. Unfortunately, there are some potential problems for tissues which use this approach to maintaining a high apical driving force for sodium. If the driving force for apical sodium entry is high, sodium will accumulate within the cell and stimulate basolateral sodium pump activity. Because the pump is electrogenic, increased pump activity will hyperpolarize the basolateral and (through the tight junctions) the apical membrane. The additional hyperpolarization will increase sodium entry and lead to additional pump activity. The combination of sodium entry and pump electrogenicity is a positive feedback system which is only limited by the maximal basolateral pump rate and has the potential for large oscillations in apical membrane voltage and transepithelial sodium transport. Incorporation of inwardly rectifying potassium channels in the apical membrane would strongly damp the positive feedback of such a system since, if the apical membrane hyperpolarized beyond the potassium equilibrium potential, the inwardly rectifying conductance would increase dramatically and effectively clamp the apical membrane at E_K. Yet, if the membrane was at E_K or more depolarized, there would be little or no potassium conductance and therefore no potassium loss from the cell into the luminal fluid.

For potassium channels observed in cells from CCT, the physiological significance seems much more clear cut. Discretionary control of total body potassium balance takes place in the CCT and appears to be modulated by mineralocorticoid hormones [264, 275, 318]. The major cell type in mammalian CCT is the principal cell which is known to secrete potassium and to absorb sodium and chloride [172, 264, 271, 272, 274, 324, 330, 342].

Early single-channel recordings from rabbit CCT primary cultures demonstrated several different apical membrane potassium conductances [96, 186, 398], but single potassium channels were not observed which corresponded to the barium-blockable, apical K^+ conductive pathway observed in macroscopic ion transport studies of isolated rabbit CCT [75, 173, 272–274, 318]. Several groups have documented that this barium-sensitive, apical K^+ conductance is responsible for physiological potassium secretion and is aldosterone-dependent in rabbit, but not rat CCT [76, 318, 321].

Frindt and Palmer [76] first observed an apical channel which might be a candidate pathway for this secretion. The channel is inwardly rectifying with a unit conductance of 9 pS; it is not stimulated by changes in intracellular calcium or by changes in voltage but was blocked by low concentrations of barium. The surface density of the channel, the open probability (> 0.9 with no applied potential), and the relatively weak rectification are all consistent with this channel being a significant pathway for potassium secretion. Recently, other investigators have found that this rat CCT low-conductance potassium channel is also regulated by intracellular pH; open probability decreased by lowering pH [386]. This effect of intracellular H^+ concentration is also consistent with the macroscopic properties of CCT potassium secretion. In both of these cases manipulation of the mineralocorticoid status did not influence the observed frequency of these low-conductance K^+ channels. In contrast, in primary cultures of rabbit CCT cells, there is also a low-conductance potassium channel with characteristics almost exactly the same as the rat K^+ channels [218]. However, this low-conductance K^+ channel was much more likely to be found in patches formed on aldosterone-treated cells than in patches formed on cells which had been grown in the absence of mineralocorticoid hormone. This observation is consistent with several whole tissue studies which demonstrate the importance of mineralocorticoid status in determining levels of potassium secretion in rabbit CCT cells. Cortical collecting tubules from rabbits given deoxycorticosterone for more than 4 days have twice the apical K^+ conductance of tubules from untreated rabbits. This increase in K^+ conductance increases K^+ secretion even though there is little or no change in transepithelial electromotive driving force [318, 321]. The differences in the effect of mineralocorticoids on rabbit and rat CCT channels corresponds to the whole tissue observation that in rat, unlike rabbit, mineralocorticoid treatment increases potassium secretion primarily by depolarizing the CCT apical membrane via an increase in apical Na^+ conductance [321]. This increased driving force was sufficient to explain increased K^+ transport without invoking changes in apical K^+ conductance. These differences in the hormonal regulation of potassium secretion emphasize fundamentally different ion transport properties between rabbit and rat CCT principal cells.

Besides their induction by mineralocorticoids, the small-conductance potassium channels in CCT can also be induced by apical application of arachidonic acid [219]. Whether this activation is in any way related to the effects of steroid hormones is unknown.

Basolateral Potassium Channels in Epithelia

Most epithelial tissues have (Na^+, K^+)-ATPase at the basolateral membrane which serves as the exit step in transepithelial movement of sodium. Of course at the same time that ATP is used to extrude sodium from the cell, the ATPase also transports potassium into the cell. If the turnover rate of the pump is not to be rate limited by the energy required to transport potassium uphill into the cell, then there must be an exit pathway to prevent intracellular potassium from reaching very high levels. In general this is accomplished by potassium efflux through basolateral potassium channels.

Basolateral efflux of potassium could be accomplished by transport systems other than channels, but the presence of channels serves an important additional role which could not be met by other classes of transport system. Having numerous potassium conductance pathways with the relatively large unit conductance typical of channels means that the basolateral membrane potential is largely determined by the potassium reversal potential with smaller contributions from channels selective for other ions and from other electrogenic transport systems [e.g., Na/Ca exchange or the (Na^+, K^+)-ATPase]. The relatively small changes in basolateral potential imply that the contribution of potential to the energy required for ionic transport by any mechanism remains essentially constant with only differences in intracellular and, to a lesser extent, extracellular concentration producing changes in the driving force for transport.

But this constancy of potential energy is probably not of great physiological significance compared to the secondary effect of the large numbers of basolateral potassium channels. Epithelial tissues are unusual since, because of their morphology, there are two pathways for current flow across the tissue: one transcellular and one paracellular. The paracellular pathway is via the tight junctions and the lateral spaces. The resistance of this pathway connects the basolateral membrane ionic batteries with the apical membrane ionic batteries. If there is a difference in potential of the apical and basolateral batteries, then current flows through the tight junctional resistance with the resistive potential drop across the junctional resistance being just equal to the transepithelial, open-circuit potential. The net effect of the junctional shunting is to cause current flow which will tend to cause the difference between the apical and basolateral potentials to be less than it would be if the tight junctions were not conductive. In the case of a relatively hyperpolarized basolateral membrane (since the basolateral battery should be near E_K which is normally between -60 and -75 mV) and a relatively depolarized apical membrane (because of the various conductive sodium entry pathways), then junctional current flow will depolarize the basolateral membrane and hyperpolarize the apical. The magnitude of the potential change in each membrane is just the amount of junctional current divided by the respective membrane conductance. Since the conductance of the basolateral membrane is very large (since there are many potassium channels) then the change in basolateral potential is relatively small (usually only 5–10 mV), but the change in the apical potential is usually quite large since its conductance is generally much lower. The net effect is to strongly hyperpolarize the apical membrane and provide significant potential energy for any sodium entry process which is not electroneutral (e.g., Na/nonelectrolyte cotransport or Na transport through Na channels).

A variety of different types of potassium channels have been identified in the basolateral membranes of epithelial cells. The evidence for many of these has been reviewed recently in several places [43, 45–47]. One of the earliest descriptions of single basolateral potassium channels did not provide much information other than the unit conductance of the channel (30–40 pS) and some qualitative kinetic information [98]. Nonetheless, this work did demonstrate the feasibility of patch recording from the basolateral surface of renal epithelial cells and set the stage for more comprehensive subsequent work.

As with the apical membrane, potassium channels in the basolateral membrane can generally be divided in to those activated by intracellular calcium and those which do not respond to calcium. Although most Ca^{2+}-activated K channels have been reported in the apical membranes of epithelial cells, as mentioned above, there have been a few reports of Ca-activated K channels in the basolateral membrane of a few epithelial cells.

Besides BK channels, other types of Ca-activated K channels have also been identified in the basolateral membrane of secretory epithelia. In guinea pig pancreatic acinar cells, the potassium conductance of the basolateral membrane can mostly be attributed to a population of 30-pS, Ca- and voltage-regulated K channels that are specifically activated by pentagastrin [349]. A 20-pS, Ca-activated, voltage-insensitive K channel has also been identified in the basolateral membranes of cultured tracheal epithelial cells [395]. This K channel can be activated by serosal epinephrine which may be a reflection of the fact that epinephrine increases intracellular Ca^{2+} levels.

There are several mechanisms which could couple activation of potassium channels to membrane receptors for external ligands. The effect of cAMP was mentioned above. For Ca^{2+}-activated channels, the obvious intracellular second messenger is one of the inositol polyphosphates. In fact, injection of inositol trisphosphate (IP_3) into cells does activate single Ca-dependent K channels [134]. In addition to the effect of IP_3-induced increases in intracellular calcium, the diacyl glycerols released when phospholipase C catalyzes the production of inositol phosphates may also modulate K channels. Application of phorbol esters which mimic the action of diacylglycerols on PKC inhibits basolateral potassium conductance in turtle colon [45].

The pathways that couple receptor activation to production of the second messengers like cAMP or inositol phosphates and diacylglycerols often include G proteins. G proteins are important for the activation and inhibition of adenylyl cyclase and activation of phospholipase C and may therefore indirectly influence potassium channels; as pointed out for epithelial sodium channels, these proteins may also directly modulate the activity of K channels [415].

A second mechanism to couple receptor activation to alterations in potassium channel gating would involve the channels sensitivity to receptor-induced alterations in membrane potential. All BK channel and many of the smaller conductance, Ca-activated K channels are very sensitive to changes in membrane potential. In addition, there are some basolateral potassium channels which are not sensitive to Ca^{2+}, but which have increased open probabilities in response to depolarization [287]. Voltage-sensitive, basolateral K channels have been described in *Necturus* enterocytes [333] and proximal tubule [158, 313], and in rabbit urinary bladder and proximal tubule cells [101, 197]. The voltage dependence of these K channels could to couple basolateral K recycling with electrogenic sodium pump activity such that both are increased or decreased in simultaneously [158, 313]. This would be an example of using voltage to couple K channels to a membrane process other than receptor activation.

Other cellular processes may also be coupled to K channel activation. Calcium-independent mechanisms would be important to regulate basolateral potassium channels which are not sensitive to intracellular Ca^{2+}. One possible coupling mechanism would be a sensitivity of channels to changes in cell volume. Several K channels have been described which respond to cell swelling. When cells from primary cultures of colonic epithelial cells are swollen by salt or urea entry, potassium channels appear in the basolateral membrane which are characterized by their high selectivity for K over Rb and their sensitivity to block by quinidine or lidocaine [301]. These same channels can also be detected from their lidocaine-induced noise [46]. In *Necturus* small intestine, cell swelling increases the open probability of a population of Ca-dependent, K channels in isolated enterocytes [333]; and in *Necturus* proximal tubule, there is a 21-pS Ca-independent, K channel which is also activated by cell swelling [158]. There have also been several reports of single potassium channels activated by negative pressure applied to patch pippettes [30, 36, 114, 310, 311, 351, 360]. Stretch-activated ion channels could serve an osmoregulatory function by mediating rapid KCl loss during regulatory volume decrease. Consistent with this idea, Sackin identified in the basolateral membranes of *Necturus* proximal tubule a K selective channel which was activated by both stretch [311] and cell swelling [312]. In general, however, stretch-activated channels are often poorly selective and may have significant Ca permeability; therefore, we have placed a more extensive discussion of all types of stretch-activated channels below (see "Stretch and Osmotically Activated Channels").

Pharmacology of Epithelial Potassium Channels

As we have alluded to several times above, there are numerous "blockers" of potassium channels in general and epithelial potassium channels in particular. In general, most blockers seem to block open potassium channels in a voltage-dependent manner although the half-blocking concentration of different blockers varies enormously.

Inorganic Blockers

The potassium channel blocker which has been used most extensively in studies of epithelial transport is barium ion. The efficacy of barium as a blocker of a variety of different potassium channels was first established in nonepithelial tissues [7, 8, 52, 116]. Barium's ability to block potassium channels is at least partially due to the similarity in size of the two unhydrated ions; potassium is 0.133 nm and barium is 0.135 nm. The strong coordination sites within the potassium channel pore interact even more strongly with divalent barium than with monovalent potassium. The strength of the interaction promotes an extremely long residency time for Ba^{2+} within the channel while the divalent charge inhibits the occupancy of adjacent sites by potassium. The implication of these two characteristics is that Ba^{2+} should produce a flickery block of otherwise open channels and that there should be a finite (though small) permeability of potassium channels to Ba^{2+}. Both predictions have been confirmed in several different types of potassium channels [8, 52, 252].

Since Ba^{2+} produces a flickery block of most potassium channels (see Table 3), Ba^{2+} has been used extensively to alter the noise spectra of potassium channels in epithelia (for reviews, see [379, 403, 407]). Used in this way, Ba^{2+} (and other blockers) provide a useful tool for localizing the source of the potassium channel noise to either the apical or basolateral membranes or tight junctions. On the other hand, Ba^{2+} has also often been used in a less discriminate manner assuming that the Ba^{2+} will "remove" the contribution of potassium conductance to transepithelial conductance. The usual intent of such experiments is to demonstrate that potassium conductance only contributes a fraction of the total tissue conductance. Because of the nature of Ba^{2+} block, experiments which use the ion to "eliminate" potassium conductance must be interpreted with caution. In the first place, Ba^{2+} does not block all potassium conductances, but even when applied to tissue which contains Ba-blockable conductances the block is not so simple that potassium permeability can be assumed to be zero. The major difficulty with the assumption is that Ba^{2+} block is competitive with potassium and dependent upon both voltage and current. The voltage dependence implies that Ba^{2+} block becomes less effective as the cell membrane depolarizes; depolarization being the expected consequence of blocking some potassium channels. Thus, the blocking ability of Ba^{2+} is gradually reduced as it blocks more channels. In addition, Ba^{2+} block depends upon current direction; i.e., Ba^{2+} can be "knocked-off" the channel [8]. Therefore, as the membrane depolarizes and potassium efflux from the cell is favored, Ba^{2+} block will be further reduced. The "knock-off" effect is particularly pronounced because of the competitive effect of high intracellular potassium. The net effect is to produce a potassium channel which strongly rectifies, allowing significant amounts of potassium out of the cell; but not as much into the cell. Under these conditions transepithelial conductance measurements will still have a significant component of potassium conductance. Naturally, if the Ba^{2+} concentration which is used to block K^+ channel is high enough, the potassium conductance will be very small; however, for some channel types (such as outwardly rectifying channels), the 99% blocking concentration may be ten times the more typically applied concentration of 1–5 m*M* used in many investigations. A more reasonable approach for completely blocking potassium conductance may be either to use one of the organic blockers described below or to use some mix of blockers which includes Ba^{2+} and additional organic blockers.

Cs^+ is even less useful as a pharmacological tool for identifying potassium channels since the half blocking concentration is significantly higher than Ba^{2+}, and it suffers from all of the same problems as Ba^{2+}. Cs^+ has occasionally been used as a source of induced noise for fluctuation measurements in epithelial tissues [294, 379] and the effect of Cs^+ has occasionally been used as a criteria for identification of a few potassium channels since Cs^+ and Ba^{2+} occasionally have differential effects [379]. Cs^+ does find a very useful niche for certain classes of patch clamp experiments. Under whole-cell patch clamp conditions all or most of the intracellular potassium can be replaced with Cs^+ from the pipette resulting in a viable cell with no measurable conductance through potassium channels.

For the same reason that Cs^+ blocks potassium channels, all other alkali metal cations and many other small univalent cations also block potassium channels with low affinity. The relative potency of ions varies somewhat for different potassium channels, but the order of potency is usually $Ag^+ > Rb^+ > NH_4^+ > Tl^+ > Na^+ > Li^+$. As with Ba^{2+} several of these blocking agents have finite permeabilities and only block weakly at relatively low concentrations of blocker in the presence of high concentrations of potassium. Otherwise, especially in the absence of potassium, they may carry significant amounts of charge across the membrane. In particular, Rb^+, Tl^+, and NH_4^+ are usually fairly permeable through most potassium channels.

Hydrogen ion was not placed in the sequence of blockers above even though it often blocks potassium channels in the micromolar or submicromolar range. The reason for treating H^+ separately from other univalent blocking ions is that its mechanism of block appears to be different from the open channel block of the other agents. As one might expect H^+ appears to produce its block in many epithelial potassium channels by titrating some site associated with the inner surface of the channel which alters the gating of the channel and/or alters the unit conductance. As intracellular pH is made more acidic, the potassium current (which is directly proportional to the product of open probability and unit conductance) decreases. The interesting feature of the titration curve is that significant alterations in whole cell potassium conductance can occur within the physiological range of intracellular pH in epithelial cells. Therefore, it is possible that alterations of intracellular pH may play an important role in modulating the potassium conductance of the basolateral membrane. Indeed, Oberleithner and his colleagues [276–278] have demonstrated that exposure of sodium-transporting, tight epithelia to aldosterone induces an alkalinization sufficient to significantly increase the potassium conductance of the basolateral membrane.

Thus intracellular pH may be an interesting method for coupling increases in sodium transport with an increase in basolateral potassium conductance. The increase in basolateral potassium conductance is a necessary concommitant to the increase in basolateral sodium pumping associated with increased sodium transport.

There are several additional examples of alterations in intracellular pH changing the open probability of potassium channels in epithelial cells. The open probability of BK channels in choroid plexus [31] and cultured kidney cells [38] are decreased as intracellular pH is reduced; although, such reductions are not universally true for all BK channels [180].

Organic Blockers

As mentioned above, many small univalent inorganic cations are capable of blocking potassium channels. This generalization also extends to small organic cations. All substituted ammonium compounds produce block with the block becoming more potent as the side chains are extended. The most frequently used quarternary ammonium blocker is TEA, but TEA is a relatively weak blocker of most potassium channels often requiring 30–50 m*M* concentrations for 95% block. A much more potent blocker is nonyltriethylammonium which produces greater than 90% block at concentrations of 100–200 μ*M*. Unfortunately, compounds with longer side chains are so poorly soluble in water that they are not useful blocking agents. Nonetheless, the efficacy of the various quarternary ammonium compounds does suggest that the most effective blockers are likely to be cationic compounds with a large hydrophobic residue associated with the cationic site. Most organic blockers of potassium channels fit this description. But unlike small inorganic and even small organic cationic blockers, larger organic blockers often discriminate between different potassium channels and can, therefore, be used as pharmacological tools for categorizing K^+ channels.

One group of organic compounds which has seen significant use is the aminopyridines. All of the amino-substituted pyridines are capable of blocking potassium channels. The efficacy of the mono-substituted derivatives is approximately millimolar, but varies depending upon the position of the amino substitution with the order of potency being 4-amino > 3-amino > 2-amino. On the other hand, the di-substituted derivatives are much more potent with the half block for the 3,4-derivative being about 2 μ*M*. In general, the di-substituted derivatives are much more appropriate to produce close to complete block of potassium channel than any of the mono-substituted derivatives.

Local anesthetics also block potassium in a fashion which has been called "use dependent." The characteristics of this block are that the more active the channel, the more potent the block. Local anesthetics are not very useful as pharmacological tools for distinguishing potassium channels, because they block many other types of channels with approximately the same potency. Other agents with local anaesthetic-like action also block some potassium channels. For example, yohimbine blocks BK channels in medullary thick ascending limb cells of rabbit kidney [39].

Probably the most specific of potassium channel blockers is the natural product derived from scorpion venom, charybdotoxin. This toxin binds to calcium-activated potassium channels with very high affinity and specificity. The half-maximal blocking concentration is in the nanomolar range and the toxin has not been shown to interact with any other ion channel protein. Indeed, block by charybdotoxin has become the definitve test for the presence of calcium-activated K^+ channels.

Chloride Channels

Some epithelial tissues, besides having the ability to support net movement of cations, also are capable of either absorbing or secreting Cl^-. In a generalized chloride-transporting epithelial tissue, Cl^- ion movement is accompanied by Na^+ movement leading to the net movement of salt and fluid across epithelial tissues. Therefore, an examination of Cl^- transport in epithelial tissue can provide additional information about fluid and salt transport over and above that available from an examination of cation transport alone.

In general, transepithelial net Cl^- movement like sodium transport occurs via two mechanisms; one an energy requiring one such as a gradient driven transporter (e.g., the $Na^+/K^+/2Cl^-$ co-transport) and the second a passive Cl^- conductive pathway or chloride channel. Epithelial tissues are polarized, and like many other epithelial proteins those responsible for chloride transport contribute to the polarization by having chloride channels on one surface of the epithelial tissue while the membrane on the opposite cellular surface contains the energy-requiring transporter. In chloride-secreting epithelia, chloride channels are located in the apical membrane while chloride/cation symporters or chloride/anion antiporters are located in the basolateral membrane. With this arrangement, chloride is actively accumulated in the cell until its concentration increases enough to favor passive movement of chloride out of the cell down its electrochemical gradient through the apical membrane chloride channels. Panel A of Fig. 2 schematically illustrates a model for chloride secretion. Ultimately, the energy for chloride transport is provided by the ATP hydrolysis associated with the (Na, K)-ATPase at the basolateral membrane. Panel B of Fig. 2 shows the characteristics of the other class of Cl^--transporting epithelia, those which reabsorb chloride. In contrast to chloride-secreting tissues, chloride reabsorbing epithelia have the active chloride transport mechanism in the apical membrane and the chloride channels in the basolateral membrane. In theory, regulation of transepithelial chloride transport could occur at either the energy-requiring entry step or the passive exit step. However, for both classes of chloride-transporting epithelia, the limiting step in the translocation of Cl^- across the

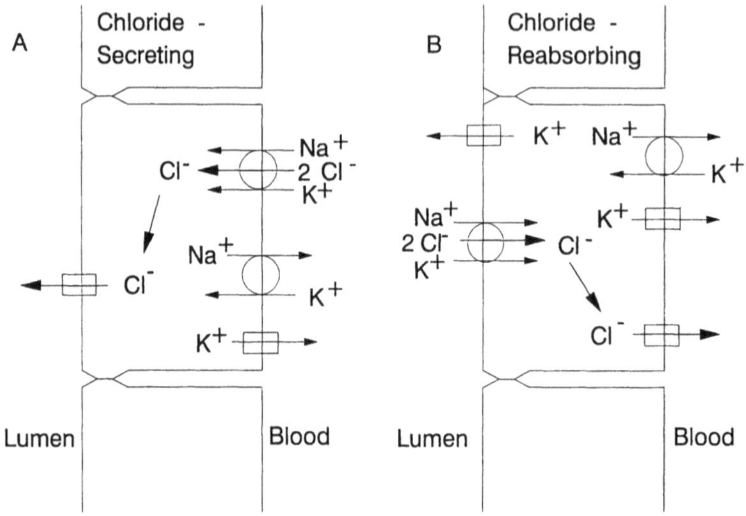

Fig. 2. Schematic diagrams of the transport machanisms present in chloride-secreting (*A*) or chloride-reabsorbing (*B*) epithelial cells

tissue is, in general, thought to be movement through Cl^- channels. This means that the control of Cl^- transport can best be accomplished by regulating the Cl^- conductive step. For example, rectal gland tubules have a very small Cl^- conductance in the apical membrane under the basal (nonstimulated) conditions which corresponds to very low basal Cl^- secretion; but an increase in cytosolic cAMP level induces an increase in the Cl^- conductance of the apical membrane resulting in an increase in the Cl^- secretion without any significant change in the activity of the basolateral $Na^+/K^+/2Cl^-$ cotransporter. Thus, examining Cl^- conductance in epithelial tissues can provide insight into the regulation of chloride transport and, secondarily, into regulation of tissue salt and water transport.

Epithelial Chloride Conductances

As with other epithelial ion channels, chloride channels have been investigated with various techniques (e.g., flux measurement, microelectrode, and single-channel recording, etc.), and Cl^- conductance has been reported in many epithelia tissues. Prior to the advent of patch clamp techniques, almost all studies of Cl^- conductance in epithelial tissue were done with conventional intracellular microelectrodes or with Cl^--selective microelectrodes. The changes in the membrane potential after application of Cl^- channel blockers or after varying the Cl^- concentration of the bathing solution were usually taken as evidence for Cl^- conductance in a particular epithelium. Additional supportive evidence was some change in the membrane resistance associated with changes in bath chloride concentration or after application of blockers. For the same reasons we did not describe potassium conductances in great detail, we will not emphasize reports which rely solėly on measurements of macroscopic conductance. Rather we will concentrate on the properties of single channels measured with patch clamp methods or fluctuation analysis.

Single Chloride Channels

While macroscopic measurements of chloride conductance imply the presence of chloride channels, they provide little specific information about individual chloride channel proteins. Recently, patch clamp methods have also been applied to an examination of single chloride channels in epithelial tissues. These methods allow a relatively easy determination of a variety of characteristics which uniquely define chloride channel proteins. In particular, chloride channels can be categorized in terms of their single-channel conductance, their mean open and closed times, and, most importantly, those factors which modify or regulate chloride flux through the channel. All of these can be used as criteria for distinguishing different chloride channels and can also provide insight into the molecular mechanisms by which chloride transport is regulated.

Basolateral Chloride Channels

As mentioned above, Cl^--reabsorbing epithelial tissues generally have Cl^- conductive pathways (presumably channels) in their basolateral membrane. But while there have been frequent reports of single chloride channels found in the apical membrane of chloride secreting epithelia, there have been only a few isolated reports of single chloride channels observed in the basolateral membrane of epithelial cells (see Table 4). This discrepancy in the number of reports probably does not reflect an intrinsic difference in the distribution of chloride channels; but rather, is merely a measure of the difficulty in gaining access to the basolateral surface of epithelial cells with patch electrodes, since access to the basolateral membrane is generally restricted by connective tissue in native epithelial tissues or by the culture support in tissue culture systems. For native tissues, access to the basolateral membrane can occasionally be achieved after treatment with collagenase. Chloride channel activity has been reported from the basolateral membrane of rabbit urinary bladder treated with collagenase [125]. In the study, two different types of chloride channels were reported; one with a unit conductance of 64 pS and the other 360 pS. The 360-pS chloride channel is not active at physiological potentials (about -50 mV). In contrast, the 64-pS chloride channel is active at physiological potentials, but the channel closes when the membrane is hyperpolarized more negative than -80 mV. However, step depolarization from -80 to -50 mV reactivates the channel. Also the 64-pS channel inwardly rectifies even when bathed in symmetric solutions containing 150 m*M* chloride.

Table 4. Chloride Channels

Epithelium	Method[a]	Conductance (pS)	I-V[b]	Cl^- Channel blockers[c]	Voltage activated[d]	Characteristics	References
Apical							
Trachea (human)	PC	25	Outward	DPC, 9-AC	No	Stimulated by isoproternol, $P_{Cl}/P_{Na} = 7$	[393]
	PC	26	Outward	DPC		Stimulated by isoproternol, epinephrine, cAMP; no change by $[Ca^{2+}]_i$ (1 nM–1 m*M*); no response to these agents in cystic fibrosis cells	[395]
Trachea (canine)	PC	29	Outward	DPC, 9-AC	Depol		[392]
	PC	30–50	Outward		Hyperpol	A purely anion-selective channel (30–50 pS)	[335]
		20	Linear (?)		Depol	A highly anion-selective channel (20 pS)	
		10	Linear (?)		Depol		
		250	Outward (?)			$P_{Cl}/P_{Na} = 4$	
Trachea (bovine)	RC	71	Rect			Stimulated by PKA	[365]
	RC	100	Linear		(−)	$P_{Cl}/P_K = 13$, one subconductive state	[181]
Alveolar (rat)	PC	300–400	Linear			$P_{Cl}/P_{Na} = 67$, Ca^{2+} decreases single channel conductance; permeability: I (1.5) > Br (1.02) = Cl (1.00) > NO_3 (0.9)	[326]
Alveolar (mouse)	PC	400				Unit conductance (or subconductance) 60–70 pS	[176]
Airway (human)	PC	20	Linear			Stimulated by high $[Ca^{2+}]_i$ in normal and CF cells	[81]
		50	Outward			Stimulated by epinephrine in normal cells; no response to epinephrine in cyctic fibrosis cells	
Airway (human)	PC	45			Depol	Stimulated by C subunit of PK-A in normal cells; no response to C subunit of PK-A in cystic fibrosis cells; stimulated by depolarization in normal and cystic fibrosis cells	[327]
Airway (human)	PC	23–65	Outward			Stimulated by PK-A, PK-C in normal cells; not stimulated by PK-A, PK-C in cystic fibrosis cells	[149, 205, 206]

Airway cell line (human)	RC	13–29	Outward		Depol	Stimulated by PK-A in normal; not stimulated by PK-A in cystic fibrosis cells	[152]
Renal cell line (MDCK)	PC	460					[173]
Renal collecting duct	PC	120				I-V; nonlinear but symmetrical	[32]
Renal distal nephron cell line (A6)	PC	300–400	Linear	SITS		$P_{Cl}/P_{Na} = 9$	[268]
		8	Linear		No	$P_{Cl}/P_{Na} = 6$, one subconductive state	
		3	Outward (strong)		Depol	Stimulated by cAMP, Ca^{2+}, $P_{Cl}/P_{Na} > 17$	[230, 231]
						I-V linear after insulin, phosphatase, Ca^{2+}	[230]
Kidney cells	RC	26	Linear		(+)	$P_{Cl}/P_K = 20$	
		400	Linear		(−)	$P_{Cl}/P_K = 2$-5, multiple subconductive states; open probability was near 1.0; (both types of Cl^- channels, 26 and 400 pS, were unaffected by the inhibitors, IAA-94, stilbene isothiocyanates or anthranilic acid analogs)	[181]
Gastric cells (*Necturus*)	PC	13	Inward		Hyperpol	Stimulated by cAMP, isobutylmethylxanthine	[49]
Colonic cell line (T84) (human)	PC	40–50	Outward		Depol	Stimulated by PGE_2, forskolin, cAMP; $P_{Cl}/P_{Na} = 50$; permeability: I (1.8) > Br (1.4) > cl (1.0) > F (0.4)	[118]
(HT29) (human)	PC	15	Linear	NPPAB			[128]
		32–50	Outward	NPPAB	Depol	Permeability: Cl = Br = I = NO_3; impermeable to gluconate	
Colon cells (rat)	RC	50	Rect			Permeability: I (1.7) > Br (1.3) > Cl(1.0) > F (0.4) = HCO_3 (0.4) > SO_4 (0.16)	[300]
						Permeability: Cl = Br = I = NO_3 = SCN > F (0.5) > acetate (0.3) > gluconate (0.7)	
						$P_{Cl}/P_K = 17$, activated by large voltage (ca. ± 50 mV)	

Table 4. Contd.

Epithelium	Method[a]	Conductance (pS)	I-V[b]	Cl^- Channel blockers[c]	Voltage activated[d]	Characteristics	References
Rectal gland (shark)	PC	40–50	Linear	DPC, NPPBAB	Depol	Permeability: $NO_3 > Cl > Br$; induced by cAMP	[110, 111]
		11	Linear	Not NPPBAB	Hyperpol	Permeability: $Cl > Br \gg I$, sulfate, gluconate	[103]
Lacrimal glands (rat)	Noise	1–2				Ca^{2+} dependent	[228]
Choroid plexus (bovine)	RC	31					[427]
Basolateral							
Urinary bladder (rabbit)	PC	64	Inward			$P_{Cl}/P_K = 23$ Permeability: $Cl = Br = I = NO_3 = SCN > F$ (0.6) > Acetate (0.3) > gluconate (0.07)	[125]

[a] PC, Patch clamp; RC, reconstitution; noise, spontaneous or blocker-induced fluctuation analysis.
[b] Outward, increasing unit current with depolarization; inward, increasing unit current with hyperpolarization; rect, nonlinear I-V with no additional characteristic described.
[c] DPC, Diphenylamine-2-carboxylate; 9-AC, anthracene-9-carboxylic acid; NPPAB, 5-nitro-2-(3-phenylpropylamino)-benzoic acid; SITS, 4-acetamido-4′-isothiocyanostilbene-2, 2′-disulfonic acid; IAA-94, indanyloxyacetic 94; PK-A, cyclic AMP-dependent protein kinase; PK-C, protein kinase C.
[d] Depol, Open probability increases with depolarization; hyperpol, open probability increases with hyperpolarization.

Current Rectification of Single Chloride Channels

Rectification of the unit conductance is an interesting property of many chloride channels. The rectification can either be inward (like the 64-pS channel described above) or outward (as is characteristic of many apical chloride channels). In general, unit current through a channel is dependent on the concentration of the charge carrier in the conductive pathway (or pore) within the channel protein. Since the intracellular Cl^- concentration is much smaller (about 25–30 m*M*; e.g., see [201]) than the extracellular Cl^- concentration (about 100–140 m*M*), there are fewer charge carriers for outward chloride movement (inward current) than for chloride influx (outward current). This implies that under normal conditions of low-intracellular and high-extracellular chloride there should be somewhat more outward current than inward current (Goldman rectification). The inward rectification of the 64-pS channel in rabbit urinary bladder cells is exactly opposite that expected for Goldman rectification. Moreover, in inside-out patches with symmetrical NaCl concentrations on both the extracellular and intracellular surfaces of the channel, outward rectification of single channel current is still present, but inconsistent with simple Goldman rectification. Both observations suggest that many Cl^- channels and the 64-pS chloride channel, in particular, have some specific mechanism to produce rectification of the unit conductance which has to do with the basic properties of the channel protein, itself, rather than concentration-dependent (Goldman) mechanisms. As we shall see, rectification appears to be a distinguishing characteristic of many chloride channels both in the basolateral as well as the apical membrane.

Apical Chloride Channels

Contrary to the paucity of single-channel data on basolateral chloride channels, there are numerous reports of chloride channels in the apical membranes of epithelial tissues. As for cation channels, there is a wide range of unit conductances for different chloride channels from a few picosiemens to several hundred picosiemens. Besides characterizing channels by their single-channel conductance, channels can also be distinguished on the basis of their current-voltage relationships, open probabilities, mean open and closed times, and their voltage dependencies. in addition, the sensitivity to various anion inhibitors is also another useful method for characterizing some chloride channels.

Current-Voltage Relationships and Single-Channel Rectification. Apical chloride channels generally have either a linear or outwardly rectifying current-voltage relationship. In a human colonic cell line (HT 29), two types of chloride channels have been reported [129]. One type has a relatively small single-channel conductance (15 pS) and a linear current-voltage relationship. On the other hand, the other type has a relatively large single-channel conductance (50 pS) and a current-voltage relationship which outwardly rectifies even in inside-out patches with symmetrical NaCl concentrations on the extracellular and intracellular surfaces of the channel. Thus, the outward rectification cannot be due to Goldman rectification. In addition to the colonic cells, many other reports of single chloride channels describe some outward rectification which cannot be attributed to a concentration-dependent effect [81, 118, 129, 334, 391, 392, 394]. A possible explanation of the rectification is suggested by recent reports that a chloride channel in a distal nephron cell line also outwardly rectifies [231], but that the extent of rectification depends upon the cytosolic Ca^{2+} concentration. As the cytosolic calcium concentration is increased, the outward rectification decreases (inward current increases) until at a cytosolic Ca^{2+} concentration of 0.8 m*M* the current-voltage relationship becomes linear. This suggests the possibility that the rectification is due to the effect of a strong fixed negative charge at the inner mouth of many Cl^- channels which would produce rectification by reducing the local concentration of Cl^- near the channel inner mouth, resulting in the outward rectification. Screening or binding of calcium to the strong anionic site would then allow normal chloride concentrations at both inner and outer mouths of the channel which would produce a linear I-V relationship.

Voltage Dependence of Open Probability. Since chloride channel proteins span the cell membrane, they are exposed to the entire membrane potential field. Therefore, it seems possible that the transition rates of the channel protein from one conformational state (open) to another (closed) might be voltage dependent. However, only a few reports suggest any effect of voltage on one measure of transition rates of chloride channels, the open probability [111, 118, 129, 231]. The change in the open probability is due to either a change in the opening rate or a change in closing rate of the channel (or both); for example, an increase in the open probability of the channel could be due to an increase in the opening rate and/or a decrease in the closing rate. Unfortunately, while there is some information about overall open probability, there are few reports which provide specific information about the individual transition rates. Greger et al. [111] indicate that the increase in the open probability of the chloride channel in the rectal gland with depolarization is due to an increase in the opening rate coupled with a decrease in the closing rate. In renal cells, an increase in open probability of a chloride channel after depolarization is mainly due to a decrease in the closing rate [231]. Thus, there is little basis for generalization about the effects of voltage on the kinetics of different chloride channels.

Regulation of Chloride Channels

Chloride channels in both chloride-reabsorbing and chloride-secreting epithelia represent the rate limiting step for transepithelial chloride transport. Therefore, they also represent the point at which most regulation of chloride transport takes place. Although regulation of transepithelial cation transport may have received more attention, anion transport appears to respond to a very diverse complement of hormonal agents and intracellular factors. With the recognition that at least one, and possibly the major, defect associated with cystic fibrosis is related to the regulation of chloride secretion through chloride channels, many reports using various patch clamp methods have recently appeared. While both apical and basolateral chloride channels are likely to be regulated, very little is known about the regulation of basolateral Cl^- channels. There are a few reports that cAMP increases basolateral Cl^- conductance; for example, in the thick ascending limb of Henle's loop of the mouse [323].

On the other hand, there are numerous reports of regulation of apical chloride conductances. Both vasoactive intestinal peptide (VIP) and catecholamines are extracellular agents which increase Cl^- secretion by stimulating the activity of Cl^- channels in the apical membranes of many epithelial tissues. In particular, catecholamines increase chloride channel activity (or Cl^- conductance) in intestine [95], colon [80, 118], airway epithelium [3, 80, 81, 326, 364, 388, 390, 392, 394, 396, 397], elasmobranch rectal gland [110, 337], renal epithelial cells [184, 231, 329, 412], and pancreatic duct cells [270].

Role of cAMP

VIP and catecholamines apparently stimulate chloride secretion by increasing intracellular cAMP levels. Thus, intracellular cAMP must have an effect on total chloride conductance, but since total conductance is the product of single-channel conductance times, the number of channels times the mean open probability of a single channel, the specific mechanism by which cAMP affects total conductance is not clear from macroscopic conductance measurements. Examination of single chloride channels exposed to cAMP suggests that one effect is to increase the open probability of chloride channels [110, 334]. However, the increase in open probability is probably not large enough to explain the increase in total tissue conductance in several tissues. This implies that cAMP must be increasing the unit conductance or increasing the number of functional chloride channels in the apical membranes of secretory epithelial cells. There are reports that cAMP does not alter the unit conductance of chloride channels [231, 364] so it appears that cAMP is likely to have a dual effect on chloride channels: first, increasing the open probability and, second, increasing the number of functional channels in the apical membrane. Of course, both of these effects will lead to increased chloride conductance of the whole tissue. While there have been several reports suggesting that cAMP increases the open probability of chloride channels in epithelial tissues, there are only a few reports which actually examine the mechanism by which the open probability is altered. That is, does cAMP increase the open probability by increasing only the opening rate or by decreasing only the closing rate (or by altering both simultaneously). In distal nephron cell cultures, cAMP increases the opening rate of Cl^- channels without any significant effect on the closing rate [231]. This means that cAMP increases the probability that the Cl^- channel will open, but the probability that the Cl^- channel will lose is not changed.

The mechanism by which cAMP increases the number of chloride channels is not clear, but the direct effect of cAMP on the channel open probability is apparently related to the phosphorylation of the channel. Schoumacher et al. [326] and Valdivia et al. [364] have reported that cAMP-dependent protein kinase (PKA) increases the open probability of Cl^- channels in the apical membrane of airway epithelium. This suggests that the open probability of Cl^- channels is increased by a series of steps starting with hormonal interaction with the basolateral membrane followed by activation of adenylate cyclase and the production of intracellular cAMP. The cAMP activates PKA which phosphorylates the channel protein. The phosphorylated channel has a higher open probability than the dephosphorylated channel.

Effects of Intracellular Calcium

Another major factor which modulates the activity of many chloride channels is the level of intracellular calcium. Ca^{2+} -activated Cl^- conductances have been reported in a variety of cell types [11, 249, 250], including lacrimal gland cells [58, 59, 67, 229], renal cells [231], and airway epithelial cells [81]. Intracellular calcium appears to have at least two separate actions on single chloride channels. First, increased levels of intracellular calcium increase the open probability of epithelial chloride channels; second, in at least one preparation, increasing intracellular calcium also changes the extent of channel rectification in way that increases chloride secretion (inward current) [231].

In lacrimal gland cells, a relatively high concentration of acetycholine (0.5–1 μM) activates a Ca^{2+} -dependent Cl^- channel; but if a nonhydrolyzable analogue GTP, guanosine 5'-γ-Thioltriphosphate (GTPγS), is added, then a much lower concentration of acetylcholine (1 n*M*) can stimulate the Ca^{2+} -activated Cl^- current [58]. The stimulation by acetylcholine is blocked by increasing the intracellular Ca^{2+} buffering capacity and is not dependent on the extracellular Ca^{2+} concentration. Further, intracellular IP_3 (20 μ*M*) also stimulates the Ca^{2+} -activated Cl^- current. These observations suggest that, in lacrimal gland cells, acetylcholine stimulates the Ca^{2+} -activated Cl^- current by increasing the intracelcular Ca^{2+} concentration released from intracellular Ca^{2+} stores in response to increases in IP_3. A similar acetylcholine-induced increase in the intracellular Ca^{2+} mediated by IP_3 is also observed in pancreatic acinar cells [344].

In general, the results of the above experiments and the sensitivity of chloride channels to intracellular calcium suggests that calcium plays a physiological role in regulating chloride channels. The sensitivity of chloride channels to intracellular calcium is an interesting feature of the channel since it appears that regulation of the sensitivity may be an additional point at which the activity of some chloride channels can be controlled. The Ca^{2+} -activated Cl^- channels in cultured renal cells are not stimulated by physiological concentrations of cytosolic Ca^{2+} (0.1–1 μ*M*), but only by high cytosolic Ca^{2+} (> 10 μ*M*) [231]. However, after treatment with insulin or CTX, the sensitivity of the Cl^- channels to Ca^{2+} is increased so that the channels can be activated by low intracellular calcium levels in the physiological range (0.1–1 μ*M*). The result of the insulin-induced change in sensitivity is an increase of the Cl^- inward current (Cl^- secretion) and an increase in the open probability of the channel.

The relationship between effects of Ca^{2+} and cAMP on the Cl^- channel is not completely clear. However, recently Frizzell et al. [81] have reported that in normal airway cells epinephrine and Ca^{2+} have stimulatory actions on Cl^- channels. However, in airway cells of cystic fibrosis patients, epinephrine cannot activate the Cl^- channel even though there is no difference of the stimulatory action of Ca^{2+} in normal and cystic fibrosis airway cells. This observation suggests that the PKA phosphorylation site is different from the Ca^{2+}-activated site, and that therefore cAMP and Ca^{2+} have different regulatory pathways.

Protein Kinase C

Besides activation by PKA-mediated phosphorylation, PKC phosphorylation also appears to have a stimulatory action on Cl^- channels in transformed human fetal tracheal cells [150]. Cl^- channels in inside-out patches from the normal tissue are activated by PKC in the presence of ATP, a diacylglycerol analogue, and 1.2 μ*M* cytosolic Ca^{2+}. Even in the presence of a PKA inhibitor, PKC is still capable of activating Cl^- channels. This observation suggests that the pathway for activation by PKC is different than that for PKA. In addition to an examination of normal tissue, the action of PKC on cystic fibrosis tissues was also examined. Like PKA, PKC also produced no activation Cl^- channels [150, 206, 207]. However, it is unclear whether PKC treatment alters the sensitivity of the Cl^- channel to cytosolic Ca^{2+}.

Other Modulatory Agents

Prostaglandins stimulate Cl^- secretion in tracheal epithelial cells [3]. In colonic cells, PGE_2 also activates Cl^- channels [80, 118]. Indomethacin, which inhibits the synthesis of prostaglandins from arachidonic acid, decreases the Cl^- secretion in tracheal tissues [3, 389, 397]. This suggests that endogenous prostaglandin can activate Cl^- secretion probably through activation of the Cl^- channel in the apical membrane.

Adenosine increases Cl^- current in a colonic cell line (T84) [34] and in isolated colonic epithelial cells [13]. Before addition of adenosine, a tenfold decrease in the extracellular Cl^- concentration depolarizes the membrane only by 3 mV. On the other hand, after addition of adenosine, a tenfold decrease in the extracellular Cl^- concentration depolarizes the membrane by 48 mV. This suggests that adenosine induces a large increase in Cl^- conductance, and the increase in Cl^- current depends upon the applied dose of adenosine or 2-chloro-adenosine (a more potent agonist). The adenosine-induced increase in Cl^-

current can be blocked by the specific adenosine receptor antagonist, 8-phenyltheophylline, indicating that adenosine stimulates Cl^- currents through a receptor-mediated mechanism. Adenosine agonists also increase arachidonic acid release from the cell without any significant change in the intracellular concentrations of Ca^{2+}, cAMP or cGMP, suggesting that the intracellular concentration of arachidonic acid is increased by adenosine agonists. This would suggest that adenosine produces an increase in arachidonic acid concentration followed by an increase in prostaglandins, which stimulates Cl^- channels in the apical membrane. This phenomenon is consistent with the stimulatory action of endogenous prostaglandins on Cl^- channels.

Chloride Channels and Cell Volume

Cl^- channels may play an important role in maintaining cell volume. Exposure to hypotonic media increases the Cl^- conductance in the colonic cell line (T84) [411] and also increases the Cl^- secretion in other epithelial tissues [29]. Associated with exposure to hypotonic media and the increase in secretion, there is also an increase in intracellular Ca^{2+} concentration. Apparently, T84 epithelial cells have a Ca^{2+} -activated Cl^- channel, which is activated by an increse in cytosolic Ca^{2+} concentration induced by the exposure to hypotonic media. The activation of the Ca^{2+} -activated Cl^- channels causes Cl^- efflux which results in the recovery of the cell volume from the hypotonic-induced cell swelling. Based on these observations, Cl^- channels may have a role in maintaining cell volume in response to changes in intracellular Ca^{2+}.

Nonselective Cation Channels

Nonselective cation channels have been observed in a variety of epithelial and nonepithelial tissues (Table 5). Most nonselective cation channels discriminate only poorly among monovalent cations with typical permeability ratios among the alkaline earth cations of less than 3. Nonselective cation channels, with a few notable exceptions, have a relatively low single-channel conductance and also have a variety of characteristics which are often attributed exclusively to highly selective sodium, potassium, and calcium channels. In particular, the pharmacology of nonselective cation channels often resembles that of more selective channels to the extent that it is possible to categorize nonselective cation channels, as we have done below, on their pharmacological resemblance to highly selective channels. Nonselective cation channels are present in many different epithelial tissues but differ enormously in their response to voltage, hormonal stimulation, intracellular second passangers and mechanical/osmotic

Table 5. Nonselective cation channels

Epithelium	Method[a]	Permeability or P_{Na}/P_K[b]	Conductance (pS)	Ca_i Activated	Voltage Activated	Amiloride Sensitive	Inhibition/Stimulation[c]	References
Apical membrane								
Lens (frog, rat, mouse)	PC	2.0–4.0	25–30	Yes	Depol	Inhibited	Stretch stimulates; Ca^{2+} inhibits	[36, 151, 293, 296]
		1.2–1.8	50–60					
		Cations, Cl^-	100					
Lens (frog, chick)	PC		10–12					[293, 296]
Lens (frog, mouse, rat, bovine)	PC	Cations, Cl^-	380	–	–	–		[293, 296]
			1900					
Cornea (frog)	PC		25–30		Depol		Stimulated by stretch	[293, 296]
			10					
Choroid plexus (*Necturus*); ventricular epithelium	PC		29–54	No	No		Stimulated by stretch	[30]
Pancreas duct (rat)	PC		23	Yes	No			[6]
Sweat gland (human)	PC		5		Yes	Inhibited		[175]
			18	Yes	Yes			
Tracheal cell (human)	PC		15–25	Yes				[82, 394]
Frog skin (larval)	Noise					Stimulated	Inhibited by Ca^{2+}, Ba^{2+}, TEA, amiloride (> 100 μ*M*); stimulated by BIG	[135]
Frog skin (adult)	Noise			No			Stimulated by alkaline pH; inhibited by acid pH, divalent ions	[135, 373]
Urinary bladder (toad)	PC		5	No			Stimulated by AVP, cAMP, IBMX	[78]
			17				Stimulated by AVP, cAMP, IBMX	
			90–490				Stimulated by cytoplasmic fluoride	
Urinary bladder (toad)	Noice			No			Inhibited by polyvalent ions, nitrendipine, nicardipine; stimulated by cAMP, theophylline, forskolin, oxytocin	[2, 371, 372]

Table 5. Contd.

Epithelium	Method[a]	Permeability or P_{Na}/P_K[b]	Conductance (pS)	Ca_i Activated	Voltage Activated	Amiloride Sensitive	Inhibition/Stimulation[c]	References
Urinary bladder (rabbit)	PC & Noise	0.54	6	No				[428]
Urinary bladder (rabbit)	Noise	0.5–0.7	6	No			Appears after kallikrein, urokinase; stimulated, inhibited by serotonin	[195, 197]
Proximal tubule, culture (rat	PC	1–5	50	No	No			[226]
Proximal tubule, S1 (rabbit)	PC		25	Yes	No			[246]
Proximal tubule (*Necturus*)	PC	1	18				Activated by stretch, inhibited by gadolinium	[65]
Proximal tubule, S1 (rabbit)	PC		25	Yes	No			[246]
Distal tubule culture (rat)	PC	1.0/0.97	18	Yes	No			[177]
CCT culture (rabbit)	PC		40					[399]
Distal nephron (A6)	PC	1	51				Activated by stretch, inhibited by gadolinium	[158]
Distal nephron (A6)	RC	2	4–80					[275, 321]
Distal nephron culture on plastic (A6)	PC	2–5	9	No		Inhibited		[53, 121]
Distal nephron culture on glass and permeable supports (A6)	PC	1.9	20.9	Yes				[119]
IMCD (rat)	PC	Cations	27.5	–	No	Inhibited	Stimulated by G_i $alpha_{i\text{-}3}$ subunit, GTP-gamma-S; inhibited by ANP, cGMP, PTX, GDP-beta-S	[208, 210, 211]
			11.5			Inhibited		
Renal cortical vesicles (rabbit)	Dye					Inhibited	Inhibited by Ba^{2+}	[50]

Basolateral membrane								
Pancreas acini (rat, mouse)	PC		30–35	Yes	No		Stimulated by cholecystokinin, acetylcholine, SITS, DIDS, and DNDS; inhibited by DPC, DCDPC, and quinine	[86, 102, 240]
Lacrimal gland (rat)	PC		25	Yes				[228]
Thyroid follicular cells (rat)	PC		35	Yes	No			[234]
Proximal tubule culture, S2 or S3 (rabbit)	PC	Cations, Cl^-	25		Depol		Inhibited by DPC and SITS	[100]
cTALH (mouse)	PC		26.8	Yes	No		Inhibited by ATP, ADP, AMP, GTP	[359]
Membrane unknown								
Urinary bladder (toad); isolated cells	PC		24–157	No	No	No	Inhibited by quinidine and, divalent ions	[41]
Opossum kidney culture; subconfluent	PC	Cations, Cl	22				Stimulated by hypoosmolarity and stretch	[360]

[a] PC, Patch clamp; RC, reconstitution in lipid blayers; noise, spontaneous or blocker-induced fluctuation analysis, Dye, ion sensitive fluorescence dye.

[b] P_{Na}/P_K, The ratio of the sodium to potassium permeability.

[c] CCT, Cortical collecting tubule; IMCD, inner medullary collecting duct; cTALH, cortical thick ascending loop of Henle; DPC, diphenylamine-2-carboxylate; DCDPC, 3, 5-dichlorodiphenylamine-2-carboxylate; BIG, benzimidazolyl-2-guanidine; IBMX, isobutylmethyl-xanthine; NPPB, 5-nitro-2-(3-phenylpropylamino)-benzoic acid; SITS acetamido-4′-isothiocyano-stilbene-2, 2′-disulfonic acid; DIDS, 4, 4′-di isothiocyano-stilbene-2, 2′-disulfonic acid; DNDS, 4,4′-dinitro-stilbene-2,2′-disulfonic acid; TEA, tetraethylammonium; PTX, pertussis toxin.

stress. This section describes the features of nonselective cation channels in epithelial tissues.

Ca^{2+}-Activated Nonselective Cation Channels

Several nonselective cation channels which are all activated by elevated levels of cytoplasmic Ca^{2+}, and which have other characteristics in common are found in the basolateral membrane of mouse and rat pancreatic acini [86, 102, 240], rat lacrimal gland [228], rat thyroid follicular cells [235], and mouse cortical thick ascending limb [358]; and the apical membrane of lens epithelium [36, 152, 293], rat pancreatic duct cells [6], human sweat gland cells from normal and cystic fibrosis patients [176], cystic fibrosis tracheal cells [82, 293], isolated rabbit proximal convoluted tubules [247], and rat distal tubule culture [178]. These channels all have a single-channel conductance of 18–35 pS, a linear I-V relationship, no voltage sensitivity of their open probability, an essentially equal permeability to all of the alkali metal cations with little or no permeability to anions or divalent cations, and all are activated by micromolar cytoplasmic Ca^{2+}. These characteristics are similar to those of other nonselective cation channels which are found in nonepithelial cells including neutrophils [383], adipose tissue cells [235], *Helix* neurons [289], mast cells [213], macrophages [221], insulinoma cells [346], neuroblastoma [416], cultured cardiac cells [266], aortic endothelial cells [64], and neonate dorsal root ganglion cells [338].

Some investigators have questioned whether these channels have a normal functional role in epithelial cells since the cytoplasmic Ca^{2+} levels required for activation (greater than 1 μM) are above normal physiological concentrations. However, in inside-out patches when many cytosolic components have been disrupted, nonselective cation channels may not be as sensitive to intracellular calcium as the same channels in intact cells. In fact, cytosolic humoral factors may play a significant role in modulating the activity of these channels. For instance, open channel probability (P_0) for the Ca^{2+} -sensitive nonselective cation channel of mouse cortical thick ascending limb of Henle's loop is reduced by cytoplasmic application of adenosine derivatives to membranes of excised patches [357] similar to the response of nonepithelial, Ca^{2+} -activated, nonselective cation channels found in pancreatic insulinoma cells [346]. Additional support for a physiological role of Ca^{2+} -activated, nonselective cation channels is that both inward macroscopic current and the nonselective cation channel in pancreatic acini are activated by cholesytokinin and acetylcholine, probably by raising cytoplasmic Ca^{2+} [238]. Therefore, it seems that ion channel activation may be an important regulatory step in exocrine gland fluid and solute secretion. In rat thyroid follicular cells, Ca^{2+} activation of nonselective cation channel could explain agonist (i.e., norepinephrine, thyroid-stimulating hormone) evoked membrane depolarization.

In renal epithelia, one postulated physiologic role for these channels includes Na^+ reabsorption. However, the contribution to total transcellular Na^+ reabsorption would be negligible compared to that of Na^+ cotransporters, Na/H exchanger, or the amiloride-blockable, Na^+ selective channel. Alternatively, these nonselective cation channels could contribute to cell volume regulation. Apical nonselective cation channels could allow cation entry (Na^+) and, thus, mediate volume regulatory increase in shrunken cells while basolateral nonselective cation channels could provide a pathway for cation loss (K^+) associated with regulatory volume decrease in swollen epithelium (see "Stretch and Osmotically Activated Channels").

Role of Nonselective Cation Channels in Renal Sodium Reabsorption

Another well characterized nonselective cation channel that may contribute to electrogenic Na^+ reabsorption is present in both primary cultures and isolated rat IMCD [211]. This channel has a unit conductance of 27.5 pS and does not discriminate between Na^+, K^+ and NH_4^+. It is not voltage or pH dependent and does not respond to the potassium channel blockers, Ba^{2+} or TEA, or the chloride channel blocker DPC. However, like the highly sodium-selective channels of tight epithelia, the mean open time is decreased and mean closed time increased when exposed to submicromolar concentrations of amiloride. Increasing cytoplasmic Ca^{2+} produces no consistent effect on the channel. The channel is usually closed cell-attached patches but opens when patches are excised, suggesting the presence of some intracellular inhibitory substance. ANP and its second messenger cGMP inactivate the channel [209] consistent with the ANP-induced inhibition of macroscopic Na^+ reabsorption seen in IMCD cells [365, 417]. Also, a PTX-sensitive, G-protein in the IMCD apical membrane directly activates this amiloride sensitive nonselective cation channel [208, 209, 212]. In excised patches, PTX or GDPβS decreases, and GTPγS or the activated $\alpha_{i\text{-}3}$ subunit of G_i increases the open probability of the channel. Besides Na^+ reabsorption, the IMCD is capable of either K^+ reabsorption or secretion in response to changes in total body potassium load. The nonselective cation channel described above could provide an ion pathway for these macroscopically observed potassium movements.

Renal medullary accumulation of ammonia increases during metabolic acidosis and is a major determinant of inner collecting duct ammonia secretion [123]. Since the nonselective cation channel described above is quite permeable to NH_4, transport of medullary interstital ammonium through this nonselective, cation channel in IMCD cells could play a role in acid/base homeostasis.

Relationship of Nonselective Cation Channels to Other Epithelial Channels

As we mentioned above, there are pharmacological similarities between nonselective cation channels and other more selective channels. In addition there are often other similarities which suggest some functional or even evolutionary relationship between nonselective cation channels and other channels.

Nonselective Cation Channels that Have Similarities to Epithelial Sodium Channels

Net transport Na^+ across tight epithelial tissue follows the classic model of Koefoed-Johnsen and Ussing [362]: basolateral membrane Na^+-K^+ -ATPase creates an electrochemical gradient for the passive movement of Na^+ across the apical membrane via a Na^+ selective pathway. There is some experimental evidence to suggest a relationship between amiloride-blockable, nonselective cation channels and amiloride-blockable, highly sodium-selective, low-conductance, apical Na^+ channel [53]. Lewis and his colleagues [194, 196, 198, 427] postulate that an intramembranous pool (presumably cytoplasmic vesicles) containing highly Na^+ selective channels is inserted into the apical membrane of rabbit urinary bladder and that the insertion is mediated through interaction of the vesicle with cytoplasmic microfilaments. The newly inserted Na^+ -selective channels then undergo hydrolytic degradation by renal tubular urokinase or kallikrein resulting in a channel with progressive loss of amiloride sensitivity and Na^+ selectivity until the only conductance remaining is the so-called "leak" pathway. This nonselective, partially degraded channel finally becomes unstable in the apical membrane and is released into the luminal fluid and possibly the cytosol.

This hypothesis is supported by the experimental observation that, after stimulation of vesicle fusion by bladder stretching [196], fluctuation analysis revealed that the newly inserted, amiloride-blockable conductance starts with a ratio of sodium permeability to potassium permeability of about 30. As the channels degrade they lose this selectivity with partially degraded channels having a Na to K permeability ratio of between 2.6 and 9.0. Finally, the fully degraded channel has little, if any, Na to K selectivity and is not blockable by amiloride [198]. Zweifach and Lewis [427] have successfully inserted a nonselective, 6-pS "leak" channel from rabbit urinary bladder washings into lipid bilayers. Patch clamp measurements of this latter channel yield a Na to K permeability ratio of 0.5, very close to the selectivity estimtes obtained from noise analysis of the urokinase/kallikrein-degraded, "leak" channel described above. Besides these functional relationships which may imply some specific relationship between the selective Na^+ channel and some cation nonselective channels, both the Na^+-selective and nonselective, cation channel are inhibited by the indolamine, serotonin, with similar half-maximal inhibitory concentrations. Consistent with this similarity in properties and possible common evolutionary ancestry is the identification using patch clamp methods of an amiloride-blockable, Na^+-selective channel and an amiloride-insensitive, nonselective cation channel possessing the same conductance (5 pS) and gating kinetics in the apical membrane of toad urinary bladder [78]. A 17-pS nonselective cation channel has also been identified in the same tissue. In cultured sweat gland secretory epithelium from normal and cystic fibrosis patients, an amiloride-blockable 5-pS nonselective cation channel has also been described [176]. Amiloride-blockable Na^+ channels from the *Xenopus* distal nephron cell line,

A6, reconstituted into lipid bilayers have consistently low selectives (P_{Na}/P_K = 2–3) with a wide range of unit conductances (4–80 pS) [279, 320]. High-conductance (40-pS) nonselective cation channels have also been described in rabbit cortical collecting tubule primary cultures [398]. That these channels may represent a continuum of differentiation or degradation states may explain the identification of both high selectivity, low conductance and low selectivity (P_{Na}/P_K3–4), high conductance (7–10 pS) Na^+ channels and 21-pS nonselective cation channels in A6 cells depending on their transporting state or permeability of the cells growth supports [119, 120]. Evidence for the existence of more than one set of distinguishable channel characteristics (either due to different channel proteins or different forms of the same protein) depending on exposure to aldosterone has also been demonstrated by Lewis and Wills [200] in rabbit urinary bladders. Measuring apical membrane potential, they calculated a P_{Na}/P_K of 2.6 for animals maintained on a normal diet (low circulating levels of aldosterone) and 9 for animals on a low Na^+ diet (high levels of circulating aldosterone). Eaton and Hamilton [53] postulated a scheme for the regulation of apical Na^+ permability in which exposure to aldosterone switches existing sodium channels from a nonconducting or "cryptic" state to a functional, conducting state with no change in the number of apical membrane channel proteins. On the other hand, in this scheme, exposure to vasopressin does not alter the properties of preexistent channels but rather promotes the insertion of new channels from a vesicular pool. In addition, besides the functional appearance and disappearance of channels in response to changes in exposure to aldosterone, the scheme also suggested that, associated with aldosterone removal there was a change in the selectivity of apical channels from being highly selective for sodium to being much less selective for sodium over potassium. There is some evidence for aldosterone-exposure changing the functional conductance of preexistant channels (see Benos et al., this volume); however, aldosterone-induced transitions from sodium-selective to cation non-selective channels are still hypothetical since there have been no reported single-channel recordings documenting the spontaneous conversion of a highly sodium selective channel to a poorly selective channel.

Another link between amiloride-blockable, poorly selective channels and highly Na^+ selective channels is a common response to several hormones and second messenger agents known to regulate sodium reabsorption. Nonselective cation channels are activated by vasopressin, cAMP analogues, phosphodiesterase inhibitors, and PKA in toad urinary bladder [2.78].

Nonselective Cation Channels that Have Similarities to Potassium Channels

Some nonselective cation channels exhibit behavior more typical of potassium-selective channels. Hillyard and Van Driesche [136] found an amiloride-blockable, nonselective cation channel in larval bullfrog skin, which did not discriminate between monovalent K^+ -like (K^+ NH_4^+, Rb^+, Cs^+) or Na^+ -like

(Na^+, Li^+) cations. This cation channel was blocked by the K^+ channel blockers Ba^{2+} and TEA (see "Inorganic Blockers"). Examining fluctuations in short circuit current, these investigators also observed that, unlike sodium-selective channels, the apparent open probability of this nonselective cation channel was increased rather than decreased by amiloride, but that, like sodium-selective channels, the open probability was increased by benzimidazole-2-guanidine. Mucosal Ca^{2+} inhibited the channel and interfered with the agonist effect of amiloride. In single-channel recordings from frog tadpole skin cells, amiloride stimulated 25-pS and 100-pS cation channels which might be responsible for the amiloride-stimulated macroscopic short-circuit current [40].

Nonselective cation channels are also present in postlarval frog skin cells, but the properties of these channels are different from those of channels in larval skins described above. These differences may not represent genetically distinct channel proteins, but rather altered posttranslational modification of the larval channel proteins associated with the radically different hormonally environment after metamorphosis. There are several examples of these differences. In skin from molting toads, amiloride reduces transepithelial conductance even when potassium is the main mucosal cation [155]. The rate of occupancy of the amiloride-receptor by amiloride decreases during bullfrog maturation [137]. Noise analysis of apical adult bullfrog skin revealed a nonselective monovalent cation channel that was also blocked by mucosal divalent cations, but produced Lorentzian noise only in the presence of both mucosal Na^+ and amiloride [136]. Amiloride-induced fluctuations in mucosal Ca^{2+}-sensitive, short-circuit current in frog skin also shows activation by hypotonicity and forskolin [372].

There are other similarities between some nonselective cation channels and some types of potassium channels. Many nonselective cation channels are voltage dependent and activated by intracellular calcium (Ca_i^{2+}) in a manner reminiscent of the voltage-dependent, Ca_i^{2+}-activated K^+ channels found in the luminal membranes of cortical collecting tubules [75,148], proximal tubules [98, 156]. *Amphiuma* early distal tubule [146], and cultured renal cells [19, 96, 112, 146]. However Ca_i^{2+}-activated nonselective cation channels are, in general, much less sensitive to the levels of intracellular calcium than Ca_i^{2+}- activated potassium channels. Ca_i^{2+}- activated nonselective cation channels generally activate only when intracellular calcium exceeds 1 μM and are closed at intracellular calcium concentrations which normally activated Ca_i^{2+}- activated K^+ channels (0.1 μM) [358].

Nonselective Cation Channels that Have Similarities to Calcium Channels

Still other nonselective cation channels exhibit characteristics similar to calcium channels. The so-called CAT-50 channel, a nonselective channel in frog lens epithelia [36], and nonselective cation channels in *Necturus* choroid plexus epithelia [30] and toad urinary bladder [2, 370] are capable of carrying Ca^{2+} current under certain conditions. The nonselective cation channel in toad urinary bladder is also inhibited by dihydropyridine Ca^{2+} channel blockers (see "Calcium-permeable Channels").

Epithelial Channels that Are Permeable to both Anions and Cations

A report by Gogelein and Greger [101] describes a 25-pS channel in the basolateral membrane of freshly isolated rabbit late proximal tubules which was permeable to both cations and anions. Like many epithelial chloride channels (see "Chloride Channels"), the channel is voltage dependent, activating at depolarizing potentials. In addition, it was inhibited by chloride channel blockers such as DPC and 4-acetamido-4′-isothiocyano-stilbene-disulfonic acid (SITS). Opossum kidney cell cultures possess a similar 22-pS anion- and cation-permeable, nonselective channel that is activated by stretch and hypotonicity [360]. Gogelein and Pfannmuller [102] have also demonstrated that DPC and the related compound, 5-nitro-2-(3-phenylpropylamino)-benzoic acid (NPPB) and 3′, 5-dichlorodiphylamine-2-carboxylic acid (DCDPC) increase the closed time of the Ca^{2+} activated, nonselective channel in the basolateral membrane of rat exocrine pancreas. However, the half-blocking concentration of these substances is different than for typical Cl^- channels and the long closed intervals produced by these blockers contrasts with the flickery block produced by the same blockers when applied to Cl^- selective channels. Surprisingly, this nonselective cation channel was activated rather than blocked by SITS and its analogues 4, 4′-diisothiocyano-stibene-disulfonic acid (DIDS) and 4,4′-dinitro-stilbene-disulfonic acid (DNDS). These stibene disulfonates acted only from the cytoplasmic side and increased mean channel open time, similar to their effect on the Na^+ channel of excitable tissues [290]. Patch clamp studies of apical lens epithelia in frog, rat and mouse reveal two types of nonselective cation channel with conductances of 100 pS and 380 pS–1.9 nS that are also permeable to chloride ions [293]. High conductance (> 300 pS) chloride-selective channels have been demonstrated in cultured epithelial cells [174, 268] and rabbit urinary bladder basolateral membrane [125].

Summary

In summary, nonselective cation channels are ubiquitous, being found in most cells and tissues which have been examined. Nonselective channels have a chameleonlike quality with characteristics which often mimic those of more selective channels. Whether this reflects a common evolutionary origin for many of these channels is only conjecture at this point and must await studies examining the amino acid sequences and structure of the channel proteins. Functionally, nonselective channels may be involved in Na^+, K^+ and acid/base homeostasis, cytosolic Ca^{2+} regulation, metamorphic changes, cell volume regulation, and possibly pathophysiological responses of epithelial cells to large changes in intracellular calcium or membrane potential.

Calcium-Permeable Channels

The kidneys play a major role in total body calcium homeostasis; 98% of filtered calcium load is reabsorbed, with reabsorption in the proximal tubular accounting for 60%, loop of Henle 20%, distal tubule 10%–15%, and collecting tubules 5%. As with renal Na^+ handling less than 2% of the filtered load of calcium is excreted and calcium does not exhibit a tubular maximum for reabsorption. Friedman [74] has proposed a model for calcium absorption in epithelial cells which parallels many characteristics of Na^+ reabsorption. Active cellular extrusion of Ca^{2+} occurs at the basolateral membrane via either the Na-Ca exchanger (driven by the Na^+ gradient created by Na-K-ATPase) or Ca-ATPase. This creates an electrochemical gradient for the passive influx of calcium across the apical membrane, presumably via calcium permeable channels or a carrier mediated process. In this scheme, regulation of calcium reabsorption would be hormonally modulated by parathyroid hormone (PTH), intracellular cAMP, vitamin D, and calcitonin. Hormones which target renal cells such as vasopressin and PTH are known to activate non-epithelial, receptor-dependent, calcium-permeable channels in smooth muscle and osteosarcoma cells, respectively [139]. Despite this transcellular flux of millimolar quantities of Ca^{2+} intracellular calcium concentration must be maintained in the submicromolar range, suggesting regulation of calcium transport which closely matches basolateral extrusion with apical influx similar to the case for transepithelial sodium transport [327, 328]. In addition, apical calcium permeability via channels may be important to produce subtle changes in cytosolic Ca^{2+} that can act as a second messenger to modulate a variety of cellular processes in epithelial tissues. To be similar to our definition of other ion channels, our definition of a "calcium channel" should be a channel which is predominantly permeable to calcium. There are few channels in epithelial tissues which strictly fit this definition (possibly only the channels described by Van Driessche [370] [Table 6]). However, many channels have a high enough permeability to calcium that calcium entry through these channels may be significant in a modulating cellular proceses. Because of this potentially important contribution to cellular regulation, we have decided to gather together all channels which have a large calcium permeability regardless of their permeability to other ions. This means that many of the references to specific channels may be mentioned in other places within this review, particularly in the section on nonselective cation channels.

Bacskai and Freidman [10] have demonstrated using FURA-2 fluorescence measurements that PTH-stimulated increases in cytoplasmic calcium are mediated by dihydropyridine-sensitive calcium channels in the apical membranes of cultured mouse cortical thick ascending limb and distal convoluted tubule cells. These channels appear to be inserted from a subapical vesicular pool since their appearance can be inhibited by the microtubule-disrupting agent, colchicine. Patch clamp studies have identified stretch-activated cation channels which

Table 6. Calcium-permeable channels

Epithelium	Method[a]	Conductance (pS)	CA^{2+} Channel Blockers	Voltage Activated	Inhibition/Activation	References
Lens (frog)	PC	4				[293, 296]
Lens (frog)	PC	28		No	Stimulated by stretch; inhibited by amiloride	[36, 151, 293, 296]
Choroid plexus (*Necturus*); ventricular epithelium	PC	29–54		Depol	Stimulated by hypoosmolarity, stretch	[30]
Urinary bladder (toad)	Noise		La^{3+}, Cd^{2+}, verapamil, nicardipine		Stimulated by Ag^{+}, Cu^{+}, Hg^{+}	[2, 371, 372]
Proximal tubule (*Necturus*)	PC	18			Stimulated by stretch, inhibited by gadolinium	[65]
Cortical thick ascending limb and distal convoluted tubule primary cultures	Fura-2	–	Nifedipine	–	Stimulated by parathyroid hormone, inhibited by colchicine	[251]
Distal nephron (A6)	PC	51			Stimulated by stretch, inhibited by gadolinium	[158]

[a] PC, Patch clamp; Noise, spontaneous or blocker-induced fluctuation analysis; Fura-2, measurements of calcium flux using the fluorescent dye, FURA-2.

carry inward calcium current in the apical membrane of isolated *Necturus* proximal tubules [65] and A6 cells [159]. The unit conductances were 18 pS and 51 pS, respectively. Both channels were inhibited by 10 μ*M* gadolinium and discriminated poorly between Na^+, K^+, and Ca^{2+}.

Van Driessche and his coworkers [2, 370, 371] have described an oxytocin-stimulated channel in the apical membrane of toad urinary bladder in which the conductance depends upon extracellular Ca^{2+}. In the absence of competing polyvalent cations (Ca^{2+}, Cd^{2+}, Ba^{2+}, Sr^{2+} Mg^{2+} La^{3+}), this channel conducts monovalent cations with no apparent selectivity (see "Nonselective Cation Channels"). This channel is not blocked by amiloride but is activated by oxytocin, cAMP, theophylline, and forskolin. Nanomolar concentrations of mucosal Ag^+, Cu^+, and Hg^+ (probably by interacting with channel sulfhydryl groups) increase the total conductance of this channel to divalent cations (Ca^{2+}, Ba^{2+}, and Mg^{2+}). Inward divalent current which is induced by oxytocin or Ag^+ can be blocked by the calcium channel blockers, La^{3+}, Cd^{2+}, nicardipine, or verapamil. Inward monovalent current stimulated by oxytocin is also blocked by the dihydropyridine calcium channel blockers nitrendipine and nicardpine. Features which this cation channel in epithelial tissue shares with classic Ca^{2+} channels in excitable tissues include: (a) monovalent cation current is carried by these channels in the absense of polyvalent cations; (b) monovalent cation current is blocked by micromolar concentrations of polyvalent cations; (c) these channels are permeable to Ca^{2+}, Ba^{2+}, and Mg^{2+}; (d) they respond to organic Ca^{2+} channel antagonists: dihydropyridines selectively block the L-type (long-lasting, high-threshold) calcium channels, and phenylalkylamines (verapamil) block both L-type and T-type (transient, low-threshold) calcium channels; (e) some Ca^{2+} channels are activated by adenine nucleotide derivatives; and (f) sulfhydryl agents activate the Ca^{2+} channel protein [370].

There are other reports in larval and adult frog skin of channels which carry monovalents in the absence of any mucosally applied polyvalent cations [136, 370, 381]. However, Ca^{2+} current through these pathways has not been observed and neither the dihydropyridine, nicardipine, nor the benzothiazepine, diltiazem, have any effect. However, binding of Ca^{2+} channel blockers to calcium channels has a complicated dependence on external divalent cations [139]. Micromolar divalent cations are required for dihydropyridine binding to L-type channels, while inhibition of binding can be seen with higher concentrations. Divalent cations can also inhibit phenylalkylamine and benzothiazepine binding to skeletal muscle calcium channels. Allosteric interactions between the high-affinity Ca^{2+} binding sites of the channel protein and the receptors for channel antagonists may explain such findings.

Evidence for the presence of calcium channels in other renal epithelial tissues is based on Ca^{2+}-dependent physiological responses to calcium channel blockers. Volume regulatory decrease in isolated proximal S_2 renal tubules is inhibited by verapamil and restored by subsequent application of a calcium ionophore [269]. It has been suggested that Ca^{2+} influx, presumably through Ca^{2+} channels, regulates volume compensation via a Ca^{2+} activated K^+

conductance. Similarly, verapamil inhibits K^+ efflux in chicken small intestine epithelia [260]. In contrast, verapamil induced decrease in Ca^{2+} influx is associated with a transient reduction in cell K^+ content and volume in MDCK cells [4]. Verapamil also blocks the antidiuretic hormone stimulated hydro-osmotic response is toad urinary bladder, a response modulated by cytosolic Ca^{2+} [142]. Reduction of basolateral Ca^{2+} or application of La^{3+} irreversibly inhibits cAMP-induced water transport in isolated rabbit cortical collecting tubule [51].

Cooper et al. [36] have described a stretch-activated, 28-pS, calcium-permeable, cation-selective channel in apical frog lens epithelia (CAT-50). In the absence of mucosal Ca^{2+}, it carries monovalent cations with $P_K > P_{Na}$. In the presence of physiological calcium concentrations (2 mM), the channel has a linear current-voltage relationship and does not discriminate between Na^+ and K^+ (see "Nonselective Cation Channels"). Divalent cations in the mucosal solution inhibit monovalent current through this channel; and, when divalents (Ca^{2+}, Ba^{2+}) are the predominant cation in the mucosal solution, divalent current is carried in a voltage-independent manner. The channel is not affected by the K^+ channel blocker TEA. This channel is probably the same as that described in frog lens [152]. This latter study demonstrated an amiloride block-with marked reduction in channel activity at doses of 50 µM. Low concentrations of amiloride are usually considered to act as a blocker for highly selective Na^+ channels in epithelial tissue, but amiloride also rapidly blocks T-type calcium channels [350] with K_i of 30 µM in mouse neuroblastoma, guinea pig atrial myocytes, and chick dorsal root ganglion neurons. Thus low-threshold, T-type calcium channels may have structural homologies with the amiloride-blockable epithelial Na^+ channel and this frog lens nonselective cation channel. Physiologically, the CAT-50 channel may be a pressure sensor activated by mechanical stress to control cell volume by altering cytoplasmic Ca^{2+} (see "Stretch and Osmotically Activated Channels"). Alternatively, it may play a role in cataractogenesis and cell death by allowing pathophysiological increases cellular Na^+ and Ca^{2+}.

Frog lens epithelium also possesses a rare 4-pS channel that is possibly selective for calcium but which has not been completely characterized [296].

Single-channel studies have been performed on the ventricular epithelium of Necturus choroid plexus using patch clamp technology [30]. The choroid plexus is responsible for cerebrospinal fluid formation via Na^+, Cl^- and HCO_3^- secretion. In single-channel studies of choroid plexus, a stretch-activated channel with unit conductance of 29–54 pS was discovered that discriminates poorly between Na^+ and K^+ and carries a significant Ca^{2+} current. In the cell attached mode with 75 mM $CaCl_2$ in the patch pipette, the unit conductance of the channel was 20 pS and the I-V relationship was linear. In addition, a 200 pS, Ca^{2+}-activated, voltage dependent K^+ channel was found in this tissue. K^+ channel activity increased when the tissue was exposed to hypotonic medium, but only in the presence of external Ca^{2+}. The suggestion was that Ca^{2+} influx through the stretch-activated channel occurs with cellular swelling and activates the K^+ channel thus mediating volume regulatory decrease (see "Stretch and Osmotically Activated Channels").

Our knowledge of epithelial calcium transport pales in comparison to the work done on calcium channels in excitable tissues. Further work using fluctuations analysis and patch clamp of renal epithelial tissues pretreated with parathyroid hormone, vitamin D, or calcitonin may provide more information on apical calcium conductance and renal calcium homeostasis [74]. Again the importance of cytosolic Ca^{2+} as a modulator of many epithelial transport processes underscores the application and importance of such investigation. Also evidence suggests that Ca^{2+} influx through Ca^{2+} channels may modulate cell volume regulation.

Stretch and Osmotically Activated Channels

Most cells have the ability to regulate their volume when exposed to an anisotonic environment through activation of a variety of ion transport mechanisms. Regulatory volume decrease (RVD) describes the shrinking of hypotonically swollen cells and regulatory volume increase (RVI) refers to the swelling of hypertonically shrunken cells. In RVD, cells generally experience cytosolic loss of K^+, Cl^- and water. Although several different types of ion transport mechanisms have been implicated in the final volume decrease (for a review, see [60]), stretch-activated channels often seem to play a significant role in the initiation of RVD. A one milliosmolar change in the transmembrane osmotic gradient producing a cellular volume change of approximately 1% will produce membrane tension in the range of a few millibars [311], which is sufficient pressure to activate stretch-activated channels. Thus, stretch-activated ion channels (SAC) could play an important role in volume regulation. Pressure-induced increases in membrane tension generally increase the probability of SAC opening with the increased openings usually occurring in bursts. At the same time, unit conductance and ion selectivity remain unchanged. Stretch-activated channels appear to be ubiquitous, occurring in an enormous variety of cells. Stretch-activated channels have been described in oocytes [248], red blood cells [414], snail heart muscle [336], toad smooth muscle [165], chick skeletal muscle [114], tobacco cells [62], hair cells [14], vascular endothelium [185] and neuroblastoma cells [63], fish embryos during cleavage [245], and yeast [115].

Because of the widespread distribution of stretch-activated channels, it is not surprising to find many examples of them in epithelial tissues [Table 7]. But in epithelial cells, stretch-activated channels may play an additional role beyond merely promoting cellular volume regulation in response to changes in cell size. Because of the large transported load of ions across epithelial tissues, transporting cells must respond rapidly to cell volume changes due to relative imbalances in solute influx and efflux [328]. In this section we will review the evidence for stretch and osmotically activated ion channels mediating epithelial cell volume regulation.

Many of the stretch-activated channels are nonselective cation channels, particularly those observed in the apical membranes of epithelial cells. Examples of such stretch-activated, nonselective cation channels are found in the apical membranes of *Necturus* proximal tubule and *Rana* diluting segment [65, 94, 143]. Stretch-activated, nonselective cation channels are not found exclusively in apical membranes; Hunter [143] has described a similar channel in the basolateral membrane of frog proximal tubule cells. All of these channels are insensitive to intracellular calcium and range in conductance from 18 to 29 pS. Christensen [30] has described a 30-pS channel which is activated by membrane stretch (i.e., negative pipette suction or exposure to hypotonic media) in *Necturus* choroid plexus epithelium. This channel is capable of carrying inward Ca^{2+} currents but also carries Na^+ and K^+ current equally well. In addition, a large conductance (about 200-pS), voltage-sensitive K^+ channel opens at the same time, but only in the presence of external Ca^{2+}. Both channels are normally closed when there is no mechanically or osmotically induced tension. Christensen suggests that stretch in response to hypotonic cellular swelling activates the Ca^{2+} -permeable channel and that the subsequent increase in intracellular calcium activates Ca^{2+}-activated K^+ channels, thereby promoting RVD.

The lens of the eye is a specialized epithelial tissue responsible for refracting light to form a retinal image. A low osmolarity bathing solution is required to maintain lens transparency. Such a bathing solution would result in cellular swelling without the capacity for RVD [36]. In frog lens epithelial cells, patch clamp studies of the apical membrane reveal a cation-selective, Ca^{2+}-permeable, 28-pS channel, the open probability of which increases in response to any mechanical pressure (i.e., negative or positive pipette suction). The characteristics are very similar to the stretch-activated channel described above [30]. It may be that increased intracellular pressure associated with hypotonic swelling induces calcium entry via this stretch-activated channel which activates Ca^{2+} and /or voltage-activated K^+ channels responsible for RVD. Such calcium entry may also be involved with the process of cataractogenesis. Ca^{2+}-activated K^+ channels similar to those associated with osmotically activated SACs in *Necturus* choroid plexus have also been described in lens epithelia [293].

Davis and Finn [42] have noted that apically applied amiloride decreases basolateral K^+ conductance and inhibits RVD in frog urinary bladder. Calcium ionophores restore the amiloride-inhibited RVD response. They suggest that blockage of apical Na^+ entry reduces intracellular Ca^{2+}, presumably by increasing the sodium gradient driving the basolateral Na-Ca exchanger, and that the subsequent reduction in intracellular calcium reduces basolateral K^+ conductance. However, the Ca^{2+}-permeable, stretch-activated channel in frog lens [152] and T-type calcium channels in excitable tissue [350] are both inhibited by amiloride suggesting that Ca^{2+} channels themselves may also play a modulatory role in RVD. Cytosolic Ca^{2+}, as measured by Ca^{2+}-sensitive

Table 7. Stretch and osmotically activated channels

Epithelium	Method[a]	Selectivity[b]	Conductance (pS)	Ca_i Activated[c]	Voltage Activated	Inhibition/Stimulation	References
Apical membrane							
Lens (frog, rat, mouse)	PC	NSCC	25–28	No	No	Stimulated by stretch; inhibited by amiloride	[36, 151, 293]
Choroid plexus (*Necturus*);	PC	NSCC	29–54	No	No	Stimulated by stretch	[30]
ventricular epithelium		K	200	Yes	Hyperpol	Stimulated by hypo-osmolarity	[30]
Urinary bladder (toad)	PC	Cation	30–60			Stimulated by stretch	[292]
Proximal tubule (*Necturus*)	PC	NSCC	18			Stimulated by stretch, inhibited by gadolinium	[65]
Proximal tubule cell culture	PC	K	195	Yes	Hyperpol	Stimulated by hypo-osmolarity	[287]
		anion	41		Hyperpol	Stimulated by hypo-osmolarity	[287]
Cortical collecting tubule cell culture	PC	Cl	300			Stimulated by hypo-osmolarity	[341]
Frog skin	Noise	Na				Amiloride blockable; inhibited by hyperosmolarity	[423]
Distal nephron (A6)	PC	NSCC	51			Stimulated by stretch; inhibited by gadolinium	[158]
Basolateral Membrane							
Proximal tubule (*Necturus*)	PC	K	26–47	No	Hyperpol	Stretch, hypo-osmolarity stimulates	[65, 311, 312]
	PC	K	21	No		Stimulated by stretch	[157]
Proximal tubule (*Xenopus*)	PC	K	27	No		Stimulated by stretch	[156]
Proximal tubule (frog)	PC	NSCC	29	No	Hyperpol	Stimulated by stretch	[142]
Proximal tubule (rabbit)	EMF	K				Stimulated by hypo-osmolarity; inhibited by Ba^{2+} and quinine	[165, 323, 388]
		Cl				Stimulated by hypo-osmolarity	[165, 323, 388]
Diluting segment (frog)	PC	NSCC	21	No	Hyperpol	Stimulated by stretch	[143]
Colon (turtle)	Noise	K	20			Stimulated by hypo-osmolarity; lidocaine, quinidine inhibit	[46, 93]

Small intestine (*Necturus*)	EMF	K				Stimulated by hypo-osmolarity; inhibited by Ba^{2+} and metabolic inhibitors	[187]
Urinary bladder (rabbit)	EMF	Na				Hypo-osmolarity, stretch stimulates	[194]
Urinary bladder (toad)	EMF	K				Stimulated by hypo-osmolarity; inhibited by Ba^{2+} and hyperosmolarity	[195]
Urinary bladder (frog)	EMF	K		Yes		Stimulated by hypo-osmolarity; inhibited by amiloride	[42]
Medullary TAL (mouse)	EMF	Cl				Inhibited by hyperosmolarity	[259]
Renal medullary vesicles (rabbit)	Flux	Cl				Inhibited by hyperosmolarity; valinomycin	[16]
Membrane polarity unknown							
Colon cell line (T84)	PC	Cl	75		Yes	Stimulated by cell swelling; inhibited by depolarization, DIDS	[411]
Opossum kidney cell	PC	NSCC, Cl	22			Stimulated by hypo-osmolarity; stretch	[360, 361]
culture; subconfluent		K	15	Yes	No	Stimulated by hypo-osmolarity	[360, 361]
Proximal tubule culture (rabbit)	PC	K	180			Stimulated by hypo-osmolarity, inhibited by barium	[158]

[a] PC, Patch clamp; EMF, microelectrode measurements; Noise, spontaneous or blocker-induced fluctuation analysis.
[b] NSCC, Nonselective cation channel.
[c] Ca_i activation, activation by intracellular clacium.

dye fluorescence, increases during hypotonic induced cell swelling in toad urinary bladder [29, 409], and reduction in extracellular Ca^{2+} abolished this increase in internal calcium. While these responses might be related to alternations in the rate of Na/Ca exchange, the calcium channel blockers verapamil and lanthanum inhibit the RVD in isolated S_2 proximal tubule [244, 269]. Calcium entry through gadolinium-inhibitable, stretch-activated cation channels has been demonstrated in *Necturus* proximal tubules [65] and apical membranes of A6 cells [159]. Therefore, it may be that the rate of extracellular Ca^{2+} entry (via either Ca^{2+} channels or Na-Ca exchange) controls intracellular Ca^{2+}. Increases in intracellular calcium would inhibit Na^+ influx and stimulate K^+ efflux promoting RVD [353]. Cytoplasmic calcium could directly open Ca^{2+}-activated, voltage-dependent K^+ channels; indirectly modulate them via Ca^{2+}-dependent enzymes (ie., PKC, calmodulin, proteases); or promote the fusion of ion channel containing vesicles with the epithelial cell plasma membrane [409]. Lewis and de Moura [196] have noted that insertion of membrane vessicles containing amiloride-sensitive Na^+ channels into the apical membrane of rabbit urinary bladder occurs when the tissue is mechanically, hydrostatically or hypotonically stretched. Cytochalasin B, a blocker or microfilament aggreation, disruptes the hypotonically mediated exocytosis of the vesicles and the endocytosis induced by hypertonic media. Cytochalasin B also inhibit RVD in *Necturus* gallbladder epithelium [69]. Activation of stretch-activated channels can also be blocked by application of cytochalasin, and it has been suggested that channel gating occurs when membrane stretch is applied directly to the channel protein or the surrounding plasma lipids via cytoskeletal strands [226, 310]. In support of this idea, the activity of most stretch-mediated channel insertion, stretch-activated channels rapidly decreases when cell attached patches are excised. Separation from an intracellular microfilmentous gating system or loss of some other cytoplasmic component has been proposed to explain such decreases in activity. Thus, stretch-mediated channel insertion, stretch-activated channels and regulatory volume decrease are all dependent on the cytoskeletal microfilament network.

Besides the stretch-activated channels described above, which are generally nonselective, channels which are selective for potassium have also been implicated in the control of cellular volume in epithelial tissues. Many of these stretch-activated potassium channels occur in the basolateral membranes of epithelial cells. Stretch-activated potassium channels have been observed in proximal tubule cells from *Necturus* and *Xenopus* [65, 158, 159, 311, 312]. In *Necturus*, Sackin [312] has also demonstrated that the channel activated by negative pressure in the patch pipette is also activated by osmotic swelling. Hypotonic cell swelling activates a Ba^{2+} -inhibitable K^+ conductance in the basolateral membrane of *Necturus* small intestine [188]. In addition, exposure of this epithelia to absorbable sugars and amino acids, thus raising intracellular osmolarity (and presumably volume), similarly increases this K^+ conductance. Consistent with the involvement of this K^+ channel, RVD in *Necturus* gallbladder epithelia is inhibited by serosal $BaCl_2$ and phencyclidine (a blocker of

N-methyl-*D*-apartate-activated K^+ channels), [84]. The same investigators also noted that replacement of apical and basolateral Cl^- with NO_3^- or SCN^- did not inhibit RVD. NO_3^- and SCN^- cannot be transported by the bumetanide-blockable KCl cotransporter but do pass through bumetanide-blockable anion channels. Mathematical models of volume regulation predict that such volume activated Cl^- fluxes should accompany outward K^+ conductance during RVD [345]; (see also "Chloride Channels and Cell Volume"). Using fluctuation analysis, Dawson et al. [46] have examined the effect of osmotic changes on the basolateral membrane of turtle colon in which the apical membrane has been permeabilized with amphotericin. They used the open-channel blocker lidocaine to induce measurable fluctuations of basolateral potassium channels. Swelling induced by application of mucosal KCl produced easily measurable lidocaine-induced noise compared with that observed under nonswelling conditions. The increase in fluctuations was attributed to the activation of a 20-pS potassium channel. Previous work had established that RVD activates quinidine and lidocaine-sensitive K^+ conductances in this tissue [93]. Patch clamp studies in turtle colon epithelial cells revealed a 17-pS K^+ channel which could be blocked by lidocaine, quinidine, or barium; and which is probably responsible for this K^+ conductance capable of inducing RVD [301].

Renal epithelia also exhibit regulatory volume compensation via ion conductive pathways [60]. Sackin [312, 313] has identified a 42-pS short open time, voltage-sensitive potassium channel in the basolateral membrane of *Necturus* proximal tubule. The open probability and density of these channels increases in response to either membrane stretch or hypotonic swelling. Calculations suggest that about a 1% change in cell volume would activate significant numbers of K^+ channels. Contrary to the Ca^{2+}-activated K^+ channels which we described earlier in conjunction with RVD, this channel appears to be insensitive to even large changes in intracellular Ca^{2+}. Tanner et al. [352] have shown that RVD in *Necturus* late proximal tubule is inhibited by $BaCl_2$ and SITS, suggesting RVD in this tissue is produced by coordinated activation of both a K^+ and Cl^- conductance respectively. Although the stretch-activated channel described in *Necturus* proximal tubule is Ca_i^{2+}-insensitive [311], the Ba^{2+}-inhibitable RVD in isolated rabbit proximal S_2 segment is slowed by verapamil or the removal of extracellular Ca^{2+} and restored by application of a calcium ionophore suggesting an important role for Ca_i^{2+} [269]. Whole cell patch clamp studies in rabbit proximal tubule have shown a 180-pS Ca^{2+}-activated K^+ channel and a Cl^- channel activated by cell swelling [160]. Other investigators have also demonstrated evidence for an increase in basolateral Cl^-conductance and quinine or Ba^+-inhibitable K^+ conductances in response to cellular swelling in perfused and nonperfused isolated rabbit proximal tubule [166, 322, 387]. Cultured opossum kidney (OK) cells demonstrate many proximal tubule characteristics. Ubl and associates [359, 360] describe a small-conductance Ca^{2+} and depolarization-sensitive K^+ channel (15-pS) activated by cellular swelling in subconfluent OK cells. These investigators also found a 22-pS channel activated by stretch and exposure to hypotonic media which is also activated by depolarization and is permeable to Na^+, K^+, and Cl^-. This latter channel inactivates when the patch is excised unlike the stretch- activated channel described in *Necturus* proximal tubule [311]. Gogelein and Greger [100] suggest a role in volume regulation for a similar 28-pS voltage-dependent channel found in cell-attached patches from the basolateral membrane of rabbit late proximal tubule that conducts outward Na^+, K^+, and Cl^- current. Additional studies on OK cells reveal that RVD is inhibited by the potassium channel blocker quinine and the chloride channel blocker DIDS [171]. Hyperpolarization by valinomycin inhibited RVD in OK cells, suggesting that increased Cl^- conductance might lead to voltage activation of K^+ channels similar to those previously described [359, 360]. Most models for regulatory volume decrease evoke basolateral mechanisms for KCl efflux followed by compensatory water extrusion [60]. However, patch clamp studies of a Ca^+ and voltage-activated K^+ channel (195-pS) and an anionic channel (41-pS) in the apical membrane of proximal tubule primary cultures shows that they are both activated by hypoosmotic challenge [228]. Cultured rabbit cortical collecting tubule intercalated cells have an apical, DIDS-sensitive, 300-pS Cl^- channel which is activated by hypotonic cell swelling and stretch [340]. Cultured rabbit medullary thick ascending limb cells contain an apical Ca^{2+}-activated potassium channel activated by hypoosmotic swelling [261].

Up till now we have focused on RVD since there is little evidence for direct ionic conductance involvement in the volume regulatory response to hyperosmolar conditions (RVI). In apical membrane of frog skin, Zieske and Van Driessche [422] found that hypertonicity dramatically reduces macroscopic short circuit current and the density of single Na^+ channels and magnitude of single Na^+ channel current based on amiloride induced noise analysis. The authors suggest that the ensuing cellular water loss would increase the cytosolic Na^+ and Ca^{2+} concentrations. The evidence for cytosolic Ca^{2+} inhibiting apical Na^+ permeability at a macroscopic and single-channel level is well established [217, 285, 354]. Although single-channel density and current subsequently returned to normal values, no role for augmented Na^+ channel activity with RVI was seen.

In addition to cell volume regulation, stretch activation may play an important role in membrane trafficking of ion channel proteins. As mentioned above, insertion of cytoplasmic membrane vesicles containing amiloride-blockable, Na^+-selective channels into rabbit urinary bladder apical membranes occurs during membrane stress induced by mechanical, hydrostatic, or hypotonic stretch [196]. This inserted channel becomes progressively less selective for Na^+ entry with time due to endogenous enzyme degradation and eventually dissociates from the membrane entirely. Also, in toad urinary bladder, successful apical membrane patch seals demonstrating Na^+-selective (5-pS) and-nonselective cation channels (5-pS and 17-pS) required tightly stretching the tissue preparation [78]. Post et al. [292] have obtained cell attached patch recordings of inwardly rectifying stretch-activated channels in the apical membrane of a toad urinary bladder cell line that conduct cations (30–60 pS) and is amiloride insensitive. Whether these stretch-activated Na^+-conductive channels play a direct role in cell volume regulation is doubtful since the increase in Na^+ entry would be counterproductive to RVD.

Conclusion

In this chapter, we have examined all types of epithelial ion channels. We have briefly described the methods which have allowed investigators to gather information about the properties of single ion channels and have stressed such information if it is available for a particular channel type. The message we are left with is that epithelial tissues have an enormously diverse population of channels some of which serve a clear purpose in the overall transport function of epithelial tissue and some whose purpose in a physiological context remains unclear. Possibly, as more structural information becomes available about various epithelial ion channels better correlations with function and relationships among channels will become clear.

References

1. Abramcheck FJ, Van Driessche W, Helman SI (1985) Autoregulation of apical membrane Na^+ permeability of tight epithelia. Noise analysis with amiloride and CGS 4270. J Gen Physiol 85:555–582
2. Aelvoet I, Erlij D, Van Driessche W (1988) Activation and blockage of a calcium-sensitive cation-selective pathway in the apical membrane of toad urinary bladder. J Physiol (Lond) 398:555–574
3. Al-Bazzaz F, Yadava VP, Westenfelder C (1981) Modification of Na and Cl transport in canine tracheal mucosa by prostaglandins. Am J Physiol 240:F101–F105
4. Allen JC, Mills JW (1989) Calcium flux, volume, and ions in MDCK cells treated with verapamil and dibutryl cyclic AMP. FASEB J 3:A863
5. Alvarez-Leefmans FJ, Gamino SM, Giraldez F, Gonzalez-Serratos H (1986) Intracellular free magnesium in frog skeletal muscle fibres measured with ion-selective micro-electrodes. J Physiol (Lond) 378:461–483
6. Argent BE, Arkle S, Gray MA, Greenwell JR (1987) Two types of calcium-sensitive cation channels in isolated rat pancreatic duct cells. J Physiol (Lond) 386:82P–82P
7. Armstrong CM, Swenson RP, Taylor SR (1982) Block of squid axon K channels by internally and externally applied barium ions. J Gen Physiol 80:663–682
8. Armstrong CM, Taylor SR (1980) Interaction of barium ions with potassium channels in squid giant axons. Biophys J 30:473–488
9. Ausiello DA, Sorscher EJ, Harlin C, Benos DJ (1989) Subunits of the epithelial sodium channel may be G-proteins. FASEB J 3:A228
10. Bacskai BJ, Friedman PA (1990) Parathyroid hormone activates apical membrane calcium channels in cultured distal kidney cells. Kidney Int 37:205–205
11. Barish ME (1983) A transient calcium-dependent chloride current in immature Xenopus oocyte. J Physiol (Lond) 342:309–325

12. Barrett JN, Magleby KL, Pallotta BS (1982) Properties of single calcium-activated potassium channels in cultured rat muscle. J Physiol (Lond) 331:211–230
13. Barrett KE (1989) A role for arachidonic acid in the induction of epithelial chloride secretion. FASEB J 3:A281
14. Baxendale LM (1988) Insulin increases apical sodium channel density in A6 epithelia. FASEB J 2:A748–A748
15. Baxendale LM, Helman SI (1986) A three state model for regulation of apical membrane Na^+ transport of epithelial cells. Fed Proc 45:516–516
16. Bayliss JM, Reeves WB, Andreoli TE (1989) Hypertonicity collapses Cl^- channels in rabbit renal medullary vesicles. Clin Res 37:485A
17. Benos DJ, Saccomani G, Sariban-Sohraby S (1987) The epithelial sodium channel. Subunit number and location of the amiloride binding site. J Biol Chem 262:10613–10618
18. Bertorello A, Aperia A (1989) Regulation of Na^+-K^+-ATPase activity in kidney proximal tubules: involvement of GTP binding proteins. Am J Physiol 256:F57–F62
19. Bolivar JJ, Cereijido M (1987) Voltage and Ca^{2+} -activated K^+ channel in cultured epithelial cells (MDCK). J Membr Biol 97:43–51
20. Bonvalet JP, Pradelles P, Farman N (1987) Segmental synthesis and actions of prostaglandins along the nephron. Am J Physiol 253:F377–F387
21. Brand K, Spokes K, Silva P, Epstein FH (1990) Atrial natiuretic peptide activates phospholipase C (PLC) in rat papilla. Kidney Int 37:336
22. Bridges RJ, Benos DJ (1990) Reconstitution of epithelial ion channels In: Helman SI, Van Driessche W (eds) Current topics in membranes and transport: channels and noise in epithelial tissues. Academic, New York. pp 283–312
23. Brown AM, Yatani A, Imoto Y, Kirsch G, Hamm H, Codina J, Mattera R, Birnbaumer L (1988) Direct coupling of G proteins to ionic channels. Cold Spring Harbor Symp Quant Biol 53:365–374
24. Brown PD, Loo DD, Wright EM (1988) Ca^{2+}-activated K^+ channels in the apical membrane of Necturus choroid plexus. J Membr Biol 105:207–219
25. Cantiello HF, Patenaude CR, Ausiello DA (1989) G protein subunit α_{i-3} activates a pertussis toxin-sensitive Na^+ channel from the epithelial cell line, A6. J Biol Chem 264:20867–20870
26. Cantiello HF, Patenaude CR, Ausiello DA (1990) G-protein activation of an epithelial sodium channel is mediated via phospholipid metabolites. Kidney Int 37:213
27. Catterall WA (1988) Structure and function of voltage-sensitive ion channels. Science 242:50–61
28. Chang D, Dawson DC (1988) Digitonin-permeabilized colonic cell layers. Demonstration of calcium-activated basolateral K^+ and Cl^- conductances. J Gen Physiol 92:281–306
29. Chase H, Wong S (1989) Exposure to hypotonic media results in a large increase in intracellular free [Ca] in epithelial cells. FASEB J 3:A983–A983
30. Christensen O (1987) Mediation of cell volume regulation by Ca^{2+} influx through stretch-activated channels. Nature 330:66–68
31. Christensen O, Zeuthen T (1987) Maxi K^+ channels in leaky epithelia are regulated by intracellular Ca^{2+}, pH and membrane potential. Pflugers Arch 408:249–259
32. Christine CW, Laskowski FH, Gitter AH, Gross P, Fromter E (1987) Chloride-selective single ion channels in the apical membrane of cultured collecting duct principal cells. Pflugers Arch 408:R32
33. Chu LLH, Edelman IS (1972) Cordycepin and alpha-amanitin: inhibitors of transcription as probes of aldosterone action. J Membr Biol 10:291–310
34. Cliff WH, Frizzell RA (1989) Adenosine increases Cl^- conductance in T84 epithelial cells. FASEB J 3:A1148
35. Colquhoun D, Hawkes AG (1983) The principles of the stochastic interpretation of ion channel mechanisms. In: Sakmann B Neher E (eds) Single channel recording. Plenum, New York, pp 135–175
36. Cooper KE, Tang JM, Rae JL, Eisenberg RS (1986) A cation channel in frog lens epithelia responsive to pressure and calcium. J Membr Biol 93:259–269
37. Cornejo M, Guggino SE, Guggino WB (1987) Modification of Ca^{2+}-activated K^+ channels in cultured medullary thick ascending limb cells by N-bromoacetamide. J Membr Biol 99:147–155
38. Cornejo M, Guggino SE, Guggino WB (1989) Ca^{2+}-activated K^+ channels from cultured renal medullary thick ascending limb cells: effects of pH. J Membr Biol 110:49–55
39. Cornejo M, Guggino SE, Sastre, A, Guggino WB (1989) Isomeric yohimbine alkaloids block calcium-activated K^+ channels in medullary thick ascending limb cells of rabbit kidney. J Membr Biol 107:25–33

40. Cox T (1989) Single channel records from dissociated frog tadpole skin cells. FASEB J 3:A861
41. Das S, Palmer LG (1989) Extracellular Ca controls outward rectification by apical cation channels in toad urinary bladder: patch-clamp and whole-bladder studies. J Membr Biol 107:157–168
42. Davis CW, Finn AL (1985) Cell volume regulation in frog urinary bladder. Fed Proc 44:2520–2525
43. Dawson DC (1987) Properties of epithelial potassium channels. Curr Top Memb Transp 28:41–71
44. Dawson DC (1987) Cellular mechanisms for K transport across epithelial cell layers. Semin Nephrol 7:185–192
45. Dawson DC, Richards NW (1990) Basolateral K conductance: role in regulation of NaCl absorption and secretion. Am J Physiol 259:C181–C195
46. Dawson DC, Van Driessche W, Helman SI (1988) Osmotically basolateral induced K^+ conductance in turtle colon: lidocaine-induced K^+ channel noise. Am J Physiol 254:C165–C174
47. Dawson DC, Wilkinson DJ, Richards NW (1990) Basolateral K channel noise: signals from the dark side. In Curr Top Membr Transp Channels and noise in epithelial tissues, Helman SI, Van Oriessche W (eds), Academic, New York, pp 191–212
48. de Sousa RC, Grosso A (1981) The mode of action of vasopressin: membrane microstructure and biological transport. J Physiol (Paris) 77:643–669
49. Demarest JR, Loo DD, Sachs G (1989) Activation of apical chloride channels in the gastric oxyntic cell. Science 245:402–404
50. Desir GV (1989) Reconstitution and partial purification of an amiloride-sensitive non-selective cation channel from the rabbit kidney. Clin Res 37:488A
51. Dillingham MA, Dixon BS, Anderson RJ (1987) Calcium modulates vasopressin effect in rabbit cortical collecting tubule. Am J Physiol 252:F115–F121
52. Eaton DC, Brodwick MS (1980) Effects of barium on the potassium conductance of squid axon. J Gen Physiol 75:727–750
53. Eaton DC, Hamilton KL (1988) The amiloride-blockable sodium channel of epithelial Tissue. In: Narahashi T (ed) Ion channels. Plenum, New York, pp 251–282
54. Eaton DC, Ling BN (1989) The effect of cholera toxin, pertussis toxin, and GTP-S on highly selective sodium channels in A6 cells. Kidney Int 37:373a
55. Eaton DC, Marunaka Y (1988) Comparison between the effects of amiloride and CDPC, an amiloride analogue, on amiloride-sensitive sodium channels from cultured renal cells. FASEB J 2:A750
56. Eaton DC, Marunaka Y (1990) Ion channel fluctuations: "noise" and single channel measurements. In: Helman SI, Van Driessche W (eds) Current topics in membranes and transport: channels, noise and impedance in epithelia. Academic, New York pp 61–114
57. Eaton DC, Marunaka Y (1990) Effects of cyclic AMP and cholera toxin on an amiloride-blockable sodium channel in renal cells. FASEB J 4:A549
58. Evans MG, Marty A (1986) Potentiation of muscarinic and alphaadrenergic responses by an analogue of guanosine 5-triphosphate. Proc Natl Acad Sci USA 83:4099–4103
59. Evans MG, Marty A (1986) Calcium-dependent chloride currents in isolated cells from rat lacrimal glands. J Physiol (Lond) 378:437–460
60. Eveloff JL, Warnock DG (1987) Activation of ion transport systems during cell volume regulation. Am J Physiol 252:F1–F10
61. Ewald DA, Williams A, Levitan IB (1985) Modulation of single Ca-dependent K-channel activity by protein phosphorylation. Nature 315:503–506
62. Falke L, Edwards KL, Pickard BG, Misler S (1987) A stretch-activated anion channel in cultured tobacco cells. Biophys J 51:251a
63. Falke LC, Misler S (1989) Activity of ion channels during volume regulation by clonal N1E115 neuroblastoma cells. Proc Natl Acad Sci USA 86:3919–3923
64. Fichtner H, Frobe U, Busse R, Kohlhardt M (1987) Single non-selective cation channels and Ca^{2+} -activated K^+ channels in aortic endothelial cells. J Membr Biol 98:125–133
65. Filipovic D, Sackin H (1990) Stretch-activated calcium currents in proximal tuble. FASEB J 4:A447
66. Findlay I (1984) A patch-clamp study of potassium channels and whole-cell currents in acinar cells of the mouse lacrimal gland. J Physiol (Lond) 350:179–195
67. Findlay I, Petersen OH (1983) Acetylcholine stimulates a Ca-dependent Cl^- conductance in mouse lacrimal acinar cells. Pflugers Arch 403:328–330
68. Fishman HM, Lenchtag HR (1990) Electrical noise in physics and biology In: Helman SI, Van

Driessche W (eds) Current topics in membranes and transport: channels, noise and impedance in epithelia. Academic, New York pp 4–35
69. Foskett JK, Spring KR (1985) Involvement of calcium and cytoskeleton in gallbladder epithelial cell volume regulation. Am J Physiol 248:C27–C36
70. Fox JA, Pfeffer BA, Fain GL (1988) Single-channel recordings from cultured human retinal pigment epithelial cells. J Gen Physiol 91:193–222
71. Frace AM, Eaton DC (1987) Chemical modification of amino groups on calcium activated K channels of GH3 cells. Biophys J 51:55a
72. Frehland E (1979) Theory of transport noise in membrane channels with open-closed kinetics. Biophys Struct Mech 5:91–106
73. Frehland E, Hoshiko T, Machlup S (1983) Competitive blocking of apical sodium channels in epithelia. Biophys Acta 732: 636–646
74. Friedman PA (1988) Renal calcium transport: sites and insights. News Physiol Sci 3:17–21
75. Frindt G, Palmer LG (1987) Ca-activated K channels in apical membrane of mammalian CCT, and their role in K secretion. Am J Physiol 252:F458–F467
76. Frindt G, Palmer LG (1989) Low-conductance K channels in apical membrane of rat cortical collecting tubule. Am J Physiol 256:F143–F151
77. Frindt G, Sackin H, Palmer LG (1990) Whole-cell currents in rat cortical collecting tubule: low-Na diet increases amiloride-sensitive conductance. Am J Physiol 258:F562–F567
78. Frings S, Purves RD, MacKnight ADC (1988) Single-channel recordings from the apical membrane of the toad urinary bladder epithelial cell. J Membr Biol 106:157–172
79. Frizzell RA, Halm DR (1990) Chloride channels in epithelial cells, In: Helman SI, Van Driessche W (eds) Current topics in membranes and transport: channels, noise, and impedance in epithelia. Academic, New York pp 248–282
80. Frizzell RA, Halm DR, Rechkemmer G, Shoemaker RL (1986) Chloride channel regulation in secretory epithelia. Fed Proc 45:2727–2731
81. Frizzell RA, Rechkemmer G, Shoemaker RL (1986) Altered regulation of airway eptihelial cell chloride channels in cystic fibrosis. Science 233:558–560
82. Frizzell RA, Schoumacher RA, Halm DR (1987) Chloride channel regulation in cystic fibrosis epithelia. Prog Clin Biol Res 254:101–113
83. Fuchs W, Larsen EH, Lindemann B (1977) Current-voltage curve of sodium channels and concentration dependence of sodium permeability in frog skin. J Physiol (Lond) 267:137–166
84. Furlong TJ, Spring KR (1989) Mechanisms underlying volume regulatory decrease by Necturus gallbladder epithelium. Kidney Int 35:155
85. Furuya K, Enomoto K, Furuya S, Yamagishi S, Edwards C, Oka T (1989) Single calcium-activated potassium channel in cultured mammary epthelial cells. Pflugers Arch 414:118–124
86. Gallacher DV, Maruyama Y, Petersen OH (1984) Patch-clamp study of rubidium and potassium conductances in single cation channels from mammalian exocrine acini. Plfugers Arch 401:361–367
87. Gallacher DV, Morris AP (1986) A patch-clamp study of potassium currents in resting and acetylcholine-stimulated mouse submandibular acinar cells. J Physiol (Lond) 373:379–395
88. Garty H (1986) Mechanisms of aldosterone action in tight epithelia. J Membr Biol 90:193–205
89. Garty H, Benos DJ (1988) Characteristics and regulatory mechanisms of the amiloride-blockable Na^+ channel. Physiol Rev 68:309–373
90. Garty H, Edelman IS (1983) Amiloride-sensitive trypsinization of apical sodium channels: analysis of hormonal regulation of sodium transport in toad bladder. J Gen Physiol 81:785–803
91. Garty H, Yeger O, Yanovsky A, Asher C (1988) Guanosine nucleotide-dependent activation of the amiloride-blockable Na^+ channel. Am J Physiol 256:F965–F969
92. George AL Jr, Staub O, Geering K, Rossier BC, Kleyman TR, Kraehenbuhl J-P (1989) Functional expression of the amiloride-sensitive sodium channel in *Xenopus* oocytes. Proc Natl Acad Sci USA 86:7295–7298
93. Germann WJ, Ernst SA, Dawson DC (1986) Resting and osmotically induced basolateral K conductances in turtle colon. J Gen Physiol 88:253–274
94. Giebisch G, Hunter M, Kawahara K (1990) Apical potassium channels in Amphiuma diluting segment: effect of barium. J Physiol (Lond) 420:313–323
95. Giraldez F, Sepulveda FV, Sheppard DN (1988) A chloride conductance activated by adenosine 3′,5′ -cyclic monophosphate in the apical membrane of necturus enterocytes. J Physiol (Lond) 395:597–623

96. Gitter AH, Beyenbach K, Christie WC, Gross P, Minuth WW, Fromter E (1987) High-conductance K^+ channel in apical membranes of principal cells cultured from rabbit renal cortical collecting duct anlagen. Pflugers Arch 408:282–290
97. Gogelein H (1990) Ion channels in mammalian proximal renal tubules. Renal Physiol Biochem 13:8–25
98. Gogelein H, Greger R (1984) Single channel recordings from basolateral and apical membranes of renal proximal tubules. Pflugers Arch 401:424–426
99. Gogelein H, Greger R (1986) Na^+ selective channels in the apical membrane of rabbit late proximal tubules (pars recta). Pflugers Arch 406:198–203
100. Gogelein H, Greger R (1986) A voltage-dependent ionic channel in the basolateral membrane of late proximal tubules of the rabbit kidney. Pflugers Arch 407 [Suppl 2]:S142–S148
101. Gogelein H, Greger R (1987) Properties of single K^+ channels in the basolateral membrane of rabbit proximal straight tubules. Pflugers Arch 410:288–295
102. Gogelein H, Pfannmuller B (1989) The nonselective cation channel in the basolateral membrane of rat exocrine pancreas (inhibition by 3′,5-dichlorodiphenylamine-2-carboxylic acid (DCDPC) and activation by stilbene disulfonates). Pflugers Arch 413:287–298
103. Gogelein H, Schlatter E, Greger R (1987) The small conductance chloride channel in the luminal membrane of the rectal gland of the dogfish (Squalus acanthias). Pflugers Arch 409:122–125
104. Gogelein H, Van Driessche W (1981) Noise analysis of the K^+ current through the apical membrane of Necturus gallbladder. J Membr Biol 63:243–254
105. Gogelein H, Van Driessche W (1981) Capacitive and inductive low frequency impedances of Necturus gallbladder epithelium. Pflugers Arch 389:105–113
106. Gogelein H, Van Driessche W (1981) The effect of electrical gradients on current fluctuations and impedance recorded from Necturus gallbladder. J Membr Biol 60:199–209
107. Greenwald JE, Needleman P, Wilkins M, Schreiner GF (1990) Renal synthesis of atriopeptin in physiology and pathophysiology. Kidney Int 37:339
108. Greger R, Gogelein H (1987) Role of K^+ conductive pathways in the nephron. Kidney Int 31:1055–1064
109. Greger R, Oberleithner H, Schlatter E, Cassola AC, Weidtke C (1983) Chloride activity in cells of isolated perfused cortical thick ascending limbs of rabbit kidney. Pflugers Arch 399:29–34
110. Greger R, Schlatter E, Gogelein H (1985) Cl^- channels in the apical cell membrane of the rectal gland "induced" by cAMP. Pflugers Arch 403:446–448
111. Greger R, Schlatter E, Gogelein H (1987) Chloride channels in the luminal membrane of the rectal gland of the dogfish (Squalus acanthias): properties of the "large" conductance channel. Pflugers Arch 409:114–121
112. Guggino SE, Guggino WB, Green N, Sacktor B (1987) Ca^{2+}-activated KT^+ T channels in cultured medullary thick ascending limb cells. Am J Physiol 252:C121–C127
113. Guggino SE, Suarez-Isla BA, Guggino WB, Sacktor B (1985) Forskolin and antidiuretic hormone stimulate a Ca-activated K channel in cultured kidney cells. Am J Physiol 249:F448–F455
114. Guharay F, Sachs E (1984) Stretch-activated single ion channel currents in tissue-cultured embryonic chick skeletal muscle. J Physiol (Lond) 352:685–701
115. Gustin MC, Zhou XL MB, Culbertson MR, Kung C (1987) Stretch-activated cation channel in yeast. Biophys J 51:251a
116. Hagiwara S, Fukuda J, Eaton DC (1974) Membrane currents carried by Ca, Sr, and Ba in barnacle muscle fiber during voltage clamp. J Gen Physiol 63:565–578
117. Halm DR, Dawson DC (1984) Control of potassium transport by turtle colon: role of membrane potential. Am J Physiol 247:C26–C32
118. Halm DR, Rechkemmer GR, Schoumacher RA, Frizzell RA (1988) Apical membrane chloride channels in a colonic cell line activated by secretory agonists. Am J Physiol 254:C505–C511
119. Hamilton KL, Benos DJ (1988) A non-selective cation channel identified from the apical membrane of A6 kidney cells. FASEB J 2:A749
120. Hamilton KL, Eaton DC (1985) Single-channel recordings from amiloride-sensitive epithelial sodium channel. Am J Physiol 249:C200–C207
121. Hamilton KL, Eaton DC (1986) Single-channel recordings from two types of amiloride-sensitive epithelial Na^+ channels. Membr Biochem 6:149–171
122. Hamilton KL, Eaton DC (1986) Cyclic AMP-induced potassium channel activity in the apical membrane of A6 kidney cells. Fed Proc 450:516

123. Hamm LL, Simon EE (1987) Roles and mechanisms of urinary buffer excretion. Am J Physiol 253:F595–F605
124. Hanrahan JW, Alles WP, Lewis SA (1984) Basolateral anion and K^+ channels from rabbit urinary bladder epithelium. J Gen Physiol 84:30a
125. Hanrahan JW, Alles WP, Lewis SA (1985) Single anion-selective channels in basolateral membrane of a mammalian tight epithelium. Proc Natl Acad Sci USA 82:7791–7795
126. Hanrahan JW, Wills NK, Phillips JE, Lewis SA (1986) Basolateral K channels in an insect epithelium. Channel de conductance, and block by barium. J Gen Physiol 87:443–466
127. Harris HW Jr, Wade JB, Handler JS (1986) Fluorescent markers to study membrane retrieval in antidiuretic hormone-treated toad urinary bladder. Am J Physiol 251:C274–C278
128. Hays SR, Baum M, Kokko JP (1987) Effects of protein kinase C activation on sodium, potassium, chloride, and total CO_2 transport in the rabbit cortical collecting tubule. J Clin Invest 80:1561–1570
129. Hayslett JP, Gogelein H, Kunzelman K, Greger R (1987) Characteristics of apical chloride channels in human colon cells (HT29). Pflugers Arch 410:487–494
130. Hebert RL, Fredin D, Jacobson HR (1990) PGE_2 inhibits sodium transport in the rabbit cortical collecting duct by increasing intracellular calcium. Clin Res 38:3
131. Helman SI, Cox TC, Van Driessche W (1983) Hormonal control of apical membrane Na transport in epithelia Studies with fluctuation anlysis. J Gen Physiol 82:201–220
132. Helman SI, Kizer NL (1990) Apical sodium ion channels of tight epithelia as viewed from the perspective of noise analysis. In: Helman SI, Van Driessche W (eds) Current topics in membranes and transport: channels and noise in epithelial tissues. Academic, New York 117–155
133. Helman SI, Koeppen BM, Beyenbach KW, Baxendale LM (1985) Patch clamp studies of apical membranes of renal cortical collecting ducts. Pflugers Arch 405:71–76.
134. Higashida H, Brown DA (1988) Ca^{2+}-dependent k^+ channels in neuroblastoma hybrid cells activated by intracellular inositol trisphosphate and extracellular bradykinin. FEBS Lett 238:395–400
135. Hille B (1984) *Ionic channels of excitable membranes.* Sinauer, Sunderland, pp1–426
136. Hillyard SD, Van Driessche W (1989) Effects of amiloride on the poorly selective cation channel of larval bullfrog skin. Am J Physiol 256:C168–C174
137. Hillyard SD, Zeiske W, Van Driessche W (1982) Poorly selective cation channels in the skin of the larval frog (stage less than or equal to XIX). Pflugers Arch 394:287–293
138. Hinton CF, Eaton DC (1989) Expression of amiloride-blockable sodium channels in Xenopus oocytes. Am J Physiol 257:C825–C829
139. Hosey MM, Lazdunski M (1988) Calcium channels: Molecular pharmacology, structure and regulation. J Membr Biol 104:81–105
140. Hoshiko T, Grossman RA, Machlup S (1988) Effects of basolateral ouabain, amphotericin B, cyanide and potassium on amiloride noise during voltage clamp of Rana pipiens skin support sodium-amiloride competition. Biochim Biophys Acta 942:186–198
141. Hudspeth AJ (1985) The cellular basis of hearing: the biophysics of hair cells. Science 230:745–752
142. Humes HD, Simmons CF, Brenner BM (1980) Effect of verapamil on the hydroosmotic response to antidiuretic hormone in toad urinary bladder. Am J Physiol 239:F250–F257
143. Hunter M (1988) Pressure-induced activation of channels in the basolateral membrane of frog proximal tubule cells. J Physiol 403:24P
144. Hunter M (1989) Stretch-activated channels in single cells of the frog diluting segment. J Physiol 416:30P
145. Hunter M, Giebisch G (1988) Calcium-activated K-channels of Amphiuma early distal tubule: inhibition by ATP. Pflugers Arch 412:331–333
146. Hunter M, Kawahara K, Giebisch G (1986) Potassium channels along the nephron. Fed Proc 45:2723–2726
147. Hunter M, Kawahara K, Giebisch G (1988) Calcium-activated epithelial potassium channels. Miner Electrolyte Metab 14:48–57
148. Hunter M, Lopes AG, Bouilpaep EL, Giebisch GH (1984) Single channel recordings of calcium-activated potassium channels in the apical membrane of rabbit cortical collecting tubules. Proc Natl Acad Sci USA 81:4237–4239
149. Hunter M, Lopes AG, Boulpaep E, Giebisch G (1986) Regulation of single potassium ion channels from apical membrane of rabbit collecting tubule. Am J Physiol 251:F725–F734

150. Hwang T-C, Lu L, Zeitlin PL, Gruenert DC, Huganir R, Guggino WB (1989) Cl^- channels in CF: lack of activation by protein kinase C and cAMP -dependent protein kinase. Science 244:1351–1353
151. Jacob TJC (1983) Spontaneous single channel events in Rana pipiens lens epithelial cells. J Physiol 343:99–100
152. Jacob TJC, Bangham JA, Duncan G (1985) Characterization of a cation channel on the apical surface of the frog lens epithelium. Q J Exp Physiol 70:403–421
153. Jetten AM, Yankaskas JR, Stutts MJ, Willumsen NJ, Boucher RC (1989) Persistence of abnormal chloride conductance regulation in transformed cystic fibrosis epithelia. Science 244:1472–1475
154. Josephson IR, Brown AM (1986) Inwardly rectifying single-channel and whole cell K^+ currents in rat ventricular myocytes. J Membr Biol 94:19–35
155. Katz U (1978) Changes in ionic conductances and in sensitivity to amiloride during the natural moulting cycle of toad skin (Bufo viridis, L). J Membr Biol 38:1–9
156. Kawahara K (1985) Ba^{2+}-sensitive potassium permeability of the apical membrane in newt kidney proximal tubule. J Membr Biol 88:283–292
157. Kawahara K (1990) A stretch-activated K^+ channel in the basolateral membrane of Xenopus kidney proximal tubule cells. Pflugers Arch 415:624–629
158. Kawahara K, Hunter M, Giebisch G (1987) Potassium channels in Necturus proximal tubule. Am J Physiol 253:F488–F494
159. Kawahara K, Matsuzaki K, Giebisch GH (1990) Stretch-activated ion channels in the apical membrane of A6 cells. J Am Soc Nephvol 1:686
160. Kawahara K, Ogawa A, Suzuki M (1990) Regulatory volume decrease in cultured rabbit proximal tubule cells: evidence for activation of Ca-dependent potassium currents. Kidney Int 37:564
161. Kemendy A, Eaton DC (1990) Aldosterone-induced sodium transport in A6 epithelia is blocked by 3 deazaadenosine, a methylation blocker. FASEB J 4:A445
162. Kemendy AE, Eaton DC (1989) Aldosterone affects apical sodium channel density and open probability in A6 epithelia. FASEB J 3:A861
163. Kim D, Clapham DE (1989) Potassium channels in cardiac cells activated by arachidonic acid and phospholipids. Science 244:1174–1176
164. Kim D, Lewis DL, Graziadei L, Neer EJ, Bar-Sagi D, Clapham DE (1989) G-protein β,γ-subunits activate the cardiac muscarinic K-channel via phospholipase A2. Nature 337:557–560
165. Kirber MT, Walsh JV Jr, Singer JJ (1988) Possible role of ion channels in stretch-induced contraction of smooth muscle. Biophys J 53:411a
166. Kirk KL, DiBona DR, Schafer JA (1987) Regulatory volume decrease in perfused proximal nephron: evidence for a dumping of cell K^+ . Am J Physiol 252:F933–F942
167. Kishimoto A, Mikawa K, Hashimoto K, Yasuda I, Tanaka S, Tominaga M, Kuroda T, Nishizuka Y (1989) Limited proteolysis of protein kinase C subspecies by calcium-dependent neutral protease (calpain). J Biol Chem 264:4088–4092
168. Kleyman TR, Cragoe EJ Jr (1988) The mechanism of action of amiloride. Semin Nephrol 8:242–248
169. Kleyman TR, Cragoe EJ Jr (1988) Amiloride and its analogs as tools in the study of ion transport. J Membr Biol 105:1–21
170. Kleyman TR, Cragoe EJ Jr, Kraehenbuhl J-P (1989) The cellular pool of Na^+ channels in the amphibian cell line A6 is not altered by mineralocorticoids. Analysis using a new photoactive amiloride analog in combination with anti-amiloride antibodies. J Biol Chem 264:11 995–12 000
171. Knoblauch C, Montrose MH, Murer H (1989) Regulatory volume decrease by cultured renal cells. Am J Physiol 256:C252–C259
172. Koeppen BM (1987) Electrophysiological identification of principal and intercalatex cells in the rabbit outer medullary collecting duct. Pflugers Arch 409:138–141
173. Koeppen BM, Giebisch G, Biagi BA (1983) Electrophysiology of mammalian renal tubules: inferences from intracellular microelectrode studies. Annu Rev Physiol 45:497–517
174. Kolb HA, Brown CD, Murer H (1985) Identification of a voltage-dependent anion channel in the apical membrane of a Cl-secretory epithelium (MDCK). Pflugers Arch 403:262–265
175. Kolb HA, Brown CD, Murer H (1986) Characterization of a Ca-dependent maxi K channel in the apical membrane of a cultured renal epithelium (JTC-12.P3). J Membr Biol 92:207–215

176. Krouse ME, Hagiwara G, Chen J, Lewiston NJ, Wine JJ (1989) Ion channels in normal human and cystic fibrosis sweat gland cells. Am J Physiol 257:C129–C140
177. Krouse ME, Schneider GT, Gage PW (1986) A large anion-selective channel has seven conductance levels. Nature 319:58–60
178. Kukuljan M, Araya N, Caviedes R, Stutzin A (1989) Characterization of a calcium-activated nonselctive cationic channel in a renal cell line. Biophys J 55:493a–493a
179. Kunzelmann K, Pavenstadt H, Beck C, Unal O, Emmrich P, Arndt HJ, Greger R (1989) Characterization of potassium channels in respiratory cells. I General properties. Pflugers Arch 414:291–296
180. Kunzelmann K, Pavenstadt H, Greger R (1989) Characterization of potassium channels in respiratory cells. II. Inhibitors and regulation. Pflugers Arch 414:297–303
181. Lahav M, Dietz T, Edelman IS (1973) The action of aldosterone on sodium transport: further studies with inhibitors of RNA and protein synthesis. Endocrinology 92:1685–1699
182. Landry DW, Reitman M, Cragoe EJ Jr, Al Awqati Q (1987) Epithelial chloride channel. Development of inhibitory ligands. J Gen Physiol 90:779–798
183. Lang DG, Ritchie AK (1987) Large and small conductance calcium-activated potassium channels in the GH3 anterior pituitary cell line. Pflugers Arch 410:614–622
184. Lang F, Defregger M, Paulmichl M (1986) Apparent chloride conductance of subconfluent Madin Darby canine kidney cells. Pflugers Arch 407:158–162
185. Lansman JB, Hallam TJ, Rink TJ (1987) Single stretch-activated ion channels in vascular endothelial cells as mechanotransducers? Nature 235:811–813
186. Laskowski FH, Christine CW, Gitter AH, Beyenbach KW, Gross P, Fromter E (1990) Cation channels in the apical membrane of collecting duct principal cell epithelium in culture. Renal Physiol Biochem 13:70–81
187. Latorre R, Miller C (1983) Conduction and selectivity in potassium channels. J Membr Biol 71:11–30
188. Lau KR, Hudson RL, Schultz SG (1984) Cell swelling increases a barium-inhibitable potassium conductance in the basolateral membrane of Necturus small intestine. Proc Natl Acad Sci USA 81:3591–3594
189. Lauger P (1978) Transport noise in membranes: current and voltage fluctuations at equilibrium. Biochim Biophys Acta 507:337–349
190. Lester DS, Asher C, Garty H (1988) Characterization of cAMP-induced activation of epithelial sodium channels. Am J Physiol 254:C802–C808
191. Levitan IB (1985) Phosphorylation of ion channels. J Membr Biol 87:177–190
192. Levitan IB, De, Peyer JE, Cachelin AB, Reuter H (1982) Ca^{2+}-activated K^+ conductance in internally perfused snail neurons is enhanced by protein phosphorylation. Proc Natl Acad Sci USA 79:4207–4211
193. Lewis SA (1983) Conttrol of Na^+ and water absorption across vertebrate "tight epithelia by adh and aldosterone. J Exp Biol 106:9–24
194. Lewis SA, Alles WP (1986) Urinary kallikrein: a physiological regulator of epithelial Na^+ absorption. Proc Natl Acad Sci USA 83:5345–5348
195. Lewis SA, Butt AG, Bowler MJ, Leader JP, MacKnight ADC (1985) Effects of anions on cellular volume and transepithelial Na transport across toad urinary bladder. J Membr Biol 83:119–137
196. Lewis SA, deMoura JLC (1982) Incorporation of cytoplasmic vesicles into apical membrane of mammalian urinary bladder epithelium. Nature 297:685–688
197. Lewis SA, Hanrahan JW (1985) Apical and basolateral membrane ionic channels in rabbit urinary bladder epithelium. Pflugers Arch 405:S83–S88
198. Lewis SA, Ifshin MS, Loo DD, Diamond JM (1984) Studies of sodium channels in rabbit urinary bladder by noise analysis. J Membr Biol 80:135–151
199. Lewis SA, Wills NK (1982) Electrical properties of the rabbit urinary bladder assessed using gramicidin D. J Membr Biol 67:45–53
200. Lewis SA, Wills NK (1983) Apical membrane permeability and kinetic properties of the sodium pump in rabbit urinary bladder. J Physiol (Lond) 341:169–184
201. Lewis SA, Wills NK, Eaton DC (1978) Basolateral membrane potential of a tight epithelium: ionic diffusion and electrogenic pumps. J Membr Biol 41:117–148
202. Li JH, Cragoe EJ Jr, Lindemann B (1985) Structure-activity relationship of amiloride analogs as blockers of epithelial Na channels. I. Pyrazine-ring modifications. J Membr Biol 83:45–56
203. Li JH, Cragoe EJ Jr, Lindemann B (1987) Structure-activity relationship of amiloride analogs as blockers of epithelial Na channels. II. Side-chain modifications. J Membr Biol 95:171–185

204. Li JH, Lindemann B (1983) Competitive blocking of epithelial sodium channels by organic cations: the relationship between macroscopic and microscopic inhibition constants. J Membr Biol 76:235–251
205. Li JH, Lindemann B (1983) Chemical stimulation of Na transport through amiloride-blockable channels of frog skin epithelium. J Membr Biol 75:179–192
206. Li M, McCann JD, Anderson MP, Clancy JP, Liedtke CM, Nairn AC, Greengard P, Welsh MJ (1989) Regulation of chloride channels by protein kinase C in normal and cystic fibrosis airway epithelia. Science 244:1353–1356
207. Li M, McCann JD, Liedtke CM, Nairn AC, Greengard P, Welsh MJ (1988) Cyclic AMP-dependent protein kinase opens chloride channels in normal but not cystic fibrosis airway epithelium. Nature 331:358–360
208. Light D, Ausiello D, Stanton BA (1989) G-protein regulation of a cation channel in renal epithelial cells. Biophys J 55:604a
209. Light DB, Ausiello DA, Stanton BA (1989) Guanine nucleotide-binding protein, $\alpha^{*}_{i\text{-}3}$ directly activates a cation channel in rat renal inner medullary collecting duct cells. J Clin Invest 84:352–356
210. Light DB, Corbin JD, Stanton BA (1990) Dual ion-channel regulation by cyclic GMP and cyclic GMP-dependent protein kinase. Nature 344:336–339
211. Light DB, McCann FV, Keller TM, Stanton BA (1988) Amiloride-sensitive cation channel in apical membrane of inner medullary collecting duct. Am J Physiol 255:F278–F286
212. Light DB, Schwiebert EM, Karlson KH, Stanton BA (1989) Atrial natriuretic peptide inhibits a cation channel in renal inner medullary collecting duct cells. Science 243:383–386
213. Lindau M, Fernandez JM (1986) A patch-clamp study of histamine-secreting cells. J Gen Physiol 88:349–368
214. Lindemann B (1980) The beginning of fluctuation analysis of epithelial ion transport. J Membr Biol 54:1–11
215. Lindemann B (1984) Fluctuation analysis of sodium channels in epithelia. Annu Rev Physiol 46:497–515
216. Lindemann B, Van Driessche W (1977) Sodium-specific membrane channels of frog skin are pores: current fluctuations reveal high turnover. Science 195:292–294
217. Ling BN, Eaton DC (1989) Effects of luminal Na^+ on single Na^+ channels in A6 cells, a regulatory role for protein kinase C. Am J Physiol 256:F1094–F1103
218. Ling BN, Hinton CF, Eaton DC (1991) Amiloride-sensitive sodium channels in rabbit cortical collecting tubule primary cultures. AM J Physiol 261:F933–F944
219. Ling BN, Eaton DC (1990) Arachidonic acid modulates apical potassium channels in A6 cells. Clin Res 38:468A
220. Ling BN, Kemendy AE, Kokko KE, Hinton CF, Eaton DC (1990) Regulation of the amiloride-blockable sodium channel from epithelial tissue. Mol Cell Biochem 99:141–150
221. Lipton SA (1986) Antibody activates cationic channels via second messenger Ca^{2+}. Biochim Biophys Acta 856:59–67
222. Loo DD, Kaunitz JD (1989) Ca^{2+} and cAMP activate K^+ channels in the basolateral membrane of crypt cells isolated from rabbit distal colon. J Membr Biol 110:19–28
223. Loo DD, Lewis SA, Ifshin MS, Diamond JM (1983) Turnover, membrane insertion, and degradation of sodium channels in rabbit urinary bladder. Science 221:1288–1290
224. Loo DDF, Diamond JM (1985) Crystal accumulation and very high short-circuit currents in rabbit urinary bladder. Am J Physiol 248:F70–F77
225. Loo DDF, Kaunitz JD (1989) Ca^{2+} and cAMP activate K^+ channels in the basolateral membrane of crypt cells isolated from rabbit distal colon. J Membr Biol 110:19–28
226. MacGregor RD, Sachs F, Hunt CA (1989) Mechanisms for opening stretch-activated ion channels. Biophys J 55:245a–245a
227. Marom S, Dagan D, Winaver J, Palti Y (1989) Brush-border membrane cation conducting channels from rat kidney proximal tubules. Am J Physiol 257:F328–F335
228. Marty A, Evans MG, Tan YP, Trautman A (1986) Muscarinic responses in rat lacrimal glands. J Exp Biol 124:15–32
229. Marty A, Tan YP, Trautman A (1984) Three types of calcium-dependent channels in rat lacrimal glands. J Physiol (Lond) 357:293–325
230. Marunaka Y, Eaton DC (1988) Effects of cdpc on single channel currents of the amiloride-sensitive sodium channel from cultured renal cells. Biophys J 530:522a
231. Marunaka Y, Eaton DC (1990) The effects of insulin and phosphatase on a calcium-dependent Cl channel in a distal nephron cell line (A6). J Gen Physiol 950:773–789

232. Marunaka Y, Eaton DC (1990) Chloride channels in the apical membrane of a distal nephron A6 cell line. Am J Physiol 258:C352–C368
233. Maruyama Y, Gallacher DV, Petersen OH (1983) Voltage and Ca^{2+}-activated K^+ channel in basolateral acinar cell membranes of mammalian salivary glands. Nature 302:827–829
234. Maruyama Y, Matsunaga H, Hoshi T (1986) Ca^{2+} -and voltage activated K^+ channel in apical cell membrane of gallbladder epithelium from Triturus. Pflugers Arch 406:563–567
235. Maruyama Y, Moore D, Petersen OH (1985) Calcium-activated cation channel in rat thyroid follicular cells. Biochim Biophys Acta 821:229–232
236. Maruyama Y, Nishiyama A, Izumi T, Hoshimiya N, Petersen OH (1986) Ensemble noise and current relaxation analysis of K current in single isolated salivary acinar cells from rat. Plfugers Arch 406:69–72
237. Maruyama Y, Nishiyama A, Teshima T (1986) Two types of cation channels in the basolateral cell membrane of human salivary gland acinar cell. Jpn J Physiol 36:219–223
238. Maruyama Y, Petersen OH (1982) Cholecystokinin activation of single-channel currents is mediated by internal messenger in pancreatic acinar cells. Nature 300:61–63
239. Maruyama Y, Petersen OH (1982) Single-channel currents in isolated patches of plasma membrane from basal surface of pancreatic acini. Nature 299:159–161
240. Maruyama Y, Petersen OH (1984) Control of K^+ conductance by cholecystokinin and Ca^{2+} in single pancreatic acinar cells studied by the patch-clamp technique. J Membr Biol 79:293–300
241. Maruyama Y, Petersen OH (1984) Single calcium-dependent cation channels in mouse pancreatic acinar cells. J Membr Biol 81:83–87
242. Maruyama Y, Petersen OH, Flanagan P, Pearson GT (1983) Quantification of Ca^{2+}-activated K^+ channels under hormonal control in pig pancreas acinar cells. Nature 305:228–232
243. Marver D (1990) PGE_2 and PMA inhibit rabbit CCT (Na, K)-ATPase activity by a primary effect on luminal membrane Na permeability. Kidney Int 37:349
244. McCarty NA, O'Neil RG (1989) Time-dependent action of calcium for regulatory volume decrease (RVD) in proximal straight tubule (PST): the calcium window. FASEB J 3:A977–A977
245. Medina IR, Bregestovski PD (1988) Stretch-activated ion channels modulate the resting membrane potential during early embryogenesis. Proc R Soc Lond [Biol] 235:95–102
246. Mene P, Simonson MS, Dunn MJ (1989) Phospholipids in signal transduction of mesangial cells. Am J Physiol 256:F375–F386
247. Merot J, Bidet M, Gachot B, Le Maout S, Tauc M, Poujeol P (1988) Patch clamp study on primary culture of isolated proximal convuloted tubules. Pflugers Arch 413:51–61
248. Methfessel C, Witzemann V, Takahashi T, Mishina M, Numa S, Sakmann B (1986) Patch clamp measurements on Xenopus laevis oocytes: currents through endogenous channels and implanted acetylcholine receptor and sodium channels. Pflugers Arch 407:577–588
249. Miledi R (1982) A calcium-dependent transient outward current in Xenopus laevis oocytes. Proc R Soc Lond [Biol] 215:491–497
250. Miledi R, Parker I (1984) Chloride current induced by injection of calcium into Xenopus oocytes. J Physiol (Lond) 357:173–183
251. Miledi R, Parker I (1989) Latencies of membrane currents evoked in Xenopus oocytes by receptor activation, inositol trisphosphate and calcium. J Physiol (Lond) 415:189–210
252. Miller C, Latorre R, Reisin I (1987) Coupling of voltage-dependent gating and Ba block in the high-conductance, Ca-activated K channel. J Gen Physiol 90:427–449
253. Miller C, Moczydlowskki E, Latorre R, Phillips M (1985) Charybdotoxin, a protein inhibitor of single Ca-activated K channels from mammalian skeletal muscle. Nature 313:316–318
254. Miller RJ (1987) Multiple calcium channels and neuronal function. Science 235:46–52
255. Miller RT, Moe O, Tejedor A, Pang I-J, Stermweis PC (1990) Distribution of G proteins and adenylyl cyclase in renal cortical brush border and basolateral membranes. Kidney Int 37:215–215
256. Mircheff AK, Lu CC (1984) A map of membrane populations isolated from rat exorbital gland. Am J Physiol 247:G651–G661
257. Moczydlowski E, Latorre R (1983) Gating kinetics of Ca-activated K channels from rat muscle incorporated into planar lipid bilayers. J Gen Physiol 82:511–542
258. Mohrmann M, Cantiello HF, Ausiello DA (1987) Inhibition of epithelial Na^+ transport by atriopeptin, protein kinase C, and pertussis toxin. Am J Physiol 253:F372–F376
259. Molony DA, Andreoli TE (1988) Diluting power of thick limbs of Henle I. Peritubular hypertonicity blocks basolateral Cl^- channels. Am J Physiol 255:F1128–F1137
260. Montero MC, Bolufer J, Ilundain A (1988) Potassium transport in epithelial cells isolated from small intestine of the chicken. Pflugers Arch 412:422–426

261. Montrose-Rafizedah C, Taniguchi J, Guggino WB (1990) Activation of apical calcium-activated potassium channels in volume regulation in the thick ascending limb. Kidney Int 37:567
262. Morris AP, Fuller CM, Gallacher DV (1987) Cholinergic receptors regulate a voltage-insensitive but Na^+-dependent calcium influx pathway in salivary acinar cells. FEBS Lett 211:195–199
263. Morris AP, Gallacher DV, Lee JA (1986) A large conductance, voltage- and calcium-activated K^+ channel in the basolateral membrane of rat enterocytes. FEBS Lett 206:87–92
264. Muto S, Giebisch G, Sansom S (1987) Effects of adrenalectomy on CCD: evidence for differential response of two cell types. Am J Physiol 253:F742–F762
265. Neer EJ, Kim S-Y, Ang S-L, Bloch DB, Bloch KD, Kawahara Y, Tolman C, Lee R, Logothetis D, Kim D, Seidman JG, Clapham DE (1988) Functions of G-protein subunits. Cold Spring Harbor Symp Quant Biol 53:241–246
266. Neher E, Colquhoun D, Reuter H, Stevens CF (1981) Inward current channels activated by intracellular Ca in cultured cardiac cells. Nature 294:752–754
267. Neher E, Stevens CF (1977) Conductance fluctuations and ionic pores in membranes. Annu Rev Biophys Bioeng 6:345–381
268. Nelson DJ, Tang JM, Palmer LG (1984) Single-channel recordings of apical membrane chloride conductance in A6 epithelial cells. J Membr Biol 80:81–89
269. Neufeld T, Terreros D, Grantham J (1983) Critical role of calcium in the regulation of intracellular volume of isolated proximal S2 renal tubules in hypotonic medium. Kidney Int 23:255
270. Novak I, Greger R (1988) Properties of the luminal membrane of isolated perfumed rat pancreatic ducts. Effects of cyclic AMP and blockers of chloride transport. Pflugers Arch 411:546–553
271. O'Neil RG, Hayhurst RA (1985) Functional differentiation of cell types of cortical collecting duct. Am J Physiol 248:F449–F453
272. O'Neil RG, Sansom SC (1984) Electrophysiological properties of cellular and paracellular conductive pathways of the rabbit cortical collecting duct. J Membr Biol 82:281–295
273. O'Neil RG, Sansom SC (1984) Electrophysiological properties of cellular and paracellular conductive pathways of the rabbit cortical collecting duct. J Membr Biol 82:281–295
274. O'Neil RG, Sansom SC (1984) Characterization of apical cell membrane Na^+ and K^+ conductances of cortical collecting duct using microelectrode techniques. Am J Physiol 247:F14–F24
275. O'Neil RG, Sansom SC (1984) Characterization of apical cell membrane Na^+ and K^+ conductances of cortical collecting duct using microelectrode techniques. Am J Physiol 247:F14–F24
276. Oberleithner H, Kersting U, Gassner B (1988) Aldosterone-controlled linkage between Na^+/H^+ exchange and K^+ channels in fused renal epithelial cells. Ciba Found Symp 139:201–219
277. Oberleithner H, Kersting U, Hunter M (1988) Cytoplasmic pH determines K^+ conductance in fused renal epithelial cells. Proc Natl Acad Sci USA 85:8345–8349
278. Oberleithner H, Weigt M, Westphale H-J, Wang W (1987) Aldosterone activates Na/H exchange and raises cytoplasmic pH in target cells of the amphibian kidney. Proc Natl Acad Sci USA 84:1464–1468
279. Olans L, Sariban-Sohraby S, Benos DJ (1984) Saturation behavior of single, amiloride-sensitive Na^+ channels in planar lipid bilayers. Biophys J 46:831–835
280. Pallotta BS (1985) N-Bromoacetamide removes a calcium-dependent component of channel opening from calcium-activated potassium channels in rat skeletal muscle. J Gen Physiol 86:601–611
281. Palmer LG, Corthesy-Theulaz I, Gaeggeler H-P, Kraehenbuhl J-P, Rossier BC (1990) Expression of epithelial sodium channels in oocytes. J Gen Physiol 96:23–46
282. Palmer LG, Frindt G (1986) Epithelial sodium channels: characterization by using the patch-clamp technique. Fed Proc 45:2708–2712
283. Palmer LG, Frindt G (1986) Amiloride-sensitive Na channels from the apical membrane of the rat cortical collecting tubule. Proc Natl Acad Sci USA 83:2767–2770
284. Palmer LG, Frindt G (1987) Ca ionophore and phorbol ester inhibit Na channels in rat cortical tubules. Fed Proc 46:495
285. Palmer LG, Frindt G (1987) Effects of cell Ca and pH on Na channels from rat cortical collecting tubule. Am J Physiol 253:F333–F339

286. Palmer LG, Li JH, Lindemann B, Edelman IS (1982) Aldosterone control of the density of sodium channels in the toad urinary bladder. J Membr Biol 64:91–102
287. Parent L, Cardinal J, Sauve R (1988) Single-channel analysis of a K channel at basolateral membrane of rabbit proximal convoluted tubule. Am J Physiol 254:F105–111
288. Parent L, Dube L, Roy G, Sauve R (1989) Patch-clamp studies of K^+ and Cl^- channels at the apical membrane of kidney proximal tubule cells in primary culture. Biophys J 55:161a
289. Partidge LD, Swandulla D (1988) Calcium-activated non-specific cation channels. TINS 11:69–72
290. Patlak J, Horn R (1983) Effect of N-bromoacetamide on single sodium channel currents in excised membrane patches. J Gen Physiol 79:333–351
291. Petersen OH, Findlay I, Iwatsuki N, Singh J, Gallacher DV, Fuller CM, Pearson GT, Dunne MJ, Morris AP (1985) Human pancreatic acinar cells: studies of stimulus-secretion coupling. Gastroenterology 89:109–117
292. Post MA, Richards NM, Ernst SA, Dawson DC (1989) Apical stretch-activated cation channels in a toad urinary bladder cell line (TMB). FASEB J 3:A861
293. Rae JL (1985) The application of patch clamp methods to ocular epithelia. Curr Eye Res 4:409–420
294. Rae JL (1986) Potassium channels from chick lens epithelium. Fed Proc 45:2718–2722
295. Rae JL, Dewey J, Cooper K (1986) Properties of single potassium-selective ionic channels from the apical membrane of rabbit corneal endothelium. Exp Eye Res 49:591–610
296. Rae JL, Levis RA (1984) Patch clamp recordings from the epithelium of the lens obtained using glasses selected for low noise and improved sealing properties. Biophys J 45:144–146
297. Rae JL, Levis RA (1984) Patch voltage clamp of lens epithelial cells: theory and practice. Mol Physiol 6:115–162
298. Rae JL, Levis RA, Eisenberg RS (1988) Ionic channels in ocular epithelia. In: Narahashi T (ed) *Ion channels*, Plenum, New York, pp 283–327
299. Reif MC, Troutman SL, Schafer JA (1986) Sodium transport by rat cortical collecting tubule. Effects of vasopressin and desoxycorticosterone. J Clin Invest 77:1291–1298
300. Reinhardt R, Bridges RJ, Rummel W, Lindemann B (1987) Properties of an anion-selective channel from rat colonic enterocyte plasma membranes reconstituted into planar phospholipid bilayers. J Membr Biol 95:47–54
301. Richards NW, Dawson DC (1986) Single potassium channels blocked by lidocaine and quinidine in isolated turtle colon epithelial cells. Am J Physiol 251:C85–C89
302. Richards NW, Dawson DC (1987) Two types of Ca-activated channels in isolated turtle colon epithelial cells. Biophys J 51:344a
303. Richards NW, Dawson DC (1989) N-phenylanthranilic acid blocks specific classes of K-conducting channels in colonic epithelial cells. FASEB J 3:A1149
304. Richards NW, Lowy RJ, Ernst SA, Dawson DC (1989) Two K^+ channel types, muscarinic agonist-activated and inwardly rectifying, in a Cl^- secretory epithelium: the avian salt gland. J Gen Physiol 93:1171–1194
305. Ritchie A (1987) Thyrotropin-releasing hormone stimulates a calcium-activated potassium current in a rat anterior pituitary cell line. J Physiol (London) 385:611–625
306. Ritchie A (1987) Two distinct calcium-activated potassium currents in a rat anterior pituitary cell line. J Physiol (London) 385:591–609
307. Rossier BC, Gaeggeler H-P, Rossier M (1978) Effects of 3′deoxyadenosine and actinomycin D on RNA synthesis in toad bladder: analysis of response to aldosterone. J Membr Biol 41:149–166
308. Rossiter KA, Liang CT, Kinsella JL, Austin HA, Spiegel AM (1990) Polar distribution and heterogenous localization of G-s-alpha and other G-proteins in a mammalian kidney. Kidney Int 37:216
309. Rousseau G, Crabbe J (1972) Effects of aldosterone on RNA and protein synthesis in the toad bladder. Eur J Biochem 25:550–559
310. Sachs F (1987) Baroreceptor mechanisms at the cellular level. Fed Proc 46:12–16
311. Sackin H (1987) Stretch-activated potassium channels in renal proximal tubule. Am J Physiol 253:F1253–F1262
312. Sackin H (1989) A stretch-activated K^+ channel sensitive to cell volume. Proc Natl Acad Sci USA 86:1731–1735
313. Sackin H, Palmer LG (1987) Basolateral potassium channels in renal proximal tubule. Am J Physiol 253:F476–F487

314. Sadoshima J, Akaike N, Kanaide H, Nakamura M (1988) Cyclic AMP modulates Ca-activated K channel in cultured smooth muscle cells of rat aortas. Am J Physiol 255:H754–H759
315. Sakmann B, Trube G (1984) Voltage-dependent inactivation of inward-rectifying single-channel currents in the guinea-pig heart cell membrane. J Physiol (Lond) 347:659–683
316. Sakmann B, Trube G (1984) Conductance properties of single inwardly rectifying potassium channels in ventricular cells from guinea-pig heart. J Physiol (Lond) 347:641–657
317. Salkoff LB, Tanouye MA (1986) Genetics of ion channels. Physiol Rev 66:301–329
318. Sansom SC, O'Neil RG (1985) Mineralocorticoid regulation of apical cell membrane Na^+ and K^+ transport of the cortical collecting duct. Am J Physiol 248:F858–F868
319. Sariban-Sohraby S, Benos DJ (1986) The amiloride-sensitive sodium channel. Am J Physiol 250:C175–C190
320. Sariban-Sohraby S, Latorre R, Burg M, Olans L, Benos D (1984) Amiloride-sensitive epithelial Na^+ channels reconstituted into planar lipid bilayer membranes. Nature 308:80–82
321. Schafer JA, Troutman SL, Schlatter E (1990) Vasopressin and mineralocorticoid increase apical membrane driving force for K^+ secretion in rat CCD. Am J Physiol 258:F199–F210
322. Schild L, Aronson PS, Giebisch G (1989) Changes in cell volume induced by K and Cl in the isolated and prefused rabbit proximal tubule. Kidney Int 35:488
323. Schlatter E, Greger R (1985) cAMP increases the basolateral Cl^--conductance in the isolated perfused medullary thick ascending limb of Henle's loop of the mouse. Pflugers Arch 405:367–376
324. Schlatter E, Schafer JA (1987) Electrophysiological studies in principal cells of rat cortical collecting tubules. ADH increases the apical membrane Na^+-conductance. Pflugers Arch 409:81–92
325. Schneider GT, Cook DL, Gage PW, Young JA (1985) Voltage sensitive, high-conductance chloride channels in the luminal membrane of cultured pulmonary alveolar (type II) cells. Pflugers Arch 404:354–357
326. Schoumacher RA, Shoemaker RL, Halm DR, Tallant EA, Wallace RA, Frizzell RA (1987) Phosphorylation fails to activate chloride channels from cystic fibrosis airway cells. Nature 330:752–754
327. Schultz SG (1981) Homocellular regulatory mechanisms in sodium-transporting epithelia: avoidance of extinction by "flush-through". Am J Physiol 241:F579–F590
328. Schultz SG, Hudson RL (1986) How do sodium-absorbing cells do their job and survive. News Physiol Sci 1:185–189
329. Schuster VL (1986) Cyclic adenosine monophosphate-stimulated anion transport in rabbit cortical collecting duct. J Clin Invest 78:1621–1630
330. Schuster VL, Stokes JB (1987) Chloride transport by the cortical and outer medullary collecting duct. Am J Physiol 253:F203–F212
331. Sepulveda FV, Mason WT (1985) Single channel recordings obtained from basolateral membranes of isolated rabbit enterocytes. FEBS Lett 191:87–91
332. Shearman MS, Sekiguchi K, Nishizuka Y (1989) Modulation of ion channel activity: a key function of the protein kinase C enzyme family. Pharmacol Rev 41:211–237
333. Sheppard DN, Giraldez F, Sepulveda FV (1988) K^+ channels activated by L-alanine transport in isolated *Necturus* enterocytes. FEBS Lett 234:446–448
334. Shoemaker RL, Frizzell RA, Dwyer TM, Farley JM (1986) Single chloride channel currents from canine tracheal epithelial cells. Biochim Biophys Acta 858:235–242
335. Sieman D, Reuhl T (1987) Non-selective cationic channel in primary cultured cells of brown adipose tissue. Pflugers Arch 408:534–536
336. Sigurdson WJ, Morris CE, Brezden BL, Gardner DR (1987) Stretch activation of a K^+ channel in molluscan heart cells. J Exp Biol 127:191–209
337. Silva P, Stoff J, Field M, Fine L, Forrest JN, Epstein FH (1977) Mechanism of active chloride secretion by shark rectal gland: role of Na/K-ATPase in chloride transport. Am J Physiol 233:F298–F306
338. Simmonneau M, Distasi C, Tauc L, Barbin G (1987) Potassium channels in mouse neonatal root gangelion cells: a patch-clamp study. Brain Res 412:224–232
339. Snider RM, Roland RM, Lowy RJ, Agranoff BW, Ernst SA (1986) Muscarinic receptor-stimulated $Ca2^+$ signaling and inositol lipid metabolism in avian salt gland cells. Biochim Biophys Acta 889:216–224
340. Stanton BA, Dietl S, Schwiebert E (1991) Cell volume regulation in the cortical collecting duct (CCD): stretch-activated Cl channels. Kidney Int

341. Stetson DL, Lewis SA, Alles W, Wade JB (1982) Evaluation by capacitance measurements of antidiuretic hormone induced membrane area changes in toad bladder. Biochim Biophys Acta 689:267–274
342. Stokes JB (1982) Ion transport by the cortical and outer medullary collecting tubule. Kidney Int 22:473–484
343. Strange K, Willingham MC, Handler JS, Harris HW Jr (1988) Apical membrane endocytosis via coated pits is stimulated by removal of antidiuretic hormone from isolated, perfused rabbit cortical collecting tubule. J Membr Biol 103:17–28
344. Streb H, Irvine RF, Berridge MJ, Schulz I (1983) Release of Ca from a nonmitochondrial intracellular store in pancreatic acinar cells by inositol-1, 4, 5,-triphosphate. Nature 306:67–69
345. Strieter J, Stephenson JL, Palmer LG, Weinstein AM (1989) Volume-activated chloride permeability can mediate cell volume regulation in a mathematical model of toad bladder epithelium. FASEB J 3:A1150
346. Sturgess NC, Hales CN, Ashford MLJ (1986) Inhibition of a calcium-activated, non-selective cation channel, in a rat insulinoma cell line, by adenine derivates. FEBS Lett 208:397–400
347. Sugimoto T, Tanabe Y, Shigemoto R, Iwai M, Takumi T, Ohkubo H, Nakanishi S (1990) Immunohistochemical study of a rat membrane protein which induces a selective potassium permeation: its localization in the apical membrane portion of epithelial cells. J Membr Biol 113:39–47
348. Suzuki K, Petersen CC, Petersen OH (1985) Hormonal activation of single K^+ channels via internal messenger in isolated pancreatic acinar cells. FEBS Lett 192:307–312
349. Suzuki K, Petersen OH (1988) Patch-clamp study of single-channel and whole-cell K^+ currents in guinea pig pancreatic acinar cells. Am J Physiol 255:G275–G285
350. Tang C-M, Presser F, Morad M (1988) Amiloride selectivity blocks the low threshold (T) calcium channel. Science 240:213–214
351. Taniguchi J, Guggino WB (1989) Membrane stretch: a physiological stimulator of Ca^{2+}-activated K^+ channels in thick ascending limb. Am J Physiol 257:F347–F352
352. Tanner GA, Horisberger JD, Giebisch G (1988) Cell volume regulation in late proximal tubule of necturus kidney. Kidney Int 35:428–428
353. Taylor A (1979) Possible role of crytosolic calcium & Na-Ca exchange in regulation of transepithelial sodium transport. Am J Physiol 236:F505–F512
354. Taylor A, Lee CO, Windhager EE (1980) Cytosolic calcium ion activity in epithelial cells of Necturus kidney. Nature 287:859–861
355. Tempel BL, Jan YN, Jan LY (1988) Cloning of a probable potassium channel gene from mouse brain. Nature 332:837–839
356. Tempel BL, Papazian DM, Schwarz TL, Jan YN, Lily YJ (1987) Sequence of a probable potassium channel component encoded at Shaker locus of Drosophilia. Science 237:770–775
357. Teulon J, Paulais M, Anagnostopoules T (1988) ATP-, ADP- and AMP-inhibition of a Ca^{2+}-activated non-selective cation channel from the basolateral membrane of the cortical thick ascending limb of Henle's loop (CAL). Kidney Int 35:428–428
358. Teulon J, Paulais M, Bouthier M (1987) A Ca^{2+}-activated cation-selective channel in the basolateral membrane of the cortical thick ascending limb of Henle's loop of the mouse. Biochim Biophys Acta 905:125–132
359. Ubl J, Murer H, Kolb H-A (1988) Hypotonic shock evokes opening of Ca^{2+}-activated K channels in opossum kidney cells. Pflugers Arch 412:551–553
360. Ubl J, Murer H, Kolb H-A (1988) Ion channels activated by osmotic and mechanical stress in membranes of opossum kidney cells. J Membr Biol 104:223–232
361. Ueda S, Loo DD, Sachs G (1987) Regulation of K^+ channels in the basolateral membrane of Necturus oxyntic cells. J Membr Biol 97:31–41
362. Ussing HH, Koefoed-Johnsen V (1958) The nature of the frog skin potential. Acta Physiol Scand 42:298–308
363. Ussing HH, Zerahn K (1951) Active transport of sodium as the source of electric current in short-circuited isolated frog skin. Acta Physiol Scand 23:110–127
364. Valdivia HH, Dubinsky WP, Coronado R (1988) Reconstitution and phosphorylation of chloride channels from airway epithelium membranes. Science 242:1441–1444
365. Van de Stolpe A, Jamison RL (1988) Micropuncture study of the effect of ANP on the papillary collecting duct in the rat. Am J Physiol 254:F477–F483
366. van der Ploeg I, Celsi G, Bertorello A, Holtback U, Fredholm B, Aperia A (1990) G-protein subunits distribution in the luminal and basolateral membrane of proximal tubular cell.

Kidney Int 37:217
367. Van Driessche W (1980) Spontaneous and blocker-induced K-channel noise in frog skin (Rana temporaria). Biophys Struct Mech 6:5–8
368. Van Driessche W (1980) Ba-induced conductance fluctuations of spontaneously fluctuating K channels in the apical membrane of frog skin (Rana temporaria). J Membr Biol 56:31–42
369. Van Driessche W (1984) Physiological role of apical potassium ion channels in frog skin. J Physiol (Lond) 356:79–95
370. Van Driessche W (1987) Ca^{2+} channels in the apical membrane of the toad urinary bladder. Pflugers Arch 410:243–249
371. Van Driessche W, Aelvoet I, Erlij D (1987) Oxytocin and cAMP stimulate monovalent cation movements through a Ca-sensitive, amiloride-insensitive channel in apical membrane of toad urinary bladder. Proc Natl Acad Sci USA 84:313–317
372. Van Driessche W, Desmedt L, Simaels J (1989) Forskolin and serosal hypotonicity activate a calcium-sensitive pathway in the apical membrane of frog skin. FASEB J 3:A983–A983
373. Van Driessche W, Erlij D (1983) Noise analysis of inward and outward Na^+ currents across the apical border of ouabain-treated frog skin. Pflugers Arch 398:179–188
374. Van Driessche W, Gogelein H (1978) Potassium channels in the apical membrane of the toad gallbladder. Nature 275:665–667
375. Van Driessche W, Hillyard SD (1985) Quinidine blockage of K channels in the basolateral membrane of larval bullfrog skin. Pflugers Arch 405:S77–S82
376. Van Driessche W, Wills NK, Hillyard SD, Zeiske W (1982) K^+ channels in an epithelial "single membrane" preparation. Arch Int Physiol Biochim 90:P12–P14
377. Van Driessche W, Zeiske W (1980) Spontaneous fluctuations of potassium channels in the apical membrane of frog skin. J Physiol (Lond) 299:101–116
378. Van Driessche W, Zeiske W (1985) Apical K^+ channels in frog skin: a pathway for K^+ excretion. In: Gilles R, Gilles-Baillien M (eds) Transport processes, iono- and osmoregulation. Springer, Berlin Heidelberg New York, pp 40–55
379. Van Driessche W, Zeiske W (1985) Ionic channels in epithelial cell membranes. Physiol Rev 65:833–903
380. Van Driessche W, Zeiske W (1985) Apical K^+ channels in frog skin: a pathway for K^+ excretion. In: Gilles R, Gilles-Baillien M (eds) Transport processes, iono- and osmoregulation. Springer, Berlin Heidelberg New York, pp 40–55
381. Van Driessche W, Zeiske W (1985) Ca^{2+} -sensitive, spontaneously fluctuating, cation channels in the apical membrane of the adult frog skin epithelium. Pflugers Arch 405:250–259
382. Vergara C, Latorre R (1983) Kinetics of Ca^{2+} -activated K^+ channels from rabbit muscle incorporated into planar bilayers. J Gen Physiol 82:543–568
383. von Tscharner V, Prod'hom B, Baggiolini M, Reuter H (1986) Ion channels in human neutrophils activated by a rise in free cytosolic calcium concentration. Nature 324:369–372
384. Wade JB (1986) Role of membrane fusion in hormonal regulation of epithelial transport. Annu Rev Physiol 48:213–223
385. Wade JB (1989) Dynamics of apical membrane responses to ADH in amphibian bladder. Am J Physiol 257:R998–R1003
386. Wang WH, White S, Geibel J, Giebisch G (1990) A potassium channel in the apical membrane of rabbit thick ascending limb of Henle's loop. Am J Physiol 258:F244–F253
387. Welling PA, O'Neil RG (1987) Cell swelling increases basolateral membrane Cl and K conductances of the rabbit proximal straight tubule (PST). Kidney Int 31:452–452
388. Welsh MJ (1983) Evidence for basolateral membrane potassium conductance in canine tracheal epithelium. Am J Physiol 244:C377–C384
389. Welsh MJ (1984) Anthracene-9-carboxylic acid inhibits an apical membrane chloride conductance in canine tracheal epithelium. J Membr Biol 78:61–71
390. Welsh MJ (1985) Ion transport by primary cultures of canine tracheal epithelium: methodology, morphology, and electrophysiology. J Membr Biol 88:149–163
391. Welsh MJ (1986) Single apical membrane anion channels in primary cultures of canine tracheal epithelium. Pflugers Arch 407:S116–S122
392. Welsh MJ (1986) An apical-membrane chloride channel in human tracheal epithelium. Science 232:1648–1650
393. Welsh MJ (1988) Defective regulation of ion transport in CF airway epithelia. In: Mastella G, Quinton PM (eds) *Cellular and molecular basis of cystic fibrosis.* San Francisco Press, San Francisco, pp 321–332

394. Welsh MJ, Liedtke CM (1986) Chloride and potassium channels in cystic fibrosis airway epithelia. Nature 322:467–470
395. Welsh MJ, McCann JD (1985) Intracellular clacium regulates basolateral potassium channels in a chloride-secreting epithelium. Proc Natl Acad Sci USA 82:8823–8826
396. Welsh MJ, Smith PL, Frizzell RA (1982) Chloride secretion by canine tracheal epithelium. II. The cellular electrical potential profile. J Membr Biol 70:227–238
397. Welsh MJ, Smith PL, Frizzell RA (1983) Chloride secretion by canine tracheal epithelium. III. Membrane resistances and electromotive forces. J Membr Biol 71:209–218
398. White SJ, Henderson RM, Boulpaep EL, Giebisch GH (1989) Ion channels in cultured rabbit renal collecting duct cells. Kidney Int 35:491–491
399. Wills NK (1984) Mechanisms of ion transport by the mammalian colon revealed by frequency domain analysis techniques. In: Wade JB, Lewis SA (eds) *Current topics in membranes and transport*. Academic, New York, pp 61–85
400. Wills NK (1985) Apical membrane potassium and chloride permeabilities in surface cells of rabbit descending colon epithelium. J Physiol (Lond) 358:433–445
401. Wills NK, Alles WP, Sandle GI, Binder HJ (1984) Apical membrane properties and amiloride binding kinetics of the human descending colon. Am J Physiol 247:G749–G757
402. Wills NK, Biagi BA (1982) Active potassium transport by rabbit descending colon epithelium. J Membr Biol 64:195–203
403. Wills NK, Clausen C, Clauss WC (1987) Electrophysiology of active potassium transport across the mammalian colon. In: Giebisch G (ed) *Current topics in membranes and transport*. Academic, New York, pp 441–456
404. Wills NK, Eaton DC, Lewis SA, Ifshin MS (1979) Current-voltage relationship of the basolateral membrane of a tight epithelium. Biochim Biophys Acta 555:519–523
405. Wills NK, Lewis SA, Eaton DC (1979) Active and passive properties of rabbit descending colon: a microelectrode and nystatin study. J Membr Biol 45:81–108
406. Wills NK, Zeiske W, Van Driessche W (1982) Noise analysis reveals K^+ channel conductance fluctuations in the apical membrane of rabbit colon. J Membr Biol 69:187–197
407. Wills NK, Zweifach A (1987) Recent advances in the characterization of epithelial ionic channels. Biochim Biophys Acta 906:1–31
408. Wolf ID, Van Driessche W (1986) Voltage-dependent Ba block of K channels in apical membrane of frog skin. Am J Physiol 251:C696–C706
409. Wong SME, Chase HS Jr (1986) Role of intracellular calcium in cellular volume regulation. Am J Physiol 250:C841–C852
410. Worrell RT, Butt AG, Cliff WH, Frizzell RA (1989) A volume-sensitive chloride conductance in human colonic cell line T84. Am J Physiol 256:C1111–C1119
411. Worrell RT, Frizzell RA (1989) DIDS inhibits a volume-sensitive epithelial chloride conductance in the colonic cell line. FASEB J 3:A1148–A1148
412. Yanase M, Handler JS (1986) Adenosine 3′, 5′-cyclic monophosphate stimulates chloride secretion in A6 epithelia. Am J Physiol 251:C810–C814
413. Yanase M, Handler JS (1986) Activators of protein kinase C inhibit sodium transport in A6 epithelia. Am J Physiol 250:C517–C522
414. Yang XC, Sachs F (1987) Stretch-activated channels in several tissues. Biophys J 51:252a
415. Yatani A, Codina J, Brown AM, Birnbaumer L (1987) Direct activation of mammalian atrial muscarinic potassium channels by GTP regulatory protein G-K. Science 235:207–211
416. Yellen G (1982) Single Ca^{2+} -activated nonselective cation channels in neuroblastoma. Nature 296:357–359
417. Zeidel ML, Kikeri D, Burrows M, Silva P; Brenner BM (1988) Atrial natriuretic peptide (ANP) inhibits ^{22}Na uptake in rabbit inner medullary collecting duct (IMCD) cells. Kidney Int 33:291–291
418. Zeiske W, Machen TE, Van Driessche W (1983) Cl^--and K^+-related fluctuations of ionic current through oxyntic cells in frog gastric mucosa. Am J Physiol 245:797–807
419. Zeiske W, Van Driessche W (1979) Saturable K^+ pathway across the outer border of frog skin: kinetics and inhibition by Cs^+ and other cations. J Membr Biol 47:77–96
420. Zeiske W, Van Driessche W (1981) Cationic permeability of an epithelial K^+ channel. Arch Int Physiol Biochim 89:55–59
421. Zeiske W, Van Driessche W (1981) Apical K^+ channels in frog skin (Rana temporaria): cation adsorption and voltage influence gating kinetics. Pflugers Arch 390:22–29
422. Zeiske W, Van Driessche W (1984) The sensitivity of apical Na^+ permeability in frog skin to hypertonic stress. Pflugers Arch 400:130–139

423. Zeiske W, Van Driessche W (1986) Impairment of Na transport across frog skin by Tl^+: effects on turnover, area density and saturation kinetics of apical Na channels. Pflugers Arch 407:145–152
424. Zeiske W, Van Driessche W, Ziegler R (1986) Current-noise analysis of the basolateral route for K^+ ions across a K^+-secreting insect midgut epithelium (Manduca sexta). Pflugers Arch 407:657–663
425. Zeiske W, Wills NK, Van Driessche W (1982) Na^+ channels and amiloride-induced noise in the mammalian colon epithelium. Biochim Biophys Acta 688:201–210
426. Zeuthen T, Christensen O, Cherksey B (1987) Electrodiffusion of Cl^- and K^+ in epithelial membranes reconstituted into planar lipid bilayers. Pflugers Arch 408:275–281
427. Zweifach A, Lewis SA (1988) Characterization of a partially degraded Na^+ channel from urinary tract epithelium. J Membr Biol 101:49–56

Amiloride-Sensitive Na^+ Transport Mechanisms *

D.J. Benos, D.G. Warnock, and J.B. Smith

Introduction

The major extracellular cation in animals is Na^+. Although intracellular ionic composition varies depending upon cell type and origin, in general the major intracellular fluid cations are K^+ and Mg^{2+}, with Na^+ kept at relatively low levels. The maintenance of these differences in ionic activities across cell membranes results from active accumulation and extrusion of specific ions, as well as from the selective permeability characteristics of the membranes themselves. In addition, the interior of most animal cells is electrically negative with respect to the bathing medium. Thus, Na^+ diffuses from the extracellular fluid, where its electrochemical potential is high, into cells, where there is a much lower Na^+ electrochemical potential energy state. This tendency for net cellular accumulation is opposed by the actions of the ($Na^+ + K^+$)-activated ATPase pump.

Cells have evolved numerous mechanisms by which to transport Na^+. These mechanisms include active transport pumps, ion channels, and co- and counter-transport systems. Certain nonexcitable tissues, like the kidney and intestine, among whose functions include the regulation of whole-body Na^+, contain Na^+ transport systems susceptible to blockade by the synthetic drug amiloride (3,5-diamino-6-chloropyrazinoylguanidine; Fig. 1). This drug was first synthesized by chemists at Merck, Sharp, and Dohme pharmaceutical laboratories in the middle 1960s [21, 48, 125]. This drug potently and specifically inhibits Na^+ transport in a wide variety of cellular and epithelial preparations ranging from erythrocytes to the tongue, to kidney tubules, and to the rectum, although typically not in nerve (see [33] for a review). Clinically, amiloride interferes with late distal and collecting tubule Na^+ reabsorption, secondarily producing an appreciable K^+ sparing action through hyperpolarization of the cell membrane. In addition, this hyperpolarization effect can diminish or offset the K^+ loss

* This research was supported by NIH grants DK37206, DK19407, DK39256, and DK39258. Research support is also acknowledged from the Veterans Administration Research Service (D.G.W.), a Research Career Development Award HL01671 (J.B.S.) from the National Institutes of Health, and a grant-in-aid from the American Heart Association with funds contributed by the American Heart Association Central Florida Region (J.B.S.).

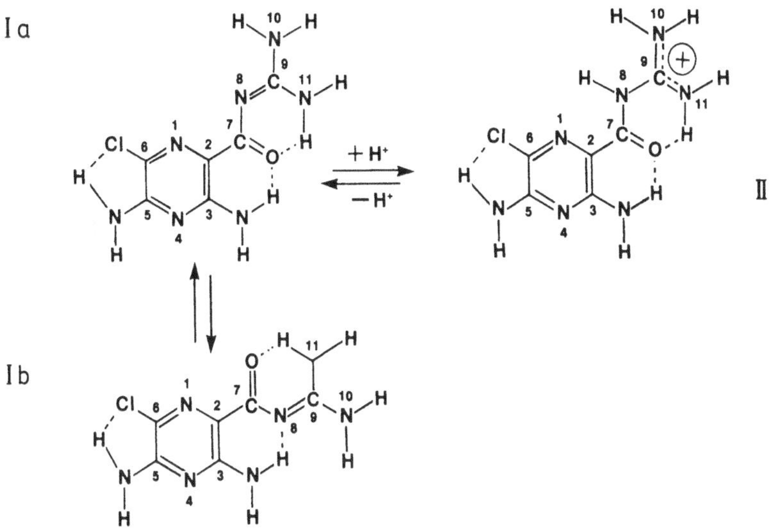

Fig. 1. The structure of amiloride tautomers in aqueous solution. The planar tautomers are stabilized by three hydrogen bonds (*dotted lines*). Tautomer II is the most prevalent in aqueous solution. (Adapted from [35,331])

induced by more potent diuretics. Thus, this drug is used primarily in combination with hydrochlorothiazide for treatment of hypertension and congestive heart failure [12, 272]. The fact that a single synthetic molecule can inhibit Na^+ transport in such a large number of diversified tissues obtained from nearly every phylum of the animal kingdom (with the exception of the sponges and coelenterates) underscores the similarity between basic Na^+ transport mechanisms that exist in the plasma membrane of most animal cells.

In this review, we will discuss the physiology, kinetics, biochemistry and molecular biology, and the immunocytochemistry of the three major classes of amiloride-sensitive Na^+ transport processes: the epithelial Na^+ channel, the Na^+/proton exchanger, and the Na^+/Ca^{2+} antiport system. Throughout, we will draw heavily from experiments designed specifically to elucidate the molecular characteristics of these three transport systems. We conclude by summarizing the status of the overall field of Na^+ transport, and we will suggest areas that we feel concentrated experimental effort should be focused.

General Description of Amiloride-Sensitive Transport Systems

Before the late 1970s, most investigators thought that amiloride only blocked Na^+ entry into so-called high electrical resistance Na^+ transporting epithelia, such as frog skin, toad urinary bladder, and mammalian colon. There were a few papers demonstrating that amiloride could inhibit Na^+ transport in certain nonepithelial systems, such as erythrocytes and sea urchin eggs, but the significance of these observations was not realized [3, 164]. Since the late 1970s, hundreds of papers have been published that describe amiloride-sensitive Na^+ transport in many different nonepithelial cells, as well as in electrically low resistance epithelia, such as renal proximal tubule and gall bladder. These amiloride-sensitive Na^+ transport systems have been implicated in important physiological processes, such as salt balance, pH regulation, fertilization, growth and differentiation, and cell volume regulation.

In general, amiloride acts rapidly and reversibly. A survey of the literature reveals that amiloride-sensitive Na^+ transport processes can be divided arbitrarily into three main groups based upon their sensitivity to the drug (Fig. 2). The first group comprises those transport systems in which the apparent equilibrium dissociation constant (K_i) for amiloride is less than 1 μM. The second group consists of those systems in which the K_i for amiloride is between 1 μM–1 mM, and the third group consists of those transporters in which the K_i for amiloride is greater than 0.3 mM. These K_i values are arbitrarily defined as the concentration of amiloride that inhibits Na^+ transport by 50%, and are determined from dose response curves at physiological external Na^+ concentrations (i.e., > 100 mM). The Na^+ transport mechanism in the first group (high amiloride affinity) consists of a conductive pathway, i.e., an ion channel. The Na^+ transport mechanism comprising the second group (intermediate amiloride affinity) consists of electroneutral exchange processes, i.e., Na^+/H^+ or Na^+/Na^+ exchange. The last group, i.e., the low amiloride affinity group, consists of a Na^+/Ca^{2+} antiport process. In general, the Na^+ channel is found primarily in high resistance epithelia although, it has been noted in low resistance epithelia, e.g., proximal straight tubule [127] and LLC/PK_1 cells

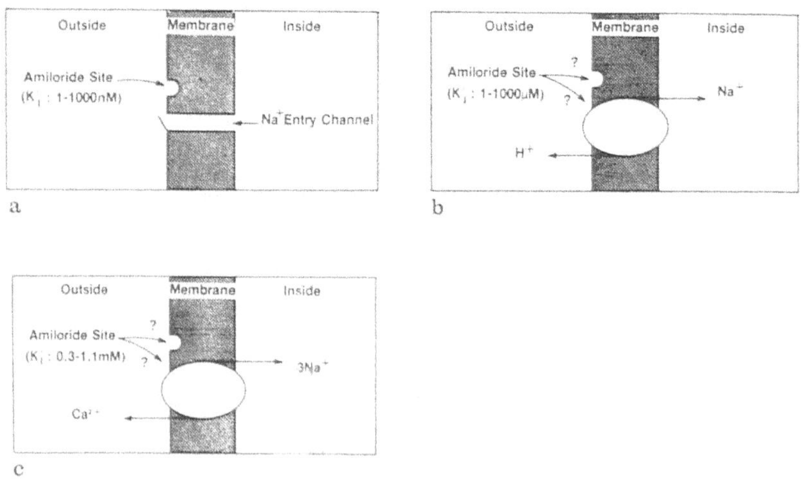

Fig. 2. Amiloride-sensitive Na^+ translocation pathways in biological membranes. (Adapted from [33])

[224]. Na^+/H^+ exchangers are found in low resistance epithelia and in most cellular systems. The electrogenic Na^+/Ca^{2+} exchange system has been described in a variety of cell types including myocardium [171, 285, 311], nerve [22, 23], smooth muscle cells [320], red cells (see [180]), neutrophils [313], and platelets [301], as well as in membrane vesicles from cardiac [281] and smooth muscle [136, 263], brain [55, 340], a pituitary tumor cell line (GH_3) [167], pancreas [31], photoreceptors [305], parotid [343], kidney cortex [126] and distal tubules [271], toad urinary bladder [76], and intestinal epithelium [152]. Exceptions to this generalization are extant; for example, Vigne et al. [363] have described an amiloride-sensitive cation channel in brain microvessels having a K_i for amiloride of 10 μM at physiological salt concentrations.

Although amiloride can inhibit three entirely distinct transport systems, albeit with quite disparate affinities, the pharmacological inhibitory profile of each of these transport systems by amiloride analogs is very distinct. Figure 3 shows several of the more commonly used analogs of amiloride. The pattern of inhibition by amiloride and its analogs is similar for the Na^+ channel and Na^+/Ca^{2+} exchange system. Inhibition of both systems require the presence of a halogen at pyrazine ring position 6 as well as an acylguandinium side chain. Aryl or arylalkyl substitutions on the terminal guanidine nitrogen (phenamil and benzamil) are much better inhibitors than amiloride. However, alkane

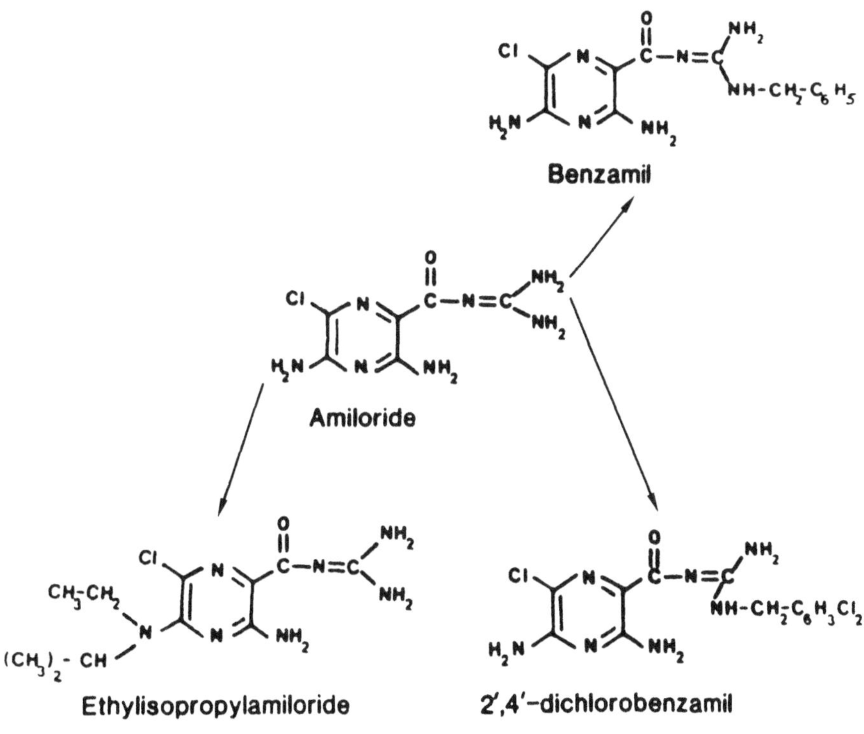

Fig. 3. Structural Formulae of amiloride and amiloride analogs. (Adapted from [110])

replacement of one or both of the 5-position amino hydrogens essentially eliminates inhibition of the Na^+ channel but maintains the inhibitory ability of the analog towards the Na^+/Ca^{2+} exchanger. 2,4-dichlorobenzamil is the most potent inhibitor of Na^+/Ca^{2+} exchange in membrane vesicles from heart and nerve tissues.

Unlike the Na^+ channel and Na^+/Ca^{2+} exchanger, derivatization of the terminal guanidinium nitrogen produces compounds lacking the ability to inhibit Na^+/H^+ exchange. On the other hand, derivatizations of the 5-position amino group with hydrocarbon side chains (e.g., dimethylamiloride and ethylisopropyl amiloride) greatly increase activity towards Na^+/H^+ exchange, while lowering or abolishing inhibition of the Na^+ channel, and moderately increasing inhibition of the Na^+/Ca^{2+} exchanger.

At concentrations greater than 10 μ*M*, amiloride has been shown to cause additional effects, including inhibition of protein synthesis [190, 201, 382], inhibiting specific ligand binding to α_1, α_2, and β adrenergic receptors [47, 154], binding to DNA [83] and guanine nucleotide regulatory proteins [13, 232], and the inhibition of the enzymatic activities of the Na^+/K^+ ATPase [334, 335], kallikrein [207], tyrosine kinase [88], monoamine oxidase [245], adenylate cyclase [202], acetylcholinesterase [87], and protein kinase C [46]. Penetration of amiloride into intracellular compartments occurs easily because the molecule is a weak base with a pKa of 8.7 with the nonprotonated moiety having a membrane permeability coefficient of 10^{-7} cm/s [40, 104, 313]. From planar lipid bilayer measurements, the maximal permeability of the charged amiloride molecule is 10^{-8} cm/s [40], but it is likely that, at least in some cells, the protonated moiety is also highly permeable. For example, Leffert et al. [190], using cultured rat hepatocytes, found that amiloride could be intracellularly accumulated 20- to 25-fold after 20 min when present at 40 μ*M* external concentration. These investigators showed that amiloride uptake consists of two components, a Na^+-independent linear component (approximately 10% of total uptake) and a Na^+-dependent, temperature-sensitive, saturable component with a K_i for amiloride of 21 m*M*. Also, amiloride has been shown to inhibit Na^+-dependent aminoisobutyric acid uptake in cultured chicken myotubes [356], Na^+-dependent phlorizin-sensitive hexose uptake by a clone (LLC-PK_1/CL_4) of pig kidney cells [146], and can inhibit the Na^+-D-glucose, Na^+-l-alanine, and Na^+-phosphate symporters at concentrations > 1 m*M* [145]. Recently, amiloride and several of its analogs have also been shown to inhibit the veratridine-activated $^{22}Na^+$ influx across membrane vesicles obtained from both synaptosomes and heart membranes. The K_i for inhibition of this flux by amiloride was 600 μ*M* [358]. Amiloride also inhibits the binding of [^{3}H]batrachotoxin and [^{3}H]tetracaine to these membranes. Amiloride has also been shown to block the low-threshhold (T) Ca^{2+} channel in whole cell patch clamp studies of mouse (N18 cells) neuroblastoma [346]. Amiloride analogs have been shown to bind to a cation site in the pore of the Ca^{2+} channel and allosterically alter the interactions of the dihydropyridines and benzothiazepines [118]. Electrophysiological studies within intact frog atrial myocytes have shown that another analog of amiloride, namely, 3,4-dichlorobenzamil, inhibits a voltage-gated Ca^{2+} channel with a K_i of 0.8 μM, and is a more potent inhibitor of this Ca^{2+} channel than the Na^+/Ca^{2+} exchanger [49]. These investigators have also shown that this same analog may inhibit the delayed rectifier K^+ channel with a 30%–40% inhibition of the K^+ current at a concentration of 5 μM. Thus, although amiloride was initially considered to be a very specific inhibitor of Na^+ transport processes, recent investigations have definitively shown that amiloride and its analogs can have many untoward and far reaching physiological and biochemical consequences. Nonetheless, when used judiciously, this class of drugs can provide useful molecular probes with which to dissect, both pharmacologically and biochemically, various transport processes of interest.

The Chemistry of Amiloride

Amiloride is a pyrazine-ring based molecule with two amino groups attached at ring positions 3 and 5, a chloride at ring position 6, and an acylguanidinium group at ring position 2. Because of the guanidinium moiety, amiloride is a weak base with a pKa of 8.7 in a 30% ethanolic solution [331]. Protonation occurs on the guanidinium part of the molecule. Thus in the physiological pH range, amiloride exists primarily as a monovalent cation with a positive charge resonating between the terminal amidinium fragment. It is this charged form of amiloride that is the biologically active one for the Na^+ channel [42], the Na^+/H^+ exchanger [187], and for the Na^+/Ca^{2+} exchanger [167]. Figure 1 shows the different tautomeric forms of amiloride in aqueous solution. The free base (i.e., unprotonated form of amiloride, Ia and Ib) exists primarily as the acylamino tautomer (double bond between N8 and C9). The protonated form of amiloride (II) assumes the acylamino form (resonating double bond between C9 and N10 and N11). There are two potential acylamido configurations that the uncharged amiloride molecule may assume (Ia and Ib). The higher dipole moment of Ib compared to Ia would imply that Ib predominates in aqueous solution. Both the acylamido and acylamino confirmations are stabilized by two internal hydrogen bonds. A much weaker hydrogen bond exists between the 6-chloro and 5-amino hydrogen. In aqueous solution, amiloride exists in a coplanar configuration with molecular dimensions of 8 × 11.5 Å [314].

Amiloride has three absorption peaks: 211, 284, and 361 nm. The molecule does not absorb light of wavelengths 410–900 nm, even when dissolved in water at 10 mM. Amiloride is, however, a highly fluorescent molecule. When excited at 316 nm, the drug emits a broad, blue fluorescence in the wavelength range 370–500 nm with a peak emission of 410 nm. The intensity of fluorescence emission by amiloride is diminished if the molecule is excited at 286 nm, although the emission spectrum is similar [61].

The Amiloride-Sensitive Na^+ Channel

The apical membrane of many sodium-reabsorbing epithelia, such as renal distal and collecting tubules, descending colon, and trachea, contain sodium-specific ion channels that are inhibited with high affinity by the diuretic compound amiloride. These channels mediate the cellular entry of sodium from the luminal contents during the initial stages of active transepithelial sodium reabsorption. The channels are not voltage-dependent, but their gating properties are regulated by the mineralocorticoid hormone aldosterone and the peptide hormone vasopressin; by intracellular calcium, hydrogen, and sodium; and by guanine nucleotide regulatory proteins. The high amiloride affinity sodium channel protein has been isolated and purified to homogeneity. It consists of six subunits, ranging in molecular masses from 300 kDa to 40 kDa. The molecular mass of the native channel protein is 730 kDa. Another amiloride-sensitive channel has also been purified [26], but it is not clear whether this protein has a high or low affinity for amiloride. This protein has a native molecular mass of 185 kDa, and consists of a dimer of identical 105 kDa polypeptides. Single-site turnover numbers deduced from current noise experiments (10^6 ions/s) are consistent with a channel or pore-type mechanism of transport [196]. Fuchs et al. [116] and Van Driessche and Lindemann [357], found that, under their experimental conditions, sodium flow through these channels could be adequately described by an electrodiffusion model in which the passive movement of sodium obeys the independence principle.

Flux Ratio Analysis of the Amiloride-Sensitive Sodium Channel

One necessary criterion of independent ion movement through a particular transport pathway is that the relationship between unidirectional sodium influx and efflux obeys the Behn-Teorell-Ussing flux ratio equation. Ussing and Zerahn [354] first proposed that the ratio of unidirectional transepithelial sodium fluxes through the active transport pathway provided a measure of the effect of driving force of the sodium pump in frog skin. However, the use of the transepithelial flux ratio is problematic because a significant fraction of flux is made to traverse the epithelium via paracellular shunt pathways. In fact, because of the relatively high degree of basolateral membrane impermeability to sodium and the very small intracellular sodium transport compartment and the electrochemical gradient across the apical membrane favoring sodium entry, a tracer flux ratio analysis through the amiloride-sensitive sodium channel in intact epithelia has been accomplished only recently [36, 246]. Both groups of investigators found that there was an excellent agreement between the measured and predicted flux ratio over a very wide range of electrochemical potential energy gradients, indicating that sodium movement through the high affinity amiloride-sensitive channel obeys independence under these experimental conditions.

Voltage Dependence of the Amiloride Block

Previous studies have indicated that differences exist between species of frog skin as well as other epithelial preparations with regard to the interaction between sodium ions and amiloride. For example, amiloride has been found to be a competitive inhibitor of sodium entry in *Rana temporaria* skin [86, 192], rabbit descending colon [131], toad urinary bladder (see [33]), the skin of the Mexican axolotl [85], and the chicken chorioallantoic membrane [85], and noncompetitive in the skins of *Rana pipiens* and *Rana catesbeiana* [38], the hen coprodeum (see [180]), and the toad urinary bladder [43]. Mixed inhibition has been reported in toad skin [38], frog skin [293], and toad urinary bladder [341].

A major criticism of all the studies cited above is that changes in apical membrane potential, when external sodium and amiloride are varied, were not considered. The electric field across the apical membrane could easily influence the binding of amiloride, especially because amiloride must carry a net positive charge in order to inhibit sodium entry in frog skin. Palmer [248] and Hamilton and Eaton [144] have found that the magnitude of the apparent amiloride inhibitory dissociation constant in frog skin and toad bladder depends upon voltage. They also estimated that the fraction of the electrical potential felt by amiloride varied from 0.14 to 0.40. By incorporating these parameters into a purely noncompetitive inhibitory model, it can be shown that previously observed deviations from inhibitory kinetics can be accounted for [34]. The competitive inhibitory kinetics between amiloride and sodium in European frogs as compared to the noncompetitive behavior in American frog species may result because of differences in the magnitude of the change in apical membrane potential difference brought about by reducing external sodium, hence resulting in larger differences in the apparent K_i. It is important to realize, however, that this entire voltage-dependent amiloride block analysis, as well as the determination of macroscopic inhibitory constants [192] assumes that the sodium channel density and single channel currents remain constant. Recent experiments based upon current fluctuation analysis indicate that this assumption may not be valid [2, 150].

Interfacial Potentials and Apical Sodium Entry Channels

Benos et al. [37] investigated the effects that alterations in interfacial potentials have on sodium entry in bullfrog skin epithelia. The results of their study demonstrate that changes in the ionic strength of the external solution do not affect the rate of sodium transport through amiloride-sensitive entry channels. Also they found that uranyl ions do not change sodium transport through these channels, in spite of the fact that this molecule, at 2.5 m*M*, can induce a 145 mV increase in the interfacial potential of phosphatidylserine monolayers. Therefore, it is likely that the surface charge density in the vicinity of the sodium selective channel located in the apical membrane of frog skin is small (i.e.,

< 1 $e^-/600$ Å^2), and that this sodium entry channel is insensitive to changes in apical surface potential.

Single Channel Measurements

With the advent of patch clamp and planar lipid bilayer recording techniques, it has become possible within the past 8 years to observe amiloride-inhibitable sodium conductive events at the single channel level. Benos and collaborators first recorded single amiloride-sensitive sodium channels in bilayers [240, 299]. These channels were observed after incorporation of apical membrane vesicles derived from cultured A6 toad kidney cells grown on Millipore filters into the bilayer. The range of single channel conductances was 4–80 pS under symmetrical salt solutions (200 m*M* NaCl). Both the open channel conductance and the probability of a channel being in the open or closed state were independent of voltage up to $\pm$ 100 mV. From dilution potential measurements, the channels were found to be perfectly cation selective. However, the ability of these channels to discriminate between sodium and potassium was low. The measured permeability ratio of sodium to potassium estimated from bionic potential or conductance measurements was 2–3:1. These results were at variance with predictions of sodium/potassium selectivity ratios estimated in frog skin and toad urinary bladder epithelia, where Na^+/K^+ ratios of at least 100–1000:1 have been reported [39, 247], assuming only a single class of high amiloride affinity epithelial Na^+ channels with no multiple conductance states.

Amiloride interacts with this channel when present either in the *cis* or *trans* compartment of the bilayer. Addition of amiloride to the *trans* side produced a flickering block, where *cis* addition of amiloride reduced the apparent open state conductance, suggesting that the block was short-lived with respect to the amplifier response time. The *cis* block by amiloride was concentration-dependent, and the open state conductance was reduced to 50% of its initial value by 0.1 μ*M* amiloride. Olans et al. [240] analyzed the variation of conductance displayed by the epithelial sodium channel in bilayers. Two extremes of channel conductances were studied. Concentration versus sodium activity curves for both populations of channels saturated with increasing sodium concentration. For the low conductance channel, the half-maximal sodium concentration (K_{Na}) was 17 m*M*, while for the high conductance (44 pS channel), K_{Na} was 47 m*M*. In spite of these differences, both the small and large conductance channels displayed identical pharmacological profiles and ion selectivity. Both were equally amiloride sensitive; both were perfectly cation selective; both showed a Na^+/K^+ selectivity ratio of 2–3:1; and both saturated with increasing sodium concentration.

Hamilton and Eaton [144] initially described amiloride blockable channels in A6 cells using the patch clamp technique, while Palmer and Frindt [249] measured single sodium channels in rat cortical collecting ducts. More recently Light et al. [194] have monitored a moderately high conductance (20 pS), poorly cation discriminatory, amiloride-sensitive channel from primary cultures of rat inner medullary papillary collecting duct cells. Culture of A6 cells on impermeable supports led to the appearance of channels with low ion selectivity and a relatively short mean open time (50 ms; [144]). However, if A6 cells were grown on permeable supports, the sodium channels observed were comparable to those measured by Palmer and Frindt [249] in the collecting duct. In both cases, the selectivity of the channel was high (> 20:1 sodium over potassium); the channel conductance low, i.e., about 4–5 pS; and the channel has long open and closed periods (on the order of seconds). The current-voltage curves obtained for these channels showed rectification expected for a channel obeying the Goldman–Hodgkin–Katz constant field current equation. Further, Palmer and Frindt [249] found that the single channel currents saturated with increasing sodium, with K_{Na} on the order of 20 mM. These channels had a unit conductance of 5–8 pS, had long lived open and closed states (seconds), and linear current-voltage curves. These single amiloride-sensitive sodium channels were influenced by pH or Ca^{2+} [250, 251]. Reducing cytoplasm pH from 7.4 to 6.9 or 6.4 at constant calcium (10^{-7} M) decreased the probability of a single channel being open from 0.41 to 0.19 and 0.05, respectively. Single channel conductance was not affected by changes in pH. The effects of cytoplasmic calcium ion concentration on sodium flow through these amiloride-inhibitable channels are not yet clear. In vesicle ion flux studies [121], calcium also inhibited sodium movement in a pH-dependent fashion. At an intravesicular calcium ion concentration of 10^{-7} M, maximal inhibition of sodium channels occurred at pH > 7.5, but there was no effect at pH values less than 7. In contrast, calcium had no direct effect on single sodium channels studied in excised patches of rat cortical collecting tubule apical membrane [250, 251].

Ling and Eaton [197] examined the effect of external Na^+ concentration, amiloride, and phorbol esters on single Na^+ channels in A6 cells using the cell-attached configuration of the patch clamp technique. They found that reduction of external [Na^+] and/or application of amiloride increased open channel probability (P_o) and increased the number of Na^+ channels in the patch. The calcium ionophore A23187 prevented this rise in channel activity. Likewise, this increase in open state probability and the appearance of new channels was prevented by phorbol myristate acetate or oleyl-acetyl-glycerol. Sphingosine, a protein kinase C inhibitor, increased P_o and produced new channels. Thus, these authors conclude that the modulation of apical membrane Na^+ permeability by external Na^+ is not a direct effect of Na^+ on the channel, but occurs by way of an intracellular regulatory pathway involving cytosolic Ca^{2+} and protein kinase C.

Gogelein and Greger [127] have also reported the presence of amiloride-sensitive sodium channels in the pars recta portion of the rabbit proximal tubule. These channels had a unit conductance of 12 pS, had linear current voltage curves, and the probability of the channel being open or closed was found to be independent of voltage. Amiloride induced flickering with no effect on single channel conductance. However, the sensitivity of this sodium channel to amiloride was much lower than is commonly observed for a high affinity channel. It is important to point out that the patches used were inside-out, and amiloride was added to the cytoplasmic side. These channels also displayed a high sodium to potassium selectivity ratio, namely > 19. The channels spent most of their time in a closed state and, when open, were open for only a short period of time, i.e., less 500 ms.

Amiloride inhibited sodium flow through single channels either by changing the probability of a channel being in the open versus closed state or by decreasing current flow through single open channels. As indicated earlier, in general two different types of high amiloride affinity inhibitable channels have been described. One type has a very low sodium/potassium selectivity; the other a much higher one. In addition, either the low or high selectivity amiloride-sensitive sodium channel can open and close with relatively fast or slow time courses, depending on the conditions of measurement and tissue source. The open/closed kinetics do not correlate with selectivity or conductance. There also appears to be another class of sodium selective channels which are inhibited by amiloride with a much lower affinity, i.e., at concentrations > 1 μM. However, these channels appear to have a very different pharmacological profile than do the more typical high amiloride affinity channel types. An example of this type of channel comes from the work of Vigne et al. [363] in which they measured the properties of an amiloride-sensitive cation channel in endothelial cells of rat and pig brain cortex. These channels had a conductance of 23 pS in 140 mM NaCl, a $Na^+:K^+$ selectivity ratio of 1.5 as determined from reversal potential measurements, and an apparent K_i for amiloride inhibition of 10 μM. Phenamil also inhibits this channel with an apparent K_i of 15 nM. These compounds produced a flickering block. Although the order of inhibitory potency is the same as for the high amiloride affinity channel, this channel's kinetic, ion selectivity, and pharmacological characteristics differ so much from the high affinity channel that it is unlikely that the proteins comprising these channels are identical (see below). It is not yet clear whether the large or moderately large nonselective, amiloride inhibitable cation channel observed in the apical membrane of the rabbit renal inner medullary cells belongs to this category because amiloride dose response curves or amiloride analog studies have not yet been done on this channel. It may be, however, that the highly selective sodium channel is restricted to cortical nephron segments and other electrically tight or high transepithelial resistance epithelia, and the lower selectivity channel to the medullary regions of the kidney. These low amiloride affinity channels may also occur in other sodium transporting epithelia like proximal tubule, perhaps in the basolateral membrane, but a structural, functional, and molecular biological comparison between these different amiloride-sensitive forms of epithelial sodium channels has yet to be made.

Hormonal Regulation of Amiloride-Sensitive Sodium Channels

Amiloride-sensitive sodium channels are regulated by a number of different mechanisms and, in particular, by the mineralocorticoid hormone aldosterone and the peptide hormone vasopressin (antidiuretic hormone or ADH). Most of our knowledge about the properties and the regulatory mechanisms on these apical sodium channels comes from electrophysiological measurements on model epithelia, such as frog skin and toad urinary bladder. These studies have shown that both aldosterone and ADH cause a two- to fourfold increase in the density of apical sodium channels without affecting the single channel current [120, 123]. In addition, there exist intracellular feedback inhibition mechanisms involving increases in cell Ca^{2+} and/or Na^+.

Although the hormonal and intracellular regulation of sodium channels is well known, the molecular events are far from understood. Aldosterone definitely causes de novo protein synthesis, but it is not clear whether the newly synthesized proteins are sodium channels. However, recent evidence obtained using polyclonal antibodies raised against purified renal papillary Na^+ channel protein shows that the aldosterone-induced 70-kDa protein is a component of the renal Na^+ channel (Blazer-Yost, Palevsky, Szerlip, and Cox, personal communication). On the other hand, data from several different laboratories indicates that the aldosterone-induced proteins activate preexisting apical channels [17, 60]. Vasopressin was shown to increase sodium transport by increasing cell cAMP levels [241]. Again, the mediating events distal to the increase in the intracellular cAMP levels are not known. It is apparent that both ADH and aldosterone act via different mechanisms, yet their actions impinge upon the same transport protein. Lester et al. [191] found that the vasopressin-induced stimulation of sodium transport across intact toad urinary bladders is preserved in apical membrane vesicles derived from it. The effect of ADH could be mimicked by 1 mM cAMP and is further enhanced by the addition of the phosphodiesterase inhibitor 3-isobutyl-1-methylxanthine (10 μM). Li et al. [193] showed the response to the hormone can be fully accounted for by an increase in the density of apical sodium conductance sites with no change in the single channel current or in K_{Na}. This result was confirmed by Helman et al. [151] using vasopressin in frog skins. In addition, Garty and Edelman [122] found that apical trypsinization, which blocks most of the baseline sodium transport, does not impair the vasopressin-induced increase in sodium transport. This result indicates that the hormone stimulates transport by recruiting channels from an intracellular pool, or from apical membrane channels not accessible to trypsin proteolysis. On the other hand, pretreating the apical surface of toad urinary bladder epithelium with tyrosine-specific reagents inhibited to the same extent the baseline and antidiuretic hormone-induced increase in sodium transport. While these results are difficult to reconcile, there may be several explanations. First, the ADH-induced channels may be continuously present in the apical surface in a conformation not accessible to trypsin. Alternatively, the group-specific reagent used in these studies may also have nonspecific effects which can directly or indirectly inhibit channel insertion or modification by other apical membrane proteins.

A major question to be addressed is how does cAMP activate or induce sodium channels. Because cAMP activates a cAMP-dependent protein kinase, it is likely that the next step in the cascade is the activation of this kinase and subsequent protein phosphorylation. A vasopressin-activated, cAMP-dependent protein kinase is found in kidney medulla, toad urinary bladder, and cultured LLC-PK_1 cells [19, 101, 304]. Perhaps the most parsimonius mechanism would be that the protein kinase directly phosphorylates sodium channels and thereby converts them from an inactive to an active or open state. Another possibility is that cAMP-dependent protein kinase modifies another protein which in turn either activates dormant channels or induces the insertion of new channels into the apical surface. Lester et al. [191] recently tested these hypotheses using apical membrane vesicles derived from toad urinary bladder epithelium. Preincubating the tissues with vasopressin or 8-bromo-cAMP increased the channel-mediated Na^+ fluxes. Incorporating cAMP-dependent protein kinase, ATP and cAMP in vesicles derived from untreated cells, or trapping phosphatases in vesicles derived from cAMP-activated cells failed to alter channel activity. On the other hand, recent work on the effects of cAMP phosphorylation by Sariban-Sohraby et al. [300] has shown that one of the components (namely, the 300-kDa subunit) of the purified amiloride-sensitive sodium channel can be phosphorylated by cAMP-dependent protein kinase. Frings and Macknight (personal communication) report that, in isolated patches of toad urinary bladder apical membranes, exposure of the cytoplasmic surface to the catalytic subunit of cAMP-dependent protein kinase plus ATP activates quiescent amiloride-sensitive Na^+ channels.

Aldosterone Regulation of Sodium Channels

Aldosterone, like antidiuretic hormone, also increases the transepithelial transport of sodium, primarily by increasing apical sodium permeability. The characteristic response of in vitro epithelia to aldosterone added exogeneously is composed of three distinct phases: first, a latent period of 30–90 min in which cellular events take place; second, an earlier response of up to 6 h during which a two- to fourfold increase in sodium transport is observed; and third, a late response of up to 24 h in which sodium transport continues to increase mainly because of de novo synthesis of $(Na^+ + K^+)$ATPase pumps. In addition, an increase in the amount of membrane protein and lipid methylation has also been reported [298, 380]. This pleotropic action of aldosterone serves several purposes. First, it enables the elevation of transepithelial transport by several fold without reaching limitation by any of the individual processes involved. Second, it prevents the acute increase in cell sodium which would accompany augmentation of sodium transport induced by an increase in apical membrane permeability alone. Third, it is a long-term effect by which the epithelium can maintain whole-body sodium homeostasis.

The early effect of aldosterone can be fully accounted for by an increase in luminal sodium entry. Current fluctuation analysis supports the idea that this enhanced sodium entry is produced by an increase in apical sodium channel density with no change in single channel current. Thus, like ADH, there are two potential mechanisms of action: one, induction of the synthesis and/or insertion of new sodium channels in the apical membrane, or two, opening preexisting apical nonconducting channels. Because the hormone mediates its action by protein synthesis, the first possibility has been favored by a number of investigators. However, recent experiments provide strong evidence in favor of the second hypothesis. Group-specific reagents and trypsin were used to inhibit irreversibly basal channel activity, and the effect of such pretreatment on the tissue response to aldosterone was examined. In all cases, it was found that partial inhibition of basal sodium transport blocked the ability of aldosterone to increase transport rate. Thus, aldosterone increases sodium entry by activating channels that have previously been accessible to permeable protein-modifying reagents. In addition, recent studies with antibodies made against the sodium channel protein ([338]; see below) have shown that aldosterone treatment does not cause the movement of intracellular channels to the apical surface.

Sariban-Sohraby et al. [298] have suggested a mechanism by which aldosterone activates sodium channels. These authors suggest that transmethylation of either the channel protein or surrounding lipids produce the early effect. The role of aldosterone in this process may be to induce a specific methyl transferase or to elevate the intracellular concentration of S-adenosyl-L-methionine by affecting other enzymes. However, because these experiments are subject to a number of criticisms, this interpretation should therefore be viewed cautiously.

Regulation of Sodium Channels by G Proteins

Guanine nucleotide binding proteins (G proteins) are a family of proteins composed of three subunits (α, β, γ) which couple plasma membrane receptors to a variety of enzymes and ion channels (63). Several ion channels are regulated by G proteins (see [194, 195] for discussion). G_i activates potassium channels and isolated atrial cells and the GH_3 pituitary cell line, G_s stimulates calcium channels in skeletal muscle and cardiac cells, and G_o activates calcium channels in neuronal cells and potassium channels in hippocampal neurons. Previous studies by Mohrmann et al. [218] and Light et al. [194, 195] have shown that electrogenic sodium absorption by LLC-PK_1 cells, a renal epithelial cell line, may also be regulated by a G protein. Pertussis toxin (PTX), a compound that prevents receptor-dependent activation of G_i and G_o, reduced sodium absorption across this epithelia. In recent papers, Light et al. [194, 195], reported that the α subunit of G_i directly activates a nonselective, amiloride-sensitive cation channel from cultured rabbit inner medullary collecting duct cells and, thus, regulates sodium entry across the apical membrane. In addition, Ausiello et al. [20] report that the purified epithelial sodium channel from A6 cells may, in fact, contain two subunits which are themselves G proteins. This group proposes that in epithelia, the G protein–ion channel complex in the apical membrane, because it is geographically separate from the G protein receptor complexes in the basolateral membrane, may have a rather unique mechanism of G protein regulation. Because hormones bind to receptors in the basolateral membrane and activate the ion channel in the apical membrane, the effector activated by the G protein complex in the basolateral membrane may serve as an intracellular mediator and either bind to and directly activate the G protein in the apical membrane or activate the G protein within the ion channel by phosphorylation reactions. PTX, then, would block the interaction between the intracellular effector and the G-protein(s) contained within the amiloride-sensitive cation channel. Tests of this hypothesis must await future experimentation.

Purification and Biochemical Characterization of Sodium Channel Proteins

There have been relatively few attempts at the isolation of the epithelial sodium channel using various radiolabeled amiloride analogs. In general, suitable high affinity radiolabeled probes with high specific activity were unavailable so that binding could not be accurately measured. In earlier studies the amiloride analog [^{3}H]benzamil inhibited a homogeneous class of binding sites in membrane vesicles obtained from either bovine or rat kidney cortex [105, 182, 183] with an equilibrium dissociation constant ranging from 5–40 n*M*. Treatment of bovine renal cortex homogenate with the detergent deoxycholate resulted in the solubilization of over 70% of the total protein, but with recovery of only 20% of the specific binding sites. In these studies sodium transport function of the solubilized protein was not reported. In the rat kidney cortex the technique of radiation inactivation analysis estimated the molecular mass of a high amiloride affinity sodium transport site of 650 kDa.

Benos and colleagues have recently been successful in isolating and purifying to homogeneity an amiloride-sensitive sodium transport system from bovine renal papillary collecting tubules and from the amphibian A6 kidney cell line [41, 297]. In these studies an enriched apical membrane preparation from the A6 cells or bovine papilla was first prepared. These vesicles were enriched in apical membrane markers approximately ten fold and contained an amiloride blockable sodium conductance as assayed by a radioactive uptake technique. Sodium channels were solubilized from these membranes by treatment with the zwitterionic detergent 3-[(cholamidopropyl)dimethylammononio]-1-propanesulfonate. This detergent step solubilized approximately 50% of the total protein with 100% recovery of the high affinity binding sites. Following detergent solubilization, the sodium channel protein was further purified by a combination of conventional biochemical methods, including wheat germ agglutinin chromatography, and gel filtration using size exclusion high performance liquid chromatography. This final purified material was reconstituted into planar lipid bilayers (Figs 4, 5) and was shown to display amiloride-sensitive, single sodium channel activity. The native sodium channel from both amphibian and mammalian kidney consists of a heterohexameric complex of subunits ranging

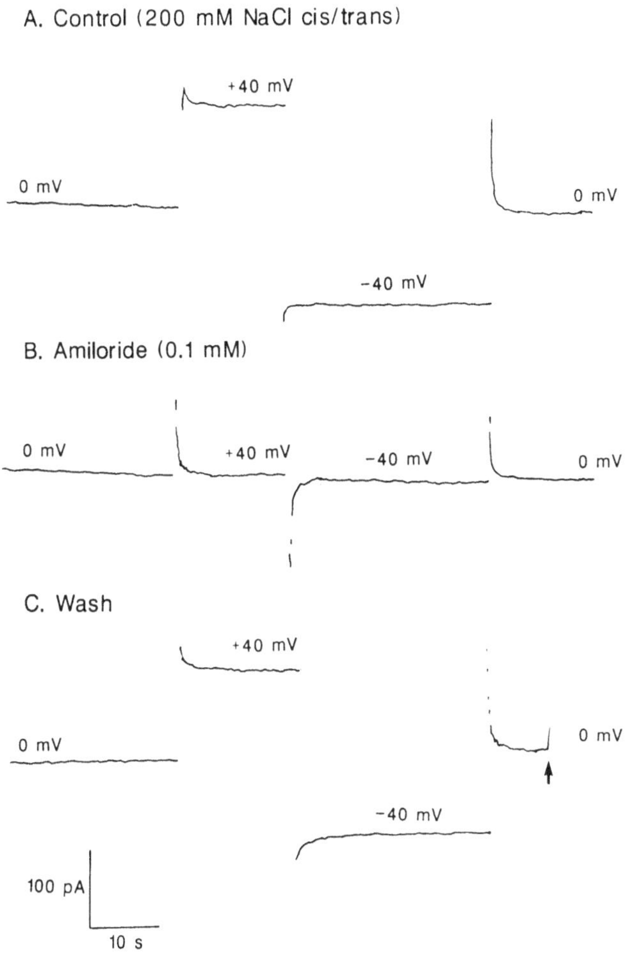

Fig. 4. Macroscopic Na^+ current measurements in a planar lipid bilayer (7:3 phosphatidylethanolamine: phosphatidyl serine, 20 mg/ml in n-decane) using purified bovine renal papillary Na^+ channel protein

in molecular size from 300 kDa to 40 kDa. These subunits are most likely covalently attached to each other by disulfide bridges because in the absence of any reducing agents, the native complex runs as a single broad band of 730 kDa. The inclusion of 6 *M* urea (a chaotropic agent that minimizes hydrogen bonding and potential hydrophobic interactions) fails to disrupt the large protein complex. Thus, it appears that the native protein contains several non-identical subunits. At a minimum, the three highest molecular mass subunits, i.e., of 300, 150, and 95 kDa, are heavily glycosylated, containing sialic acid homopolymers. The large molecular size of the purified epithelial sodium channel protein agrees well with the earlier results concerning the molecular mass of the native membrane protein from rat kidney deduced by target analysis.

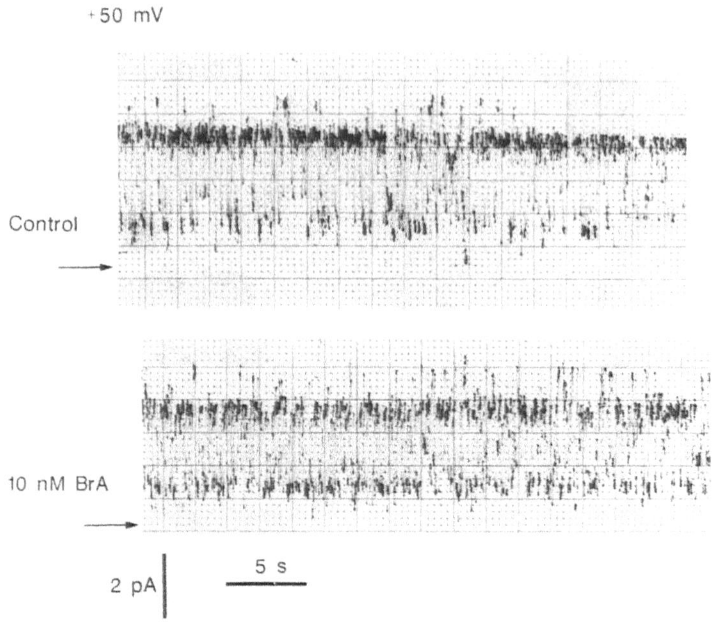

Fig. 5. The effect of 6-bromoamiloride (*BrA*) on single Na^+ channel current produced by HPLC-purified Na^+ channel protein from bovine renal papilla incorporated into a planar lipid bilayer. There was 200 m*M* symmetrical NaCl (pH 7.0) on both sides of the bilayer. *Arrows*, zero current level

The discovery of irreversible photosensitive amiloride analogs with subsequent radioactive synthesis has made possible the identification of amiloride binding subunit. Tritiated methylbromoamiloride specifically incorporates into a minor component contained in a solubilized bovine renal papillary homogenate with an apparent molecular mass of 730 kDa under nonreducing conditions and 150 kDa under reducing conditions. These proteins clearly correspond to those present in the purified epithelial sodium channel preparation. Kleyman and colleagues [182, 183], using [^{3}H]bromobenzamil, have found that this high affinity analog of amiloride covalently incorporates into three polypeptides of a bovine kidney cortex preparation with apparent molecular masses of 176, 77, and 46 kDa. In a subsequent study, Kleyman and coworkers [181] used another high affinity photoreactive amiloride analog, 2′-methoxy-5′-nitrobenzamil, and anti-amiloride antibodies (see below) to label a 130- to 180-kDa polypeptide in homogenates of A6 cells and bovine kidney cortex. In addition, 75- to 80-kDa and 50-kDa peptides were periodically labeled. These results suggest either the existence of multiple amiloride binding sites in the channel complex, or that the lower molecular weight polypeptides may be degradation products of the larger ones. This issue is unresolved at present.

Recently another putative amiloride-sensitive sodium channel from pig kidney cortex has been isolated [26]. This protein may exist in a native configuration as a dimer, composed of two identical subunits of 90- to 100-kDa size. It is most probable that this protein is not identical to the more typical high amiloride affinity epithelial sodium channel discussed above. The reasons for this are that there are major differences in affinity between amiloride and a number of its structural analogs. Amiloride inhibits the high affinity channel with a K_i ranging anywhere from 50–100 nM. Under the same conditions amiloride inhibits sodium uptake into the pig kidney cortex vesicles with a K_i on the order of 10 μM. 250 times less than what is observed for the high affinity channel [26]. In addition another analog of amiloride called ethylisopropyl-amiloride (EIPA) can inhibit this pig kidney cortex channel with the same effectiveness as amiloride. EIPA, in essence, has very little effect on the high amiloride affinity sodium channel. It may be that this protein represents a low amiloride affinity conductive pathway, which has been observed in vesicle uptake studies and in the basolateral membrane of certain epithelia.

Antibodies and Immunocytochemical Localization of Epithelial Sodium Channel Proteins

Using A6 cells, Turner and collaborators [353] raised over 100 monoclonal antibodies against the apical membranes of cells grown on plastic, but none of these antibodies inhibited sodium transport. Likewise, Moberly and Fanestil [217] also raised a number of antibodies against the apical membrane proteins from the A6 cells. In these experiments the cells were grown on plastic, but these antibodies also did not inhibit function. Kleyman et al. [182] have synthesized an amiloride analog with a caproic acid group covalently bound to the guanidinium moiety of the amiloride. This analog was in turn coupled to bovine serum albumin and used as a haptan to raise anti-amiloride antibodies. These investigators found that the anti-amiloride antibodies could reverse the inhibition of amiloride on sodium transport across toad urinary bladder. It is possible that these anti-amiloride antibodies may structurally resemble the amiloride binding site on the sodium channel. Subsequently, these authors have raised anti-idiotypic antibodies that may prove useful in the identification of the amiloride binding site.

Verrey and coworkers [360] have produced an antibody originally directed against the α subunit of the sodium/potassium ATPase that specifically labels the apical membrane of both the A6 cell line and toad urinary bladder. Immunoprecipitation followed by SDS gel electrophoresis of iodinated apical membrane proteins from A6, toad urinary bladder, and TMB monolayers (TMB cells are a cultured cell line originally derived from toad urinary bladder) yielded a series of proteins with molecular weights closely corresponding to the subunits of the epithelial sodium channel reported by Benos et al. [41]. Furthermore, this antibody, in Western blots and enzyme-linked immunoabsorbent enzyme assays (ELISA), will recognize purified epithelial sodium channel protein, specifically the 150-kDa subunit. Sorscher et al. [338] have produced several hybridoma cell lines secreting antibodies against the purified amiloride-inhibited channel protein from bovine renal papilla. In addition polyclonal antibodies were raised in rabbits against this channel protein. These antibodies show considerable specificity in ELISA and dot blot assays and can immunoprecipitate radiolabeled channel protein. The polyclonal antibodies bind at least four sodium channel subunits on Western blots, namely the 300-, 150-, 95-, and 40-kDa polypeptides, and cross-react with channel protein isolated from the A6 cells. The antibodies also have been used to affinity purify the channel in relatively high yield from crude detergent-solubilized kidney homogenates. The monoclonal antibodies, on the other hand, are very specific in their interaction with the sodium channel, i.e., they only react with a 300-kDa subunit, and do not cross-react with proteins isolated from amphibian A6 cell line.

The localization of amiloride-sensitive sodium channels in papillary collecting tubules of bovine kidney and cultured amphibian A6 cells was examined using these antibodies [348]. The distribution of the channel protein was measured in thick frozen sections at the light microscopic level using indirect immunofluorescence, and at the electron microscopic level using immunogold labeling. In cells of both the intact bovine collecting tubule and A6 confluent monolayers, only the luminal or apical-facing surface membranes showed staining. Sodium channel protein was characteristically localized on microvillar domains on the apical plasma membrane. Little or no basolateral membrane staining was evident. Channel protein was also absent from subapical membrane vesicles, tight junctions, and was not found in nonepithelial Chinese hamster ovary cells and skin fibroblasts. In addition the apical membrane of the collecting duct principle cells obtained from rat renal medullary collecting ducts was heavily stained by the anti-sodium channel antibody, whereas the intercalated cells were unstained [64]. Basolateral membranes of both cell types were unstained, as were subapical membrane vesicles in the cytoplasm of these cells. These results clearly show that amiloride-sensitive sodium channels are located predominately on the apical membrane of principal cells in the medullary collecting duct. The principal cells are involved in sodium homeostasis in this region of the kidney.

As indicated earlier, various hypotheses concerning the mechanism of sodium channel regulation by aldosterone abound. One possibility is that aldosterone induces the synthesis of new channel protein because of the effects of aldosterone on sodium entry are blocked by inhibitors of mRNA transcription and protein synthesis. With the availability of an anti-sodium channel antibody, this possibility was tested by comparing the staining patterns of untreated A6 cells or those treated for 16 h with $0.1 \mu M$ aldosterone. The data obtained from these experiments indicate that latent sodium channels are always present at the cell surface, and can be activated by an effector protein which may be induced by

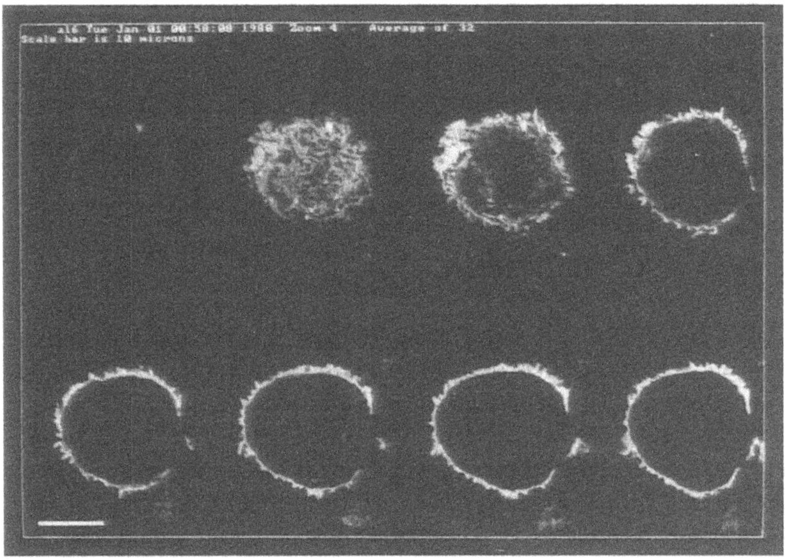

Fig. 6. En face view of an aldosterone-treated, filter-grown amphibian A6 kidney cell stained with polyclonal anti-Na^+ channel antibodies. The cell was optically scanned with a confocal fluorescent microscope at 0.5-μM intervals. Staining is restricted to the apical membrane surface. *Bar*, 10 μM. (Micrographs courtesy of A. Toussan and R. Zinkowski)

aldosterone or by post-translational modifications. No evidence of channel recruitment from subapical membrane vesicles or other cytoplasmic pools by aldosterone was observed (Fig. 6).

Because of the availability of anit-sodium channel antibodies, the examination of intracellular synthesis, processing and regulation of the epithelial sodium channel is now feasible. In addition, these antibodies are currently being used to screen libraries of complementary DNA clones prepared from poly A^+ messenger RNA, so that eventually the primary amino acid sequence of the components of this important channel protein can be determined.

The Na^+/H^+ Antiport System

Introduction

Kinetic studies of the Na^+/H^+ antiporter in epithelial cells have defined its basic properties, regulatory features, and responses to various activators and inhibitors. A special feature of epithelial cells relates to the distribution of the Na^+/H^+ antiporter in the plasma membranes with respect to its specialized role in vectorial transport of protons across the epithelium. The net effects of vectorial Na^+/H^+ exchange may include titration of extracellular buffer, cell volume regulation, as well as participation in the absorptive flux of Na^+. In general, the basic properties of the Na^+/H^+ antiporter are similar in a variety of epithelial tissues [372]. The best characterized preparation is the brush border membrane vesicle from the renal cortex, which will serve as the primary focus for this review.

Basic Properties of Na^+/H^+ Exchangers

The Na^+/H^+ antiporter is a mechanism for proton extrusion and sodium uptake that is driven by the prevailing Na^+ and/or H^+ electrochemical energy gradient in epithelial cells. This process does not directly require ATP or other forms of energy [229]. As such, the Na^+/H^+ antiporter is an example of a "secondary" active transport process by which uphill Na^+ influx is driven by an outwardly directed proton gradient, or by which an outwardly directed Na^+ gradient can cause intracellular or intravesicular accumulation of protons.

Stoichiometry

The Na^+/H^+ antiporter is an electrically neutral exchanger through which equal but opposite fluxes of Na^+ and protons proceed. This characteristic was established by studies in which the net fluxes of both cations were measured in proximal tubule cells, by the lack of its effect on transmembrane electrical potentials [204, 205] and by the thermodynamic demonstration that oppositely directed Na^+ and proton gradients reduce the net flux through the exchanger to zero [177]. In addition, Hill analyses of the brush border Na^+/H^+ antiporter are also consistent with a stoichiometric ratio of 1:1 [204]. A recent study in human leukocytes indicates that internal pH may affect the stoichiometry of the exchanger. Ng and Bomford [231] found a Hill ratio of 1:1 for Na^+ and H^+ fluxes when the internal pH was 6.9, but an activation of the proton efflux process occurred with a Hill coefficient of 2.0 when the internal pH was set at 6.0. These results suggest that the stoichiometry of the Na^+/H^+ antiporter is not absolute, and that the exchanger may operate in an electrogenic mode (e.g., $2\,H^+/1\,Na^+$) under certain experimental conditions.

Pre-steady-state kinetic studies of the rabbit renal brush border Na^+/H^+ antiporter have been performed by Otsu et al. [242]. Utilizing ^{22}Na uptake measurements made at 0°C, they were able to distinguish an early exponential increase in the uptake rate before the steady-state rate was established. As shown in Fig. 7, extrapolation of the steady-state rate to zero time provides an estimate of the "burst amplitude" of the pre-steady-state process. The burst amplitude (*y* intercept) is clearly a function of the Na^+ concentration. The inset in Fig. 7 demonstrates, however, that the rate of the early exponential burst is independent of the Na^+ concentration. An apparent turnover number (s^{-1}) can be obtained by dividing the steady-state rate (nmol mg^{-1} protein s^{-1}) by the

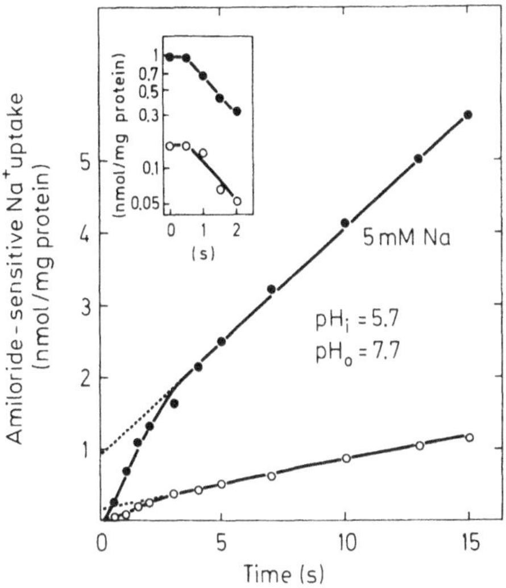

Fig. 7. Amiloride-sensitive $^{22}Na^+$ uptake (nmol/mg protein) versus time in rabbit renal brush border vesicles at 0°*C*. *Inset*, semilogarithmic plots of Na$^+$ uptake during the burst phasey- obtained by subtracting the pre-steady-state time points from the extrapolated steady-state balue at time zero. Open symbols, 1 m*M* Na$^+$; *closed symbols*, 5 m*M* Na$^+$. (From [242])

burst amplitude (nmol mg^{-1}protein). The advantage of this approach is that it is independent of any estimate of number of binding sites from a Scatchard analysis which may greatly overestimate the number of functional transporters [99].

These results have been interpreted by Otsu et al. [242] to indicate that at least two Na$^+$ transport sites must be occupied to activate the exchanger in the pre-steady-state. The transition to steady-state Michaelis-Menten kinetics was explained by a "flip-flop," or alternating site mechanism in which the functional unit of the Na$^+$/H$^+$ antiporter is made up of monomers. At the start of the initial burst phase, the monomers would be outwardly oriented as a result of the initial pH gradient and the lack of external Na$^+$, but following the first turnover, only half of the monomers per transport cycle would form a translocation complex with external Na$^+$ [242]. Beliveau et al. [32] have estimated the molecular size of the rat renal Na$^+$/H$^+$ exchanger to be 321 kDa with radiation activation analysis; this estimate would include all of the monomeric units which are functionally linked and will therefore exceed the size estimate for the individual transporter unit.

Initial reports by Murer et al. [229] and Kinsella and Aronson [174] did not demonstrate any effect of imposed electrical potential on the rate of Na$^+$/H$^+$ exchange, consistent with the electrically neutral character of this process. These studies were done, however, before it was appreciated that a significant proton leak pathway is present in brush border membrane vesicles [65, 275, 290]. In fact, an effect of interior-negative electrical potential on the rate of $^{22}Na^+$ uptake has been observed in renal brush border vesicles by Warnock and Yee [375]. The conclusion that the Na$^+$/H$^+$ antiporter is an electrically neutral process rests most firmly upon the thermodynamic analysis utilized by Kinsella and Aronson [177], but may depend upon the precise experimental conditions [231, 375].

Substrates

In addition to Na^+/H^+ exchange with Na^+ moving in the influx direction, the Na^+/H^+ antiporter is reversible, and can also mediate Li^+/H^+ exchange, NH_4^+/H^+ exchange, and self-exchange for Na^+, Li^+, and ammonium. As would be expected, the alternative substrates for the Na^+/H^+ antiporter compete at the cation transport site and thereby inhibit Na^+ influx. Direct measures of Li^+/H^+ exchange [109, 160, 161] demonstrate that Li^+ is an alternative substrate for the Na^+/H^+ antiporter. Of note, there is an increase in the apparent cation affinity when Li^+ is compared to Na^+ in rabbit renal cortical vesicles, and no significant change in the colonic brush border vesicle system (Table 1).

Similar studies have not examined NH_4^+/H^+ exchange because of technical limitations. The acridine orange technique has not been used because transmembrane entry of NH_3 buffers internal protons and thereby collapses the preformed pH gradient. Although this result would mimic NH_4^+/H^+ exchange, the actual mechanism would simply represent permeation of a weak base (e.g., NH_3). Competition studies have been done using $^{22}Na^+$ uptake. These studies indicate that NH_4^+ competes for Na^+ at the external cation binding site [138, 175]. In addition, H^+ itself has an appreciable affinity for the external cation binding site at which a competitive interaction occurs with Na^+ [15]. Based upon these kinetic and competition studies, it appears that the cation specificity of the external binding site is $H^+ \gg Li^+ > Na^+ > NH_4^+$. In addition, other cations including K^+, Rb^+, Cs^+, choline, tetramethylammonium and *N*-methylglucamine have a negligible affinity at the external cation binding site [175, 373].

Inhibition by Amiloride

Reversible inhibition by amiloride is a nearly universal characteristic of the Na^+/H^+ antiporter. For the epithelial form, the inhibition constant for amiloride is approximately 15 μM [176] and is several orders of magnitude greater than that for inhibition of the classical Na^+ channel in tight epithelia [33] and

Table 1. Kinetic features of Na^+/H^+ and Li^+/H^+ exchange in rabbit brush border membrane vesicles

	Renal Cortes[a]	Colon[b]
Na^+/H^+ exchange		
Maximum rate (FV/mg protein s^{-1})	7.1	2.5
Cation equilibrium constant (mM)	14.3	20.0
Li^+/H^+ exchange		
Maximum rate (FV/mg protein s^{-1})	2.4	2.4
Cation equilibrium constant (mM)	1.2	29.0

[a] FV, Arbitrary fluorescence units: from [160, 161].
[b] Protein content assumed to be 125 μg: from [109].

the Na^+/H^+ antiporter which is expressed in nonepithelial cells [134]. More importantly, entirely different patterns of inhibition are obtained when various analogs are tested against the various transport systems. As indicated earlier, substitutions on the acylguanidinium group of amiloride produces much more potent inhibition of the Na^+ channel and Na^+/Ca^{2+} exchanger, while alkyl-substitutions on the 5-amino group produce more potent inhibition of the Na^+/H^+ exchanger [167, 180, 313, 364]. In addition to its rapid onset and reversible inhibition of the Na^+/H^+ exchanger, there appears to be a sidedness to the effect of amiloride. Preloading studies do not demonstrate an effect of amiloride on the internal aspect of the Na^+/H^+ antiporter [176, 373], and a transinhibitory effect is apparent during Na^+ efflux studies [373]. A similar asymmetry has been observed in canine red blood cell ghosts [135]. Therefore, it can be concluded that the inhibition of the Na^+/H^+ antiporter is mediated by an interaction of amiloride at an external site (or sites) on the Na^+/H^+ antiporter. The functional asymmetry with respect to amiloride implies that the transport site undergoes a conformational change upon translocation across the membrane or, alternatively, that a second site required for amiloride binding is accessible only from the outside [135].

The kinetic details of amiloride inhibition of the exchanger have proven to be controversial. Mahnensmith and Aronson [205] have clearly demonstrated simple competitive inhibition of the Na^+/H^+ antiporter by relatively low concentrations of amiloride in rabbit renal brush border membrane vesicles, a result which is similar to findings in other systems. In contrast, the report by Ives et al. [160] clearly documented a mixed inhibitory pattern in which the maximal transport rate was reduced and the apparent affinity for Na^+ was increased by amiloride. A simple technical difference appears to account for this discrepancy. The studies of Ives et al. [160] were performed with buffers that lacked chloride (replaced by gluconate), while the studies which report simple competitive inhibition had at least 50 m*M* chloride present in the uptake medium. A recent report by Warnock et al. [372] demonstrated that either competitive or mixed inhibition could be obtained with the same vesicle preparation when chloride versus gluconate buffers were used for the amiloride studies. In addition, other studies have recently reported a non-competitive effect of amiloride or a 5-amino substituted analogue on the Na^+/H^+ antiporter [84, 99, 223]. The report by Costa-Casnellie et al. [84] demonstrated that the properties of the Na^+/H^+ exchanger in a granulocyte cell line (HL-60) are influenced by the cell maturation stage and the intracellular pH. It appears that in most systems, a clear cut competition exists between amiloride and Na^+ at the external cation binding site. It also appears that a separate site exists through which amiloride exerts a noncompetitive effect on the external aspect of the Na^+/H^+ antiporter. It is also possible that the noncompetitive effect of micromolar concentrations of Li^+ on Na^+/H^+ exchange [160] is exerted on this second site on the external aspect of the Na^+/H^+ antiporter. Also, the noncompetitive Na^+ interaction with amiloride may occur at a second external site [135, 160] which may also be the locus at which ambient chloride interacts with the Na^+/H^+ exchanger [374].

As a generalization, the pharmacological inhibition profile typical for the Na^+/H^+ exchanger is 5′-*N* substituted analogs > amiloride > guanidinium-substituted analogs [180]. Several exceptions are worth noting. Kapus et al. [169] examined the mitochondrial inner membrane Na^+/H^+ exchanger and found that benzamil derivatives were the most potent inhibitors, while 5′-*N* substituted analogues were very weak inhibitors. Takaichi et al. [342] have recently described the endogenous Na^+/H^+ exchanger in oocytes of *Xenopus laevis* and found that amiloride was a better inhibitor than 5′-*N* substituted analogues, while benzamil was a very weak inhibitor. Moran et al. [222] have found an electrically conductive pathway in vesicles prepared from cultured renal epithelial cells and rat renal brush border membrane vesicles. Unlike other conductive Na^+ channels [33], this system was inhibited by 5′-*N* substituted analogues of amiloride better than benzamil analogs.

Amiloride sensitivity is nearly a universal characteristic of the Na^+/H^+ antiporter. Two exceptions from mammalian systems are worth noting. Grinstein et al. [133] have described an electrically neutral Na^+/H^+ exchanger in rat thymocytes which is not inhibited by amiloride but shares the same substrate specificity as the more usual form. Gurich and Warnock [139] have described an electrically neutral Na^+/H^+ exchanger in endosomes prepared from rabbit renal cortex which is completely insensitive to amiloride or its more potent analogues. Furthermore, Li^+/H^+ exchange was not evident in this preparation, even though Li^+ completely inhibited Na^+/H^+ exchange [139]. Further study is required to determine whether these systems are entirely different forms of the Na^+/H^+ antiporter, or perhaps represent precursors and/or degradation products in the biosynthetic and processing pathways for the Na^+/H^+ exchanger.

Internal Modifier Site

Acidification of the cytosolic aspect of the Na^+/H^+ antiporter activates the exchanger at an allosteric site such that the rate of Na^+/H^+ exchange exceeds what would be predicted by the change in the proton gradient. [14]. Similarly, in renal proximal tubule vesicles and cells [57, 235, 365] proton efflux showed steeper dependence on the intracellular pH than would be predicted by a simple Michaelis-Menten relationship. This finding, and the observation that internal acidification accelerated Na^+/Na^+ exchange, stand in marked contrast to the competitive inhibition observed when external pH is lowered [15, 235, 365]. These observations formed the basis for the proposal that an allosteric modifier site exists on the cytosolic aspect of the Na^+/H^+ antiporter [14].

The internal modifier site plays a central role in the rapid regulation of the Na^+/H^+ antiporter. The “set point” [134] is an issue which warrants special consideration. Since the Na^+/H^+ exchange is driven by the respective proton and Na^+ gradients, a steady-state should be achieved in which both gradients are balanced. The resulting intracellular pH can be viewed as the set point and as such would determine the intracellular pH, especially in the simplified case where bicarbonate has been omitted so as to inactivate other mechanisms which regulate intracellular pH (e.g., Cl^-/HCO_3^- exchange, Na^+/HCO_3^- cotransport). This set point idea was tested in a cultured renal cell line in which steady-state intracellular pH (in the absence of bicarbonate) was measured with the dimethyloxazolidine-dione technique [264]. If Na^+/H^+ antiporter activity was inhibited with amiloride, the steady-state pH fell. If Na^+/H^+ exchange activity was increased by introducing monensin, the steady-state intracellular pH rose. These results demonstrate that the regulation of the Na^+/H^+ antiporter and the apparent zero net flux at the set-point is due to kinetic rather than thermodynamic factors. Because increasing (monensin) or decreasing (amiloride) the rate of Na^+/H^+ exchange altered intracellular pH, the system was not at thermodynamic equilibrium, but was at a kinetically defined steady-state. This sort of regulation also explains the sensitivity of the Na^+/H^+ antiporter to changes in the intracellular pH around the set point. This finding is entirely consistent with the allosteric, internal modifier site in which decreases in cell pH below the set point would activate the Na^+/H^+ antiporter, and alkalinization inhibits its activity. Numerous examples [134] demonstrate that rapid activation of the Na^+/H^+ antiporter by hormones, growth factors, etc. is accompanied by an adjustment of the set point in the alkaline direction by 0.1to 0.3 pH units, such that the Na^+/H^+ antiporter is more readily activated by cytosolic acidification and less readily inhibited by cytosolic alkalinization. Furthermore, inhibition of the Na^+/H^+ antiporter by parathyroid hormone or cAMP adjusts the set point in the acid direction such that the Na^+/H^+ antiporter becomes less sensitive to activation by cytosolic acidification. [213]. Finally, Cassel et al. [71] have suggested that ATP may play a regulatory role at the inner aspect of the Na^+/H^+ exchanger. Thus, the internal modifier site provides a locus for the short-term regulation of the Na^+/H^+ antiporter activity by stimulatory and inhibitory factors.

Other cations appear to interact at the internal modifier site. Li^+ has an inhibitory effect in renal brush border membrane vesicles, while Na^+ has relatively little effect [160]. The effects of Na^+ on the internal modifier site are influenced by the prevailing local pH, suggesting that protons may compete with Na^+ for binding to the modifier site [128, 369]. Low concentrations of ammonia also stimulate the initial rate of $^{22}Na^+$ uptake [371], an effect which could be attributed to a change at the modifier site. Cell shrinkage and swelling can activate the Na^+/H^+ exchanger [99] and also affect the affinities of the intracellular cation binding site. Further studies are required to define the ionic selectivity of the internal modifier, and the effects of pH on selectivity. At present, it appears that the internal modifier site may be a locus where the cations which are substrates of the Na^+/H^+ antiporter may also allosterically regulate its activity.

Biochemistry of the Na^+/H^+ Antiporter

Recent studies have used site specific reagents to characterize the functional groups involved in the Na^+/H^+ antiporter. The pKa for the interaction of external protons with the Na^+/H^+ antiporter is approximately 7.4, consistent with titration of an imidazolium group [132]. Furthermore, a histidine-reactive reagent (diethylpyrocarbonate) inhibits the Na^+/H^+ antiporter in renal brush border membrane vesicles [132]. Carboxy-reactive reagents (*N*, *N'*-dicyclohexylcarbodiimide, DCCD; *N*-ethoxycarbonyl-2-ethoxy-1, 2-dihydroquinoline, EEDQ) also irreversibly inhibit the Na^+/H^+ antiporter [65, 115, 158, 288]. Amiloride protects the Na^+/H^+ antiporter against inactivation by carboxy-reactive reagents. It is not known whether this critical carboxy group is involved at the substrate binding site or some other external aspect of the Na^+/H^+ antiporter. The fact that amiloride and the 5-amino substituted analogs do not protect against EEDQ attack, suggests that this critical carboxy group interacts with the 5-amino group on amiloride [288]. There also appear to be different critical carboxy groups in the Na^+/H^+ antiporter because different reagents have different effects on the kinetics of the exchanger. For example, DCCD decreases the affinity for substrates, but does not affect the maximal rate of transport of the exchanger [159]. Furthermore, substrates such as Na^+ or Li^+ protect the exchanger against attack by DCCD. In contrast, EEDQ does not affect the affinity of the exchanger for substrates, but does reduce its maximal rate of transport, and Na^+ and Li^+ do not protect the exchanger against attack by EEDQ [288]. The Na^+/H^+ antiporter in vesicles prepared from rat renal cortex appear to be more sensitive to DCCD than the exchanger in brush border vesicles prepared from rabbit renal cortex [114].

Recent studies have suggested that isothiocyanates can inhibit the Na^+/H^+ antiporter in rabbit brush border membrane vesicles [155, 156]. The alkaline conditions used in these studies suggest that the nucleophile undergoing attack is a critical lysine group in the exchanger. Substrates do not protect against attack, but amiloride analogs do protect. A number of isothiocyanates will attack the Na^+/H^+ antiporter and reduce its maximal transport rate without changing the affinity for substrates, but the most potent analog identified by Huang et al. [155] is a 5'-(3-isothio-cyanatophenyl) derivative of amiloride synthesized by Cassel et al. [70].

To summarize the results obtained with group-specific reagents, there appears to by a critical carboxy group (attacked by EEDQ), a critical histidine group (attacked by diethylpyrocarbonate), and a critical lysine group (attacked by isothiocyanate derivatives) in a hydrophobic region of the Na^+/H^+ antiporter. All of these reagents irreversibly reduce the maximal transport rate but do not affect substrate affinities. Amiloride and its analogs protect the exchanger from attack by these reagents, but substrates do not provide any protection. Of interest, Kaback [166] has recently described a similar arrangement of critical amino acid side chains in a hydrophobic membrane spanning region of another proton translocating system, the *lac* permease of *Escherichia coli*. Molecular modelling by McDaniel et al. [210] has developed evidence consistent with this arrangement of functional side chains in the Na^+/H^+ antiporter.

Labeling Experiments

Recent studies have used amiloride analogs and group specific reagents to characterize the renal Na^+/H^+ antiporter in brush border preparations. Studies with [^{14}C]DCCD, a carboxy-reactive reagent, have provided molecular mass estimates of 65 kDa for the rat renal cortical Na^+/H^+ exchanger [113, 114] and 100 kDa for the rabbit renal Na^+/H^+ exchanger [159]. Cross-linking studies with DCCD and [^{3}H]MBA have demonstrated two bands in rabbit renal brush border vesicles [158].

Burckhardt and colleagues have undertaken photoaffinity labeling of the rat renal Na^+/H^+ exchanger, using a ^{14}C labeled analog of amiloride. The photolabel was mainly incorporated into a band at 65 kDa, confirming their previous findings with [^{14}C]DCCD [113–115]. A similar approach was used by Wu and Lever [381] to label a number of bands in brush border vesicles from hog kidney. When labeling was combined with purification and reconstitution studies, the major peaks of incorporation migrated at 81 kDa and 107 kDa [381]. Recent studies with a novel photoaffinity derivative of amiloride and anti-amiloride antibodies have demonstrated labeling at 77 kDa in the rabbit brush border system [370]. Similar results have been obtained with an isothiocyanato derivative of amiloride [155, 156]. If a preliminary estimate of approximately 75 kDa is accepted for the epithelial form of the Na^+/H^+ exchanger, then an interesting interpretation can be made of the pre-steady-state kinetic studies of Otsu et al. [242] and the radiation inactivation experiments of Beliveau et al. [32]. The functional size of the Na^+/H^+ exchanger appears to be 321 kDa from the studies of Beliveau et al. [32], and the functional complex would then be made up of four monomeric subunits of approximately 80 kDa each. Otsu et al. [242] have hypothesized such an oligomeric structure from the studies of pre-steady-state kinetics.

Solubilization and Affinity Chromatography

Weinman and colleagues have recently reconstituted amiloride-sensitive Na^+/H^+ exchange activity into asolectin vesicles [376–379]. Even though the efficiency of reconstitution is relatively modest, this group has demonstrated regulation of the reconstituted Na^+/H^+ exchange activity by cAMP-dependent protein kinase and protein kinase C [376–379]. The reconstitution experiments of Wu and Lever [381] have been combined with photoaffinity labeling studies, and at present must be regarded as the most convincing studies to date.

Another approach has utilized an amiloride affinity column with ligand specific elution [157]. Even though these workers were able to isolate a 25-kDa protein from rabbit renal cortex with this approach, the functional characteristics of this protein has not yet been established. It would seem that functional characterization is absolutely essential because there are other membrane transport systems in the proximal tubule brush border which can be inhibited by amiloride and could therefore be purified with an amiloride affinity column [80, 92, 216, 222].

Molecular Biology of the Na^+/H^+ Antiporter

Pouyssegur and coworkers [296] have made a tremendous advance in the cloning and sequencing of the Na^+/H^+ exchanger from human lymphocytes. Because of the lack of appropriate antibodies and affinity ligands, conventional purification and cloning strategies could not be used. Utilizing the unique properties of the Na^+/H^+ exchanger, mutant mouse L cells were developed which lacked Na^+/H^+ exchange activity and could therefore survive being Na^+ loaded in an acid environment. After transfection with human DNA from lymphocytes, these cells were screened for the revertant phenotype: transformed cells were loaded with acid and rescued by Na^+ in the external medium if they had been transfected with a stable gene for the human Na^+/H^+ exchanger.

Once stable transformants were created, a tedious but ultimately successful cloning effort permitted the identification of a full-length cDNA clone for the human Na^+/H^+ exchanger. The coding region contains 2445 nucleotides which code for a putative 815 amino acid protein. Hydropathy plots suggest that there are at least ten membrane spanning regions with a large cytoplasmic tail of at least 400 amino acids.

The functional domains of the Na^+/H^+ exchanger, including the substrate binding sites, the amiloride binding sites, the internal modifier sites, etc., have not yet been identified, but one can anticipate very rapid progress once the first representative of this class of transporters has been identified. The putative molecular mass of the Na^+/H^+ exchanger, ignoring glycosylation, is 100 kDa [296] which exceeds most previous estimates of the mass of the Na^+/H^+ exchanger in brush border vesicle systems. It should be borne in mind that Sardet et al. [296] have probably cloned the "housekeeping" form of the Na^+/H^+ exchanger, at least judging from its amiloride sensitivity. If the ten membrane spanning regions constitute the majority of the ion translocation machinery, then the large cytosolic tail could be truncated without adversely affecting the basic transport characteristics of the system. It will be of great interest to discover the regions of homology between the epithelial forms of the exchanger and the housekeeping form. It is not known at present whether these forms are derived from the same mRNA with alternative splicing or if entirely different gene products are involved.

Physiology of the Renal Na^+/H^+ Antiporter

Proximal Tubule

It appears certain that the Na^+/H^+ antiporter is the predominant, though not the sole, mechanism of proton secretion in the proximal convoluted tubule. This conclusion was initially based on the observations that inhibition of Na^+ reabsorption by several different procedures (Na^+ removal, inhibition of Na^+/K^+ ATPase with ouabain, or potassium removal from the peritubular compartment) inhibited bicarbonate reabsorption by over 80% in vitro and in vivo studies [224]. However, this interpretation has been reexamined in view of the demonstration of H^+-ATPase activity in the proximal tubule [139] and, more importantly, that a Na^+-dependent process accounts for bicarbonate exit across the basolateral membrane [5, 9, 10, 58]. Removal of sodium or depolarization of the basolateral membrane limits bicarbonate exit across the basolateral membrane and thereby inhibits bicarbonate reabsorption, regardless of the mechanism of proton secretion (Na^+/H^+ exchange or H^+-ATPase) across the apical membrane. Furthermore, the resulting cell alkalinization when the Na^+/HCO_3^- exit step is inhibited would be expected to reduce the rate of Na^+/H^+ exchange by reducing the intracellular substrate concentration, and via the allosteric internal modifier site on the Na^+/H^+ antiporter [14].

Recent studies have used potent amiloride analogs and appropriate controls to determine directly that the apical Na^+/H^+ exchanger plays a predominate role in proximal bicarbonate reabsorption and transcellular NaCl reabsorption in the proximal tubule [266, 267]. Nevertheless, an additional component of bicarbonate absorption is mediated by an amiloride-insensitive, Na^+-independent process which is presumably the H^+-ATPase in the luminal membrane.

These considerations are most relevant to the mammalian proximal tubule where bicarbonate is reabsorbed preferentially with respect to chloride [274]. In contrast, in the amphibian proximal tubule, preferential bicarbonate reabsorption does not occur, suggesting that the net rate of proton secretion is less in the amphibian than in the mammalian system. The demonstration of Na^+/H^+ antiporter activity in both basolateral and apical membrane of the salamander proximal tubule [57] could account for this difference; proton extrusion occurs across both membranes and thereby reduces the net rate of transepithelial proton secretion. It seems likely that a vectorial arrangement of H^+/HCO_3^- transporters would be found in the mammalian proximal tubule because of the more efficient rate of proton secretion, as compared with other nephron segments. Cell fractionation studies by Ives et al. [161] and Sabolic and Burckhardt [290] confirmed this expectation; the Na^+/H^+ antiporter was found to be confined to the apical membrane of the rabbit and rat proximal tubule. These findings have been confirmed by in vivo studies of the rat proximal tubule, using fluorescent probes of intracellular pH [9]. Furthermore, fractionation studies of brush border membrane preparations demonstrated that the minor contamination with Na^+/K^+-ATPase activity is associated with markers for the Golgi apparatus and is completely devoid of Na^+/H^+ antiporter activity [214].

Another aspect of the distribution of the Na^+/H^+ antiporter correlates with the rates of net transepithelial proton secretion in the proximal tubule. As described by Jacobsen et al. [162], the V_{max} for the Na^+/H^+ antiporter is nearly four times greater in early segments (i.e., outer cortex) of the proximal tubule than in later segments (outer medulla). Similar findings have been observed with intact, perfused tubules by Kurz [185], in which the activity of the Na^+/H^+ exchanger was localized to the S_3 segment of the proximal tubule. Also of note is the finding that the adaptive increase in Na^+/H^+ antiporter induced by several days of metabolic acidosis is limited to the early segments of the proximal tubule [162]. This finding indicates that the characteristics and regulation of Na^+/H^+ antiporter activity in different segments of the proximal tubule may be heterogeneous. The differences in maximal transport rate appear to reflect differences in the total Na^+/H^+ activity between outer cortical and outer medullary segments. Mircheff et al. [214] found that more than 80% of the Na^+/H^+ antiporter activity comigrated with maltase activity in studies of a brush border membrane preparation from the entire rabbit renal cortex which were purified with a phase separation technique. The remaining activity comigrated with leucine amino-peptidase activity, which may be a marker for later segments of the proximal tubule [352].

A variety of factors have been discovered which increase the maximal transport rate of the Na^+/H^+ exchanger in the proximal tubule. Chronic adaptation has been described in metabolic acidosis [6, 79, 178, 265, 351], respiratory acidosis [184, 345], and increases in glomerular filtration rate [78, 146, 236]. In addition, recent studies have identified several factors which can regulate the apical membrane Na^+/H^+ exchanger in a very acute fashion, including parathyroid hormone and cAMP [79, 264, 379], angiotensin II [198, 291], $alpha_2$ adrenergic agonists [124, 237], and acute uninephrectomy [294]. It remains to be determined whether the acute regulatory effects are accomplished by phosphorylation events or membrane recycling while longer term adaptations may well represent synthesis and insertion of new membrane transport proteins.

Distal Tubule

Amiloride-sensitive, electrically neutral Na^+/H^+ exchange has been reported in the cortical thick ascending limb from mouse [112] and rat [130] kidney. These studies utilized in vitro perfusion of isolated nephron segments, and could therefore localize the Na^+/H^+ exchange to the apical membrane. Similar findings have been reported for the amphibian diluting segment and late distal tubule [238, 339, 383]. Both bicarbonate reabsorption and net NaCl absorption (via parallel Na^+/H^+ and Cl^-/HCO_3^- exchangers) have been demonstrated in these studies, consistent with the apical localization of the Na^+/H^+ exchanger. Cell fractionation studies have not yet been undertaken, due to the lack of sufficient amounts of appropriate cells. There also appears to be an array of Na^+/H^+ and Cl^-/HCO_3^- exchangers in the mouse medullary ascending limb which participates in cell volume regulation rather than net NaCl absorption, and is localized in the basolateral membrane [148].

Basolateral Na^+/H^+ exchange has been demonstrated in isolated, perfused rabbit cortical collecting tubules by Chaillet et al. [74] and Breyer and Jacobson [59]. The physiologic role of this system has not been defined, but presumably it is localized in cells which also have basolateral Na^+,K^+-ATPase rather than the mitochondrial-rich cells. Inhibitor studies by Laski [188] are consistent with the basolateral localization of the Na^+/H^+ exchanger, and also suggests that these cells also have a sodium-independent mechanism for proton extrusion across the apical membrane. It is also possible that this system may regulate intracellular pH and account for the effects of changes in intracellular pH on the hydroosmotic response to vasopressin [11] and for the effects of amiloride on apical K^+ conductance in the frog diluting segment [239].

Na^+/H^+ exchange activity, and H^+-ATPase activity have been described in cultured cells from the renal papilla [179, 308, 369, 384]. These studies have not localized the Na^+/H^+ antiporter to the apical or basolateral membrane. It seems unlikely that the Na^+/H^+ exchanger is located on the apical surface because the low Na^+ concentration of the final urine would not provide a favorable driving force for H^+ secretion. It seems more likely that the Na^+/H^+ exchanger plays a role in regulation of intracellular pH or cell volume regulation than in net transepithelial Na^+ transport in the papilla.

Cultured Epithelial Cells

Amiloride-sensitive $^{22}Na^+$ uptake was clearly demonstrated in the Madin-Darby canine kidney (MDCK) cell line by Saier and coworkers [286, 287]. This early description included most of the salient features of the Na^+/H^+ exchanger, including Na^+/Na^+ exchange, inhibition by amiloride, competitive inhibition by Li^+ with an apparent affinity for lithium which was greater than Na^+. This system is presumably localized to the apical membrane in confluent MDCK cells [292].

The LLC-PK_1 cell line was derived from pig kidney, and possesses Na^+/H^+ exchange activity as assayed in studies in monolayers and in membrane vesicles prepared from monolayers [66, 25, 141, 219, 221, 223]. The kinetic features of the Na^+/H^+ exchanger are similar to those obtained in previous studies in renal brush border membrane vesicles, including Km for Na^+, and apparent decrease in maximal transport rate when Li^+/H^+ exchange was compared to Na^+/H^+ exchange in membrane vesicles prepared from confluent monolayers by the magnesium aggregation technique [223]. The steady-state intracellular pH was found to be near the "set-point" as determined from pH-recovery studies in suspended LLC-PK_1 cells [219]. A distinction was noted in the studies with amiloride; noncompetitive inhibition was observed [221, 223] in contrast to the mixed inhibition [160, 161] or competitive inhibition observed in previous studies [176, 365]. Furthermore, changes in external pH affected the apparent affinity for Na in a competitive fashion, similar to that described for renal brush border membrane vesicles [15] and also decreased the maximal transport rate for the Na^+/H^+ exchanger [221], an effect which was not observed with the renal brush border system.

Recent studies in LLC-PK_1 cells appear to localize the Na^+/H^+ antiporter in the basolateral membrane [219, 220], despite cell fractionation studies from the same laboratory [223] which examined Na^+/H^+ exchange activity in membrane vesicles which were presumed to be of apical origin because they were enriched in an apical membrane marker (trehalose; 7.8-fold enrichment), and deficient in Na^+/H^+-ATPase (relative specific activity < 1.0), a basolateral membrane marker. Measurements of $^{22}Na^+$ uptake in confluent monolayers by Cantiello et al. [66] were also consistent with the apical localization of the antiporter. This discrepancy has been resolved by the studies of Haggerty et al. [140], who found pharmacologically distinct Na^+/H^+ exchangers in the apical and basolateral membranes LLC-PK_1 cells grown on permeable supports. The apical form of the exchanger has a relatively low affinity for amiloride (> 1 μM) which is typical for the epithelial form [234], while the basolateral form of the exchanger had a higher affinity for amiloride which is characteristic of the "housekeeping" form of the Na^+/H^+ exchanger found in fibroblasts and other nonpolarized cells [134].

The opossum kidney (OK) cell line is another continuous renal epithelial cell line which has a well-defined Na^+/H^+ exchange system [213, 264]. A distinguishing feature of this cell line is its response to parathyroid hormone [347], and the striking regulation of the Na^+/H^+ exchange system by cAMP. This effect could be discerned at 10 pM parathyroid hormone and was limited to the amiloride-sensitive component of $^{22}Na^+$ exchange. Kinetic analysis indicates that parathyroid hormone acutely decreases the maximal rate of transport and shifted the internal pH sensitivity of the Na^+/H^+ exchanger in the acidic (e.g., less sensitive) direction [213]. The regulation of membrane transport in response to parathyroid hormone appears to be a unique feature of the OK cell line [206].

All of these cell lines (MDCK, LLC-$PK_1$1, and OK) have been considered to be examples of established cell lines with characteristics of the proximal tubule because of the presence of the Na^+/H^+ exchanger, and other Na^+-coupled cotransport systems (including Na^+/phosphate, Na^+/glucose, and various Na^+/amino acid transporters) [80, 223, 270]. Each of these cell lines has undergone some sort of alteration which converted them into a continuous cell line. It would not be surprising if such an alteration were also associated with the appearance of transport systems which were under-expressed in the original cell type [50, 108]. Rather than functional identification based upon the presence of transporters, the precise membrane localization of a given transport system and/or hormonal responsiveness of the cell line may prove to be of greater value. For example, the responsiveness of calcitonin [129] and vasopressin [89] of LLC-PK_1 cells, and basolateral localization of a Na^+/H^+ exchanger [220], if this finding can be confirmed, may be more typical of cells derived from the thick ascending limb or other distal segments than of proximal tubule cells. MDCK cells also have a hormone response profile which is more similar to distal tubule cells [292], while the hormone responses of the OK cell line is quite characteristic of the proximal tubule [347]. Careful functional localization of the Na^+/H^+ antiporter is required in each of these continuous cell lines to see if this functional characteristic is similar to that described for the proximal tubule (i.e., apical), or if a basolateral localization can be demonstrated, similar to that in the mouse thick ascending limb and segments of the collecting system.

Na^+/Ca^{2+} Exchange

Introduction

Historical Background

Certain cells contain a protein called the Na^+/Ca^{2+} exchanger or antiporter that catalyzes the obligatory transfer of Na^+ and Ca^{2+} in opposite directions through the plasma membrane. The exchanger was discovered independently in nerves [22, 23] and cardiac muscle [285]. Most studies of the exchanger have used dialyzed squid axons [62] or cardiac membrane vesicles [281]. Several reports prior to 1987 concluded that Na^+/Ca^{2+} exchange played little or no role in smooth muscle function [1, 72, 102, 153, 225, 336, 355]. However, the preponderance of data reported since 1986 indicates that the exchanger does play a prominent role in smooth muscle function. In this section, we will concentrate on the recent progress on Na^+/Ca^{2+} exchange in smooth muscle. Excellent reviews are available on Na^+/Ca^{2+} exchange in squid axon [24, 97] and cardiac muscle [253, 276, 309]. It should also be mentioned that Na^+ discharges Ca^{2+} from mitochondria isolated from cardiac muscle and nerve tissues via an exchanger about which little is known [67].

General Description of the Exchanger

The magnitude and direction of net Ca^{2+} movement via the exchanger are governed by kinetic and thermodynamic factors. The driving force for Na^+/Ca^{2+} (in millivolts) is the difference between the membrane potential ($\Delta\psi$) and the reversal potential of the exchanger ($E_{Na/Ca}$). $E_{Na/Ca}$ is the membrane potential at which there is no net Na^+ or Ca^{2+} movement via the exchanger.

$E_{Na/Ca} = n\,E_{Na} - 2\,E_{Ca}$ where n is the stoichiometry of $Na^+:Ca^{2+}$ exchange, n − 2

$$E_{Na} = -RT/F \ln([Na^+]_i/[Na^+]_o) \tag{1}$$

$$E_{Ca} = -RT/2F \ln([Ca^{2+}]_i/[Ca^+]_o) \tag{2}$$

and $[Na^+]$ and $[Ca^{2+}]$ are the free-ion concentrations in the extracellular (o) and intracellular (i) fluids, R is the gas constant, T is absolute temperature, and F is Faraday's number. Net Ca^{2+} movement via the exchanger, J_{Ca}, is the product of the kinetic and thermodynamic terms,

$$J_{Ca} = k\left[\Delta\psi - \frac{nE_{Na} + 2E_{Ca}}{n-2}\right] \tag{3}$$

The stoichiometry of the exchanger appears to be $3\,Na^+:1Ca^{2+}$ (see next section), so

$$J_{Ca} = k(\Delta\psi - E_{Na/Ca}) \text{ where } E_{Na/Ca} = 3\,E_{Na} - 2\,E_{Ca} \tag{4}$$

as indicated previously [27, 52, 93, 226, 278]. The kinetic term k is a complex parameter that depends on such factors as the number of exchangers, the degree of saturation of substrate and effector sites, membrane potential, and temperature.

If the $\Delta\psi$ is more negative than $E_{Na/Ca}$, the exchanger may cause Ca^{2+} efflux (negative J_{Ca}). On the other hand, if $\Delta\psi$ is more positive than $E_{Na/Ca}$, the exchanger may catalyze Ca^{2+} influx (positive J_{Ca}). The calculated value of $E_{Na/Ca}$ is − 19 mV in aortic myocytes with 0.11 μ*M* free Ca^{2+} and 6.7 m*M* Na^+ at 2 and 140 m*M* external Ca^{2+} and Na^+, respectively [329]. Because $\Delta\psi$ is usually between − 30 and − 60 mV, the driving force poises the exchanger in the forward mode under physiological conditions; $\Delta\psi$ would have to be more positive than −19 mV, cell Na^+ increase by about two fold, or some combination of depolarization and an increase in cell Na^+, for the driving force to switch the exchanger to the reverse mode. Because the exchanger in nonstimulated aortic myocytes is latent [329], the contribution of the exchanger to net Ca^{2+} flux appears to be governed primarily by kinetic factors rather than the driving force.

Kinetic Properties and Stoichiometry of Na^+/Ca^{2+} Exchange

Kinetics of Exchange

Na^+/Ca^{2+} exchange activity is high in smooth muscle cells from rat aorta [320]. Reverse (Ca^{2+} influx) mode activity can be readily observed as multifold increases in unidirectional $^{45}Ca^{2+}$ influx [320] or $[Ca^{2+}]_i$ [328] that occur when extracellular Na^+ is removed from Na^+-loaded cells. The K_m and V_{max} of reverse mode exchange are 100 μ*M* and approximately 30 nmol min^{-1} mg^{-1}

cell protein [320]. These values are similar to the values (100 μM and 77 nmol min^{-1} mg^{-1} cell protein) obtained by Wakabayashi and Goshima [366] in myocytes from mouse heart. The minimal turnover number of impure exchanger from cardiac plasmalemma is ~ 1000/s after reconstitution in phospholipid vesicles [77], which is similar to turnover number of the Na^+/H^+ exchanger [99, 365]. From the turnover number, the V_{max}, and about 2×10^6 cells per mg cell protein, it appears that there are approximately 150000 Na^+/Ca^{2+} exchange molecules per smooth muscle cell (Smith et al., manuscript sumitted).

The apparent K_m for Ca^{2+} is 2.5 μ*M* and 15 μ*M* for the exchanger in sarcolemmal vesicles from canine mesenteric arteries [208] and bovine aorta [316], respectively. The apparent K_m for Na^+ is 45 m*M* and 15 ± 7 m*M* in vesicles from canine mesenteric arteries [208] and bovine aorta [316], respectively. The K_m values for aortic vesicles are similar to those reported for the exchanger in cardiac sarcolemmal vesicles [317, 318]. The differences in the K_m values for the exchanger in mesenteric versus aortic vesicles may be caused by differences in technique, associated modulators of exchange activity, or exchanger structure.

The K_m for Ca^{2+} in the sarcolemmal vesicles is considerably lower than the K_m for Ca^{2+} of reverse mode exchange in intact aortic or cardiac myocytes. A number of factors may affect the kinetics of Na^+/Ca^{2+} exchange such as asymmetries of exchanger structure, vesicle size and sidedness, and the presence or absence of endogenous effectors. The Ca^{2+} affinities of the forward and reverse modes of exchange may differ by orders of magnitude because of asymmetries in the exchanger and its associated modulators.

The V_{max} of the exchanger in mesenteric sarcolemmal vesicles is 13 nmol min^{-1} mg^{-1} vesicle protein [208]. The V_{max} of aortic plasmalemmal vesicles differed by almost an order of magnitude between different preparations by the same laboratory: 24 and 205 nmol min^{-1} mg^{-1} vesicle protein [316]. Assuming that the sarcolemma comprises under 5% of total cell protein [7, 349], the V_{max} of the exchanger in intact aortic myocytes is at least 600 nmol min^{-1} mg^{-1} sarcolemmal protein. Hence, large decreases in specific exchange activity apparently occur during the preparation of the sarcolemmal vesicles. Conceivably, dephosphorylation of the exchanger during vesicle preparation may decrease specific exchange activity because the aortic myocytes depletion of ATP decreased specific exchange activity by 80% as discussed below [326]. Studies of exchange activity in vesicles prepared from cultured myocytes with known exchange activity may help to develop methods for preventing decreases in specific exchange activity during membrane isolation.

Stoichiometry of Na^+/Ca^{2+} Exchange

The exchanger is electrogenic (> 2 Na^+ : 1 Ca^{2+}) as indicated by measurements of putative Na^+/Ca^{2+} exchange currents in isolated heart cells [27, 106, 173,

211]. Manipulation of $\Delta\psi$ by varying the K^+ gradient in the presence of valinomycin affects Na^+ gradient-dependent Ca^{2+} fluxes in cardiac [45, 69, 257, 278, 282, 283], mesenteric [208, 209], tracheal [318], and aortic [316] sarcolemmal vesicles. Conversely, Ca^{2+} accumulation driven by an imposed Na^+ gradient produces a membrane potential, interior negative, as indicated by the transient accumulation of the lipophilic cation, tetraphenylphosphonium, by sarcolemmal vesicles [69, 282].

Pitts [263] obtained a coupling ratio of $\sim 3\ Na^+:1\ Ca^{2+}$ in cardiac sarcolemmal vesicles by measurements of $^{22}Na^+$ influx and $^{45}Ca^{2+}$ efflux. Rasgado-Flores and Blaustein [273] obtained a stoichiometry of $3\ Na^+:1\ Ca^{2+}$ from measurements of $^{45}Ca^{2+}$ influx and $^{22}Na^+$ efflux that were activated by internal Ca^{2+} and dependent on the concentration of internal Na^+ in perfused giant barnacle muscle cells. Wakabayashi and Goshima [366] obtained the same coupling ratio from flux measurements in myocytes cultured from mouse heart. The flux ratio method may be subject to large errors, however, because of difficulties in estimating the contributions of other transporters to the fluxes. Thus, coupling ratios obtained from flux measurements have to be interpreted with caution.

Reeves and Hale [278] devised a thermodynamic method to estimate the stoichiometry of exchange. With equal concentrations of intra- and extravesicular Ca^{2+} (i.e., J_{Ca} and $E_{Ca} = 0$), Eq. 3 (above) reduces to $(n-2)\ \Delta\psi = n\ E_{Na}$. E_{Na} is calculated from Eq. 1 (above) using the predetermined intra- and extravesicular concentrations of Na^+. The Na^+ gradient which just opposes the membrane potential and prevents net Ca^{2+} movement via the exchanger is experimentally determined. This procedure is done in the presence of valinomycin and known intra- and extravesicular concentrations of K^+ so the membrane potential equals $-60\ mV \times \log [K^+]_i/[K^+]_o$. Knowing $\Delta\psi$ and experimentally estimating the E_{Na} that just opposes a given $\Delta\psi$, the equation for n can be solved. The method gave a Na^+/Ca^{2+} stoichiometry of 2.97 ± 0.03 for cardiac sarcolemmal vesicles [278] and 2.9 ± 0.2 for the exchanger from renal basolateral membranes reconstituted into liposomes [344].

Measurements of Na^+ and Ca^{2+} activities in cardiac muscle with microelectrodes gave a value of 2.5 for the ratio of Na^+ to Ca^{2+} activities under a variety of conditions [310]. This method gives a true stoichiometry only when the electrochemical Na^+ and Ca^{2+} gradients are in equilibrium, and it is subject to limitations in the sensitivity and selectivity of the electrodes [309]. Recently Ehara and colleagues [106] measured the reversal potential of the Na^+/Ca^{2+} exchange current in single guinea pig ventricular cells using the technique of whole-cell voltage clamp and internal perfusion. Under a variety of conditions the reversal potentials were consistent with a $3\ Na^+:1\ Ca^{2+}$ exchange.

Na^+/Ca^{2+} exchange in rod outer segments from bovine photoreceptors shares some kinetic properties with exchange activity in other cell types [73, 306]. However, the transporter in rod outer segments appears to exchange $4\ Na^+$ for $1\ Ca^{2+}$ and $1\ K^+$ [73, 307], so it is unclear how structurally similar it is to the Na^+/Ca^{2+} exchanger in cardiac and smooth muscle.

Ca^{2+}/Ca^{2+} and Na^+/Na^+ Exchanges

The Na^+/Ca^{2+} exchanger will also catalyze Ca^{2+}/Ca^{2+} and Na^+/Na^+ exchanges [97, 276]. Ca^{2+}/Ca^{2+} exchange becomes apparent when the concentration of extracellular Na^+ is low or absent [51, 189, 258, 317]. Low Na^+ stimulates, high Na^+ inhibits, Ca^{2+}/Ca^{2+} exchange in cardiac sarcolemmal vesicles [317]. Monovalent cations (K^+, Rb^+, Li^+, Cs^+) increase the V_{max} of Ca^{2+}/Ca^{2+} exchange and have little effect on the K_m for Ca^{2+} [317]. The stoichiometry of Ca^{2+}/Ca^{2+} exchange is a 1:1 [28]. Measurements of $^{86}Rb^+$ fluxes showed that Rb^+-stimulated Ca^{2+}/Ca^{2+} exchange does not involve Rb^+ cotransport [317].

The Na^+/Ca^{2+} exchanger in cardiac sarcolemmal vesicles [276] and dialyzed squid axons [97] is also capable of Na^+/Na^+ exchange. Na^+/Na^+ exchange is inhibited by Ca^{2+} as well as inhibitors of Na^+/Ca^{2+} exchange such as La^{3+}, quinacrine, and amiloride analogs [276]. In squid axons internal Ca^{2+} and ATP activate Na^+/Na^+ exchange [96].

Inhibition of Exchange by Divalent Metals

Extracellular magnesium competitively inhibits reverse mode Na^+/Ca^{2+} exchange in aortic myocytes [320]. The K_i of Mg^{2+} is 100 μM, the same as the K_m for Ca^{2+} [320]. Mg^{2+} is not transported by the exchanger because Mg^{2+} does not stimulate $^{22}Na^+$ efflux. Moreover, Mg^{2+} inhibits $^{22}N^+$ efflux produced by external Ca^{2+} [320]. Other divalent cations are more potent than Mg^{2+} as inhibitors of reverse mode exchange activity. The order of effectiveness of the cations is: $Cd^{2+} > Mn^{2+} > Co^{2+} > Ba^{2+} > Mg^{2+}$ [320]. About 2 μM Cd^{2+} inhibits exchange by 50% at 100 μM Ca^{2+} [320]. Inhibitory potency correlates well with the closeness of the crystal ionic radius of the metal to that of Ca^{2+} [320]. Divalent cations also inhibit Na^+/Ca^{2+} exchange activity in cardiac sarcolemmal vesicles [350]. The order of effectiveness of the divalent cations towards the cardiac exchanger is similar, but not identical, to that in aortic myocytes. Mn^{2+} and Ba^{2+} are equipotent towards exchange activity in cardiac vesicles, whereas Mn^{2+} was about eight times more potent than Ba^{2+} towards exchange activity in aortic myocytes [320]. Additionally, all the divalents appear to be considerably more potent towards aortic myocytes compared to cardiac vesicles. It is unclear whether assay conditions or differences in exchanger structure cause the differences in sensitivity to the divalent cations.

Mg^{2+} is the only divalent metal that is present endogenously at concentrations that might be sufficient to inhibit exchange. Intracellular free Mg^{2+} is ~ 0.4 mM in smooth [186] and cardiac [228] muscle. Mg^{2+} appears to be about tenfold more potent towards exchange activity in aortic than in cardiac myocytes [367] or cardiac sarcolemmal vesicles [350]. Decreasing extracellular K^+ from 120 to 5 mM increased the potency of Mg^{2+} about sevenfold [320, 329]. Decreasing extracellular K^+ increased the potency of the other divalent cations by four- to about ninefold depending on the cation [320].

Two-Site Model of Exchange

Reeves [276, 279] and others [54, 119] have discussed a two-site model of the exchanger. According to the model [276] the exchanger has a common (A) site that binds 1 Ca^{2+} or, after changing configuration, 2 Na^+. A separate monovalent (B) site binds one Na^+ or another monovalent cation. The rate-limiting step in the Na^+/Ca^{2+} exchange cycle may be the postulated conformational change in the A site, which would be necessary to accommodate 1 Ca^{2+} or 2 Na^+ [276]. This suggestion is supported by the finding that Na^+/Ca^{2+} exchange is much more sensitive to temperature than Ca^{2+}/Ca^{2+} exchange [28]. Slaughter et al. [317] proposed that the B site is highly specific for Na^+ when 2 Na^+ occupy the A site and that occupation of the B site by Na^+ is necessary for Na^+ translocation, all 3 Na^+ ions being translocated per cycle. On the other hand, when Ca^{2+} occupies the A site, the B site is less specific and accepts K^+, Li^+, or Rb^+. The binding of K^+ to the B site [320], or to another monovalent site [119], may increase the selectivity of the exchanger for Ca^{2+} which could explain the effect of K^+ on the potency of the divalent inhibitors of exchange. Because intracellular free Mg^{2+} is about three orders of magnitude greater than free Ca^{2+}, high intracellular K^+ may be an important modulator of the specificity of the A site. Alternatively, K^+ may indirectly influence divalent inhibition of exchange activity by altering the membrane potential, however, this possibility seems less likely because internal K^+ activates Na^+/Ca^{2+} exchange in squid axon under constant voltage clamp [98].

Inhibition of Exchange by Amilorides

Smith et al. [332] found that amiloride inhibited Ca^{2+} fluxes in murine erythroleukemia cells and suggested that the amiloride-sensitive fluxes were caused by Na^+/Ca^{2+} exchange. Schellenberg et al. [303] showed that amiloride and amiloride congeners [302] inhibit Na^+/Ca^{2+} exchange in synaptosomal vesicles. Substitution of either the terminal guanidino or 5-amino nitrogen atoms of amiloride increased inhibitory potency towards exchange activity in membrane vesicles from bovine brain, porcine heart, or GH_3 cells, a line of rat pituitary tumor cells [167, 302, 311]. By contrast, alkyl or phenyl substitution on the terminal guanidino nitrogen usually decreases potency towards Na^+/H^+ exchange, while similar substitutions on the 5-amino nitrogen decrease potency towards Na^+ channels (as already indicated in this chapter). Replacement of the 6-chloro atom with H, F, Br, or I decreased amiloride potency [302].

Amiloride data on reverse mode Na^+/Ca^{2+} exchange activity in aortic myocytes agree well with the data on cardiac and pituitary membrane vesicles. Amiloride itself was the least potent inhibitor of Na^+/Ca^{2+} exchange in aortic myocytes (Fig. 8). EIPA, t-butylamiloride, phenamil, and benzamil were 15–22 times more potent in inhibiting Na^+/Ca^{2+} exchange than amiloride. 3′,4′-Dichlorobenzamil (DCB) and 2′,4′-dimethylbenzamil (DMB) were 59 and 534 times more potent than amiloride, respectively. Therefore, derivatizing either the

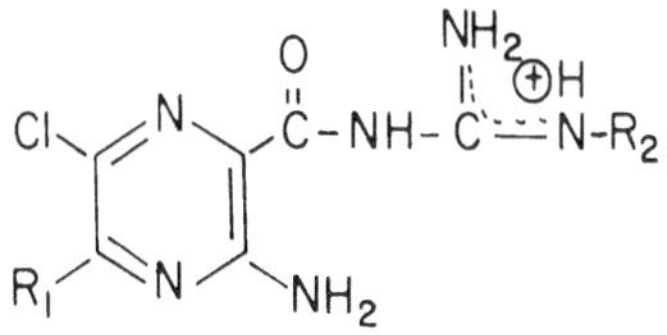

Compound	R_1	R_2	I_{50} (μM) Na/Ca	Na/H
Amiloride	$-NH_2$	H	5880	1.2
EIPA	$-N(CH_2CH_3)(CH(CH_3)_2)$	H	261	0.019
Butyl	$-NH-C(CH_3)_3$	H	361	0.034
Phenamil	$-NH_2$	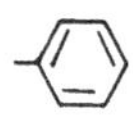	268	78
Benzamil	$-NH_2$	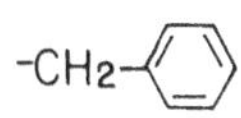	398	90
DMB	$-NH_2$	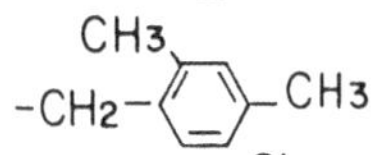	11	98
DCB	$-NH_2$	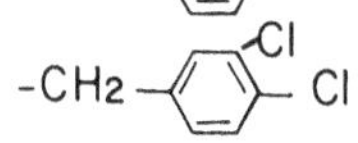	99	59

Fig. 8. Summary of inhibitory potencies of amiloride congeners towards Na^+/Ca^{2+} and Na^+/H^+ exchange in aortic myocytes. Drug concentrations that inhibited exchange by 50% (I_{50}) were obtained from log-log plots of the data. Na^+/Ca^{2+} exchange was assayed in Na^+- loaded cells as described [320, 325]. Na^+/H^+ exchange was measured as the initial rate of $^{22}Na^+$ uptake in acid-loaded cells as described [111]

5-amino-pyrazine-nitrogen or the guanidino-nitrogen increases potency towards Na^+/Ca^{2+} exchange. DMB was the only compound tested that was more potent (about tenfold) towards Na^+/Ca^{2+} than Na^+H^+ exchange. EIPA was the most potent inhibitor of Na^+/H^+ exchange that we tested. EIPA was more than 10000 times more potent towards Na^+/H^+ than Na^+/Ca^{2+} exchange (Fig. 9).

Amiloride and DCB appear to be mechanism-based inhibitors of Na^+/Ca^{2+} exchange [90, 315]. Protonation of the guanidinium group is essential for activity [90, 167, 302]. The protonated acylguanidinium moiety appears to bind mostly to the Na^+ specific (B) site because inhibition by lower concentrations of DCB or amiloride is competitive versus Na^+ and noncompetitive versus Ca^{2+}

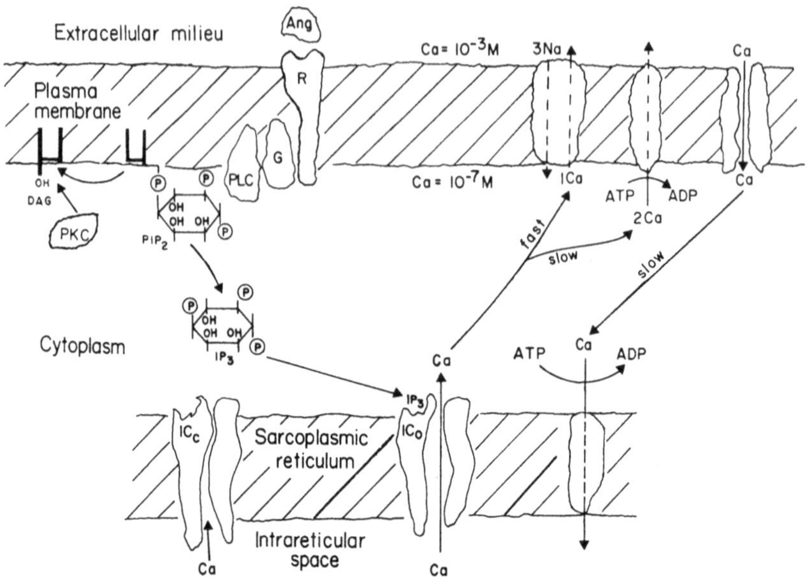

Fig. 9. Scheme of Ca^{2+} regulation in smooth muscle. *R*, Transmembrane receptor; *PLC*, phospholipase C; *G*, GTP-binding protein; IP_3, inositol triphosphate; *DAG*, diacylglycerol; *ANG*, angiotensin II; IC_c, closed intracellular Ca^{2+} channel; IC_o, open intracellular Ca^{2+} channel; *PKC*, protein kinase C

[190, 315]. At higher concentrations DCB and amiloride also interact with a site that is common for Na^+, K^+, and Ca^{2+} [315]. A specific interaction with the Na^+ specific site may explain the observation that defined structural modifications of the guanidino substituents are required for increasing potency [302]. With Na^+ and amiloride on opposite sides of the sarcolemma, however, they inhibit exchange synergistically rather than interacting in a competitive manner [90]. By contrast, merely increasing the hydrophobicity of the 5-amino substituents increases potency of inhibition of Na^+/Ca^{2+} exchange [302]. Increasing the hydrophobicity of amiloride by substitutions that are distal to the guanidino moiety would increase the partitioning of the drug into the lipid bilayer of the plasma membrane and, hence, increase its ability to compete with Na^+ for binding to the exchanger [302]. Alternatively, interaction with a hydrophobic domain of the exchange protein may account for the increase in potency with greater hydrophobicity of the 5-amino substituents.

DCB and DMB are equipotent inhibitors of exchange in membrane vesicles from bovine brain and guinea pig and porcine heart [302, 311]. In aortic myocytes, however, DMB is tenfold more potent than DCB towards Na^+/Ca^{2+} exchange (Table 1). Differences in assay conditions or in exchanger properties between intact cells versus isolated membranes probably do not cause the differences in DMB versus DCB potency. First, DMB was almost equipotent towards exchange activity in aortic myocytes (Table 1) and membrane vesicles from heart, brain, or GH_3 cells [167]. Second, Kaczorowski et al. [167] found that DMB was threefold more potent than DCB towards exchange activity in vesicles from GH_3 cells. Species and/or cell type differences in exchanger structure may account for the differences in DMB versus DCB potency. The exchanger in squid axon is not inhibited by 50 μ*M* DCB [8]. As indicated earlier in this chapter, the amiloride analogs must be used with caution because of their potential to inhibit most pathways of Na^+ and Ca^{2+} transport as well as a variety of other cellular processes. Benzamil and some amiloride compounds inhibit Na^+-dependent Ca^{2+} efflux from heart mitochondria [165, 337]. Half-maximal inhibition occurred between 100 and 400 μ*M* for the effective analogs [165]. DCB is about five times more potent as an inhibitor of the light-activated Ca^{2+} channel than of the Na^+/Ca^{2+} exchanger in rod outer segments [232].

Other Inhibitors of Na^+/Ca^{2+} Exchange

La^{3+} [281] and other lanthanides [350] are potent inhibitors of Na^+/Ca^{2+} exchange. Because all transmembrane Ca^{2+} pathways are inhibited by La^{3+} [100], its usefulness as a specific blocker for Na^+/Ca^{2+} exchange is limited. For example, in aortic myocytes Ca^{2+} efflux via the Ca^{2+} pump appears to be somewhat more sensitive to La^{3+} than efflux via the exchanger [117].

Bepridil blocks Ca^{2+} channels and has antiarrhythmic and antianginal activity [168]. Bepridil and DCB appear to inhibit exchange by the same mechanism except bepridil appears to interact only with the Na^+ specific B site of the exchanger [119]. Bepridil also inhibits Ca^{2+}/Ca^{2+} exchange. High K^+ decreases the potency of bepridil and causes the inhibition to become partial. Garcia and colleagues [119] suggested that K^+ may alter the interaction of bepridil with the exchanger by binding to an allosteric site which is distinct from the A or B sites.

The acridine derivative, quinacrine, inhibits Na^+/Ca^{2+} as well as Na^+/Na^+ and Ca^{2+}/Ca^{2+} exchanges in cardiac sarcolemmal vesicles with a K_i of 50 μ*M* [91]. Inhibition by quinacrine is noncompetitive with Ca^{2+}, but abated by high concentrations of Na^+. Quinacrine also stimulates Na^+/Ca^{2+} exchange when Na^+-loaded vesicles are preincubated with the compound and then assayed in quinacrine-free medium [91]. Certain lipophilic cations, like quinacrine and tetraphenylphosphonium, stimulate exchange two- to threefold by decreasing the apparent K_m for Ca^{2+} [91]. It appears that the distribution of charged groups between the inner and outer surfaces of the membrane influences exchange activity because fatty acids [16, 261] and anionic lauryl derivatives [252] also increase the apparent affinity of the exchanger for Ca^{2+} [261]. There appears to be some specificity in the interaction of quinacrine and tetraphenylphosphonium with the exchanger because a variety of similar compounds do not stimulate exchange [91]. Moreover, a series of amphiphilic cations with unbranched alkyl groups, dodecylamine for example, potently inhibit exchange in cardiac vesicles [252]. Because exchange activity is sensitive to a variety of agents that influence membrane structure including fatty acids, phospholipids, and cholesterol [262, 359], quinacrine and the amphiphiles may affect exchange activity by altering the membrane environment in which the exchanger works as well as by direct interactions with the exchanger.

A number of other agents are relatively weak inhibitors of exchange activity. These include amrinone; L-type Ca^{2+} channel blockers, e.g., diltiazem; verapamil; harmaline; and agents that alkylate sulfhydryl groups (reviewed by Kaczorowski et al. [168]). It does not appear that any of these agents will be useful for studying exchanger mechanism or structure.

Regulation of Na^+/Ca^{2+} Exchange

Dependence of Na^+/Ca^{2+} Exchange on Metabolic Energy

ATP stimulates both forward and reverse modes of Na^+/Ca^{2+} exchange in squid axons [25, 94]. ATP increases the affinities of the exchanger for external Na^+ and internal Ca^{2+} without affecting the maximal velocity of forward mode exchange [94]. ATP also increases the Ca^{2+} affinity of the internal modulatory site [96]. Only hydrolyzable analogs of ATP activate exchange [94], which apparently occurs via a Ca^{2+}-dependent protein kinase in axons [95] and sarcolemmal vesicles [68].

Haworth et al. [147] found that depleting cellular ATP strongly inhibited Na^+ gradient-dependent $^{45}Ca^{2+}$ uptake by cardiac myocytes. Smith and Smith [326] have shown that energy poisons inhibit reverse mode Na^+/Ca^{2+} exchange activity in aortic muscle cells. Rotenone rapidly and reversibly depleted cell ATP. Na^+/Ca^{2+} exchange activity decreased almost as rapidly as cell ATP. About 20% of the exchange activity remained after cell ATP was virtually eliminated. Repletion of ATP by the addition of glucose substantially restored exchange activity. Cell ATP increased shortly before exchange activity. An effect of ATP on exchange activity in membrane vesicles from vascular smooth muscle has not yet been demonstrated, and the possibility that secondary changes in cell metabolism, concurrent with ATP depletion and repletion, modulate exchange activity has not been excluded. However, if ATP modulates exchange activity in smooth muscle, the mechanism would differ kinetically from that in squid axon. ATP alters the Na^+ and Ca^{2+} affinities in squid axon, whereas ATP depletion and repletion in aortic myocytes altered reverse mode exchange activity assayed with saturating concentrations of internal Na^+ and internal and external Ca^{2+} [326]. Because ATP is not required for exchange activity in perfused squid axons or sarcolemmal vesicles, the exchanger does not split ATP with each transport cycle. The most plausible explanation for the apparent modulation of exchange by ATP is a phosphorylation reaction catalyzed by a protein kinase in the smooth muscle cells.

Activators of Na^+/Ca^{2+} Exchange in Membrane Vesicles

Na^+/Ca^{2+} exchange in cardiac sarcolemmal vesicles is activated by internal Ca^{2+} [280], external alkalinization [254], proteases [254], dodecyl sulfate [252], fatty acids [261], certain combinations of oxidizing and reducing agents [277], phospholipases [255, 260], cholesterol [359], and the lectin, concanavalin A [203]. Philipson and Nishimoto [260] showed that exchange activity in cardiac sarcolemmal vesicles increased fourfold when phospholipase D had converted about 10% of membrane phospholipid to phosphatidic acid. Phosphatidic acid probably stimulates exchange because the exchanger is optimally active after reconstitution in liposomes containing phosphatidic acid or other anionic phospholipids [359]. The possibility that phosphatidic acid activates the exchanger in vivo is especially intriguing because angiotensin II, which releases stored Ca^{2+} and markedly activates forward mode Na^+/Ca^{2+} exchange in aortic myocytes [325], also rapidly increases phosphatidic acid in these cells [324]. The exchanger in vesicles from smooth muscle, however, differs somewhat from that in cardiac vesicles because neither internal Ca^{2+}, chymotrypsin, nor redox modification affects exchange activity in vesicles from tracheal or aortic smooth muscle membranes [316, 318]. Vesicle studies may suggest candidate physiological modulators and provide biochemical insight to exchanger regulation. However, studies of exchange activity in intact cells probably offer the best opportunity for evaluating the physiological significance of a particular effector.

Hormonal Regulation of Exchange Activity

The exchanger is latent in unstimulated aortic myocytes [329]. Releasing stored Ca^{2+} with angiotensin strikingly increases forward (Ca^{2+} efflux) mode exchange [325]. The mechanism of activation is not known. Activation may be caused by phosphatidic acid (as discussed above) and/or the rise in $[Ca^{2+}]_i$ because the Ca^{2+} ionophore, ionomycin, releases stored Ca^{2+} and activates forward mode exchange similarly to angiotensin [322]. Intracellular Ca^{2+} activates forward mode Na^+/Ca^{2+} exchange in squid axon [51, 96] and forward mode Na^+/Ca^{2+} current in cardiac myocytes [211]. Reverse mode exchange in squid axons [96], barnacle muscle [273], and cardiac cells [27, 173] is also activated by intracellular Ca^{2+}. Extracellular Na^+-dependent $^{45}Ca^{2+}$ efflux increased monotonically with log $[Ca^{2+}]_i$ between 0.1 and 1 μM in aortic myocytes stimulated with ATP [117]. Intracellular Ca^{2+} may activate forward mode exchange at the substrate level by increasing the degree of saturation of the Ca^{2+} translocating site. Latency may be caused, at least in part, by the high cytosolic free Mg^{2+}: Ca^{2+} ratio ($\sim$ 4000:1) in nonstimulated cells because Mg^{2+} competitively inhibits reverse mode exchange in aortic myocytes [320]. Additionally, Ca^{2+} appears to activate exchange activity via a Ca^{2+}-dependent kinase [68, 95].

Vigne and colleagues [361] found that phorbol myristate acetate increased exchange activity about twofold in a cell line (A7r5) from embryonic rat aorta.

Phorbol esters activate protein kinase C and mimic diacylglycerol [233], which increases rapidly after stimulating Ca^{2+} mobilizing receptors [44]. Phorbol myristate acetate, however, had no effect on reverse mode exchange in early passage aortic myocytes [322]. It is not known if the activation of protein kinase C by Ca^{2+} mobilizing stimuli modulates forward mode exchange in A7r5 or other cells.

Molecular Biology of the Exchanger

Biochemical and Immunological Studies of the Exchanger

Miyamoto and Racker [215] reconstituted the Na^+/Ca^{2+} exchanger in proteoliposomes after solubilizing sarcolemmal membranes with cholate in the presence of soybean phospholipids. After reconstitution, specific exchange activity increased fivefold with no loss of total activity [215]. Several laboratories have partially purified the exchanger by assaying exchange activity after reconstitution in proteoliposomes. Some candidate proteins that copurified with exchange activity have the following molecular masses in SDS-polyacrylamide gels: <40 kDa [368]; 82 kDa [143]; 33, 70, and 140 kDa [333]; 33 and 70 kDa [30]; 55–70 kDa [289]; 70 and 120 kDa [256]. None of the proteins has been positively identified as a component of the exchanger.

Cook and Kaupp [81] have purified a 220 kDa glycoprotein from bovine photoreceptor membranes which appears to be a 4 Na^+:1 Ca^{2+}:1 K^+ exchanger [73, 307]. Exchange activity correlated with the amount of 220-kDa protein in chromatographic fractions from gel filtration and anion exchange columns as indicated by SDS gel electrophoresis and staining with Coomassie blue. It is noteworthy that Hale and colleagues [142] obtained a molecular mass of 226 kDa for the Na^+/Ca^{2+} exchanger in cardiac sarcolemmal vesicles using electron radiation inactivation and target size analysis. They [142] obtained essentially the same molecular mass (224 kDa) by molecular sieve chromatography in the presence of cholate and phospholipids.

Three laboratories [30, 199, 256] have obtained antibodies that appear to react specifically with the Na^+/Ca^{2+} exchanger. Barzilai et al. [30] obtained polyclonal antibodies to fractions comprised of 33- or 70-kDa sized proteins, that were prominent in a purified exchanger preparation from rat synaptic membranes. Previously Barzilai et al. [29] devised a clever method for purifying the exchanger. Membranes were solubilized with detergent and reconstituted into liposomes at a 1:40 protein to phospholipid ratio. Thus, each vesicle contained few or no proteins. Vesicles containing the exchanger were selectively loaded with Ca^{2+} and separated from nonloaded vesicles by centrifugation in a sucrose gradient. The 70-kDa protein was the most prominent and was consistently present in purified exchanger preparations. Antibodies to both bands bound exclusively to a 70-kDa protein in native synaptic membranes. Both antibodies, however, bound to both of the proteins in preparations of purified exchanger. Furthermore, both antibody preparations immunoprecipitated Na^+ gradient-dependent Ca^+ transport activity from detergent solubilized synaptic membranes and had no effect on ATP-dependent Ca^{2+} transport. The antiserum against the 70-kDa protein from rat synaptic membranes cross-reacted with a 70-kDa protein in SDS-gels of rat heart sarcolemma. Barzilai et al. [30] suggested that the 33-kDa protein was a subunit or fragment of the 70-kDa protein that contains the catalytic unit of the exchanger.

Soldati et al. [333] fractionated sarcolemmal proteins and found that exchange activity correlated best with the presence of a 33-kDa protein. Longoni and Carafoli [199] obtained polyclonal antibodies to about 30- or 140-kDa sized sarcolemmal proteins. Antibodies to the 30-kDa proteins reacted with proteins of 33, 70, and 140 kDa [199]. In the absence of dithiothreitol, which reduces disulfide bonds, antibodies to 140-kDa sized proteins reacted with the 33- and 140-kDa proteins. Neither antibody preparation reacted with the 140-kDa protein in the presence of dithiothreitol. The reaction of antibodies to the 140-kDa sized proteins with the 70-kDa protein was prominent when the membranes were solubilized in the presence of dithiothreitol. From these data Longoni and Carafoli [199] proposed that the exchanger is a tetramer of 33-kDa subunits. Although they also proposed that the 70-kDa protein is a dimer of 33-kDa subunits, the basis for dimerization in the presence of SDS and dithiothreitol is unclear.

Philipson et al. [256] obtained a 30-fold purification of exchange activity. Two protein bands, of 70 and 120 kDa, were prominent in silver-stained SDS-gels of the purified exchanger. A mild treatment with chymotrypsin, which caused no loss of exchange activity, eliminated the 120-kDa protein and increased the amount of the 70-kDa band. This finding suggests that the 70-kDa band is an active fragment of the 120-kDa protein. Antibodies to the purified, reconstituted exchanger immunoprecipitated exchange activity and the 70- and 120-kDa proteins.

Monoclonal antibodies produced to leukemic cell lines were reported to inhibit the Na^+/Ca^{2+} exchange in sarcolemmal vesicles from cardiac and skeletal muscle [212], however, this finding was not reproduced [168].

Expression Cloning of the Exchanger

Expression cloning circumvents protein purification or the production of a specific antibody for obtaining sequence information. Wright and colleagues [149] cloned the Na^+/glucose cotransporter by assaying for the expression of transport activity in *Xenopus laevis* oocytes injected with mRNA from rabbit intestine. Two laboratories [200, 312] have detected Na^+/Ca^{2+} exchange activity in oocytes injected with mRNA from cardiac cells. Both groups assayed exchange activity by measuring the increase in $^{45}Ca^{2+}$ uptake by Na^+-loaded oocytes that was produced by removing external Na^+. The photoprotein, aequorin, which can be injected at the same time as the mRNA, has provided a sensitive assay for the expression of Ca^{2+} mobilizing receptors in oocytes

[295]. Luminescence produced by Ca^{2+} influx via the exchanger offers an alternative to $^{45}Ca^{2+}$ uptake for assaying exchange activity in a single oocyte. Size fractionation of poly(A)$^{+}$ RNA on sucrose gradients yielded enriched fractions that provide a starting point for expression cloning of the exchanger gene [200, 312].

Physiological Role of the Exchanger

Role of Na^{+}/Ca^{2+} Exchange in Smooth Muscle Tissues

Assessment of the exchanger's contributions to Ca^{2+} regulation has been hampered by the occurrence of a parallel mechanism of Ca^{2+} transport, the sarcolemmal Ca^{2+} ATPase, and the lack of a specific inhibitor of either Ca^{2+} transporter. It appears that the principal role of the exchanger in cardiac muscle is to expel Ca^{2+} (forward mode) during the repolarizing phase of the action potential [8, 52, 276]. The exchanger may also contribute to Ca^{2+} influx (reverse mode) during the early part of the action potential [52, 226, 309].

The significance of Na^{+}/Ca^{2+} exchange in Ca^{2+} regulation in smooth muscle has been controversial [102]. Reuter et al. [284] described Na^{+} gradient dependent $^{45}Ca^{2+}$fluxes in rabbit aortic strips and attributed them to Na^{+}/Ca^{2+} exchange. However, several reports prior to 1987 concluded that the exchanger played little or no role in smooth muscle function [1, 72, 102, 103, 336, 355], although the evidence was inconclusive. For example, changing the Na^{+} gradient had no effect on the stimulation of $^{45}Ca^{2+}$ efflux from arterial strips produced by histamine or an energy poison [102]. $^{45}Ca^{2+}$ desaturation data obtained with vascular tissues are complicated by the slow exchange of a relatively large proportion of extracellular isotope. Aickin et al. [4] monitored contraction and intracellular Na^{+} activity with electrodes in guinea pig ureter. In ouabain-treated tissues either lowering external Na^{+} or raising external Ca^{2+} concomitantly decreased intracellular Na^{+} activity and elicited a transient contracture. Therefore, Na^{+}/Ca^{2+} exchange apparently plays a major role in the regulation of intracellular Na^{+} in guinea pig ureter when the Na^{+} pump is inhibited.

Grover et al. [136] demonstrated Na^{+} gradient-dependent Ca^{2+} transport in membrane vesicles from myometrium. Na^{+}-dependent Ca^{2+} transport had a lower capacity, lower affinity for Ca^{2+}, and a slower rate than ATP-dependent Ca^{2+} pump [137, 225]. However, higher Na^{+}/Ca^{2+} exchange activities were obtained in plasma membranes from tracheal muscle [318], aorta [316], and mesenteric arteries [208]. In fact, Slaughter et al. [316] found that the specific activity of Na^{+}/Ca^{2+} exchange assayed under V_{max} conditions was three- to sixfold greater than the V_{max} of ATP-dependent Ca^{2+} uptake even after correcting for vesicle sidedness. The correction is necessary because only inside-out vesicles carry out ATP-dependent Ca^{2+} uptake, whereas both inside-out and rightside-out vesicles carry out exchange because it is reversible.

Much of what is known about the role of Na^+/Ca^{2+} exchange in cells and tissues has been learned by manipulating the Na^+ gradient. Ouabain or K^+-free medium may be used to inhibit the Na^+/K^{2+} pump and increase cell Na^+. However, in intact tissues ouabain or the removal of external K^+ can produce contractions by releasing endogenous catecholamines from nerve endings [56, 170]. Although the Ca^{2+} pump appears to be unaffected by changes in the concentration of intracellular or extracellular Na^+ [97, 324], decreasing the concentration of extracellular Na^+ may release endogenous neurotransmitters in vascular tissue [244] or trigger the release of stored Ca^{2+} [172, 321, 323]. Decreasing the concentration of external Na^+ evokes the production of inositol trisphosphate and releases stored Ca^{2+} in several cell types [321, 323]. A Na^+-sensitive receptor that is activated by certain divalent cations (e.g., Cd^{2+} or Co^{2+} but not Ca^{2+} or Mg^{2+}) apparently stimulates a phosphoinositidase [322]. Extracellular Ca^{2+} or Mg^{2+} markedly potentiates receptor activation, and lanthanum prevents receptor activation (unpublished observations). Therefore, it is difficult to distinguish readily changes in cytoplasmic free Ca^{2+} that are produced by the Na^+-sensitive, divalent cation receptor from those produced by Na^+/Ca^{2+} exchange when both are present in the same cell type. Myocytes from human umbilical artery have the receptor, but rat aortic myocytes do not [323]. A thorough analysis of the effects of the Na^+ gradient on Ca^{2+} fluxes and cytosolic free Ca^{2+}, as can be only done on cultured cells, is essential for assessing the role the exchanger in Ca^{2+} regulation.

Role of Na^+/Ca^{2+} Exchange in Aortic Myocytes

Figure 9 depicts our working hypothesis for Ca^{2+} regulation in vascular smooth muscle. The diagram is modified from a review on angiotensin-receptor signaling [319]. The basic features of the model apply generally to Ca^{2+}-mobilizing hormones [44, 269]. The diagram shows a transmembrane receptor that is coupled to an enzyme, phospholipase C, via a GTP-binding protein. Two second messengers, inositol trisphosphate and diacylglycerol are produced concomitantly when an agonist, e.g., angiotensin II, binds to its receptor on the cell surface. Inositol trisphosphate opens an intracellular Ca^{2+} channel, shown in closed and open states, which releases Ca^{2+} from the sarcoplasmic reticulum. Diacylglycerol activates protein kinase C, which apparently migrates from the cytoplasm to the inner face of the plasma membrane [233]. After being released into the cytoplasm, Ca^{2+} rapidly exits the cell, which transiently lowers total cell Ca^{2+}. Rapid net Ca^{2+} efflux occurs largely via Na^+/Ca^{2+} exchange and partly via the Ca^{2+} pump ATPase. The diagram does not show the activation of Ca^{2+}/calmodulin protein kinases (e.g., myosin light chain kinase) or other metabolic intermediates caused by raising $[Ca^{2+}]_i$ and diacylglycerol. Ca^{2+} enters the cells via channels which are activated by angiotensin and other Ca^{2+} mobilizing neurohormones [107, 269, 327]. In aortic myocytes angiotensin causes a delayed, moderate increase in $^{45}Ca^{2+}$ uptake [325] that helps to refill the sarcoplasmic reticulum.

Forward (Ca^{2+} efflux) mode exchange can most readily be demonstrated as a several-fold increase in angiotensin-evoked $^{45}Ca^{2+}$ efflux that depends on extracellular Na^+ [230, 325]. Angiotensin markedly stimulates $^{45}Ca^{2+}$ efflux in the presence of extracellular Na^+ (Fig. 10). Replacing external Na^+ with other monovalent cations (choline or *N*-methyl-D-glucamine) strongly decreases the stimulation of efflux produced by angiotensin (Fig. 10). Basal efflux, on the other hand, is the same in the presence or absence of external Na^+ (Fig. 10). Angiotensin produces a spike in $[Ca^{2+}]_i$ mostly by triggering the release of stored Ca^{2+} because the hormone produces similar peak $[Ca^{2+}]_i$ increases in the presence or absence of external Ca^{2+} [319, 325, 330]. Replacing external Na^+ with other cations does not effect the release of stored Ca^{2+} because angiotensin produces similar peak $[Ca^{2+}]_i$ increases in the presence and absence of external Na^+ [325].

Measurements of total and free Ca^{2+} in the presence and absence of external Na^+ also indicate that Na^+/Ca^{2+} exchange rapidly expels Ca^{2+} after its released from the sarcoplasmic reticulum. Angiotensin decreases total cell Ca^{2+} by about 50% in 1 min [325]. Replacing external Na^+ with other monovalent cations almost prevents the decrease in total Ca^{2+} produced by a 1 min incubation with angiotensin [325]. Removing extracellular Na^+ substantially increases and prolongs the plateau phase of the increase in $[Ca^{2+}]_i$ produced by angiotensin [325], as would be expected from the efflux and total cell Ca^{2+} data. Na^+/Ca^{2+} exchange, which is latent in nonstimulated cells [328], becomes highly active after hormone stimulation and makes a major contribution to net Ca^{2+} efflux [325]. Thus, efflux, rather than re-sequestration is largely responsible for rapidly decreasing $[Ca^{2+}]_i$ from the peak produced by angiotensin [325].

Ionomycin at low concentrations mimics the effects of angiotensin on $^{45}Ca^{2+}$ fluxes, total Ca^{2+}, and $[Ca^{2+}]_i$ [328]. Low concentrations of ionomycin selectively release stored Ca^{2+} and have little effect on Ca^{2+} influx apparently because the ionophore has such a low affinity for Ca^{2+} ($K_m = 8$ mM

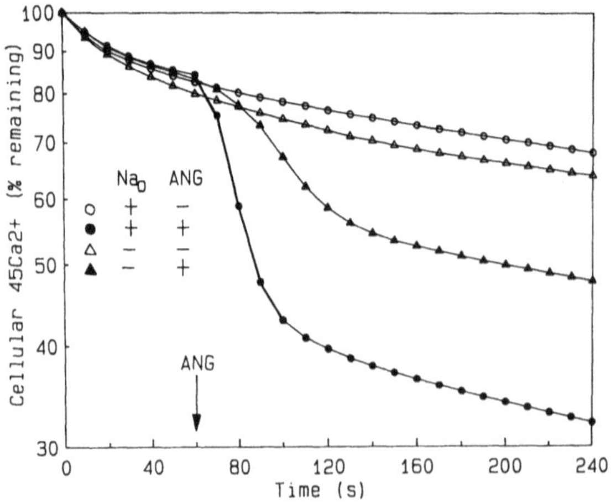

Fig. 10. Effect of removing extracellular Na^+ on basal and angiotensin-stimulated $^{45}Ca^{2+}$ efflux. Efflux was assayed as described [325]. Equimolar *N*-methyl-D-glucamine chloride replaced 140 mM NaCl as indicated

[163]). Thus, at 1 m*M* external Ca^{2+}, compared with a presumably much higher concentration in the sarcoplasmic reticulum, ionomycin selectively releases Ca^{2+} from the sarcoplasmic reticulum. Removing external Na^+ strongly inhibits net Ca^{2+} export produced by 15- to 60-s incubations with ionomycin [328]. In contrast to the brief incubations with ionomycin or angiotensin, longer ones decreased total cell Ca^{2+} even in the absence of extracellular Na^+ ([328] and unpublished data). Therefore, Na^+/Ca^{2+} exchange appears to be largely responsible for the rapid phase of Ca^{2+} export. A Na^+-gradient independent pathway, presumably the Ca^{2+} pump, apparently causes the slower component of Ca^{2+} efflux from aortic myocytes.

Vigne et al. [361, 362] have investigated the role of Na^+/Ca^{2+} exchange in a clonal cell line (A7r5) derived from embryonic rat aorta. The free Ca^{2+} transient produced by vasopressin was not affected by the removal of external Na^+. This finding is in striking contrast to the marked dependence of the changes in free and total Ca^{2+} and $^{45}Ca^{2+}$ fluxes produced by angiotensin in low passage aortic myocytes [325]. Because A7r5 cells have appreciable Na^+/Ca^{2+} exchange activity, Vigne et al. [381, 382] suggested that the exchanger may be involved in long-term Ca^{2+} homeostasis and that Ca^{2+} ATPases in the plasma membrane and sarcoplasmic reticulum determine the size and duration of the free Ca^{2+} transient. Free Ca^{2+} was measured with indo-1 acetoxymethyl ester by flow cytometry. A problem with dye-loading may account for the failure to observe an effect of removing external Na^+ on free Ca^{2+} in A7r5 cells because Cornwell and Lincoln [82] found that the amount of loading with fura-2 or indo-1 markedly influenced the duration of the hormone-evoked increase in $[Ca^{2+}]_i$ in aortic myocytes. The influence of removing external Na^+ on the $[Ca^{2+}]_i$ transient produced by angiotensin was more marked at lower concentrations of dye [82].

In contrast to the Na^+/Ca^{2+} exchanger, the Ca^{2+} pump ATPase appears to be ubiquitous in animal cells. Removing external Na^+ has no effect on basal or vasopressin-evoked $^{45}Ca^{2+}$ efflux in Swiss 3T3 cells [324]. Therefore, the plasma membrane Ca^{2+} pump, rather than Na^+/Ca^{2+} exchange, apparently mediates $^{45}Ca^{2+}$ efflux in Swiss 3T3 cells. The Ca^{2+} pump also appears to mediate net Ca^{2+} efflux produced by bradykinin in human fibroblasts [321]. Manipulating the Na^+ gradient or the membrane potential does not affect the Ca^{2+} pump in squid axons [97]. In squid axon the Ca^{2+} pump is a higher affinity, lower capacity pathway of Ca^{2+} efflux than the Na^+/Ca^{2+} exchanger [97].

Agreement between data on $[Ca^{2+}]_i$ in Cultured Myocytes and Tissue Contractility Studies

There is good agreement between the data on the effect of changing the Na^+ gradient on contraction and $[Ca^{2+}]_i$. Inhibition of the Na^+/K^+ pump with ouabain causes a very small, slow contraction of helical strips of rabbit aorta providing a Ca^{2+} channel blocker and adrenergic antagonist are present [243]. Ouabain releases endogenous catecholamines from nerve terminals and apparently activates voltage-gated Ca^{2+} channels. Superfusion of rat aortic rings with a K^+-free solution containing phentolamine and verapamil, in order to inhibit the Na^+/K^+ pump without activating adrenergic receptors or Ca^{2+} channels, produced a small increase in tension after 30–45 min [18, 329]. We [329] found that a 30-min incubation of myocytes cultured from rat aorta with ouabain slightly increased $[Ca^{2+}]_i$ from 113 ± 6 to 136 ± 2 n*M*. Thus, inhibition of the Na^+ pump produced very small increases in tension and $[Ca^{2+}]_i$.

Decreasing extracellular Na^+ from 139 to 7.5 m*M* did not increase tension in rings of rat aorta unless caffeine was present to inhibit Ca^{2+} sequestration [18]. Decreasing external Na^+ to 4 m*M* did not change $[Ca^{2+}]_i$ over short intervals (3–5 min) in aortic myocytes [329]. Mulvany and colleagues [227] found that incubation of denervated rings of rat mesenteric arteries at 25 m*M* Na^+ or with ouabain did not increase ring tension. A contractile response occurred, however, if the rings were first incubated with ouabain to increase intracellular Na^+, and then exposed to low external Na^+ [227]. A similar treatment of aortic myocytes produces a large $[Ca^{2+}]_i$ spike that is caused by Ca^{2+} influx via the exchanger [320, 329]. In addition, the rate of relaxation of rat aortic rings after inducing a contraction with caffeine was fourfold faster at 139 compared to 1 m*M* external Na^+ (18). This finding also agrees well with the $[Ca^{2+}]_i$, $^{45}Ca^{2+}$ efflux, and total cell Ca^{2+} data [325], indicating that Na^+/Ca^{2+} exchange is the major pathway for rapidly decreasing $[Ca^{2+}]_i$ from the peak produced by stimuli that release Ca^{2+} from the sarcoplasmic reticulum.

It is noteworthy that there appear to be species and/or vessel differences with regard to the effect of decreasing the concentration of external Na^+ on tissue tension. In contrast to rat aorta, decreasing the external Na^+ from 139 to 7.5 m*M* significantly increased the tonic tension of bovine tail artery [18]. Additionally Pritchard and Ashley [268] found that ouabain increased $[Ca^{2+}]_i$ from 100 ± 18 to 266 ± 83 nM in smooth muscle cells from guinea pig taenia coli, which is much greater than the increase produced by ouabain in aortic myocytes [329].

Role of Na^+/Ca^{2+} Exchange in Hypertension

Blaustein [51] proposed that a primary increase in Na^+ in vascular smooth muscle increases sarcoplasmic $[Ca^{2+}]_i$ and tone by affecting Na^+/Ca^{2+} exchange. According to the current hypothesis [53], raising cell Na^+ would prolong the $[Ca^{2+}]_i$ transient produced by vasostimulation and thereby increase the Ca^{2+} content of the sarcoplasmic reticulum. Thus, more Ca^{2+} would be released by a subsequent stimulus. This hypothesis has not yet been tested by measurements of the effects of increasing cell Na^+ on the $[Ca^{2+}]_i$ transients produced by vasoconstrictors or on the amount of Ca^{2+} in the sarcoplasmic reticulum.

Summary and Projections

Amiloride-sensitive transport systems are present in most cells of the body and thus subserve numerous physiological functions. Amiloride-sensitive Na^+ channels are undoubtedly involved in extracellular Na^+ ion homeostasis, while Na^+/H^+ and Na^+/Ca^{2+} exchangers contribute to intracellular pH and Ca^{2+} ion regulation, respectively. Abnormalities in either the operation or regulation of these proteins may underlie, at least in part, the etiology of such disease states as hypertension and cystic fibrosis. One major challenge of the future will be to deduce the functional components of these transport proteins, and to understand how they regulate channel function.

The Na^+/H^+ exchanger serves specific functions in epithelial systems beyond those encountered in nonpolarized cells. Besides cell volume regulation and defense of intracellular pH, specialized roles may be played in the net transepithelial transport of protons and Na^+. In the latter setting, the roles of parallel transport pathways (e.g., Cl^-/HCO_3^- exchange) and the precise localization of the Na^+/H^+ exchanger in apical or basolateral membranes may determine the net rate of vectorial transport. As a generalization, the kinetic features, substrate specificity and inhibitor effects are similar in all of the epithelial systems. On the other hand, subtle but definite differences are beginning to emerge in the details of inhibitor studies, substrate specificity and regulation which suggest that there may well be distinct forms of the transporter expressed in different tissues.

Studies of Na^+/Ca^{2+} exchange at the tissue level are complex and hampered by the parallel operation of the Ca^{2+} pump ATPase and the lack of a specific inhibitor of either transporter. Presently studies at the cellular level provide the best opportunity to assess definitively the contribution of the exchanger to Ca^{2+} regulation in smooth muscle. Recent studies indicate that the exchanger makes a major contribution to the regulation of Ca^{2+} in rat aortic myocytes. The majority of the data since 1986 differs from earlier studies that failed to show that the exchanger was important in smooth muscle. More work is needed to elucidate the distinctive features of the exchanger in smooth muscle and evaluate its significance in smooth muscle from different blood vessels and pathophysiological states.

The identification and functional reconstitution of each of these transporters remains a priority. It is evident that there are multiple types or isoforms of each of these proteins. With the application of modern molecular biological, immunological, and biochemical techniques, the primary structure and gene families coding for these proteins will be deduced in the near future.

"Note added in proof"

The cardiac Na^+–Ca^{2+} exchanger has been cloned and sequenced (Nicoll, Longoni, and Philipson, Science 250, 562–565, 1990) and has little homology to any other protein including the retinal rod Na^+/Ca^{2+}-K^+ exchanger (Achilles et al., Ann. N.Y. Acad. Sci., in press).

Acknowledgements. We thank Cathy Guy, Juanita Baggett, and Lucinda Smith for help in preparing the manuscript.

References

1. Aaronson P, van Breemen C (1981) Effects of sodium gradient manipulation upon cellular calcium, ^{34}Ca fluxes and cellular sodium in the guinea-pig taenia coli. J Physiol (Lond) 319:443–461
2. Abramcheck FJ, Van Driessche W, Helman SI (1985) Autoregulation of apical membrane Na^+ permeability of tight epithelia. J Gen Physiol 85:555–582
3. Aceves J, Cereijido M (1973) The effect of amiloride on sodium and potassium fluxes in red cells. J Physiol (Lond) 229:709–718
4. Aickin CC, Brading AF, Walmsley D (1987) An investigation of sodium-calcium exchange in the smooth muscle of guinea-pig ureter. J Physiol (Lond) 391:325–346
5. Akiba T, Alpern RJ, Eveloff J, Calamia T, Warnock DG (1986) Electrogenic sodium/bicarbonate cotransport in rabbit renal cortical basolateral membrane vesicles. J Clin Invest 78:1472–1478
6. Akiba R, Rocco VK, Warnock DG (1987) Parallel adaptation of the rabbit renal cortical sodium/proton antiporter and sodium/bicarbonate cotransporter in metabolic acidosis and alkalosis. J Clin Invest 80:308–315
7. Alberts B, Bray D, Lewis J, Raff M, Roberts K, Watson JD (1983) Molecular biology of the cell. Garland, New York, pp 322
8. Allen TJA, Baker PF (1986) Comparison of the effects of potassium and membrane potential on the calcium-dependent sodium efflux in squid axons. J Physiol (Lond) 378:53–76
9. Alpern RJ (1985) Mechanism of basolateral membrane $H^+/OH^-/HCO_3^-$ transport in the rat proximal convoluted tubule (a sodium-coupled electrogenic process). J Gen Physiol 86:613–636
10. Alpern RJ, Chambers M (1986) Cell pH in the rat proximal convoluted tubule. J Clin Invest 78:502–510
11. Ando Y, Jacobson HR, Breyer MD (1987) Phorbol myristate acetate, dioctanoylglycerol, and phosphatidic acid inhibit the hydroosmotic effect of vasopressin on rabbit cortical collecting tubule. J Clin Invest 80:590–593
12. Antcliff AC, Beevers DG, Hamilton M, Harpur JE (1971) The use of amiloride hydrochloride in the correction of hypokalaemic alkalosis induced by diuretics. Postgrad Med J 47:644–647
13. Anand-Srivastava MB (1989) Amiloride interacts with guanine nucleotide regulatory proteins and attenuates the hormonal inhibition of adenylate cyclase. J Biol Chem 264:9491–9496
14. Aronson PS, Nee J, Suhm MA (1982) Modifier role of internal H^+ in activating the Na^+–H^+ exchanger in renal microvillus membrane vesicles. Nature 299:161–163
15. Aronson PS, Suhm MA, Nee J (1983) Interaction of external H^+ with the Na^+–H^+ exchanger in renal microvillus membrane vesicles. J Biol Chem 258:6767–6771
16. Ashavaid TF, Colvin RA, Messineo FC, MacAlister T, Katz AM (1985) Effects of fatty acids on Na/Ca exchange in cardiac sarcolemmal membranes. J Mol Cell Cardiol 17:851–861
17. Asher C, Moran C, Rossier BC, Garty H (1988) Sodium channels in membrane vesicles from cultured toad bladder cells. Am J Physiol 254 (Cell Physiol):C512–C518
18. Ashida T, Blaustein MP (1987) Regulation of cell calcium and contractility in mammalian arteral smooth muscle: the role of sodium-calcium exchange. J Physiol (Lond) 392:617–635
19. Ausiello DA, Hall DH, Dayer JM (1980) Modulation of cyclic AMP-dependent protein kinase by vasopressin and calcitonin in cultured porcine renal LLC-PK_1 cells. Biochem J 186:773–780
20. Ausiello D, Sorscher E, Harlin C, Benos DJ (1989) Subunits of the epithelial Na^+ channel may be G-proteins. FASEB J 3(3):A228

21. Baer JE, Jones CB, Spitzer SA, Russo HF (1967) The potassium-sparing and natriuretic activity of N-amidino-3,5-diamino-6-chloropyrazinecarboxamide hydrochloride dihydrate (amiloride hydrochloride). J Pharmacol Exp Ther 157:472–485
22. Baker PF, Blaustein MP (1968) Sodium-dependent uptake of calcium by crab nerve. Biochim Biophys Acta 150:167–170
23. Baker PF, Blaustein MP, Hodgkin AL, Steinhardt RA (1969) The influence of calcium on sodium efflux in squid axons. J Physiol (Lond) 200:431–458
24. Baker PF, DiPolo R (1984) Axonal calcium and magnesium homeostasis. Curr Top Membr Transp 22:195–247
25. Baker PF, Glitsch HG (1973) Does metabolic energy participate directly in the Na^+-dependent extrusion of Ca^{2+} ions from squid giant axons. J Physiol (Lond) 233:44P–46P
26. Barbry P, Chassande O, Vigne P, Frelin C, Ellore C, Cragoe EJ Jr, Lazdunski M (1987) Purification and subunit structure of the $[^3H]$ phenamil receptor associated with the renal apical Na^+ channel. Proc Natl Acad Sci USA 84:4836–4840
27. Barcenas-Ruiz L, Beuckelmann DJ, Wier WG (1987) Sodium-calcium exchange in heart: membrane currents and changes in $[Ca^{2+}]$. Science 238:1720–1722
28. Bartschat DK, Lindenmayer GE (1980) Calcium movements promoted by vesicles in a highly enriched sarcolemma preparation from canine ventricle. J Biol Chem 255:9626–9634
29. Barzilai A, Spanier R, Rahamimoff H (1984) Isolation, purification, and reconstitution of the Na^+ gradient-dependent Ca^{2+} transporter (Na^+–Ca^{2+} exchanger) from brain synaptic plasma membranes. Proc Natl Acad Sci USA 81:6521–6525
30. Barzilai A, Spanier R, Rahamimoff H (1987) Immunological identification of the synaptic plasma membrane Na^+–Ca^{2+} exchanger. J Biol Chem 262:10 315–10 320
31. Bayerdorffer E, Haase W, Schulz I (1985) Na^+/Ca^{2+} countertransport in plasma membrane of rat pancreatic acinar cells. J Membr Biol 87:107–119
32. Beliveau R, Demeule M, Potier M (1988) Molecular size of the Na^+–H^+ antiport in renal brush border membranes, as estimated by radiation inactivation. Biochem Biophys Res Commun 152:484–489
33. Benos DJ (1982) Amiloride: a molecular probe of sodium transport in tissues and cells. Am J Physiol 242:C131–C145
34. Benos DJ (1986) In: Ionic channels in cells and model systems, Amiloride-sensitive epithelial sodium channels. Plenum, New York, pp 401–420
35. Benos DJ (1988) Amiloride: chemistry, kinetics, and structure-activity relationships. In: Grinstein S (ed) Na^+/H^+ Exchange. CRC Uniscience, Boca Raton, pp 121–136
36. Benos DJ, Hyde BA, Latorre R (1983) Sodium flux ratio through the amiloride-sensitive entry pathway in frog skin. J Gen Physiol 81:667–685
37. Benos D, Latorre R, Reyes J (1981) Surface potentials and sodium entry in frog skin epithelium. J Physiol (Lond) 321:163–174
38. Benos DJ, Mandel LJ, Balaban RS (1979) On the mechanism of the amiloride-sodium entry site interaction in anuran skin epithelia. J Gen Physiol 73:307–326
39. Benos DJ, Mandel LJ, Simon SA (1980) Cationic selectivity and competition at the sodium entry site in frog skin. J Gen Physiol 76:233–247
40. Benos DJ, Reyes J, Shoemaker DG (1983) Amiloride fluxes across erythrocyte membranes. Biochim Biophys Acta 734:99–104
41. Benos DJ, Saccomani G, Sariban-Sohraby S (1987) The epithelial sodium channel: subunit number and location of amiloride binding site. J Biol Chem 262:10 613–10 618
42. Benos DJ, Simon SA, Mandel LJ, Cala PM (1976) Effect of amiloride and some of its analogues on cation transport in isolated frog skin and thin lipid membranes. J Gen Physiol 68:43–63
43. Bentley PJ (1968) Amiloride: a potent inhibitor of sodium transport across the toad bladder. J Physiol (Lond) 195:317–330
44. Berridge MJ, Irvine RF (1984) Inositol trisphosphate, a novel second messenger in cellular signal transduction. Nature 312:315–321
45. Bers DM, Philipson KD, Nishimoto AY (1980) Sodium-calcium exchange and sidedness of isolated cardiac sarcolemmal vesicles. Biochim Biophys Acta 601:358–371
46. Besterman JM, May WS Jr, Le Vine H III, Cragoe EJ Jr, Cuatrecasas (1985) Amiloride inhibits phorbol ester-stimulated Na^+/H^+ exchange and protein kinase C. J Biol Chem 260:1155
47. Bhalla RC, Sharma RV (1986) Competitive interaction of amiloride and verapamil with α_1-adrenoceptors in vascular smooth muscle. J Cardiovasc Pharmacol 8:927–932

48. Bickling JE, Mason JW, Woltersdorf OW Jr, Jones JH, Kwon SF, Robb CM, Cragoe EJ (1965) J Med Chem 8:638
49. Bielefeld DR, Hadley RW, Vassilev PM, Hume JR (1986) Membrane electrical properties of vesicular Na–Ca exchange inhibitors in single atrial myocytes. Circ Res 59:381–389
50. Birnbaum MJ, Haspel HC, Rosen OM (1987) Transformation of rat fibroblasts by FSV rapidly increases glucose transporter gene transcription. Science 235:1495–1498
51. Blaustein MP (1977) Sodium ions, calcium ions, blood pressure regulation, and hypertension a reassessment and a hypothesis. Am J Physiol 232:C165–C173
52. Blaustein MP (1988) Sodium/calcium exchange and the control on contractility in cardiac muscle and vascular smooth muscle. J Cardiovasc Pharmacol 12 [Suppl 5]:S56–S68
53. Blaustein MP, Hamlyn JB (1984) Sodium transport inhibition, cell calcium, and hypertension. The natriuretic hormone/Na^+–Ca^{2+} exchange/hypertension hypothesis. Am J Med 77:45–59
54. Blaustein MP, Russell JM (1975) Sodium-calcium exchange and calcium-calcium exchange in internally dialyzed squid giant axons. J Membr Biol 22:285–312
55. Blaustein MP, Wiesmann WP (1970) Effect of sodium ions on calcium movements in isolated synaptic terminals. Proc Natl Acad Sci USA 66:664–671
56. Bonaccorsi A, Hermsmeyer K, Smith CB, Bohr DF (1977) Norepinephrine release in isolated arteries induced by K-free solution. Am J Physiol 232:H140–H145
57. Boron WF, Boulpaep EL (1983a) Intracellular pH regulation in the renal proximal tubule of the salamander. Na–H exchange. J Gen Physiol 81:29–52
58. Boron WF, Boulpaep EL (1983b) Intracellular pH regulation in the renal proximal tubule of the salamander. Basolateral HCO_3–transport. J Gen Physiol 81:53–94
59. Breyer MD, Jacobson HR (1989) Regulation of rabbit medullary collecting duct cell pH by basolateral Na^+/H^+ and Cl^-/base exchange. J Clin Invest 84:996–1004
60. Bridges RJ, Garty H, Benos DJ, Rummel W (1988) Sodium uptake into colonic enterocyte membrane vesicles. Am J Physiol 254 (Cell Physiol) 23:C484–C490
61. Briggman JV, Graves JS, Spicer SS, Cragoe EJ Jr (1983) The intracellular localization of amiloride in frog skin. Histochem J 15:239–255
62. Brinley FJ Jr, Mullins LJ (1987) Sodium extrusion by internally dialyzed squid axons. J Gen Physiol 50:2303–2331
63. Brown A, Birnbaumer L (1988) Signal transduction by G proteins. Am J Physiol 254:H401–H410
64. Brown D, Sorscher EJ, Ausiello DA, Benos DJ (1989) Immunocytochemical localization of Na^+ channels in rat kidney medulla. Am J Physiol (Renal) 256:F366–F369
65. Burnham C, Munzesheimer C, Rabon E, Sachs G (1982) Ion pathways in renal brush border membranes. Biochim Biophys Acta 685:260–272
66. Cantiello HF, Scott JA, Rabito CA (1986) Polarized distribution of the Na^+/H^+ exchange system in a renal cell line (LLC-PK1) with characteristics of proximal tubular cells. J Biol Chem 261:3252–3258
67. Carafoli E (1987) Intracellular calcium homeostasis. Annu Rev Biochem 56:395–433
68. Caroni P, Carafoli E (1983) The regulation of the Na^+–Ca^{2+} exchanger of heart sarcolemma. Eur J Biochem 132:451–460
69. Caroni P, Reinlib L, Carafoli E (1980) Charge movements during the Na^+–Ca^{2+} exchange in heart sarcolemmal vesicles. Proc Natl Acad Sci USA 77:6354–6358
70. Cassel D, Cragoe EJ Jr, Rotman M (1987) A dextran-bound amiloride deprivative is a selective inhibitor of Na^+/H^+ antiport. Application for studying the role of the antiporter in cellular proliferation in human fibroblasts. J Biol Chem 262:4587–4591
71. Cassel D, Katz M, Rotman M (1986) Depletion of cellular ATP inhibits Na^+/H^+ antiport in cultured human cell. Modulation of the regulatory effect of intracellular protons on the antiporter activity. J Biol Chem 261:5460–5466
72. Casteels R, van Breemen C (1975) Active and passive Ca^{2+} fluxes across cell membranes of the guinea pig taenia coli. Pflugers Arch 359:197–207
73. Cervetto L, Lagnado L, Perry RJ, Robinson DW, McNaughton PA (1989) Extrusion of calcium from rod outer segments is driven by both sodium and potassium gradients. Nature 337:740–743
74. Chaillet JR, Lopes AG, Baron WF (1985) Basolateral Na–H exchange in the rabbit cortical collecting tubule. J Gen Physiol 86:795–812
75. Chaillet JR, Amsler K, Boron WF (1986) Optical measurements of intracellular pH in single LLC-PK1 cells: demonstration of Cl–HCO_3 exchange. Proc Natl Acad Sci USA 83:522–526

76. Chase HS Jr, Al-Awqati Q (1981) Regulation of the sodium permeability of the luminal border of toad bladder by intracellular sodium and calcium. J Gen Physiol 77:693–712
77. Cheon J, Reeves JP (1988) Site density of the sodium-calcium exchange carrier in reconstituted vesicles from bovine cardiac sarcolemma. J Biol Chem 263:2309–2315
78. Cohn DE, Hruska KA, Klahr S, Hammerman MR (1982) Increased Na^+–H^+ exchange in brush border vesicles from dogs with renal failure. Am J Physiol 243:F293–F299
79. Cohn DE, Klahr S, Hammerman M (1983) Metabolic acidosis and parathyroidectomy increase Na^+–H^+ exchange in brush border vesicles. Am J Physiol 245:F217–F222
80. Cook JS, Shaffer C, Cragoe EJ Jr (1987) Inhibition by amiloride analogues of Na^+-dependent hexose uptake in LLC-PK1/C14 cells. Am J Physiol 253:C199–C204
81. Cook NJ, Kaupp UB (1988) Solubilization, purification, and reconstitution of the sodium-calcium exchanger from bovine retinal rod outer segments. J Biol Chem 263:11 382–11 388
82. Cornwell TL, Lincoln TM (1989) Demonstration of Na^+-dependent Ca^{2+} efflux using low concentrations of fluorometric Ca^{2+} probes. Cell Calcium 10:47–55
83. Costa CJ, Kirschner LB, Cragoe EJ Jr (1984) Identification of apical membranes from tight epithelia using spin-labeled amiloride and electron paramagnetic resonance spectroscopy. J Membr Biol 82:49–57
84. Costa-Casnellie MR, Segel GB, Lichtman MA (1988) The Na^+/H^+ exchanger in immature and mature granulocytic HL-60 cells. Property changes induced by intracellular acidification and cell maturation. J Biol Chem 263:11 851–11 858
85. Cuthbert AW (1981) In: SC Schultz (ed) Ion transport by epithelia. Sodium entry step in transporting epithelia: results of ligand-binding studies. Raven, New York, pp 181–195.
86. Cuthbert AW, Shum WK (1974) Amiloride and the sodium channel. Naunyn Schmiedebergs Arch Pharmacol 281:261–269
87. Dannenbaum D, Rosenheck K (1986) Amiloride, an inhibition of Na^+/H^+ exchange, inhibits also acetylcholinesterase activity. Biophys J 49:370a (abstract)
88. Davis RJ, Czech MP (1985) Amiloride directly inhibits growth factor receptor tyrosine kinase activity. J Biol Chem 260:2543–1551
89. Dayer J-M, Vassalli JD, Bobbit JL, Hull N, Reich E, Krane SM (1981) Calcitonin stimulates plasminogen activator in porcine renal tubular cells:LLC-PK1. J Cell Biol 91:195–200
90. Debetto P, Luciani S, Tessari M, Floreani M, Carpenedo F (1989) Inhibition of Na^+/Ca^{2+} exchange by amiloride acting from opposite sides of cardiac sarcolemma. Biochem Pharmacol 38:1137–1145
91. de la Pena P, Reeves JP (1987) Inhibition and activation of Na^+–Ca^{2+} exchange activity by quinacrine. Am J Physiol 252:C24–C29
92. Desir GV (1989) Reconstitution and partial purification of an amiloride-sensitive non-selective cation channel from the rabbit kidney. Clin Res 37:488a (abstract)
93. DiFrancesco D, Noble D (1985) A model cardiac electrical activity incorporating ionic pumps and concentration changes. Philos Trans R Soc Lond [Biol] 307:353–398
94. DiPolo R (1977) Characterization of the ATP-dependent calcium efflux in dialyzed squid giant axons. J Gen Physiol 69:795–813
95. DiPolo R, Beauge L (1987a) In squid axons, ATP modulates Na^+–Ca^{2+} exchange by a $Ca_i^{2\pm}$ dependent phosphorylation. Biochim Biophys Acta 897:347–354
96. DiPolo R, Beauge L (1987b) Characterization of the reverse Na/Ca exchange in squid axons and its modulation by Ca_i and ATP. Ca_i-dependent Na_i/Ca_o and Na_i/Na_o exchange modes. J Gen Physiol 90:505–525
97. DiPolo R, Beauge L (1988) Ca^{2+} transport in nerve fibres. Biochim Biophys Acta 947:549–569
98. DiPolo R, Rojas H (1984) Effects of internal and external K^+ on Na^+ Ca^{2+} exchange in dialyzed squid axons under voltage clamp conditions. Biochim Biophys Acta 776:313–316
99. Dixon SJ, Cohen S, Cragoe EJ Jr, Grinstein S (1987) Estimation of the number and turn over rate of Na^+/H^+ exchangers in lymphocytes. Effect of phorbol ester and osmotic shrinking. J Biol Chem 262:3626–3632
100. dos Remedios CG (1981) Lanthanide ion probes of calcium-binding sites on cellular membranes. Cell Calcium 2:29–51
101. Dousa TP, Barnes LD (1977) Regulation of protein kinase by vasopressin in renal medulla in situ. Am J Physiol 232(Renal Fluid Electrolyte Physiol 1):F50–F57
102. Droogmans G, Casteels R (1979) Sodium and calcium interactions in vascular smooth muscle cells of the rabbit ear artery. J Gen Physiol 74:57–70
103. Droogmans G, Himpens B, Casteels R (1985) Ca-exchange, Ca-channels and Ca-antagonists. Experentia 41:895–900

104. Dubinsky WP Jr, Frizzell RA (1983) A novel effect of amiloride on H^+-dependent Na^+ transport. Am J Physiol 245(Cell Physiol 14):C157–C159
105. Edwardson JM, Fanestil DD, Ellory JC, Cuthbert AW (1981) Extraction of a [^{3}H]benzamil binding component from kidney cell membranes. Biochem Pharmacol 30:1185–1189
106. Ehara T, Matsuoka S, Noma A (1989) Measurement of reversal potential of Na^+-Ca^{2+} exchange current in single guinea-pig ventricular cells. J Physiol (Lond) 419:227–249
107. Exton JH (1988) Mechanisms of action of calcium-mobilizing agonists: some variations on a young theme. FASEB J 2:2670–2676
108. Flier JS, Mueckler MM, Usher P, Lodish HF (1987) Elevated levels of glucose transport and transporter messenger RNA are induced by ras or src oncogenes. Science 235:1492–1495
109. Foster ES, Dudeja PK, Brasitus TA (1986) Na^+-H^+ exchange in rat colonic brush-border membrane vesicle. Am J Physiol 250:G781–G787
110. Frelin C, Barbry P, Vigne P, Chassande O, Cragoe EJ Jr, Lazdunski M (1988) Amiloride and its analogs as tools to inhibit Na^+ transport via the Na^+ channel, the Na^+/H^+ antiport and the Na^+/Ca^+ exchanger. Biochimie 70:1285–1290
111. Frelin C, Vigne P, Lazdunski M (1983) The amiloride-sensitive Na^+/H^+ antiport in 3T3 fibroblasts. J Biol Chem 258:6272–6276
112. Friedman PA, Andreoli TE (1982) CO_2-stimulated NaCl absorption in the mouse renal cortical thick ascending limb of Henle. Evidence for synchronous Na^+/H^+ and Cl^-/HCO_3^--exchange in apical plasma membranes. J Gen Physiol 80:683–711
113. Friedrich T, Sablotni J, Burckhardt G (1986) Identification of the renal Na^+/H^+ exchanger with N,N'-dicyclohexylcarbodiimide (DCCD) and amiloride analogues. J Membr Biol 94:253–266
114. Friedrich T, Sablotni J, Burckhardt G (1987a) Species differences between rat rabbit renal Na^+/H^+ exchangers. Biochem Biophys Res Commun 144:869–875
115. Friedrich T, Sablotni J, Burckhardt G (1987b) Inhibition of sodium-dependent transport systems in rat renal brush-border membranes with N,N'-dicyclohexylcarbodiimide. Biochem Biophys Res Commun 147:375–381
116. Fuchs W, Hviid Larsen E, Lindemann B (1977) Current-voltage curve of sodium channels and concentration dependence of sodium permeability in frog skin. J Physiol (Lond) 267:137–166
117. Furukawa K-I, Tawada Y, Shigekawa M (1988) Regulation of the plasma membrane Ca^{2+} pump by cyclic nucleotides in cultured vascular smooth muscle cells. J Biol Chem 263:8058–8065
118. Garcia ML, King VF, Cragoe EJ Jr, Kaczorowski GJ (1987) Interaction of amiloride derivatives with the cardiac Ca^{2+} entry blocker receptor complex. Biophys J 51:428A
119. Garcia ML, Slaughter RS, King VF, Kaczorowski GJ (1988) Inhibition of sodium-calcium exchange in cardiac sarcolemmal membrane vesicles. 2. Mechanism of inhibition by bepridil. Biochemistry 27:2410–2415
120. Garty H (1986) Mechanisms of aldosterone action in tight epithelia. J Membr Biol 90:193–205
121. Garty H, Asher C, Yeger O (1987) Direct inhibition of epithelial Na^+ channels by a pH-dependent interaction with calcium, and by other divalent ions. J Membr Biol 95:151–162
122. Garty H, Edelman S (1983) Amiloride-sensitive trypsinization of apical sodium channels. Analysis of hormonal regulation of sodium transport in toad bladder. J Gen Physiol 81:785–803
123. Garty H, Lindemann B (1984) Feedback inhibition of sodium uptake in K^+-depolarized toad urinary bladders. Biochim Biophys Acta 771:89–98
124. Gesek FA, Cragoe EJ Jr, Strandhoy JW (1989) Synergistic alpha-1 and alpha-2 adrenergic stimulation of rat proximal nephron Na^+/H^+ exchange. J Pharmacol Exp Ther 249:694–700
125. Glitzer MS, Steelman SL (1966) N-Amidino-3, 5-diamino-6-chloropyrazine-carboxamide: an active diuretic in the carboxamide series. Nature 212:191–193
126. Gmaj P, Murer H, Kinne R (1979) Calcium ion transport across plasma membranes isolated from rat kidney cortex. Biochem J 17B:549–557
127. Gogelein H, Greger R (1986) Na^+ selective channels in the apical membrane of rabbit late proximal tubules (pars recta). Pflugers Arch 406:198–203
128. Goldfarb D, Nord EP (1987) Asymmetric affinity of Na^+-H^+ antiporter for Na^+ at the cytoplasmic versus external transport site. Am J Physiol 253:F959–F968
129. Goldring SR, Dayer J-M, Ausiello DA, Krane SM (1978) A cell strain cultured from porcine kidney increases cyclic AMP content upon exposure to calcitonin or vasopressin. Biochem Biophys Res Commun 83:434–440

130. Good DW (1985) Sodium-dependent bicarbonate absorption by cortical thick ascending limb of rat kidney. Am J Physiol 248:F281–F289
131. Gottlieb GP, Turnheim K, Frizzell RA, Schultz SG (1978) p-Chloromercuribenzene sulfonate blocks and reverses the effect of amiloride on sodium transport across rabbit colon in vitro. Biophys J 22:125–129
132. Grillo G, Aronson PS (1986) Inactivating of the renal microvillus membrane Na^+-H^+ exchanger by histidine-specific reagents. J Biol Chem 261:1120–1125
133. Grinstein S, Goetz JD, Rothstein A (1984) $^{22}Na^+$ fluxes in thymic lymphocytes. I. Na^+/Na^+ and Na^+/H^+ exchange through an amiloride-insensitive pathway. J Gen Physiol 84:565–584
134. Grinstein S, Rothstein A (1986) Mechanisms of regulation of the Na^+/H^+ exchange. J Membr Biol 90:1–12
135. Grinstein S, Smith JD (1987) Asymmetry of the Na^+/H^+ antiport of dog red cell ghosts. Sidedness of inhibition by amiloride. J Biol Chem 262:9088–9092
136. Grover AK, Kwan CY, Daniel EE (1981) Na-Ca exchange in rat myometrium membrane vesicles highly enriched in plasma membranes. Am J Physiol 240:C175–C182
137. Grover AK, Kwan CY, Rangachari PK, Daniel EE (1983) Na-Ca exchange in a smooth muscle plasma membrane-enriched fraction. Am J Physiol 244:C158–C165
138. Gunther RD, Wright EM (1983) Na^+, Li^+, and Cl^- transport by brush border membranes from rabbit jejunum. J Membr Biol 74:85–94
139. Gurich RW, Warnock DG (1986) Electrically neutral Na^+-H^+ exchange in endosomes obtained from rabbit renal cortex. Am J Physiol 251:F702–F709
140. Haggerty JG, Agarwal N, Reilly RF, Adelberg EA, Slayman CW (1988) Pharmacologically different Na^+/H^+ antiporters on the apical and basolateral surfaces of cultured porcine kidney cells (LLC-PK1). Proc Natl Acad Sci USA 85:6797–6801
141. Haggerty JG, Cragoe EJ Jr, Slayman CW, Adelberg EA (1985) Na^+/H^+ exchanger activity in the pig kidney epithelial cell line, LLC-PK1: inhibition by amiloride and its derivatives. Biochem Biophys Res Commun 127:759–767
142. Hale CC, Kleiboeker SB, Carlton CG, Rovetto MJ, Jung C, Kim HD (1988) Evidence for high molecular weight. Na-Ca exchange in cardiac sarcolemmal vesicles. J Membr Biol 106:211–218
143. Hale CC, Slaughter RS, Ahrens DC, Reeves JP (1984) Identification and partial purification of the cardiac sodium-calcium exchange protein. Proc Natl Acad Sci USA 81:6569–6573
144. Hamilton KL, Eaton DC (1985) Single-channel recordings from amiloride-sensitive epithelial sodium channel. Am J Physiol 249:C200–C207
145. Harris RC, Lufburrow RA III, Cragoe EJ Jr, Seifter JL (1985) Amiloride analogs inhibit Na-glucose and alanine cotransport in renal brush border membrane vesicles (BBMV). Kidney Int 27:310 (abstract)
146. Harris RC, Seifter JL, Brenner BM (1984) Adaptation of Na^+-H^+ exchange in renal microvillus membrane vesicles. Role of dietary protein and urinephrectomy. J Clin Invest 74:1979–1987
147. Haworth RA, Goknur AB, Hunter DR, Hegge JO, Berkoff HA (1987) Inhibition of calcium influx in isolated adult rat heart cells by ATP depletion. Circ Res 60:586–594
148. Hebert SC (1986) Hypertonic cell volume regulation in mouse thick limbs. II. Na^+–H^+ and Cl(–)-HCO_3-exchange in basolateral membranes. Am J Physiol 250:C920–C931
149. Hediger MA, Coady MJ, Ikeda TS, Wright EM (1987) Expression cloning and cDNA sequencing of Na^+/glucose co-transporter. Nature 330:379–381
150. Helman SI, Baxendale LM (1990) Blocker related changes of channel density: analysis of a 3-state model for apical Na channels of frog skin. J Gen Physiol 95:647–678
151. Helman SI, Cox TC, Van Driessche W (1983) Hormonal control of apical membrane Na transport in epithelia studies with fluctuation analysis. J Gen Physiol 82:201–220
152. Hildmann B, Schmidt A, Murer H (1982) Ca^{2+}-transport across basal-lateral plasma membranes from rat small intestinal epithelial cells. J Membr Biol 65:55–62
153. Hirata M, Itoh T, Kuriyama H (1981) Effects of external cations on calcium efflux from single cells of the guinea-pig taenia coli and porcine coronary artery. J Physiol (Lond) 310:321–336
154. Howard MJ, Mullen MD, Insel PA (1987) Amiloride interacts with renal alpha- and beta-adrenergic receptors. Am J Physiol 253:F21–F25
155. Huang Z-Q, Warnock DG (1988) Evidence for a critical lysine group at the amiloride binding site of the renal Na/H antiporter. Kidney Int 33:401 (abstract)
156. Huang Z-Q, Warnock DG (1988) Studies of the critical lysine group at the amiloride binding site of the renal Na/H antiporter. Kidney Int 35:455 (abstract)
157. Huot SJ, Cassel D, Igarashi P, Cragoe EJ Jr, Slayman CW, Aronson PS (1989) Identification

and purification of a renal amiloride-binding protein with properties of the Na^+-H^+ exchanger. J Biol Chem 264:683–686
158. Igarashi P (1987) Discussion: Pouyssegur et al. Genetic approach to the Na^+/H^+ antiporter. Kidney Int 32:S150
159. Igarashi P, Aronson PS (1987) Covalent modification of the renal Na^+/H^+ exchanger by N,N'-dicyclohexycarbodiimide. J Biol Chem 262:860–868
160. Ives HE, Yee VJ, Warnock DG (1983a) Mixed type inhibition of the renal Na^+/H^+ antiporter by Li^+ and amiloride. Evidence for a modifier site. J Biol Chem 258:9710–9716
161. Ives HE, Yee VJ, Warnock DG (1983b) Asymmetric distribution of the Na^+/H^+ antiporter in the renal proximal tubule epithelial cell. J Biol Chem 258:13513–13516
162. Jacobsen C, Kragh-Hansen U, Sheikh MI (1986) Na^+-H^+ exchange in luminal-membrane vesicles from rabbit proximal convoluted and straight tubules in response to metabolic acidosis. Biochem J 239:411–416
163. Johansson JS, Haynes DH (1988) Deliberate quin2 overload as a method for in situ characterization of active calcium extrusion systems and cytoplasmic calcium binding: application to the human platelet. J Membr Biol 104:147–163
164. Johnson JD, Epel D, Paul M (1976) Intracellular pH and activation of sea urchin eggs after fertilisation. Nature 262:661–664
165. Jurkowitz MS, Altschuld RA, Brierley GP, Cragoe EJ Jr (1983) Inhibition of Na^+-dependent Ca^{2+} efflux from heart mitochondria by amiloride analogues. FEBS Lett 162:262–265
166. Kaback HR (1988) Site-directed mutagenesis and ion-gradient driven active transport: on the path of the proton. Annu Rev Physiol 50:243–256
167. Kaczorowski GJ, Barros F, Dethmers JK, Trumble MJ, Cragoe EJ Jr (1985) Inhibition of Na^+/Ca^{2+} exchange in pituitary plasma membrane vesicles by analogues of amiloride. Biochemistry 24:1394–1403
168. Kaczorowski GJ, Slaughter RS, King F, Garcia ML (1989) Inhibitors of sodium-calcium exchange: identification and development of probes of transport activity. Biochim Biophys Acta 988:287–302
169. Kapus A, Lukacs G, Cragoe EJ Jr, Ligeti E, Fonyo A (1988) Characterization of the mitochondrial Na^+-H^+ exchange. The effect of amiloride analogues. Biochim Biophys Acta 944:383–390
170. Karaki H, Urakawa N (1977) Possible role of endogenous catecholamines in the contractions induced in rabbit aorta by ouabain, sodium depletion and potassium depletion. Eur J Pharmacol 43:65–72
171. Kennedy RH, Berlin JR, Jg Y-C, Akera T, Brody TM (1986) Amiloride: effects on myocardial force of contraction, sodium pump, and Na^+/Ca^{2+} exchange. J Mol Cell Cardiol 18:177–188
172. Kim D, Okada A, Smith TW (1987) Control of cytosolic calcium activity during low sodium exposure in cultured chick heart cells. Circ Res 61:29–41
173. Kimura J, Noma A, Irisawa H (1986) Na-Ca exchange current in mammalian heart cells. Nature 319:596–597
174. Kinsella JL, Aronson PS (1980) Properties of the Na^+–H^+ exchanger in renal microvillus membrane vesicles. Am J Physiol 238:F461–F469
175. Kinsella JL, Aronson PS (1981a) Interaction of $NH4^+$ and lit with the renal microvillus membrane Na^+-H^+ exchange. Am J Physiol 241:C220–C226
176. Kinsella JL, Aronson PS (1981b) Amiloride inhibition of the Na^+-H^+ exchanger in the renal microvillus membrane vesicles. Am J Physiol 241:F374–379
177. Kinsella JL, Aronson PS (1982) Determination of the coupling ratio for Na^+-H^+ exchange in renal microvillus membrane vesicles. Biochim Biophys Acta 689:161–164
178. Kinsella J, Cujdik T, Sacktor B (1984) Na^+-H^+ exchange in isolated renal brush-border membrane vesicles in response to metabolic acidosis. Kinetic effects. J Biol Chem 259:13 224–13 227
179. Kleinman J, Blumenthal S, Weissner J, Reetz K, Leward DL, Mandel NS, Mandel GS, Garancis JC, Cragoe EJ Jr (1987) Regulation of pH in rat papillary tubule cells in primary culture. J Clin Invest 80:1660–1664
180. Kleyman TR, Cragoe EJ Jr (1988) Amiloride and its analogs as tools in the study of ion transport. J Membr Biol 105:1–21
181. Kleyman TR, Cragoe EJ Jr, Kraehenbuhl JP (1989) The cellular pool of Na^+ channels in the amphibian cell line A6 is not altered by mineralocorticoids. Analysis using a new photoactive amiloride analog in combination with anti-amiloride antibodies. J Biol Chem 264:1995–2000

182. Kleyman TR, Rajagopalan R, Cragoe EJ Jr, Erlanger BF, Al-Awqati Q (1986) New amiloride analogue as hapten to raise anti-amiloride antibodies. Am J Physiol 250 (Cell Physiol 19):C165–170
183. Kleyman TR, Yulo T, Ashbaugh C, Landry D, Cragoe EJ Jr, Al-Awqati Q (1986) Photoaffinity labeling of the epithelial sodium channel. J Biol Chem 261:2839–2843
184. Krapf R (1989) Mechanism of adaptation to chronic respiratory acidosis in the rabbit proximal tubule. J Clin Invest 83:890–896
185. Kurtz I (1987) Apical Na^+/H^+ antiporter and glycolysis-dependent H^+-ATPase regulate intracellular pH in the rabbit S3 proximal tubule. J Clin Invest 80:928–935
186. Kushmerick MJ, Dillon PF, Meyer RA, Brown TB, Krisanda JM, Sweeney HL (1986) 31P NMR spectroscopy, chemical analysis, and free Mg^{2+} of rabbit bladder and uterine smooth muscle. J Biol Chem 261:14420–14429
187. L'Allemain G, Franchi A, Cragoe EJ Jr, Pouyssegur J (1984) Blockade of the Na^+/H^+ antiport abolishes growth factor-induced DNA synthesis in fibroblasts. Structure-activity relationships in the amiloride series. J Biol Chem 259:4313–4319
188. Laski ME (1987) Total CO_2 flux in isolated collecting tubules during carbonic anhydrase inhibition. Am J Physiol 252:F322–330
189. Ledvora RF, Hegyvary C (1983) Dependence of Na^+–Ca^{2+} exchange and Ca^{2+}–Ca^{2+} exchange on monovalent cations. Biochim Biophys Acta 729:123–136
190. Leffert HL, Koch KS, Fehlmann M, Heiser W, Lad PJ, Sdelly H (1982) Amiloride blocks cell-free protein synthesis at levels attained inside cultured rat hepatocytes. Biochim Biophys Res Commun 108:738–745
191. Lester DS, Asher C, Garty H (1988) Characterization of cAMP-induced activation of epithelial sodium channels. Am J Physiol (Cell Physiol) 254:C802–C808
192. Li JH-Y, Lindemann BJ (1984) Competitive blocking of epithelial sodium channels by organic cations: the relationship between macroscopic inhibition constants. J Membr Biol 76:235–251
193. Li JH-Y, Palmer LG, Edelman IS, Lindemann B (1982) The role of sodium-channel density in the natriferic response of the urinary bladder to an antidiuretic harmone. J Membr Biol 64:77–89
194. Light DB, Ausiello DA, Stanton BA (1989) Guanine nucleotide-binding protein, alpha i-3, directly activates a cation channel in rat renal inner medullary collecting duct cells. J Clin Invest 352–356
195. Light DB, Schwiebert EM, Karlson KH, Stanton BA (1989) Atrial natriuretic peptide inhibits a cation channel in renal inner medullary collecting duct cells. Science 243:383–385
196. Lindemann B, Van Driessche W (1977) Sodium-specific membrane channels of frog skin are pores: current fluctuations reveal high turnovers. Science 195:292
197. Ling BN, Eaton DC (1989) Effects of luminal Na^+ on single Na^+ channels in A6 cells, a regulatory role for protein kinase C. Am J Physiol (Renal Fluid Electrolyte) 256:F1094–1103
198. Liu F-Y, Cogan MG (1988) Angiotensin II stimulation of hydrogen ion secretion in the rat early proximal tubule, modes of action, mechanism, and kinetics. J Clin Invest 82:601–607
199. Longoni S, Carafoli E (1987) Identification of the Na^+/Ca^{2+} exchanger of calf heart sarcolemma with the help of specific antibodies. Biochem Biophys Res Commun 145:1059–1063
200. Longoni S, Coady MJ, Ikeda T, Philipson KD (1988) Expression of cardiac sarcolemmal Na^+-Ca^{2+} exchange activity in Xenopus laevis oocytes. Am J Physiol 255:C870–C873
201. Lubin M, Cahn F, Coutermarsh BA (1982) Amiloride, protein synthesis, and activation of quiescent cells. J Cell Physiol 113:247–251
202. Mahe Y, Garcia-Romeu F, Motais R (1985) Inhibition by amiloride of both adenylate cyclase activity and the Na^+/H^+ antiporter in fish erythrocytes. Eur J Pharmacol 116:199–206
203. Makino N, Zhao D, Dhalla NS (1988) Stimulation of heart sarcolemmal Na^+Ca^{2+} exchange by concanavalin A. Biochem Biophys Res Commun 154:245–251
204. Mahnensmith RL, Aronson PS (1985a) The plasma membrane sodium-hydrogen exchange and its roles in physiological and pathophysiological processes. Circ Res 56:773–788
205. Mahnensmith RL, Aronson PS (1985b) Interrelationships among quinidine, amiloride, and lithium as inhibitors of renal Na^+-H^+ exchanger. J Biol Chem 260:12586–12592
206. Malmstrom K, Murer H (1986) Parathyroid hormone inhibits phosphate transport in OK cells but not in LLC-PK1 and JTC-12.p3 cells. Am J Physiol 251:C23–C31
207. Margolius HS, Chao J (1980) Amiloride inhibits mammalian renal kallikrein and a kallikrein-like enzyme from toad bladder and skin. J Clin Invest 65:1343–1350
208. Matlib MA (1988) Na^+–Ca^{2+} exchange in sarcolemmal membrane vesicles of dog mesenteric artery. Am J Physiol 255:C323–330

209. Matlib MA, Reeves JP (1987) Solubilization and reconstitution of the sarcolemmal Na^+–Ca^{2+} exchange system of vascular smooth muscle. Biochim Biophys Acta 904:145–148
210. McDaniel HB, Huang Z-Q, Cook WJ, Warnock DG (1990) Molecular modeling of the Na/H antiporter: The change relay hypothesis. Kidney Int 37:232
211. Mechmann S, Pott L (1986) Identification of Na–Ca exchange current in single cardiac myocytes. Nature 319:597–599
212. Michalak M, Quackenbush EJ, Letarte M (1986) Inhibition of Na^+/Ca^+ exchanger activity in cardiac and skeletal muscle sarcolemmal vesicles by monoclonal antibody 44d-D7. J Biol Chem 261:92–95
213. Miller RT, Pollock AS (1987) Modification of the internal pH sensitivity of the Na^+/H^+ antiporter by parathyroid hormone in cultured renal cell line. J Biol Chem 262:9115–9120
214. Mircheff AK, Ives HE, Yee VJ, Warnock DG (1984) Na^+/H^+ antiporter in membrane populations resolved from a renal brush border vesicle preparation. Am J Physiol 246:F853
215. Miyamoto H, Racker E (1980) Solubilization and partial purification of the Ca^{2+}/Na^+ antiporter from the plasma membrane of bovine heart. J Biol Chem 255:2656–2658
216. Miyamoto Y, Tiruppathi C,. Ganapathy V, Leiback FH (1989) Multiple transport systems for organic cations in renal brush-border membrane vesicles. Am J Physiol 256:F540–F548
217. Moberly J, Fanestil DD (1987) Heterogeneity of expression of apical membrane determinants in A6 epithelial cells. Am J Physiol C37–C44
218. Mohrmann M, Cantiello H, Ausiello DA (1987) Inhibition of epithelial Na^+ transport by atriopeptin, protein kinase C, and pertussin toxin. Am J Physiol 253:F372–376
219. Montrose M, Murer H (1986) Regulation of intracellular pH in LLC-PK1 cells by Na^+/H^+ exchange from an acid load via basolateral Na^+/H^+ exchanger. J Membr Biol 93:33–42
220. Montrose M, Friedrich T, Murer H (1987) Measurement of intracellular pH in single LLC-PK1 cells:recovery. J Membr Biol 97:63–78
221. Moran A (1987) Sodium-hydrogen exchange system in LLC-PK1 epithelium. Am J Physiol 252:C63–C67
222. Moran A, Asher C, Cragoe EJ Jr, Garty H (1988) Conductive sodium pathway with low affinity to amiloride in LLC-PK1 cells and other epithelia. J Biol Chem 263:19 586–19 591
223. Moran A, Biber J, Murer H (1986) A sodium-hydrogen exchange system in LLC-PK1 epithelia. Am J Physiol 251:F1003–1008
224. Moran A, Moran N (1984) Amiloride-sensitive channels in LLC-PK_1 apical membranes. Fed Proc 43:447
225. Morel N, Godfraind T (1984) Sodium/Calcium exchange in smooth-muscle microsomal fractions. Biochem J 218:421–427
226. Mullins LJ (1981) *Ion transport in heart.* Raven, New York, p 20
227. Mulvany MJ, Aalkjaer C, Peterson TT (1984) Intracellular sodium, membrane potential, and contractility of rat mesenteric small arteries. Circ Res 54:740–749
228. Murphy E, Freudenrich CC, Levy LA, London RE, Lieberman M (1989) Monitoring cytosolic free magnesium in cultured chicken heart cells by use of the fluorescent indicator Furaptra. Proc Natl Acad Sci USA 86:2981–2984
229. Murer H, Hopfer U, Kinne R (1976) Sodium/proton antiport in brush-border membrane vesicles isolated from rat small intestine and kidney. Biochem J 154:597–604
230. Nabel E, Berk BC, Brock TA, Smith TW (1988) Na^+–Ca^{2+} exchange in cultured vascular smooth muscle cells. Circ Res 62:486–493
231. Ng LL, Bomford J (1989) Altered stoichiometry of the human leucocyte Na^+/H^+ antiport with decreasing pH. Biochem J 259:311–314
232. Nicol GD, Schnetkamp PPM, Saimi Y, Cragoe EJ Jr, Bownds MD (1987) A derivative of amiloride blocks both the light-regulated and cyclic GMP-regulated conductances in pH. J Gen Physiol 90:651–669
233. Nishizuka Y, (1984) The role of protein kinase C in cell surface signal transdution and tumour promotion. Nature 308:693–698
234. Nord EP, Goldfarb D, Fine LG (1987a) In: Purchett JE, Greenberg A, (eds) Diuretics II: chemistry, pharmacology, and clinical applications. Relative insensitivity of renal proximal tubular Na^+/H^+ antiport activity to amiloride analogs. p. 231–234, Elsevier, New York, pp 231
235. Nord EP, Goldfarb D, Mikhail N, Moradeshagi P, Hafezi A, Vaystub S, Cragoe EJ Jr, Fine LG (1986) Characteristics of the Na^+-H^+ antiporter in the intact renal proximal tubular cell. Am J Physiol 250:F539–F550
236. Nord EP, Hafezi A, Kaunitz JD, Trizna W, Fine L (1985) pH gradient-dependent increased Na^+-H^+ antiport capacity of the rabbit remnant kidney. Am J Physiol 249:F90–98

237. Nord EP, Howard MJ, Hafezi A, Moradeshagi P, Vaystub S, Insel P, (1987b) Alpha 2 adrenergic agonists stimulate Na^+-H^+ antiport activity in the rabbit renal proximal tubule. J Clin Invest 80:1755–1762
238. Oberleithner H, Lang F, Messner G, Wang W, Deetjen F, (1983) Evidence for an amiloride sensitive Na^+ pathway in the amphibian diluting segment induce by K^+ adaptation. Pflugers Arch 399:166–172
239. Oberleithner H, Munich G, Schwab A, Dietl P (1986) Amiloride reduces potassium conductance in frog kidney via inhibition of Na^+-H^+ exchange. Am J Physiol 251:F66–F73
240. Olans L, Sariban-Sohraby S, Benos DJ (1984) Saturation behavior of single, amiloride-sensitive Na^+ channels in planar lipid bilayers. Biophys J 46:831–835
241. Orloff J, Handler J (1987) The role of adenosine 3′,5′-phosphate in the action of antidiuretic hormone. Am J Med 42:757–768
242. Otsu K, Kinsella J, Sacktor B, Froehlich JP (1989) Transient state kinetic evidence for an obligomer in the mechanism of Na^+-H^+ exchanger. Proc Natl Acad Sci USA 86:4818–4822
243. Ozaki H, Kishimoto T, Karaki H, Urakawa N (1982) Effects of the Na ionophore monensin on the contractile response and the movements of monovalent cations in the vascular smooth muscle of rabbit aorta. Naunyn Schmiedebergs Arch Pharmacol 321:140–144
244. Ozaki H, Karaki H, Urakawa N (1978) Possible role of Na-Ca exchange mechanism in the contractions induced in guinea-pig aorta by potassium free solution and ouabain. Naunyn Schmiedebergs Arch Pharmacol 304:203–209
245. Palaty V (1985) Inhibition of monoamine oxidase by amiloride. Can J Physiol Pharmacol 63:1586–1589
246. Palmer LG (1982a) Na^+ transport and flux ratio through apical Na^+ channels in toad bladder. Nature 297:688–690
247. Palmer LG (1982b) Ion selectivity of the apical membrane Na channels in the toad urinary bladder. J Membr Biol 67:91–98
248. Palmer LG (1984) Voltage-dependent block by amiloride and other monovalent cations of apical Na channels in the toad urinary bladder. J Membr Biol 80:153–165
249. Palmer LG, Frindt G (1986) Amiloride sensitive Na channels from the apical membrane of the rat cortical collecting tubules. Proc Natl Acad Sci USA 83:2767–2770
250. Palmer LG, Frindt G (1987) Ca Ionophore and phorbol ester inhibit Na channels in rat cortical collecting tubules. Fed Proc 46:4951
251. Palmer LG, Frindt G (1987) Effects of cell Ca and pH on Na channels from rat cortical collecting tubule. Am J Physiol 253(Renal Fluid Electrolyte Physiol 22):F333–339
252. Philipson KD (1984) Interaction of charged amphiphiles with Na^+-Ca^{2+} exchange in cardiac sarcolemmal vesicles. J Biol Chem 259:13999–14002
253. Philipson KD (1985) Sodium-Calcium exchange in plasma membrane vesicles. Annu Rev Physiol 47:561–571
254. Philipson KD, Bersohn MM, Nishimoto AY (1982) Effects of pH on Na^+-Ca^+ exchange in canine cardiac sarcolemmal vesicles. Circ Res 50:287–293
255. Philipson KD, Frank JS, Nishimoto AY (1983) Effects of phospholipase C on the Na^+-Ca^{2+} exchange and Ca^{2+} permeability of cardiac sarcolemmal vesicles. J Biol Chem 258:5905–5910
256. Philipson KD, Longoni S, Ward W (1988) Purification of the cardiac Na^+-Ca^{2+} exchange protein. Biochim Biophys Acta 945:298–306
257. Philipson KD, Nishimoto AY (1980) Na^+-Ca^{2+} exchange is affected by membrane potential in cardiac sarcolemmal vesicles. J Biol Chem 255:6880–6882
258. Philipson KD, Nishimoto AY (1981) Efflux of Ca^{2+} from cardiac sarcolemmal vesicles influence of external Ca^{2+} and Na^+. J Biol Chem 256:3698–3702
259. Philipson KD, Nishimoto AY (1982) Stimulation of Na^+-Ca^{2+} exchange in cardiac sarcolemmal vesicles by proteinase pretreatment. Am J Physiol 243:C191–195
260. Philipson KD, Nishimoto AY (1984) Stimulation of Na^+-Ca^{2+} exchange in cardiac sarcolemmal vesicles by phospholipase D. J Biol Chem 259:16–19
261. Philipson KD, Ward R (1985) Effects of fatty acids on Na^+-Ca^{2+} exchange and Ca^{2+} permeability of cardiac sarcolemma vesicles. J Biol Chem 260:9666–9671
262. Philipson KD, Ward R (1987) Modulation of Na^+-Ca^{2+} exchange and Ca^{2+} permeability in cardiac sarcolemmal vesicles by doxylstenric acids. Biochim Biophys Acta 897:152–158
263. Pitts BJ (1979) Stoichiometry of sodium-calcium exchange in cardiac sarcolemmal vesicles. J Biol Chem 254:6232–6235
264. Pollock AS, Warnock DG, Strewler GJ (1986) Parathyroid hormone inhibition of Na^+-H^+ antiporter activity in a cultured renal cell line. Am J Physiol 250:F217–F225

265. Preisig P, Alpern RJ (1988) Chronic metabolic acidosis causes an adaptation in the apical membrane Na/H antiporter and basolateral membrane $Na(HCO_3)3$ symporter in the rat proximal convoluted tubule. J Clin Invest 82:1445–1453
266. Preisig P, Ives HE, Cragoe EJ Jr, Alpern RJ, Rector FC Jr (1987) Role of Na^+-H^+ antiporter in rat proximal tubule bicarbonate absorption. J Clin Invest 80:970–978
267. Preisig P, Rector FC Jr (1988) Role of Na^+–H^+ antiport in rat proximal tubule NaCl absorption. Am J Physiol 255:F461–F465
268. Pritchard D, Ashley CC (1987) Evidence for Na^+/Ca^{2+} exchange in isolated smooth muscle cells: a fura-2 study. Pflugers Arch 410:401–407
269. Putney JW Jr, Takemura H, Hughes AR, Horstman DA, Thastrup O (1989) How do inositol phosphates regulate calcium signaling? FASEB J 3:1899–1905
270. Rabito CA (1986) Sodium cotransport processes in renal epithelial cell lines. Miner Electrolyte Metab 121:32–41
271. Ramachandran C, Brunette MG (1989) The renal of Na^+/Ca^{2+} exchange system is located exclusively in the distal tubule. Biochem J 257:259–264
272. Ramsay LE, Hettiarachchi J, Fraser R, Morton JJ (1980) Amiloride, spironolactone, and potassium chloride in thiazide-treated hypertensive patients. Clin Pharmacol Ther 27:533–543
273. Rasgado-Flores H, Blaustein MP (1987) Na^+/Ca^{2+} exchange in barnacle muscle cells has a stoichiometry of 3 Na^+/1 Ca^{2+}. Am J Physiol 252:C499–504
274. Rector FC Jr (1983) Sodium, bicarbonate, and chloride absorption by the proximal tubule. Am J Physiol 244:F461–F471
275. Reenstra WW, Warnock DG, Yee VJ, Fortex JG (1981) Proton gradients in renal cortex brush-border membrane vesicles. Demonstration of a rheogenic proton flux with acridine orange. J Biol Chem 256:11 663–11 666
276. Reeves JP (1985) The sarcolemmal sodium-calcium exchange system. Curr Top Membr Transp 25:77–127
277. Reeves JP, Bailey CA, Hale CC (1986) Redox modification of sodium-calcium exchange activity in cardiac sarcolemmal vesicles. J Biol Chem 261:4948–4955
278. Reeves JP, Hale CC (1984) The stoichiometry of the cardiac sodium-calcium exchange system. J Biol Chem 259:7733–7739
279. Reeves JP, Hale CC, de la Pena P, Slaughter RS, Boulware T, Froehlich J, Kaczorowski G, Cragoe E (1984) In:Bronner F, Peterlik M, (eds) Epithelial calcium and phosphate transport: molecular and cellular aspects. A model for the cardiac sodium-calcium exchange carrier. New York, pp 77–82
280. Reeves JP, Poronnik P (1987) Modulation of $Na^+ - Ca^{2+}$ exchange in sarcolemmal vesicles by intracellular Ca^{2+}. Am J Physiol 252:C17–C23
281. Reeves JP, Sutko JL (1979) Sodium-calcium ion exchange in cardiac membrane vesicles. Proc Natl Acad Sci USA 76:590–594
282. Reeves JP, Sutko JL (1980) Sodium Calcium exchange activity generates a current in cardiac membrane vesicles. Science 208:1461–1464
283. Reeves JP, Sutko JL (1983) Competitive interaction of sodium and calcium with the sodium-calcium exchange system of cardiac sarcolemmal vesicles. J Biol Chem 258:3178–3182
284. Reuter H, Blaustein MP, Haeusler G (1973) Na–Ca exchange and tension development in arterial smooth muscle. Philos Trans R Soc Lond [Biol] 265:87–94
285. Reuter H, Seitz N (1968) The dependence of calcium efflux from cardiac muscle on temperature and external ion composition. J Physiol (Lond) 195:451–470
286. Rindler MJ, Taub M, Saier MH Jr (1979) Uptake of 22 Na^+ by cultured dog kidney cells (MDCK). J Biol Chem 254:11431–11439
287. Rindler MJ, Saier MH Jr (1981) Evidence for Na^+/H^+ antiport in cultured dog kidney cells (MDCK). J Biol Chem 256:10 820
288. Rocco VK, Cragoe EJ Jr, Warnock DG (1987) N-ethoxycarbonyl-2-ethoxy-1,2-dihydroquinoline, amiloride analogues and renal Na^+/H^+ antiporter. Am J Physiol 252:F517–F524
289. Ruscak M, Juhaszova M, Orlicky J, Zachar J (1987) Reconstitution and partial purification of $Na^+ - Ca^{2+}$ exchanger from crayfish striated muscle plasma membranes. Gen Physiol Biophys 6:523–527
290. Sabolic I, Burckhardt G (1983) Proton pathways in rat renal brush-border and basolateral membranes. Biochim Biophys Acta 734:210–220
291. Saccomani G, Mitchell KD, Navar GL (1990) Angiotensin II stimulation of Na^+–H^+

exchange in proximal tubule cells. Am J Physiol 258(Renal Fluid Electrolyte Physiol 27):F1188–F1195

292. Saier MH Jr, Boerner P, Grenier FC, McRoberts JA, Rindler MJ, Taub M, Ü HS (1986) Sodium entry pathways in renal epithelial cell lines. Miner Electrolyte Metab 12:42–50

293. Salako LA, Smith AJ (1970) Changes in sodium pool and kinetics of sodium transport in frog skin produced by amiloride. Br J Pharmacol 39:99–109

294. Salihagic A, Mackovic M, Barfic H, Sabolic I (1988) Short-term and long-term stimulation of Na^+–H^+ exchange in cortical brush-border membranes during compensatory growth of the rat kidney. Pflugers Arch 413:190–196

295. Sandberg K, Markwick AJ, Trinh DP, Catt KJ (1988) Calcium mobilization by angiotensin II and neurotransmitter receptors expressed in xenopus laevis oocytes. FEBS Lett 241:177–180

296. Sardet C, Franchi A, Pouyssegur J (1989) Molecular cloning, primary structure, and expression of the human growth factor-activatable Na^+/H^+ antiporter. Cell 56:271–280

297. Sariban-Sohraby S, Benos DJ (1986) Detergent solubilization, functional reconstitution, and partial purification of epithelial amiloride-binding protein. Biochemistry 25:4639–4646

298. Sariban-Sohraby S, Burg M, Wiesmann WP, Chiang PK, Johnson JP (1984) Methylation increases sodium transport into A6 apical membrane vesicles: possible mode of aldosterone action. Science 225:745–746

299. Sariban-Sohraby S, Latorre R, Burg M, Olans L, Benos D (1984) Amiloride-sensitive epithelial Na^+ channels reconstituted into planar lipid bilayer membranes. Nature 308:80–82

300. Sariban-Sohraby S, Sorscher EJ, Brenner BM, Benos DJ (1988) Phosphorylation of a single subunit of the epithelial Na^+ channel protein following vasopressin treatment of A6 cells. J Biol Chem 263:13 875–13 879

301. Schaeffer J, Blaustein MP (1989) Platelet free calcium concentrations measured with fura-2 are influenced by the transmembrane sodium gradient. Cell Calcium 10:101–113

302. Schellenberg GD, Anderson L, Cragoe EJ Jr, Swanson PD (1985) Inhibition of synaptosomal membrane Na^+–Ca^{2+} exchange transport by amiloride and amiloride analogues. Mol Pharmacol 27:537–543

303. Schellenberg GD, Anderson L, Swanson PD (1983) Inhibition of Na^+–Ca^{2+} exchange in rat brain by amiloride. Mol Pharmacol 24:251–258

304. Schlondorff D, Franki N (1980) Effect of vasopressin on cyclic AMP-dependent protein kinase in toad urinary bladder. Biochim Biophys Acta 628:1–12

305. Schnetkamp PPM (1980) Ion selectivity of the cation transport system of isolated intact cattle rod outer segments. Evidence for a direct communication between the rod plasma membrane and the rod disk membranes. Biochim Biophys Acta 598:66–90

306. Schnetkamp PPM (1986) Sodium-calcium exchange in the outer segment of bovine rod photoreceptors. J Physiol (Lond) 373:25–45

307. Schnetkamp PPM, Basa DK, Szerencsei RT (1989) Na^+– Ca^{2+} exchange in bovine rod outer segments requires and transports K^+. Am J Physiol 257:C153–C157

308. Selvaggio AM, Schwartz JH, Bengle HH, Gordon FD, Alexander EA (1988) Mechanisms of H^+ secretion by inner medullary collecting duct cells. Am J Physiol 254:F391–F400

309. Sheu SS, Blaustein MP (1986) In:Fozzard HA, Haber E, Jennings RB, Katz AM, Morgan HE (eds) The heart and cardiovascular system. Sodium/calcium exchange and regulation of cell calcium and contractility in cardiac muscle, with a note about vascular smooth muscle. Raven, New York, pp 509–535

310. Sheu SS, Fozzard HA (1982) Transmembrane Na^+ and Ca^{2+} electrochemical gradients in cardiac muscle and their relationship to force development. J Gen Physiol 80:325–351

311. Siegl PKS, Cragoe EJ Jr, Trumble MJ, Kaczorowski GJ (1984) Inhibition of Na^+/Ca^{2+} exchange in membrane vesicle and papillary muscle preparations from guinea pig heart by analogs of amiloride. Proc Natl Acad Sci USA 81:3238–3242

312. Sigel R, Baur R, Porzig H, Reuter H (1988) mRNA-induced expression of the cardiac Na^+–$Ca_3{}^+$ exchanger in xenopus oocytes. J Biol Chem 262:14 614–14 616

313. Simchowitz L, Cragoe EJ Jr (1986) Inhibition of chemotactic factor-activated Na^+/H^+ exchange in human neutrophils by analogues of amiloride: structure-activity relationships in the amiloride series. Mol Pharmacol 30:112–120

314. Singer MA (1974) Effects of amiloride and dibucaine on a model phospholipid membrane-structure-activity relationships. Biochem Pharmacol 23:2939–2950

315. Slaughter RS, Cragoe EJ Jr, Reeves JP, Kaczorowski GJ (1988) Inhibition of sodium-calcium

exchange in cardiac sarcolemmal membrane vesicles. 1. Mechanism of inhibition by amiloride analogues. Biochemistry 27:2403–2409

316. Slaughter RS, Shevell JL, Felix JP, Garcia ML, Kaczorowski GJ (1989) High levels of sodium-calcium exchange in vascular smooth muscle sarcolemmal membrane vesicles. Biochemistry 28:3995–4002
317. Slaughter RS, Sutko JL, Reeves JP (1983) Equilibrium calcium-calcium exchange in cardiac sarcolemmal vesicles. J Biol Chem 258:3183–3190
318. Slaughter RS, Welton AF, Morgan DW (1987) Sodium-calcium exchange in sarcolemmal vesicles from tracheal smooth muscle. Biochim Biophys Acta 904:92–104
319. Smith JB (1986) Angiotensin-receptor signaling in cultured vascular smooth muscle cells. Am J Physiol 250:F759–F769
320. Smith JB, Cragoe EJ Jr, Smith L (1987) Na^+/Ca^{2+} antiport in cultured arterial smooth muscle cells. Inhibition by magnesium and other divalent cations. J Biol Chem 262:11 988–11 994
321. Smith JB, Dwyer SD, Smith L (1989a) Decreasing extracellular Na^+ concentration triggers inositol polyphosphate production and Ca^{2+} mobilization. J Biol Chem 264:831–837
322. Smith JB, Dwyer SD, Smith L (1989b) Cadium evokes inositol polyphosphate formation and calcium mobilization. Evidence for a cell surface receptor that cadium stimulates and zinc antagonizes. J Biol Chem 264:7115–7118
323. Smith JB, Dwyer SD, Smith L (1989c) Lowering extracellular pH evokes inositol polyphosphates formation and calcium mobilization. J Biol Chem 264:8723–8728
324. Smith JB, Smith L (1984) Ripid calcium mobilization by vasopressin and prostoglandin F2 alpha is independent of sodium influx in quiescent 3t3 cells. Biochem Biophys Res Commun 123:803–809
325. Smith JB, Smith L (1987) Extracellular Na^+ dependence of charges in free Ca^{2+}, $45Ca^{2+}$ efflux, and total cell Ca^{2+} produced by angiotensin II in cultured arterial muscle cells. J Biol Chem 262:17 455–17 460
326. Smith SB, Smith L (1989) Na^+/Ca^{2+} exchange in smooth muscle cells: role in Ca^{2+} regulation in human arterial muscle and energy dependence. Biophys J 55:487a
327. Smith JB, Smith L, Brown ER, Barnes D, Sabir MA, Davis JS, Farese RV (1984) Angiotensin II rapidly increases phosphatidate-phosphoinositide synthesis and phosphoinositide hydrolysis and mobilizes intracellular calcium in cultured arterial muscle cells. Proc Natl Acad Sci USA 81:7812–7816
328. Smith JB, Zheng T, Lyu R-M (1989d) Ionomycin release calcium from the sarcoplasmic reticulum and activates Na^+/Ca^{2+} exchange in vascular smooth muscle cells. Cell Calcium 10:125–134
329. Smith JB, Zheng T, Smith L (1989e) Relationship between cytosolic free Ca^{2+} and Na^+–Ca^{2+} exchange in aortic muscle cells. Am J Physiol 256:C147–154
330. Smith L, Higgins BL, Cragoe EJ Jr, Smith JB (1987a) Comparison of the effectiveness of seven amiloride congeners as inhibitors of Na/H and Na/Ca antiport in cultured smooth muscle cells. Fed Proc 46:2028
331. Smith RL, Cochran DW, Gund P, Cragoe EJ Jr (1979) Proton, carbon-13, and nitrogen-15 nuclear magnetic resonance and CNDO/2 studies on the tautomerism and configuration of amiloride, a novel acylguanidine. J Am Chem Soc 101:191–201
332. Smith RL, Macara IG, Levenson R, Housman D, Cantley L (1982) Evidence that a Na^+/Ca^{2+} antiport system regulats murine erythroleukemia cell differentiation. J Biol Chem 257:773
333. Soldati L, Longoni S, Carafoli E (1985) Solubilization and reconstitution of the Na^+/Ca^{2+} exchanger of the cardiac sarcolemma. Properties of the reconstituted system and tentative identification of the protein(s) responsible for the exchange activity. J Biol Chem 260:13 321–13 327
334. Soltoff SP, Cragoe EJ Jr, Mandel LJ (1986) Amiloride analogues inhibit proximal tubule metabolism. Am J Physiol 250:C744–C747
335. Soltoff SP, Mandel LJ (1983) Amiloride directly inhibits the Na, K-AtPase activity of rabbit kidney proximal tubules. Science 220:957–958
336. Somlyo AP, Broderick R, Somlyo AV (1986) Calcium and sodium in vascular smooth muscle. Ann NY Acad Sci 488:228–239
337. Sordahl LA, LaBelle EF, Rex KA (1984) Amiloride and diltiazem inhibition of microsomal and mitochondrial Na^+ and Ca^{2+} transport. Am J Physiol 246:C172–C176
338. Sorscher EJ, Accavitti MA, Keeton D, Steadman E, Frizzell RA, Benos DJ (1988) Antibodies against purified epithelial sodium channel protein from bovine renal papilla. Am J physiol (Cell Physiol) 24:C835–843

339. Stanton GB, Omerovic A, Koeppen B, Giebisch G (1987) Electroneytral H^+ secretion in distal tubule of amphiuma. Am J Physiol 252:F691–699
340. Swanson PD, Anderson L, Stahl WL (1974) Uptake of calcium ions by synaptosomes from rat brain. Biochim Biophys Acta 356:174–183
341. Takada M, Hayashi H (1981) Interaction of cadmium, calcium, and amiloride in the kinetics of active sodium transport through frog skin. Jpn J Physiol 31:285–303
342. Takaichi K, Wang D, Hancock J, Cragoe EJ Jr, Warnock DG (1990) Follicular cells but not oocytes themselves of xenopus laevis (african clawed frog) pooves amiloride-inhibitable sodium uptake. Kidney Int 37:547 (abstract) (in press)
343. Takuma T, Kuyatt BL, Baum BJ (1988) Calcium transport mechanisms in basolateral plasma membrane-enriched vesicles from rat parotid gland. Biochem J 227:239–245
344. Talor Z, Arruda JAL (1985) Partial purification and reconstitution of renal basolateral Na^+–Ca^{2+} exchanger into liposomes. J Biol Chem 260:15473–15476
345. Talor Z, Yang W-C, Shuffield J, Sack E, Arruda JAL (1987) Chronic hypercapnia enhances Vmax of Na-H antiporter of renal brush-border membranes. Am J Physiol 253:F394–F400
346. Tang C-M, Presser F, Morad M (1988) Amiloride selectively blocks the low threshold (T) calcium channel. Science 240:213–215
347. Teitlebaum AP, Strewler GJ (1984) Parathyroid hormone receptors coupled to cyclic adenosine monophosphate formation in an established renal cell line. Endocrinology 114:980–985
348. Tousson A, Alley CD, Sorscher EJ, Brinkley BR, Benos DJ (1989) Immunochemical localization of amiloride-sensitive sodium channels in sodium-transporting epithelia. J Cell Sci 93:349–362
349. Touster O, Aronson NN Jr, Dulaney JT, Hendrickson H (1970) Isolation of rat liver plasma membranes. Use of nucleotide pyrophosphatase and phosphodiesterase I as marker enzymes. J Cell Biol 47:604–618
350. Trosper TL, Philipson KD (1983) Effects of divalent and trivalent cations on Na^+-Ca^{2+} exchange in cardiac sarcolemmal vesicles. Biochim Biophys Acta 731:63–68
351. Tsai C-J, Ives HE, Alpern RJ, Yee YJ, Warnock DG, Rector FC Jr (1984) Increased Vmax for Na^+/H^+ antiporter activity in proximal tubule brush border vesicles from rabbits with metabolic acidosis. Am J Physiol 247:F339–F343
352. Turner RJ, Moran A (1982) Heterogeneity of sodium-dependent D-glucose transport sites along the proximal tubule: evidence from vesicles studies. Am J Physiol 242:F406–F414
353. Turner RJ, Thompson J, Sariban-Sohraby S, Handler JS (1985) Monoclonal antibodies as probes of epithelial membrane polarization. J Cell Biol 101:2173–2180
354. Ussing HH, Zerahn K (1951) Active transport of sodium as the source of electric current in the short-circuited isolated frog skin. Acta Physiol Scand 23:110–127
355. van Breemen C, Aaronson P, Loutzenhiser R (1979) Sodium-calcium interactions in mammalian smooth muscle. Pharmacol Rev 30:167–208
356. Vandenburgh HH, Kaufman S (1982) Coupling of voltage-sensitive sodium channel activity to stretch-induced amino acid transport in skeletal muscle in vitro. J Biol Chem 257:13448–13454
357. Van Driessche W, Lindemann B (1979) Concentration dependence of currents through single sodium-selective pores in frog skin. Nature 282:519–520
358. Velly J, Grima M, Decker N, Cragoe EJ Jr, Schwartz J (1988) Effects of amiloride and its analogues on [^{3}H] batrachotoximin-A 20-alpha benzoate binding, [^{3}H] tetracaine binding and ^{22}Na influx. Eur J Physiol 149:97–105
359. Vemuri R, Philipson KD (1987) Phospholipid composition modulates the Na^+–Ca^{2+} exchange activity of cardiac sarcolemma in reconstituted vesicles. Biochim Biophys Acta 937:258–268
360. Verrey F, Scharer E, Paccolat MP, Geering K, Kraehenbuhl J-P, Rossier BC (1986) loning of xenopus laevis Na,K-ATPase cDNAs: regulation of mRNA expression in cultured kidney cells. J Cell Biol 103:51a
361. Vigne P, Breittmayer J-P, Duval D, Frelin C, Lazdunski M (1988a) The Na^+/Ca^{2+} antiporter in aortic smooth muscle cells. Characterization and demonstration of an activation by phorbol esters. J Biol Chem 263:8078–8083
362. Vigne P, Breittmayer J-P, Lazdunski M, Frelin C (1988b) The regulation of the cytoplasmic free Ca^{2+} concentration in aortic smooth muscle cells (A7r5 line) after stimulation by vasopressin and bombesin. Eur J Biochem 176:47–52
363. Vigne P, Champigny G, Marsault R, Barbry P, Frelin C, Lazdunski M (1989) A new type of amiloride-sensitive cationic channel in endothelial cells of brain microvessels. J Biol Chem 264:7663–7668

364. Vigne P, Frelin C, Cragoe EJ Jr, Lazdunski M (1984) Structure-activity relationships of amiloride and certain of its analogues in relation to the blockade of the Na^+/H^+ exchange system. Mol Pharmacol 25:131–136
365. Vigne P, Jean T, Barbry P, Frelin C, Fine LG, Lazdunski M (1985) [^{3}H] Ethylpropylamiloride, a ligand to analyze the properties of the Na^+/H^+ exchange system in the membranes of normal and hypertrophied kidneys. J Biol Chem 260:14120–14125
366. Wakabayashi S, Goshima K (1981a) Kenetic studies on sodium-dependent calcium uptake by myocardial cells and neuroblastoma cells in culture. Biochim Biophys Acta 642:158–172
367. Wakabayashi S, Goshima K (1981b) Comparison of kinetic characteristics of Na^+–Ca^{2+} exchange in sarcolemma vesicles and cultured cells from chick heart. Biochim Biophys Acta 645:311–317
368. Wakabayashi S, Goshima K (1982) Partial purification of Na^+–Ca^{2+} antiporter from plasma membrane of chick heart. Biochim Biophys Acta 693:125–133
369. Wall SM, Kraut JA, Muallem S (1988) Modulation of Na^+–H^+ exchange activity by intracellular Na^+, H^+, and Li^+ in IMCD cells. Am J Physiol 255:F331–F339
370. Warnock DG, Cragoe EJ Jr, Rosster B, Kleyman TR (1989) Use of photoaffinity analogues of amiloride to characterize the renal Na/H antiporter. p. 465a Kidney Int. 35:465 (abstract)
371. Warnock DG, Lowe A, Lin H-Y, Yee VJ (1985) Regulation of the activity of the Na^+–H^+ antiporter in brush-border membrane vesicles from the proximal tubule. p. 229–231. Ann NY Acad Sci 456:229
372. Warnock DG, Pollock AS (1988) In: Grinstein S (ed) Na/H exchange, Na^+/H^+ exchange in epithelial cells. p. 77–101 CRC Press, Boca Raton
373. Warnock DG, Reenstra WW, Yee VJ (1982) Na^+/H^+ antiport of brush border vesicles: studies with acridine orange uptake. Am J Physiol 242:F733–F739
374. Warnock DG, Yang W-C, Huang Z-Q, Cragoe EJ Jr (1988) Interactions of chloride and amiloride with the renal Na^+/H^+ antiporter. J Biol Chem 263:7216–7221
375. Warnock DG, Yee VJ (1982) Sodium uptake mechanisms in brush border membrane vesicles prepared from rabbit renal cortex. Biochim Biophys Acta 684:137–140
376. Weinman EJ, Dubinsky WP, Fisher K, Steplock D, Dinh Q, Chang L, Shenolikar S (1988) Regulation of reconstituted renal Na^+/H^+ exchanger by calcium-dependent protein kinases. J Membr Biol 103:237–244
377. Weinman EJ, Dubinsky WP, Shenolikar S (1988) Reconstitutiton of cAMP-dependent protein kinase regulated renal Na^+– H^+ exchanger. J Membr Biol 101:11–18
378. Weinman EJ, Shenolikar S, Cragoe EJ Jr, Dubinsky WP (1988) Solubilization and reconstitution of renal brush border Na^+-H^+ exchanger. J Membr Biol 101:1–9
379. Weinman EJ, Shenolikar S, Kahn AM (1987) cAMP-associated inhibition of Na^+–H^+ exchanger in rabbit kidney brush-border membranes. Am J Physiol 252:F19–F25
380. Wiesmann WP, Johnson JP, Miura GA, Chiang PK (1985) Aldosterone-stimulated inhibition of Na^+-H^+ exchanger in rabbit kidney brush-border membranes. Am J Physiol 248 (Renal Fluid Electrolyte Physiol 17):F43–47
381. Wu J-SR, Lever JE (1989) Photoaffinity labeling by [^{3}H]-NS-methyl-N5-isobutylamiloride of proteins which cofractionate with Na^+/H^+ antiport activity. Biochemistry 28:2980
382. Yamaguchi DT, Sakai R, Cragoe EJ Jr, Jordan SC (1986) Amiloride inhibition of DNA synthesis and immunoglobulin production by activated human peripheral blood monoclear cells is dependent of sodium/hydrogen antiport. J Immunol 137:1300–1304
383. Yucha CB, Stoner LC (1987) Bicarbonate transport by initial collecting tubule of aquatic and land phase amphibia. Am J Physiol 253:F310–F317
384. Zeidel ML, Silva P, Seifter JL (1986) Intracellular pH regulation and proton transport by rabbit renal medullary collecting duct cells. Role of plasma membrane proton adenosine triphosphitase. J Clin Invest 77:113–120

Anion Exchange Mechanism of Band 3 and Related Proteins

R.B. Gunn

Erythroid Anion Exchange

Introduction

Band 3 is a diffuse band on an SDS-polyacrylamide gel of electrophoretically separated erythrocyte membrane proteins. Originally it was the third band from the origin, running slower in the electric field than any of the other proteins except the two spectrin bands, but refined techniques have resolved the second band into band 2 and band 2.1. Quite separately the anion exchange mechanism in erythrocytes behaves in many ways like a carrier, for example, it saturates with Michaelis-Menten kinetics as a function of substrate. The pH dependence of chloride flux is like the titration of a protein, and the flux has a high temperature coefficient like a chemical reaction. The identification of the band 3 protein with the anion exchange mechanism was first suggested by the similarity between the organic anion inhibitors of exchange and the organic phosphate anion used to show that band 3 has functional groups on both sides of the membrane and to show for the first time that a membrane protein was transmembrane [13]. This identification of the band 3 protein with anion exchange function was firmly established by the use of labeled disulfonic stilbenes ([$^3H_2$]DIDS) to both block transport and label the band 3 protein on SDS-polyacrylamide gels [17]. In this way the anion exchanger has become synonymous with band 3, though there are two problems with this nomenclature: First, in the SDS-polyacrylamide gel at the same location as the [$^3H_2$]-DIDS-labeled band there are other proteins, including other major membrane proteins such as the glycophorins (also known as PAS-I). Second, even within the proteins coded for by the gene for the anion exchanger, there may be heterogeneity. These variations of the band 3-like proteins in this region of the gel have only recently been hinted at. This is not simply a variability in glycosylation that causes the smear on the gel and range in the apparent molecular weights, but in the detailed amino acid composition and perhaps even in the function of the proteins in this band. The late Jens O. Wieth suggested calling the anion exchange protein "capnophorin," while others have used initials AE1 for the gene for the erythrocyte anion exchange protein. Despite the imprecisions in the molecular identity of band 3 as originally described, I will continue to use that term to mean that protein that mediates the rapid exchange of chloride and bicarbonate across the erythrocyte membrane.

Since the previous edition of this chapter [45] there have been several reviews that contain a wealth of information and references. These include those by Knauf [73], Passow [112], Kopito [78], and Jennings [61].

Molecular Biology of Band 3

The gene for human band 3 (hB3) is located on chromosome 17 between 17 q21 and 17 qter [124]. The coding region of murine band 3 (mB3) is spread over 17 kilobase pairs (kbp) of DNA [80, 82]. There are 20 exons interrupted by 19 introns which vary in length from 79 bp to 3.9 kbp. In the mouse gene there are multiple transcription initiation sites resulting in at least six mRNAs of different lengths but all being translated into band 3 [79]. The 5′ flanking region has no CAAT or TATA box that often indicates a promoter region. The mB3 and hB3 genes (97, 132) are highly homologous in the coding region for the protein and in both 5′ and 3′ untranslated regions. Both have out-of-frame methionine codons upstream from the initiator codons and the hB3 gene has two in-frame stop codons at bases -84 and -75 (+ 1 is defined as the adenosine in the initiator methionine codon, ATG).

The mRNA of mB3 has a long 3′ untranslated segment before the poly-adenylation sequence. Much of this portion of the hB3 mRNA has not yet been sequenced since the polyadenylation signal has yet to be identified and the 3′ untranslated region in hB3's mRNA is about 1200 base pairs smaller than that for mB3.

The cDNA for mB3 codes for 929 amino acids requiring 2787 base pairs. The cDNA for hB3 codes for 911 amino acids requiring 2730 base pairs. The N-terminal cytoplasmic domain of the protein has several regions with strong conservation of sequence including a region with 3 or 4 tryptophan (W) residues (mB3 residues 89, 95, 108 and 119; hB3 residues 75, 81, 94 and 105); chicken band 3, cB3, residues 78, 84, and 97) two proline(P)-rich regions (mB3 residues 161–204 and 272–287, hB3 residues 147–190 and 259–274); cB3 residues 149, 166, and 176 and region 266–281) and the putative ankyrin-binding region (mB3 residues 120–160, hB3 residues 106–146, cB3 residues 108–148). Other conserved sequences in the cytoplasmic domain (mB3 residues 232–260, 327–360, and 398–419; hB3 residues 219–247, 314–347, and 379–400; cB3 residues 228–254, 321–354, and 392–413) have no putative function or understandable characteristic. The primary sequence of mB3, hB3 and cB3 are shown in Fig. 1.

Structure and Chemistry of Band 3

Oligomeric Structure but Monomeric Transport

The 95-kda band 3 polypeptide is an intrinsic membrane protein that cannot be purified free from adherent lipid or detergent without irreversible aggregation.

The hydrophobic surfaces of the polypeptide that bind lipid can also interact to form oligomers. There is substantial evidence that in the membrane band 3 is a dimer or a tetramer [112]. This oligomeric structure appears to be important for the stability and conformation of band 3 since isolated and refolded monomers will not bind 4-benzamido-4′-aminostilbene-2,2′-disulfonate (BADS) but isolated and refolded dimers or in situ band 3 dimers will [11]. However the transport of anions does not seem to involve interactions between subunits of a multimeric structure. Each band 3 monomer binds a single stilbene disulfonate and there is a linear relationship between the number bound and the irreversible inhibition of transport [146]. This implies that even when a single band 3 is irreversibly inhibited the other band 3 monomers with which it is associated can transport anions normally.

Two Domains

The band 3 protein has two major domains. The N-terminal 43-kDa cytoplasmic domain consists of 423 (mB3), 404 (hB3), or 418 (cB3) amino acids most of which are rather hydrophilic. Trypsin treatment of open, porous cell membranes releases the cytoplasmic domain from the membrane that retains the rest of the protein. The cytoplasmic domain has been isolated and separately studied [92]. The functions of the cytoplasmic domain include the binding of hemoglobin and glycolytic enzymes and the attachment of cytoskeleton (see "Cytoskeletal Interactions with Band 3"). The second domain is the C-terminal 52-kDa membrane domain of 506 (cB3), 507 (hB3), or 504 (cB3) amino acids. In this region there are 10–14 strings of rather hydrophobic amino acids that might each span the membrane as an α-helix. Since the phospholipid bilayer is about 3.0 nm thick and an α-helix has 3.6 amino acids per turn and a pitch of 0.54 nm per turn along the axis of the helix, it should take 19 or 20 amino acids to span from the phospholipid head groups on one side to the phospholipid head groups on the other side. Thus a run of 19 or 20 hydrophobic aminoacids is a good candidate, but not proof, for a membrane spanning segment.

The N-terminus [127] and the C-terminus [90, 140] are both on the cytoplasmic side of the membrane and thus there must be an even number of transmembrane segments. The α-helical content of the membrane domain is high, 58% [108], whereas the α-helical content of the cytoplasmic domain is low, 37% [5] to 27% [108]. Fourteen transmembrane segments of 20 amino acids each would constitute 55% of the 507 amino acids of the membrane domain thus the α-helical content must be almost exclusively confined to the transmembrane segments. From chemical modifications and proteolysis experiments there is proof for at least eight transmembrane segments [63]; and extensive surface proteolysis and sequencing of the resistant, and presumably transmembrane, peptides indicates together with modeling that there are 14 of them [64, 131]. The functions of the protein also are separable and correspond to the two domains. The membrane domain can transport anions without the cytoplasmic domain both in red cells with the cytoplasmic domain removed by

```
mB3  MGDMRDHEEVLEIPDRDSEEELENIIGQIAYRDLTIFVTEMQDPEALPTEQTATDYVPSSTSTPHPSSGQVYVELQELMMDQRNQELQWVEAAHWIGLEENLREDGVWGR  110
hB3  .........MEELQDDYEDMMEENLEQEEYEDPDIPESQMEEPAAHDTEATATDY....HTTSHPGTHKVYVELQELVMDEKNQELRWMEAARWVQLEENLGENGAWGR  96
cB3  .........MEGPGQDTEDALRRSLDPEGYEDTKGSRTSLGTMSNPLVSDVDLEAAGSRQPTAHRDTYEGYVELHELVLD.SRKDPCWMEAGRWLHLEESMEPGGAWG.  98
CON  .........mE.p..D.Ed.le..l.qe.YeD..ip.t.m..p.a..te.tatdy..s...T.Hp.t..vYVELqELvmD..nqel.WmEAarW..LEEnl.e.GaWGr

mB3  PHLSYLTFWSLLELQKVFSKGTFLLGLAETSLAGVANHLLDCFIYEDQIRPQDREELLRALLLKRSHAEDLGNLEGVKPAVLTRSGGASEPLLPHQPSLETQLYCGQAEG  220
hB3  PHLSHLTFWSLLELRRVFTKGTVLLDLQETSLAGVANQLLDRFIFEDQIRPQDREELLRALLLKHSHAGELEALGGVKPAVLTRSGDPSQPLLPQHSSLETQLFCEQGDG  206
cB3  SHLPLLTYHSLLELHRAFAKGVVLLDVAANSLAAVAHVLLDQLIYEGQLKPQHRDDVLRALLLRHKHPSEAESVWTLPAAQLQCSDGEQKDADERALLRDQRAVEMRELH  208
CON  pHLs.LTfwSLLEL.rvF.KGtvLLdlaetSLAgVAn.LLD.fIyEdQirPQdReelLRALLLkhsHa.ele.l.gvkpAvLtrSgg.s.pllp...sletql.c.q..g

mB3  GSEGPSTSGTL......KIPPDSETTLVLVGRANFLEKPVLGFVRLKEAVPLEDLVLPEPVGFLLVLLGPEAPHVDYTQLGRAAATLMTERVFRITASMAHNREELLRSL  324
hB3  GTEGHSPSGIL.....EKIPPDSEATLVLVGRADFLEQPVLGFVRLQEAAELEAVELPVPIRFLFVLLGPEAPHIDYTQLGRAAATLMSERVFRIDAYMAQSRGELLHSL  311
cB3  GAGQSPSRAQLGPQLHQQLPEDTEATLVLVACAAFLEQPLLALVRLGAPCPDAVLAVPLPVRFVLTVLGPDSPRLSYHEIRRAAATVMADRVFRRDAYLCGGRAELLGGL  318
CON  G.eg.s.sg.L......kiPpDsEaTLVLVgrA.FLEqPvLgfVRL.ea.ple.l.lP.PvrFllvlLGPeaPh.dYtqlgRAAATlM.eRVFRidAyma..R.ELL.sL

mB3  ESFLDCSLVLPPTDAPSEKALLNLVPVQKELLRRRYLPSPAKPDPNLYNTLDLNGGKGGPGDEDDPLRRTGRIFGGLIRDIRRRYPYYLSDITDALSPQVLAAVIFIYFA  434
hB3  EGFLDCSLVLPPTDAPSEQALLSLVPVQRELLRRRYQSSPAKPDSSFYKGLDLNGG......PDDPLQQTGQLFGGLVRDIRRRYPYYLSDITDAFSPQVLAAVIFIYFA  415
cB3  QGFLEASIVLPPQEVPSEQHLHALIPIQRHAVRRRYQHPDTVRSPGGPTAPKDTGDKGQAPQDDDPLLRTRRPFGGLVRDIRRRYPKYLSDIRDALNPQCLAAVIFIYFA  428
CON  egFLdcSlVLPPtdaPSEqaLl.LvPvQrellRRRYq.spakpdp..y..ldlnGgkg.....DDPL.rTgr.FGGLvRDIRRRYPyYLSDItDAlsPQvLAAVIFIYFA

mB3  ALSPAVTFGGLLGEKTRNLMGVSELLISTAVQGILFALLGAQPLLVLGFSGPLLVFEEAFFSFCESNNLEYIVGRAWIGFWLILLVMLVVAFEGSFLVQYISRYTQEIFS  544
hB3  ALSPAITFGGLLGEKTRNQMGVSELLISTAVQGILFALLGAQPLLVVGFSGPLLVFEEAFFSFCETNGLEYIVGRVWIGFWLILLVVLVVAFEGSFLVRFISRYTQEIFS  525
cB3  ALSPAITFGGLLGEKTRGMMGVSELLLSTSVQCLLFSLLSAQPLLVVGFSGPLLVFEEAFFRFCEDHGLEYIVGRVWIGFWLILLVLLVVACEGSVLVRYLSRYTQEIFS  538
CON  ALSPAiTFGGLLGEKTRn.MGVSELLiSTaVQgiLFaLLgAQPLLVvGFSGPLLVFEEAFFsFCE.ngLEYIVGRvWIGFWLILLV.LVVAfEGSfLVryiSRYTQEIFS

mB3  FLISLIFIYETFSKLIKIFQDYPLQQTYA.PVVMKPKPQGPVPNTALFSLVLMAGTFLLAMTLRKFKNSTYFPGKLRRVIGDFGVPISILIMVLVDSFIKGTYTQKLSVP  653
hB3  FLISLIFIYETFSKLIKIFQDHPLQKTYNYNVLMVPKPQGPLPNTALLSLVLMAGTFFFAMMLRKFKNSSYFPGKLRRVIGDFGVPISILIMVLVDFFIQDTYTQKLSVP  635
cB3  FLISLIFIYETFAKLVTIFEAHPLQQSYDTDV..STEPSVPKPNTALLSLVLMAGTFFLALFLRQFKNSVFLPGKVRRLIGDFGVPISIFVMALADFFIKDTYTQKLKVP  646
CON  FLISLIFIYETFsKLikIFqdhPLQqtY...V.m.pkPqgP.PNTALlSLVLMAGTFflAm.LRkFKNS.yfPGKlRRvIGDFGVPISIliMvLvDfFIkdTYTQKLsVP

mB3  DGLKVSNSSARGWVIHPLGLYRLFPTWMMFASVLPALLVFILIFLESQITTLIVSKPERKMIKGSGFHLDLLLVVGMGGVAALFGMPWLSATTVRSVTHANALTVMGKAS  763
hB3  DGFKVSNSSARGWVIHPLGLRSEFPIWMMFASALPALLVFILIFLESQITTLIVSKPERKMVKGSGFHLDLLLVVGMGGVAALFGMPWLSATTVRSVTHANALTVMGKAS  745
cB3  RGLEVINGTARGWFIHPMGSATPFPIWMMFASPVPALLVFILIFLETQITTLIVSKPERKLVKGSGFHLDLLLIVAMGGLAALFGMPWLSATTVRTITHANALTVVGKSA  756
CON  dGlkVsNssARGWvIHPlGl...FPiWMMFAS.lPALLVFILIFLEsQITTLIVSKPERKmvKGSGFHLDLLLvVgMGGvAALFGMPWLSATTVRsvTHANALTVmGKas

mB3  GPGAAAQIQEVKEQRISGLLVSVLVGLSILMEPILSRIPLAVLFGIFLYMGVTSLSGIQLFDRILLLFKPPKYHPDVPFVKRVKTWRMHLFTGIQIICLAVLWVVKSTPA  873
hB3  TPGAAAQIQEVKEQRISGLLVAVLVGLSILMEPILSRIPLAVLFGIFLYMGVTSLSGIQLFDRILLLFKPPKYHPDVPYVKRVKTWRMHLFTGIQIICLAVLWVVKSTPA  855
cB3  VPGERAHIVEVKEQRLSGLLVAVLIGVSILMEPILKYIPLAVLFGIFLYMGVTSLFGIQLFDRILLLMPPKYHPKEPYVIRVKTWRMHIFTLTQILVVALLWGVKVSPA  866
CON  .PGaaAqIqEVKEQRiSGLLVaVLvGlSILMEPILsrIPLAVLFGIFLYMGVTSLsGIQLFDRILLLfkPPKYHPdvPyVkRVKTWRMHlFTgiQIiclAvLWvVKstPA

mB3  SLALPFVLILTVPLRRLILPLIFRELELQCLDGDDAKVTFDEENGLDEYDEVPMPV  929
hB3  SLALPFVLILTVPARRVLLPLIFRNVELQCLDADDAKATFDEEEGRDEYDEVAMPV  911
cB3  SLRCPFVLVLTVPLRRLLLPRIFSEIELKCLDIDDAVVTFEEAEGQDVYNEVQMPS  922
CON  SLalPFVLiLTVPlRRllLPlIFre [illegible]
```

proteolysis [43, 89] and in frog oocytes when only the cDNA for only the membrane domain is expressed (Gunn, Kopito, and Smith, unpublished results).

Chemical Modifications

The band 3 protein has been chemically modified with stilbene disulfonates 4,4′-diisothiocyanostilbene-2,2′-disulfonate (DIDS), 4-acetamido-4′isothiocyano-stilbene-2,2′-disulfonate (SITS), 4,4′-dinitrostilbene-2,2′-disulfate (DNDS), BADS, etc. (see [112] for review); reductive methylation [60], pyridoxal phosphate [16, 104], arginine reagents such as 1,2-cyclohexanedione [149], and phenylglyoxal [146]. These studies have shown that there are two lysines (a and b) to which DIDS can react and inhibit transport, but which may not be critical for anion transport, and that there is an arginine that must be protonated (pK ~ 12) to form a critical part of the anion binding/transport pathway [145, 146] and a lysine (hB3-K851) site at which pyridoxal phosphate reacts that is important for transport [16, 54].

Recently Jennings [62] has modified a carboxyl group on a glutamic acid that normally behaves as the principal titration site. When it is protonated, proton-sulfate cotransport is increased and halide fluxes are reduced This conversion of an extracellular glutamic acid to an alcohol also removes the intracellular pH dependence of sulfate efflux and thus the group may have access to the internal solutions [62]. From chloride kinetics alone it has been argued that the anion binding/transport site and the proton binding/transport site could be on opposite sides of the membrane at the same time [100] and thus are physically as well as chemically separable.

The molecular identity of these reactive groups along the band 3 sequence is being actively pursued by both direct chemical studies and by comparative molecular biology. For example, lysine-a that reacts with DIDS has been identified as either K-558 or K-561 of the mouse sequence. But since the chicken band 3 is missing the lysine corresponding to K-561 it is highly likely that K-558 is lysine-a [21]. When lysine-b is methylated, transport is inhibited [82]. Lysine-b was suggested by Kopito and Lodish [16, 54] to be K-657, but this lysine is also absent in the chicken sequence. Recent speculation has focused on hB3-K851 where pyridoxal reacts and inhibits transport [103].

Fig. 1. The amino acid sequence for murine band 3 (*mB*3), human band 3 (*hB*3), and chicken band 3 (*cB*3). The sequences have been aligned to provide the best consensus sequence (*CON*). The consensus sequence is given a *capital letter* if all three sequences agree, a *lower-case letter* if two of the three sequences agree, and a *dot* if there is no consensus. The band 3 sequences have *dots* inserted where gaps have been inserted to enhance the overall consensus. Right: the position number for the last amino acid on each line. The single letter code for amino acids with nonpolar side groups is alanine (*A*), isoleucine (*I*), leucine (*L*), methionine (*M*), phenylalanine (*F*), proline (*P*), tryptophan (*W*), valine (*V*); with polar uncharged side groups is asparagine (*N*), cysteine (*C*), glutamine (*Q*), glycine (*G*), serine (*S*) threonine (*T*), tyrosine (*Y*); with positively charged side groups is arginine (*R*), histidine (*H*), lysine (*K*); with negatively charged side groups is aspartic acid (*D*) and glutamic acid (*E*)

Heterogeneity of Band 3

There is only a single gene for band 3 in the genome, though of course there are two copies, one from the mother and one from the father. It is of importance to know all of the band 3 gene products in erythrocytes. Is band 3 a pure protein? The diffuse band that is produced upon SDS-polyacrylamide gel electrophoresis is mostly if not entirely due to the variable extent of glycosylation at a single asparagine (N-660 in the mouse, N-642 in the human, N-653 in the chicken) since this band becomes much sharper upon pretreatment of the protein with endoglycosidases. But it is not this heterogeneity that interests us here. Heterogeneity of the aminoacid sequence could arise from variable processing of the protein after translation or from alternate splicing of the DNA to form different mRNAs during transcription or from differences in the two DNA alleles. The first documented heterogeneity was the report of Mueller and Morrison [10, 146] who found a variant with a 3-kDa elongation in the normally 23-kDa N-terminal cytoplasmic fragment.

Indications of heterogeneity within an individual donor came from the irreversible labeling with phenyoglyoxal [145]. When resealed red cell ghosts are reacted with phenylglyoxal only 91% of the chloride exchange transport can be inhibited under a variety of conditions if incubated long enough. But if the ghosts are washed with albumin solutions that remove adsorbed but unreacted phenylglyoxal, and retreated with phenylglyoxal, the flux can be further reduced to 99% of the original value [9]. This is not consistent with a single population of band 3 transporters and the observed steady state irreversible inhibition and labeling at long times. Membranes treated with 1-ethyl-3-(4-azonia-4,4-dimethylpentyl)-carbodiimide (EAC) maximally inhibit the flux 90%, and 75% of the residual flux is inhibitable by DIDS. This may be either due to heterogeneity of band 3 or a residual transport rate 10% of normal through the EAC modified band 3 [130].

When band 3 is solubilized in Triton X-114, it partitions spontaneously into two phases, a detergent-poor phase with 35% of the band 3 and a detergent-rich phase with 65% of the band 3 [130]. The purified chymotrypsin fragment with M_r 60000 also partitions into two phases but 70% into the detergent-poor phase and 30% into the detergent-rich phase. Repeated isolation and repartition of either of the fractions of whole band 3 does not result in a 35%/65% repartitioning of the band 3 or 70%/30% repartitioning of the M_r 60000 fragment but the appearance instead of two subpopulations of band 3 or two subpopulations of fragment. Thus, band 3 is heterogeneous as determined by this procedure and the heterogeneity is present in the M_r 60000 fragment that includes the M_r 42000 N-terminal cytoplasmic domain. Deglycosylated band 3, dephosphorylated band 3, and DIDS-labeled band 3 each behave similarly to native band 3; thus the heterogeneity is not due to heterogeneity in glycosylation, phosphorylation, or the DIDS reactivity [69].

A family with acanthocytosis, an altered band 3 and increased sulfate flux, has been reported [71]. It appears to be an autosomal recessive trait with at least a 1.8-fold increase in maximal sulfate transport, a normal number of band 3s per red cell, and an abnormality in the proteolysis pattern. The discovery of natural mutants with functional differences will provide a most important line of insight into how the protein functions, and identifying which amino acids are along the transport pathway and which are involved in the conformational changes required for transmembrane translocation.

In chicken erythrocytes, a single genomic locus for band 3 (cB3) gives rise to two different mRNAs with different translational start sites [55]. On SDS-PAGE, the avian band 3 appears as a doublet with M_r 100000 and 105000, as opposed to the human and mouse band 3 which are a single band with M_r 95000. Thus in chicken red cells the heterogeneity is established. Differences in anion transport by the two proteins should be small since their difference is an absence of 33 amino acids from the N-terminal region. Both protein sequences derive from the same DNA sequence. They align with mB3, but with the N-terminal 10 amino acids missing in the longest avian sequence and the N-terminal 43 amino acids missing in the shorter avian sequence. This is consistent with the lack of glyceraldehyde-3-phosphate dehydrogenase binding to chicken band 3 [141] which binds near the N-terminus of other band 3s [119]. Whether these two mRNAs arise from alternative splicing or if they are transcribed from different promoters within the same locus is not established.

The heterogeneity of band 3 is thus established, not only by amino acid differences between individuals of the same species but also within the erythrocytes of one individual (human and chicken), and this may have functional significance (see "Anion Conductance...").

Physiology of Band 3

Early and Middle History

In 1867 Schmidt [151] and Zuntz [105] independently discovered that the amount of alkali in the serum increased when the partial pressure of CO_2 over the blood was increased. Zuntz postulated that CO_2 caused the dissociation of sodium from proteins within the cells and that some of this sodium exited from the cells with a bicarbonate ($NaHCO_3$). The converse of this bicarbonate efflux was discovered by Nasse in 1874 who found that the chlorine in serum decreased as the partial pressure of CO_2 was increased over the blood. He surmised that the chlorine was entering the cells in exchange for bicarbonate. He also found that the red cells swelled upon CO_2 treatment by taking up water. When one considers the understanding of physical chemistry at this time these notions are surprisingly elegant. It was not until 1887 that Arrhenius proposed his famous hypothesis that the ions in a solution of electrolytes are present in the free state. Previously the equivalent movement of anions and cations was thought to demonstrate their total association in solution. Nasse's published report in 1878 [143] was largely unnoticed until Warburg's thesis in 1922 in which this history was detailed [50]. The credit for discovering the chloride-bicarbonate exchange is falsely but usually given to Hamburger for his article in 1892 [101]. Textbooks continue to mention the Hamburger shift as an integral part of the uptake of CO_2 by the blood though it more properly should be the Nasse shift if due credit were given.

The important events in the next stage of study were (a) the positive fixed charge model of Mond in 1927 [29] in which the erythrocyte membrane was compared with a porous anion exchange membrane; (b) the flow tube measurements by Dirkin and Mook in 1931 [94, 95] of chloride-bicarbonate exchange with a half-life of 0.3 s; (c) the measurements of Luckner in 1939 and 1948 [134] that erroneously indicated a low activation energy of anion transport; (d) the flow tube measurements of tracer chloride exchange by Tosteson in 1959 [111]; and (e) the tracer sulfate kinetics of Passow [120] and Schnell [19, 118] to support the porous fixed charge model. Then in 1965–1968 there were several papers by mitochondriacs [24] indicating that erythrocyte chloride net transport was restricted even in cationophore treated cells. These observations were not understood by the red cell community at that time. In 1970 Dalmark and Wieth [52] showed that chloride exchange had a high activation energy and thus the feasibility of using rapid filtration methods at low temperatures. In 1971 Hunter [47] first measured the net transport permeability and showed that most of the exchange was electrically silent as suggested by Tosteson in 1959 and indirectly shown by treating red cells with ionophores in 1965–1968. In 1971 we characterized the chloride kinetics which were incompatible with the then current porous fixed charge mechanism [44] and postulated the titratable carrier model to reconcile both chloride and sulfate transport characteristics by a common molecular machinery [51].

Chloride-Bicarbonate Exchange in the Uptake and Transport of Co_2

When a red cell enters a peripheral capillary it encounters an increase in the pCO_2 from 40 torr to 46 torr. The CO_2 produced by the tissue respiration is freely and rapidly permeant across cell membranes. It diffuses rapidly over the short distance from the tissue across the capillary endothelium, across the thin layer of plasma, across the erythrocyte membrane and into the cytoplasm. There it is rapidly hydrated (or hydroxylated) in the presence of carbonic anhydrase (E.C. 4.2.1.1) to form carbonic acid, H_2CO_3 (or HCO_3 directly). The carbonic acid is almost instantaneously dissociated into a proton (more properly a hydronium ion, H_3O^+) and a bicarbonate, HCO_3^-. Thus the CO_2, the carbonic acid, and the protons with bicarbonate are in equilibrium within 250 ms in the presence of the carbonic anhydrase activity found within the erythrocyte. The equilibrium equation relating pH, pCO_2, and bicarbonate given by Henderson and Hasselbalch [148] is true for the cell interior after this transient period. $pH_i = pK + \log[(HCO_3)_i/(\alpha pCO_2)]$, where pK is about 6.1 and α is about 0.03 mM/torr and the subscript refers to the interior aqueous phase of the cells.

By mass action the CO_2 continues to react and form protons and bicarbonate so long as one or both of these products are removed. Protons are partially removed by their combination with the intracellular buffers, the principal one being hemoglobin. This titration of hemoglobin causes its affinity for oxygen to decrease (the Bohr effect) and thus enhances the extent of oxygen dissociation and the delivery of O_2 to the tissue. Bicarbonate is partially removed by the band 3 protein which exchanges it for extracellular chloride. It is in this way that much of the CO_2 transported from the tissues to the lungs comes to the plasma bicarbonate. CO_2 is also transported as carbamino groups on the N-terminal valines of the alpha and beta chains of hemoglobin. Of the total CO_2 transport about 8% is in the form of dissolved CO_2, 11% is in the form of carbamino groups (this assumes the normal complement in fresh red cells of 2,3-diphosphoglycerate which competitively prevents carbamino formation), and 81% is in the form of bicarbonate, with 67% of the total CO_2 as bicarbonate within the plasma and 14% of the total CO_2 as bicarbonate within the erythrocyte cytoplasm. This loading of the blood with CO_2 in its several forms occurs within 0.5 s which in most capillary beds is well within the average transit time of the red cells.

In the lung, the processes described above passively take place in the opposite order, driven by the gradient of pCO_2 from the blood to the alveolus. CO_2 diffuses out of the cell across the plasms and the endothelial and alveolar membranes into the alveolar space. The equilibrium between CO_2, H_2CO_3, protons and bicarbonate within the erythrocyte is disturbed, and by mass action some bicarbonate is converted back to CO_2. This lowering of the intracellular bicarbonate concentration in turn causes disequilibration across the cell membrane. The total electrochemical potential difference for chloride and bicarbonate now favors the entry of bicarbonate and the efflux of chloride on the passive anion exchange mechanism of band 3. Again this process is rapid compared with the normal transit time of the erythrocyte through the alveolar capillary bed (0.7s).

The net driving force for obligatory chloride-bicarbonate exchange is proportional to the difference in the electrochemical potential differences for each of the ions. If, as experimentally proven, the ion fluxes are strictly coupled so that one Cl ion exchanges for one HCO_3 ion, then there is no net electrical work performed and the electrical potential terms cancel. Thus the direction (but not the rate) of the exchange is strictly determined by the ratio of ion activities given by $[(Cl_0)\cdot(HCO_{3i})]/[(Cl_i)\cdot(HCO_{3o})]$. If this ratio is greater than unity, external chloride will exchange for internal bicarbonate until the ratio is equal to unity. If the ratio is less than unity, internal chloride will exchange for external bicarbonate until the ratio is unity. In the circulation this ratio is unity except in the peripheral capillaries during CO_2loading when it is transiently greater than unity and in the alveolar capillaries during CO_2 unloading when it is transiently less than unity.

Wieth and Brahm [12] have reviewed the arguments for and against the possible limitation of CO_2 transport by anion exchange. They conclude that the details of chloride and bicarbonate transport by erythrocytes are such that the physiological anion exchange after a step change in pCO_2 is half completed in about 100 ms and 90% complete in about 350 ms. Thus if red cell transit time of a coronary or pulmonary capillary is 0.3 s or less there would be incomplete equilibration. Transit times in this range are believed to be achieved during heavy exercise [107]. It seems very likely that the autotransfusions that some athletes believe help their performance may not be just due to the higher oxygen carrying capacity with the higher hematocrit but also the greater CO_2 transport that can be achieved. In a similar way the use of noncellular blood substitutes that carry oxygen may be less beneficial than anticipated because they do not enhance the removal of the CO_2 produced from that oxygen.

The importance of this mechanism to the transport of CO_2 can only be surmised by the extensive preservation of the gene and its function in the red cells of all vertebrates (with the exception of the lamprey [37]). There are no known individuals lacking band 3 and it may be surmised that if such individuals were carried to term they would be found in the new born nursery and diagnosed as "failure to thrive" in the absence of a critical assay of their red cell anion exchange function.

Anion Conductance and the Red Cell Membrane Potential

The conductance of the red cell membrane to ions is lower than that for most other cells. Unlike most other cells the anion conductance of red cells exceeds the cation conductance, specifically the potassium conductance. In human red cells the conductance for chloride is 4×10^{-6} S cm^{-2} or 6 pS per cell or 10 mmol (kg cell solid min)$^{-1}$, which is about ten fold greater than that for either Na or K (see [76] for a review). Since chloride is by far the most conductive ion, the fraction of the current carried by chloride ions is nearly unity and the membrane potential is given by the chloride equilibrium potential with almost no contribution by Na or K conductances. In isotonic chloride solutions at pH 7.4 and 37°C the membrane potential is -7 mV and $[Cl_i]/[Cl_o] = 0.76$. Some examples of abnormal conditions where cation conductances contribute to the membrane potential are when the cells have been treated with a cationophore like valinomycin or gramicidin which greatly increase the conductance of K or when the intracellular Ca activity is increased which activates a Ca-dependent K conductance native to the membrane [35, 65, 74].

The mechanism of anion conductance in human red cells is increasingly in doubt. Until recently the conductance was believed to be mediated though the band 3 protein since inhibitors of exchange such as the stilbene disulfonates (DIDS, SITS, DNDS) also inhibited the conductance of the same anions as those exchanged on band 3. The possible mechanisms on band 3 include slippage or tunneling. In the first, the conductance is conceived to be the result of the rare (once in 5×10^4 times) return of the empty carrier. According to this hypothesis, the band 3 instead of being obligated alternately to transfer anion in each direction is able to carry a conducting ion in one direction and then reset its conformation without returning with the usually obligatory bound anion. There is now significant experimental data against such a mechanism for chloride and for sulfate conductance. Separately and in complementing experiments Fröhlich, Knauf, and Kaplan with coworkers [41] showed that the predictions of the slippage hypothesis for the concentration dependence of the conductance under different exchange conditions were not met. In the tunneling hypothesis, the conductance is mediated by the passage of the anion through the band 3 protein without the usual conformational change that accompanies exchange of an anion. In this view the protein provides the pathway for anion entrance to the membrane and across it, but the anion does not contribute to the changes in the protein structure. There are two possibilities. Either the anion takes the same physical path through the protein as it would by exchange, but without the usual accompanying conformational change, or the anion takes an entirely separate or partially separate pathway from that of an exchanging anion. Gottipaty and Fröhlich [40] have evidence that the anion does not use all of the exchange pathway and specifically it does not interact with the external binding site during conductance as it does during exchange. They have found that external sulfate or external nitrate ions do not compete with the conducting chloride ion either during inward conductance or outward conductance. During exchange these anions compete with chloride at a single extracellular site with a K_i of about 5 mM for the return of the loaded conformation. During chloride conductance these anions have little effect on the net efflux or influx even when present at 150 mM [37]. It is possible that, given the heterogeneity in band 3, the anion exchanger and anion conductance mediating molecules are distinct subpopulations but that both have stilbene binding sites (see Fröhlich and King [37]).

The Equilibration of Protons and Chloride

One consequence of the rapid exchange of Cl and bicarbonate is that in a bicarbonate medium the intracellular proton activity is in electrochemical equilibrium with the extracellular proton activity [123]. This comes about because of five facts. (a) The ratio of chloride concentrations, r_{Cl}, between the cytosol, Cl_i, and the extracellular phase, Cl_o, is rapidly made equal to the ratio for bicarbonate ions ($r_{Cl} = r_{HCO3}$) by the exchange mechanism (see "Chloride-Bicarbonate Exchange. . ."). (b) The ratio of Cl activity coefficients in the two phases and the ratio of HCO_3 activity coefficients in the two phases are both very nearly equal to unity so that the ratio of concentrations is equal to the ratio of the activities for both of the species [$r(a_{Cl}) = r(a_{HCO3})$]. (c) The high conductance of Cl (and perhaps HCO_3) relative to all other ions (see "Anion conductance . . .") results in the membrane potential being the Nernst equilibrium potential for Cl and since $r(a_{Cl}) = r(a_{HCO3})$ this is also the Nernst equilibrium potential for HCO_3. (d) In the presence of intracellular carbonic anhydrase and the extracellular carbonic anhydrase either on the vascular endothelial cells of the capillaries or in the plasma from lysed red cells in the suspension, there is a rapid equilibration between CO_2, H_2CO_3, HCO_3, and H_3O^+. This assures that the Henderson-Haselbalch equation is rapidly valid for both the cytosol and extracellular fluid phases separately. (e) CO_2 is very rapidly permeant thus pCO_2 is the same in both fluid phases. Under these conditions the ratio of proton activities [$r(a_H) = a_{Ho}/a_{Hi}$; note the opposite convention from the definition of r_{Cl}] equals the ratio of chloride concentrations ($r_{Cl} = Cl_i/Cl_o$) and thus $pH_i = pH_o + r_{Cl}$ since $pH = -\log a_H$. This relationship is true in any membrane system where both Cl and protons have very high conductance since the Gibbs-Donnan equilibrium then says that $Cl_i/Cl_o = H_o/H_i$. But it is interesting that in this case the conductance of protons, and even the conductance of HCO_3 need not be high for this relationship to be true. On the other hand although protons and Cl behave as if they were both at equilibrium across the red cell membrane most of the time, there are instances in vitro when this is not true. If, for example, cells are suspended in a bicarbonate-free medium then the ratio of proton activities can be varied independently of the membrane potential or equivalently the Cl concentration ratio. If the carbonic anhydrase activity is blocked then the rate of equalization of the chloride and proton ratios (in the convention defined above) will be slowed but eventually the Henderson-Haselbalch equation will be true in both the cytosol and plasma. The importance of these exceptional conditions in vitro for non-red cells containing an anion exchange mechanism should not be underestimated.

Phosphate Transport

At 37°C and physiological orthophosphate (P_i; inorganic phosphate) concentrations (1.0–1.5 m*M*) approximately 75% of the tracer phosphate uptake is DNDS sensitive and presumed to be mediated by band 3. Approximately 20% is mediated by a newly described Na-dependent pathway that is DNDS-insensitive and the remaining 5% is independent of Na and DNDS concentrations and is linearly dependent on orthophosphate concentration [56]. Although the pH dependence and magnitude of P_i fluxes is like that of SO_4 fluxes the mechanism is somewhat different. The tracer flux of each has a symmetrical maximum around pH 6.2 when protons and total sulfate or total phosphate are at equilibrium though there is little data below pH 5.6. Sulfate is principally cotransported with a proton on band 3 [99]. The rate of cotransport being about 100 times faster than the transport of $SO_4^=$ alone [98]. The divalent sulfate anion and the proton bind in random order to the transporter, but the binding of the first enhances the affinity of the second about nine fold [122]. Thus the pH dependence of sulfate arises principally from the titration of the carrier. (The titration of the carrier decreases the maximum rate of chloride transport about 100-fold but does not have a great effect on the external affinity for chloride.) Phosphate which has a pK of 6.87 at physiological ionic strength and 37°C is present both as $H_2PO_4^-$ and $HPO_4^=$ ions in the solution. When the pH is lowered the proportion of monovalent anion increases and the band 3 is titrated as well. By measuring the flux as a function of pH_o and total P_i concentration in the external solution one finds that the absolute flux is a single Michaelis-Menten function of the monovalent form of the anion between pH values 6 and 7.5 whereas there is no simple relationship for the flux as a function of divalent concentration at different pH values [86, 87]. At very low pH_o the P_i influx (at constant pH_i) is inhibited like the chloride influx but unlike the sulfate influx into chloride containing cells. When phosphate inhibition of Cl flux is measured, monovalent phosphate is a competitive inhibitor, but divalent phosphate has little or no inhibitory effect. From these observations it seems likely that monovalent phosphate is the form that binds to the external transport site on band 3 and is transported as if it were a very slowly transported monovalent anion and not like sulfate with which it has been likened for many years. The pH dependence of P_i transport is thus probably due to the titration of the phosphate to form the proper substrate and not due to the titration of the carrier as has been shown for sulfate. The use of NMR to study the transport of phosphate analogs has shown that phosphite ($HPO_3^=$) which is structurally akin to HCO_3^- is the most rapidly transported analog [126].

Glycine Transport on Band 3

Glutathione (glutamyl-cysteinylglycine) is the major reducing substance in erythrocytes. This tripeptide is synthesized from its component amino acids in the mature erythrocyte. There is a glutathione transporter in the red cell [72] which may function mainly as an efflux leak pathway. The influx of glycine is mediated by two pathways: a Na-dependent, DNDS-insensitive pathway [33] and a Na-independent, DNDS-sensitive pathway that is probably on band 3 [8, 106]. Extracellular chloride competitively inhibits while protons activate glycine influx on the DNDS-sensitive pathway. It seems that the band 3 mechanism can carry the glycine anion though at a very slow rate.

Cytoskeletal Interactions with Band 3

Band 3 has a large cytoplasmic 404 amino acid N-terminal domain that can interact with several proteins of the cytoplasm. These interactions and the interactions between band 3 monomers affect the physical behavior of band 3. Band 3 molecules are bound to the underlying cytoskeletal network of spectrin and actin through the protein ankyrin [4, 23, 138] and through glycophorin and band 4.1 [39, 137]. There are about 10^5 ankyrin molecules per red cell with 10^6 band 3 thus only 10% of the band 3 can be directly tethered to the cytoskeleton via ankyrin. But of course by self-association band 3 dimers, tetramers and hexamers would directly and indirectly tether 20%, 40% or 60% of the band 3 using only the ankyrin connection. The measurements of rotational and translational diffusion coefficients indicate that about 20% of the band 3 are immobile at 37°C [137], and that fraction increases as the temperature is lowered. These data also indicate that the mobile band 3 monomers are limited in their rotational diffusion by association into multimers and that their lateral diffusion is limited by the underlying cytoskeleton that seems to form small domains from which the band 3 may not escape laterally. This cytoskeletal fence model [136] is consistent with the dependencies of rotational and lateral diffusion on the ionic strength and amine concentrations that alter the dimer/tetramer equilibrium of spectrin, the major cytoskeletal protein that makes up the meshwork that is believed to be the lateral barriers or fences [42].

This tethering of the cytoskeleton to the membrane through band 3 may paly an important role in the terminal differentiation of erythroblasts into mature, non-mitotic, erythrocytes. Avian (reptilian and amphibian) erythrocytes have a nucleus that is believed to have very little transcriptional activity. Normally erythroblasts divide a few times and then terminally differentiate into erythrocytes. Avian erythroblastosis virus infection results in an immortal cell line [117]. The virus contains two oncogenes, v-*erb*A that is homologous to the normal cellular gene c-*erb*A that codes for the thyroid hormone (T_4) receptor [30], and v-*erb*B that is homologous to the normal cellular gene c-*erb*B that codes for the epidermal growth factor receptor [150]. The v-*erb*A product in the infected eythroblasts presumably can bind T_4 and the complex can bind to nuclear DNA and prevent by competition with the normal receptor-T_4 complex the activation of certain genes. One of those genes is that for band 3 [116, 141]. In these infected cells the component proteins of the cytoskeleton, actin and spectrin, are synthesized but rapidly degraded throughout the erythroblast's life cycle [129]. Only when the band 3 gene product is made is the cytoskeleton assembled and, of course, tethered to the membrane. It is unknown whether it is the newly produced tether for the cytoskeleton or the new Cl/HCO_3 exchange function of the band 3 that acts to cause terminal differentiation. Or it may be fortuitous that cells only differentiate into mature erythrocytes once band 3 is inserted into the membrane.

In addition to ankyrin binding sites there are binding sites for hemoglobin [135] and the glycolytic enzymes aldolase [66], glyceraldehyde-3-phosphate dehydrogenase [27], and phosphofructokinase [93]. Each of these reversibly binds to a hydrophobic sticky region within the N-terminal 23 amino acids. The enzymes are about half bound in the intact red cell and when bound they have very low activity. Tyrosine specific protein kinases associated with the erythrocyte membrane can phosphorylate band 3 [6]. The binding of these enzymes, hemoglobin and even ankyrin is blocked by phosphorylation of tyrosine 8 and 21 (hB3). Ferricyanide increases the glycolysis of human red cells [114, 115]. A tyrosine kinase activity that phosphorylates these sites would in turn displace glyceraldehyde-3-phosphate dehydrogenase (G3PDH) from these sites and thus increase the cytosolic activity of this enzyme that partially controls the glycolytic rate. The murine band 3 does not have these tyrosines and has no ferricyanide stimulated glycolysis. Normally the phosphofructokinase reaction is rate limiting in glycolysis but in certain states (after partial depletion of energy stores and pH 8 where the glycolysis rate is two times that at pH 7.4 [67, 68, 102]) the G3PDH reaction may limit glycolysis. It is thus proposed that at least in human red cells under certain conditions, regulation of band 3 phosphorylation may be an important component in the regulation of glycolysis. The factor that binds to its receptor and stimulates this tyrosine kinase activity in vivo remains unknown.

The most interesting and yet perhaps the least understood site on the cytoplasmic domain of band 3 is the site for self-association. Band 3 molecules are mostly dimers in the intact membrane as determined by both crosslinking and rotational diffusion measurements. These are also monomers and tetramers in equilibrium with these dimers. The dimerization could have functional consequences [34]. But while the kinetics of anion exchange are not always simply that expected for a ping-pong mechanism on monomers there are only suggestive experiments that lead one to consider a scheme in which the exchange requires monomer–monomer interactions or cooperativity. These experiments which involve slowly transported anions where parallel pathways of comparable magnitude may confuse the kinetics or experiments which measure the flux as a function of the anion concentrations as they are simultaneously changed on both sides are difficult to interpret and equally difficult to reject as invalid windows on the behavior of the underlying and truly unknown molecular mechanism of exchange.

Band 3 Antibodies and the Clearance of Senescent Red Cells

About 0.83% of the normal human red cells are cleared from the blood by the reticuloendothelial system each day. The average life span of the red cell is about 120 days. The mechanisms that are involved in clearing these older and generally more dense cells are still in dispute. There are autoantibodies in normal human blood and some of these are to band 3 and spectrin and a rare glycosphingolipid. Older red cells have different creatine contents and 100–200 more IgG molecules per cell than do younger cells [70]. There are also complement components (C3 fragments) found on older red cells that might play a part in the mechanism of their phagocytosis [96]. There is also evidence that the IgG antibodies do not bind to band 3 in normal erythrocytes but only after they have been enzymatically treated or subjected to membrane oxidation [7, 38]. Oxidation causes band 3 clustering and clustering enhances binding of autologous IgG. Phagocytosis of diamide-treated (oxidant-treated) red cells is stimulated by anti-band 3 which mediates complement (C3b) binding [125, 133].

Cation Transport on Band 3

Although this chapter is devoted to anion exchange mechanisms it would be an egregious error not to mention that ion pairs between Na^+ or Li^+ and $CO_3^{=}$ form monovalent anions, $NaCO_3^-$ and $LiCO_3^-$, that enable the anion exchanger to transport cations. Although the concentrations of these ion pairs are in the micromolar range, anion exchange is very much faster than the other pathways for cation entry. At physiological concentrations of Na and HCO_3 about half of the steady state Na influx into erythrocytes is on band 3 through this mechanism. Thus if HCO_3 is removed from the medium or anion transport inhibitors are added the Na influx declines by half and the load of Na that must be pumped out by the Na pump is reduced to half. In a careful steady-state metabolic study the ATP sparing effect of these manipulations should be observed [35, 53]. In addition, heavy metals such as Zn and Pb also form complexes with bicarbonate that are transported into the erythrocyte by stilbene-sensitive pathways presumably on band 3 [25].

Kinetics of Band 3 Exchange

Stoichiometry

Band 3 mediates the obligatory exchange of small inorganic anions. The obligation is strict. The present estimate is that less than one chloride ion in 10^6 exchanges can move in one direction and be unaccompanied by the antiport of an anion in the opposite direction. Cells made highly conductive to cations by the presence of valinomycin or gramicidin also show obligatory exchange of anion equivalents [88]. Consequently the coupling is not through the electrical potential but rather is a mechanical property of the mechanism.

Reaction with Ligands

Chloride/chloride exchange is the most rapid, with a turnover of 10^4 ions/site s^{-1} at 23°C and 5×10^4 ions/site s^{-1} at 37°C. Chloride activation at the external surface when internal Cl is 140 m*M* has a K_m of 2–3 m*M*. The self-exchange has a K_m of 30–60 m*M*. The activation at the internal surface has a K_m of about 60 m*M*. Thus at physiological chloride concentrations in the absence of any other substrates the external transport site is 97% saturated and the internal transport site is about 60% saturated. This implies that in a simple mechanism the internal transport site would control the obligatory reaction since it has the greater chance of being empty and thus the greater chance of being unable to perform half of the exchange reaction. The activation energy for exchange is about 37 kcal/mol below 17°C and about 25 kcal/mol above that temperature when measured in a 140 m*M* chloride solution at pH 7.4 [49].

Bicarbonate transport has been measured directly at higher pH values where the rate of spontaneous dehydration in the absence of carbonic anhydrase activity is slow. But most measurements of bicarbonate flux are implied from the measurement of initial chloride flux in an obligatory bicarbonate/chloride exchange experiment. External bicarbonate has a K_m of less than 1 m*M* [58]. It has a maximal rate very nearly the same as chloride since the chloride efflux rate into bicarbonate media is the same as into chloride. The self-inhibition by bicarbonate is slight but not extensively studied.

Bromide transport shows a K_m on the outside of 2–3 m*M* if it is estimated from the external bromide activation of the chloride efflux. The K_m for self-exchange is about 20 mM but both calculations depend upon how the correction for external bromide self-inhibitions is made. The maximum bromide-bromide exchange is about 20% of that of chloride-chloride exchange. Heteroexchange of bromide and chloride is twice the rate of bromide-bromide exchange and thus the bromide half of the transport cycle seems to be rate limiting. Bromide-chloride heteroexchange has the same rate in both directions thus in a simple mechanism the maximal rate of bromide is the same in each direction and thus any overall asymmetry must be due to differences of affinity on the two sides [75]. A similar symmetry in sulfate transition rates has been observed [48]. But for chloride, quite the opposite conclusion was drawn by Knauf and Mann [14, 147] who used niflumic acid, a noncompetitive inhibitor that preferentially binds to the outward facing conformation. They found no differences in the chloride affinities on the two sides and surmised that the observed overall asymmetry must be principally in the two transition rates (inward > outward) for chloride ions.

Ping-Pong Versus Sequential Kinetics

Although the details of the molecular mechanism of anion transport are unknown, significant effort has been placed on trying to understand the class of behaviors that the mechanism possesses and thus place some limitation on the possible underlying molecular mechanism. One such distinction is whether the mechanism handles one anion at a time or handles two simultaneously. For the rapidly transported anions, chloride, bicarbonate, and bromide, the anions are handled one at a time. The system exhibits ping-pong kinetics, and these kinetics are most easily derived from having a single binding/transport site that has alternate access to the two sides of the membrane [36]. In such a scheme the band 3 protein binds only one transported anion at a time to form at most a binary complex with the transported anion. (This is not to mean that other sites on the protein do not exist and do not have anions bound to them, but only that there is only one anion that is on the transport pathway at any given instant). It is possible to construct multiple-transport-site models that have the mathematical form of ping-pong kinetics [113]. For example, if there is a series of sites along a single diffusion pathway through the protein, but only one of the sites may be occupied at any given moment. Thus the ping-pong kinetics restrict the model to one effective anion binding/transport site at any given moment but does not restrict the number of potential or cryptic sites. The mathematics of a single-site, obligatory exchange mechanism has been recently reviewed [48].

The alternative mechanism has sequential kinetics. It has been excluded for chloride, bicarbonate and bromide exchange in red cells but apparently is used in anion exchange in HL-60 cells [75]. This type of mechanism is most easily derived from a scheme in which there are two (or more) binding/transport sites such that a ternary complex of the protein and two transporting anions (one derived from each of the two bounding solutions) can be formed. In the transport step(s) the two anions are exchanged between their binding/transport sites.

The characteristics of ping-pong kinetics are that both the K_m and V_{max} measured by increasing the substrate on one side of the membrane are Michaelis-Menten functions of the substrate concentration on the other side of the membrane and that the binding/transport sites can be recruited from having access to ions on one side of the membrane to having access to ions on the other. By this we mean the recruitment of the ability of the site to access anions from one solution to the ability of the site to access anions from the other solution. Thus if a single-site, alternate access mechanism (which has ping-pong kinetics) has transportable anions only on one side of the membrane the mechanism will be sort of stuck with the binding/transport site having access to the solution on the other side of the membrane because there will be no anions to bind from that side. (Strictly speaking it is possible that an anion could bind to the transport site and not be transported but here by non-transportable anions we also assume non-binding of the anions to the transport site.) This is recruitment with a high fraction of the transporter molecules in one conformation: the one with the transport site facing (having access to the anions of) the solution without transportable anions and with the transport site empty (not complexed with a transportable anion). The evidence supporting a ping-pong mechanism includes: the K_m and V_{max} for chloride on one side of the membrane is a hyperbolic (Michaelis-Menten) function of the chloride concentration on the other side [59]; the recruitment of sites in inhibitor studies [20]; the measurement of the half-turnover of the transport mechanism [36]. Many other studies are consistent with this view but do not necessarily test the alternatives in an exclusive way.

There are several reports of experiments that do not show ping-pong kinetics or have cooperative effects that suggest interactions between sites. This has led to considerations of interactions between band 3 dimers or tetramers in mediating the transport of anion. All of these studies are with slowly transported anions. It might be argued that the rapidly transported anions prevent the access of the mechanism to certain conformations and that these studies with the more slowly transported anions provide an additional window onto the underlying mechanism. However there are certain limitations that come with slowly transported substrates including: parallel pathways may contribute to the total flux in unforeseen ways that make interpretation complex; the slowly transported anion is that way because it operates through a rare set of conformational changes (this is the obverse of the argument that the rapidly transported anions prevent certain conformations); the state of recruitment of a ping-pong mechanism is not well defined by the experimental conditions imposed by the slowly transported anions.

Inhibition

Anion transport in red cells is inhibited by many organic and inorganic anions. The mechanisms of inhibition can be classified like those for enzymes but with the added requirement that for nonpermeant inhibitors it is important to specify the side of the membrane to which they are added. The classification scheme that is simplest and most logical is that espoused by Cornish-Bowden [148]. There are two main classes of inhibitors: competitive and uncompetitve. Each is characterized by a single inhibitory constant, K_1 or K_2 respectively. When the characteristics of both classes are present, the inhibition is called mixed inhibition, and if K_1 equals K_2, then the mixed inhibition has a special name: noncompetitive. A starting point is to view competitive inhibition as due to the mutual exclusion of the substrate (here the transported anion like chloride or bicarbonate) and the inhibitor from the binding/transport site. That is they cannot both be bound to that site at the same time. Uncompetitive inhibition, on the other hand, is viewed as due to the binding of the inhibitor to a site separate from the substrate binding/transport site but only when the substrate is bound. In mixed inhibition the inhibitor binds at a separate site both when substrate is bound or not bound and influences the binding of the substrate to its site. Noncompetitive inhibition is a special case of mixed inhibition when the inhibitor binding at its site has no effect on the binding affinity of the substrate at the binding/transport site.

As suggested above, in transport kinetics there are two substrate and product solutions that are kept separate by the plasma membrane and thus one can add the inhibitor to the solution on one side of the membrane and systematically vary the substrate concentration either on the same side, the opposite side or both sides simultaneously. Surprisingly the kind of inhibition that a given chemical inhibitor demonstrates depends on how the experiment is performed. For example, if the inhibitor binds only the external empty conformation, T_o, the inhibition will be competitive (C) with respect to external substrate, X_o, but uncompetitive, U, with respect to changes in internal substrate concentrations, X_i; and competitive if both substrate concentrations are changed together, $X^{i=0}$. The full range of possibilities is presented in Table 1.

Table 1. Inhibition and transport activation by the substrate

Inhibitor binds to conformation (s)	Type of inhibition with respect to transport activation by the substrate		
	X_o	X_i	$X^{i=0}$
T_o	C	U	C
T_i	U	C	C
TX_o	U	U	U
TX_i	U	U	U
T_o, TX_o	M	U	M
T_i, TX_i	U	M	M
T_o, T_i, TX_o, TX_i	M	M	M

The uncomplexed transporter, T, faces either the extracellular fluid, T_o, or intracellular fluid, T_i, as does the substrate loaded transporter, TX. Extracellular substrate, X_o, or intracellular substrate, X_i, concentrations may be varied independently or together, $X^{i=o}$. The inhibition may be competitive, C, uncompetitive, U, or a mixture of these, M, depending on the conformations that the inhibitor binds to *and* the kind of experiment performed.

The mathematics for each of these situations has been derived for a single-site alternating access mechanism with ping-pong kinetics [56, 57].

The substrates for anion exchange are all mutually exclusive with each other and thus each is a competitive inhibitor of the transport of the others. For example, chloride and bicarbonate compete with each other and thus the apparent K_m for bicarbonate in plasma is much greater than the K_m in the absence of chloride which is less than 3 mM. And the K_m for chloride which is 3–5 mM on the outside of the red cell is increased in the presence of 24 mM bicarbonate. On the other hand the presence of internal bicarbonate (in the absence of any other internal substrates of the exchange mechanism) greatly increases the influx of chloride into the cell. So while cis bicarbonate inhibits, trans bicarbonate will always stimulate because it is the fastest transported substrate (equal to chloride). In contrast to this, consider the case when bromide is added and chloride-chloride exchange is being measured with tracer chloride and the cis chloride concentration is above K_m. If bromide, a slower substrate, is added to the cis side (even without removal of equal molar chloride) then the chloride exchange flux from trans to cis will be inhibited.

The substrates of the mechanism are chloride, bicarbonate, bromide, nitrate, iodide, thiocyanate, sulfate, monovalent phosphate, divalent citrate, dithionate and some organic anions such as EDTA and glycine. The first seven of these substrates have K_m values for the external binding/transport site in the 2–5 mM range and thus the mechanism is rather nonselective among them. Monovalent phosphate has a K_m value for this site in the 10–15 mM range.

The reversible inhibitors that have been used to inhibit chloride exchange (with the micromolar concentrations required to half inhibit the flux) include: DNDS (0.2), furosemide (150), ethacrynic acid (7), bumetanide (400), salicylate (900), chlorpromazine (150), probenecid (40), tetracaine (5000), benzocaine (1000), dipyridamole (50), and niflumic acid (0.6) (see [98]).

Proton-Anion Cotransport

In erythrocytes where the internal and external proton activities are rapidly brought into equilibrium in the presence of catalytic amounts of CO_2, the cotransport of protons and anions is of little physiological importance. But mechanistically the titratable carrier model has been an important element in relating the fluxes of monovalent anions and divalent anions. The protonation of a single site with a pK between 5 and 6 depending on the anionic conditions markedly changes the rates of anion transport. Protonation decreases the chloride transport rate about 100-fold from 40000 mmol/(kg cell solids·min) to 400 at room temperature, and increases the sulfate transport rate about 100-fold from 0.68 to 68 mmol/(kg cell solids · min). In addition in the strict absence of bicarbonate and CO_2 it has been shown that the proton is cotransported with the anion [15, 84]. The protonation increases the affinity of sulfate for the binding/transport site by a factor of 9, and reciprocally sulfate enhances the affinity (raises the pK) for protons by a factor of 9 [84]. The proton flux through this pathway is stilbene inhibitable, it has no voltage dependence, and it is increased as a function of proton concentration at a constant difference in proton (and hydroxyl) motive force. Thus it behaves as proton-anion cotransport and not hydroxyl-anion exchange on band 3.

In non-erythroid cells that contain anion exchangers this type of mechanism could be more important to identify and quantitate, since in those cells neither the protons nor the chloride are at equilibrium across the plasma membrane.

Nonerythroid Anion Exchange

Band 3 Protein (AE1) and Band-3-Related Protein (AE2) in the Kidney

The gene for band 3 that is on chromosome 17 (in the human) produces at least three transcripts in the mouse kidney. A minor one is identical to erythroid mB3 and the other two make what is called kidney band 3, KB3. KB3 is of at least two types in the mouse and rat. The major mouse mRNA is made from a different first exon than the erythroid exon 1, has none of the erythroid exon 2 or exon 3 but has the usual exons 4–20 as the erythroid mRNA. The minor KB3 mRNA is made using an alternative promoter and has a larger than normal exon 1 but then has the erythroid exons 2–20. The protein that is made from both of these mRNAs begins with the methionine that is the 80th amino acid in the erythroid band 3 and is coded for in exon 5. Thus the renal band 3 product made from both mRNAs is 79 amino acids shorter than the erythroid band 3 and has only 850 amino acids in the mouse and 848 amino acids in the rat [32, 121, 139]. In the rat the major transcript contains sequences derived from what corresponds to intron 3 in the mouse which together with sequences from exons 4 and 5 are 5′ untranslated sequences in rat KB3 [77, 128]. The location of KB3 in kidney tubules has been determined using antisera or monoclonal antibodies raised against proteolytic fragments of erythroid band 3 [3]. In the inner stripe of the outer medulla all of the intercalated cells are of the A-type that secrete acid. A-type intercalated cells of the distal nephron secrete protons across their apical membrane into the tubular lumen and secrete HCO_3^- across the basolateral membrane in exchange for pertubular chloride. Inhibition of basolateral anion exchange suppresses apical proton secretion, showing the importance of anion exchange in the acidification of the urine [15]. These A-type cells have an apical proton ATPase and have KB3 on their basolateral membranes though from which mRNA transcript the KB3 protein is derived is not known [22]. In the rat renal cortex about half of the intercalated cells are A-type and half are the B-type that secrete base. the B-type intercalated cells have a basolateral or diffuse cytoplasmic staining for H^+ ATPase and no KB3. Less than 1% of the cortical intercalated cells have an apical H^+ ATPase and have no KB3. KB3 has been expressed in *Xenopus laevis* oocytes [139]. Cytoplasmic injection of an in vitro synthesized mRNA from the cDNA causes a DIDS-sensitive (1 μM) ^{36}Cl influx six times the control flux which was 20 pmol/oocyte min^{-1}. This proves that this truncated form of band 3 functions as an anion transporter with pharmacology like that of band 3 in red cells. Band 3 gene products are also found in the renal tubule of the chicken [31] and probably in basolateral membranes of intercalated cells [110] and in the dark cells of the turtle urinary bladder [28].

A band 3 related protein (B3RP; also known as AE2) is encoded in a gene on chromosome 7 (in the human) at a site different than that for the cystic fibrosis defect in chloride transport [2]. The mRNA for this gene is found in many tissues and cultured cells. The cDNA was first cloned and sequenced by Demuth et al. [1]. It has about 4100 base pairs and encodes a 1237 amino acid protein. By hydropathic analysis it resembles band 3 in having an N-terminal cytoplasmic domain consisting of 703 amino acids and a C-terminal membrane domain of 534 amino acids. The membrane domain has at least 10 hydrophobic runs of 20 or more amino acids that might form transmembrane segments. There is only 33% homology with the cytoplasmic domain of band 3 at the amino acid level while there is a 67% homology with the membrane associated domain of band 3 [91]. Alper and colleagues have expressed the protein in COS cells and have showed that it is a glycoprotein and is processed from 125 kDa up to 155 kDa as it is transported from cis Golgi membranes to the plasma membrane. There are three consensus sequences for N-linked glycosylation in the B3RP whereas there are two in human and mouse band 3 and only one is glycosylated. Alper and colleagues have demonstrated functional expression B3RP mRNA by injecting it into the cytoplasm of *Xenopus laevis* oocytes [81]. The ^{36}Cl influx was increased two- to three fold above water-injected oocyte controls. There are no known inhibitors of this anion transport though stilbenes and chloride channel blockers have been tried. It is now known that this is an anion exchange protein. [Lee B.S., Gunn, R.B., Kopito R.R. (1991) J Biol Chem 266:11448–11454]. Antibodies to peptides designed from the cDNA sequence and not found in band 3 have been used to localize this band 3 related protein in the kidney. It is found on the apical membrane of principal cells of the distal convoluted tubule and the cortical collecting duct. These cells also contain the amiloride-sensitive Na channel whose function is increased by vasopressin. The rat brain AE2 cDNA is restricted to the basolateral membrane of the choroid plexus. It has been functionally expressed in COS cells [83].

Other Band-3-Related Proteins

Recently a third gene (AE3) of the band 3 family has been cloned and sequenced [81]. It is a neuronal homolog of band 3, and its mRNA is found in the cortical layers II and V, deep pontine nuclei, median raphe nucleus, hippocampal granule cells of the dentate gyrus and hypoglossal nucleus. In addition the mRNA is found in the heart [26]. While the function of the protein product is not known in these tissues it has been expressed in COS cells where it behaves as a Na-independent anion exchanger that can extrude alkali in the presence of bicarbonate [144].

Lymphocytes have an anion exchanger that extrudes bicarbonate and is stimulated by an increase in intracellular pH [18, 85] and thereby is coupled to Na/H exchange. This protein and its gene have not isolated.

Parietal cells also have a pH sensitive anion exchanger which is activated further by secretagogues for HCl secretion [142]. The function of this chloride-bicarbonate exchange is presumably to maintain intracellular pH in the face of massive secretion of acid equivalents into the stomach.

Other cells with anion exchangers that are independent of external cations include cultured cell lines [109], cardiac myocytes, and smooth muscle cells. But the most interesting cases surely remain undiscovered.

References

1. Alper SL, Brosius FCIII, Garcia AM, Gluck S, Brown D, Lodish HF (1989) Two gene products encoding putative anion exchangers of the kidney. Ann NY Acad Sci 54:102–103
2. Alper SL, Kopito RR, Libresco SM, Lodish HF (1988) Cloning and characterization of a murine band 3-related cDNA from kidney and from a lymphoid cell line. J Biol Chem 263:17092–17099
3. Alper SL, Natale J, Gluck S, Lodish HF, Brown D (1989) Intercalated cell subtypes in rat kidney collecting duct defined using antibodies against erythrocyte band 3 and renal vacuolar H^+ATPase. Proc Natl Acad Sci USA 86:5429–5433
4. Anderson RA, Lovrein RE (1984) Glycophorin is linked by band 4.1 protein to the human erythrocyte membrane skeleton. Nature 307:655–658
5. Appell KC, Low PS (1981) Partial structural characterization of the cytoplasmic domain of the erythrocyte membrane protein, band 3. J Biol Chem 256:11104–11111
6. Arese P, Bosia A, Tescarmona GP, Till U (1981) The connection between ionophore mediated calcium movements and intermediary metabolism in human red cells. Cell Calcium 2:509–524
7. Becker BF, Duhm J (1978) Evidence for anionic cation transport of lithium, sodium and potassium across the human erythrocyte membrane induced by divalent anions. J Physiol (Lond) 282:149–168
8. Bennett, V, Stenbuck PJ (1979) The membrane attachment protein for spectrin is associated with band 3 in human erythrocyte membranes. Nature 280:468–473
9. Bjerrum PJ, Andersen OS, Borders CL Jr, Wieth JO (1989) Functional carboxyl groups in the red cell anion exchange protein. Modification with an impermeant carbodiimide. J Gen Physiol 93:813–839
10. Bjerrum PJ, Wieth JO, Borders CL Jr (1983) Selective phenylglyoxalation of functionally essential arginyl residues in the erythrocyte anion transport protein. J Gen Physiol 81:453–484
11. Boodhoo A, Reithmeier RAF (1984) Characterization of matrix-bound band 3, the anion transport protein from human erythrocyte membranes. J Biol Chem 259:785–790
12. Brahm J, Gronlund J, Hlastala MP, Ohlsson J, Swenson ER (1987) Effects of in vivo inhibition of red-cell anion exchange and Haldane effect on carbon dioxide output (V CO2) in the dog lung. J Physiol (Lond) 390:245P–245P (abstract)

13. Bretscher MS (1971) A major protein which spans the human erythrocyte membrane. J Mol Biol 59:351–357
14. Brock CJ, Tanner MJA, Kempf C (1983) The human erythrocyte anion-transport protein: partial amino acid sequence, conformation and a possible molecular mechanism for anion exchange. Biochem J 213:577–586
15. Brosius FC III, Alper SL, Garcia AM, Lodish HF (1989) The major kidney band 3 gene transcript predicts an amino-terminal truncated band 3 polypeptide. J Biol Chem 264:7784–7787
16. Cabantchik IZ, Balshin M, Breuer W, Rothstein A (1975) Pyridoxal phosphate. An anionic probe for protein amino groups exposed on the outer and inner surfaces of intact human red blood cells. J Biol Chem 250:5130–5136
17. Cabantchik ZI, Rothstein A (1974) Membrane proteins related to anion permeability of human red blood cells. I. Localization of disulfonic stilbene binding sites in proteins involved in permeation. J Membr Biol 15:207–226
18. Chaillet JR, Amsler K, Boron WF (1986) Optical measurements of intracellular pH in single LLC-PK1 cells: demonstration of Cl-HCO3 exchange. Proc Natl Acad Sci USA 83:522–526
19. Chappell JB, Crofts AR (1965) Ion transport and reversible volume changes of isolated mitochondria. In: Tager JM, Papa S, Quagliariello E, Slater EC (eds) BBA Library, regulation of metabolic processes in mitochondria. Elsevier, Amsterdam, pp 302–316
20. Cornish-Bowden A (1979) Fundamentals of enzyme kinetics. Butterworth, London
21. Cox JV, Lazarides E (1988) Alternative primary structures in the transmembrane domain of the chicken erythroid anion transporter. Mol Cell Biol 8:1327–1335
22. Cox JV, Moon RT, Lazarides E (1985) Anion transporter: highly cell-type-specific expression of distinct polypeptides and transcripts in erythroid and non-erythroid cells. J Cell Biol 100:1548–1557
23. Da Silva PP, Nicolson GL (1974) Freeze-etch localization of concanavalin A receptors to the membrane intercalated particles of human erythrocyte ghost membranes. Biochim Biophys Acta 363:311–319
24. Dalmark M, Wieth JO (1970) Chloride and sodium permeabilities of human red cells. Biochim Biophys Acta 219:525–527
25. Dalmark M, Wieth JO (1972) Temperature dependence of chloride, bromide, iodide, thiocyanate and salicylate transport in human red cells. J Physiol (Lond) 224:583–610
26. De Weer P, Breitwieser GE, Smith HG, Kennedy B (1983) ADP-ATP exchange in internally dialyzed squid giant axons. Curr Top Membr Transp 19:665–669
27. Dekowski SA, Rybicki A, Drickamer K (1983) A tyrosine kinase associated with the red cell membrane phsophorylates band 3. J Biol Chem 258:2750–2753
28. Demuth DR, Showe LC, Ballantine M, Palumbo A, Fraser PJ, Cioe L, Rovera G, Curtis PJ (1986) Cloning and structural characterization of a human non-erythroid band 3-like protein. EMBO J 5:1205–1214
29. Dirken MNJ, Mook HW (1931) The rate of gas exchange between blood cells and serum. J Physiol (Lond) 73:349–360
30. Downward J, Yarden Y, Mayes E, Scrace G, Stockwell P, Ullrich A, Schlessinger J, Waterfield MD (1984) Close similarity of epidermal growth factor receptor and v-erb-B oncogene protein sequences. Nature 307:521–527
31. Drenckhahn D, Franz H (1986) Identification of actin-, alpha-actinin-, and vinculin-containing plaques at the lateral membrane of epithelial cells. J Cell Biol 102:1843–1852
32. Drenckhahn D, Schluter K, Allen DP, Bennett V (1985) Colocalization of band 3 with ankyrin and spectrin at the basal membrane of intercalated cells in the rat kidney. Science 230:1287–1289
33. Fincham DA, Teoh R, Mason DK, Young JD (1989) Glycine and L-glutamate transport by the band 3 anion-exchange transporter in human and in horse erythrocytes. Biochem Soc Trans 17:541–542
34. Freedman J (1984) Membrane-bound immunoglobulins and complement components on young and old red cells. Transfusion 24:477–481
35. Frohlich O (1984) Relative contributions of the slippage and tunneling mechanisms to anion net efflux from human erythrocytes. J Gen Physiol 84:877–893
36. Frohlich O, Gunn RB (1986) Erythrocyte anion transport: the kinetics of a single-site obligatory exchange system. Biochim Biophys Acta 864:169–194
37. Frohlich O, King PA (1988) Mechanisms of anion net transport in the human erythrocyte. In

Gunn RB, Parker JC (eds) Cell physiology of blood. Rockefeller University Press, New York, pp 181–192
38. Funder J, Tosteson DC, Wieth JO (1978) Effects of bicarbonate on lithium transport in human red cells. J Gen Physiol 71:721–746
39. Golan DE, Veatch W (1980) Lateral mobility of band 3 in the human erythrocyte membrane studied by fluorescence photobleaching recovery: evidence for control by cytoskeletal interactions. Proc Natl Acad Sci USA 77:2537–2541
40. Gottipaty V, Fröhlich O (1988) Effects of extracellular sulfonates on chloride net efflux. Biophys J 53:531a–531a (abstract)
41. Gottipaty VK (1988) Characteristics of band 3-mediated anion net fluxes in human erythrocytes. Thesis, Emory University, Atlanta
42. Graf T, Beug H (1983) Role of the v-erbA and v-erbB oncogenes of avian erythroblastosis virus in erythroid cell transformation. Cell 34:7–9
43. Grinstein S, Ship S, Rothstein A (1978) Anion transport in relation to proteolytic dissection of band 3 protein. Biochim Biophys Acta 507:294–304.
44. Gunn RB (1972) A titratable carrier model for both mono- and di-valent anion transport in human red blood cells. In: Rorth M, Astrup P (eds) Oxygen affinity of hemoglobin and red cell acid base status. Academic, New York, pp 823–827
45. Gunn RB (1979) Transport of anions across red cell membranes. In: Giebish G, Tosteson DC, Ussing HH (eds) Membrane transport in biology II. Transport across single biological membranes. Springer, Berlin Heidelberg New York, pp 59–80
46. Gunn RB (1989) Buffer equilibria in red blood cells. In: Seldin DW, Giebisch G (eds) The regulation of acid-base balance. Raven, New York, pp 57–67
47. Gunn RB, Dalmark M, Tosteson DC, Wieth JO (1973) Characteristics of chloride transport in human red blood cells. J Gen Physiol 61:185–206
48. Gunn RB, Frohlich O (1979) Asymmetry in the mechanism for anion exchange in human red blood cell membranes: evidence for reciprocating sites that react with one transported anion at a time. J Gen Physiol 74:351–374
49. Haest CWM, Kamp D, Deuticke B (1979) Formation of disulfide bonds between glutathione and membrane SH groups in human erythrocytes. Biochim Biophys Acta 557:363–371
50. Hamburger HJ (1892) Über de Einfluss der Athmung auf die Permeabilität der roten Blutkörperchen. Z Biol 28:405
51. Hasselbalch KA (1916) Die Berechnung der Wasserstoffzahl des Blutes aus der freien und gebundenen Kohlensäure desselben, und die Sauerstoffbindung des Blutes als Funktion der Wasserstoffzahl. Biochem. Z 78:112–144
52. Hunter MJ (1971) A quantitative estimate of the non-exchange-restricted chloride permeability of the human red cell. J Physiol (Lond) 218:49P–50P (abstract)
53. Hunter MJ (1977) Human erythrocyte anion permeabilities measured under conditions of net charge transfer. J Physiol (Lond) 268:35–49
54. Hwang J-K, Warshel A (1988) Why ion pair reversal by protein engineering is unlikely to succeed. Nature 334:270–272
55. Jay DG (1983) Characterization of the chicken erythrocyte anion exchange protein. J Biol Chem 258:9431–9436
56. Jennings ML (1976) Proton fluxes associated with erythrocyte membrane anion exchange. J Membr Biol 27:187–205
57. Jennings ML (1978) Characteristics of CO2-independent pH equilibration in human red blood cells. J Membr Biol 40:365–391
58. Jennings ML (1980) Apparent "recruitment" of SO4 transport sites by the Cl gradient across the human erythrocyte membrane. In: Lassen UV, Ussing HH, Wieth JO (eds) Membrane transport in erythrocytes. Munksgaard, Copenhagen, pp 450–466
59. Jennings ML (1982) Stoichiometry of a half-turnover of band 3, the chloride transport protein of human erythrocytes. J Gen Physiol 79:169–185
60. Jennings ML (1982) Reductive methylation of the two 4,4′-diisothiocyanodihydrostilbene-2,2′-disulfonate-binding lysine residues of band 3, the human erythrocyte anion transport protein. J Biol Chem 257:7554–7559
61. Jennings ML (1989) Structure and function of the red blood cell anion transport protein. Annu Rev Biophys Biophys Chem 18:397–430
62. Jennings ML, Al-Rhaiyel S (1988) Modification of a carboxyl group that appears to cross the permeability barrier in the red blood cell anion transporter. J Gen Physiol 92:161–178

63 Jennings ML, Anderson MP, Monaghan R (1986) Monoclonal antibodies against human erythrocyte band 3 protein. J Biol chem 261:9002–9010
64. Kanes KJ (1989) Band 3 structure and function: 35-Cl NMR and topographical investigations. Thesis California Institute of Technology, Pasadena
65. Kaplan JH, Pring M, Passow H (1983) Band-3 protein-mediated anion conductance of the red cell membrane. Slippage vs. ionic diffusion. FEBS Lett 156:175–179
66. Karadsheh NA, Uyeda K (1977) Changes in allosteric properties of phosphofructokinase bound to erythrocyte membranes. J Biol Chem 252:7418–7420
67. Kay MMB (1975) Mechanism of removal of senescent cells by human macrophages in situ. Proc Natl Acad Sci USA 72:3521–3525
68. Kay MMB (1978) Role of physiologic autoantibody in the removal of senescent human red cell. J Supramol Struct 9:555–567
69. Kay MMB, Bosman GJCGM, Johnson GJ, Beth AH (1988) Band-3 polymers and aggregates, and hemoglobin precipitates in red cell aging. Blood Cells 14:275–289
70. Kay MMB, Goodman JR (1984) IgG antibodies do not bind to band 3 in intact erythrocytes. Enzymatic treatment of cells is required for IgG binding. Biomed Biochim Acta 43:841–846
71. Kim H-RC, Yew NS, Ansorge W, Voss H, Schwager C, Vennstrom B, Zenke M, Engel JD (1988) Two different mRNAs are transcribed from a single genomic locus encoding the chicken erythrocyte anion transport proteins (band 3). Mol Cell Biol 8:4416–4424
72. King PA, Gunn RB (1989) Na- and Cl-dependent glycine transport in human red blood cells and ghosts. A study of the binding of substrates to the outward-facing carrier. J Gen Physiol 93:321–342
73. Knauf PA (1979) Erythrocyte anion exchange and the band 3 protein: transport kinetics and molecular structure. Curr Top Membr Transp 12:249–363
74. Knauf PA, Law F-Y, Marchant PM (1983) Relationship of net chloride flow across the human erythrocyte membrane to the anion exchange mechanism. J Gen Physiol 81:95–126
75. Knauf PA, Mann NA (1984) Use of niflumic acid to determine the nature of the asymmetry of the human erythrocyte anion exchange system. J Gen Physiol 83:703–725
76. Knauf PA, Rothstein A (1971) Chemical modification of membranes. I. Effects of sulfhydryl and amino reactive reagents on anion and cation permeability of the human red blood cell. J Gen Physiol 58:190–210
77. Koeppen BM, Helman SI (1982) Acidification of luminal fluid by the rabbit cortical collecting tubule perfused in vitro. Am J Physiol 242:F521–F531
78. Kopito RR (1991) Molecular biology of the anion exchanger gene family. Int Rev Cytol 123:177–199
79. Kopito RR, Andersson MA, Lodish HF (1987) Multiple tissue-specific sites of transcriptional initiation of the mouse anion antiport gene in erythroid and renal cells. Proc Natl Acad Sci USA 84:7149–7153
80. Kopito RR, Andersson MM, Herzlinger DA, Al-Awqati Q, Lodish HF (1988) Structure and tissue-specific expression of the mouse anion-exchanger gene in erythroid and renal cells. In: Gunn RB, Parker JC (eds) Cell physiology of blood. Rockefeller University Press, New York, pp 151–161
81. Kopito RR, Lee BS, Simmons DM, Lindsey AE, Morgans CW, Schneider K (1989) Regulation of intracellular pH by a neuronal homolog of the erythrocyte anion exchanger. Cell 59:927–937
82. Kopito RR, Lodish HF (1985) Primary structure and transmembrane orientation of the murine anion exchange protein. Nature 316:234–238
83. Kudrycki KE, Newman PR, Schull GE (1990) cDNA cloning and tissue distribution of mRNAs for two proteins that are related to the band 3 Cl^-/HCO_3^- exchanger. J Biol Chem 265:462–471
84. Kudrycki KE, Shull GE (1989) Primary structure of the rat kidney band 3 anion exchange protein deduced from a cDNA. J Biol Chem 264:8185–8192
85. Kurtz I, Golchini K (1987) Na^+-independent Cl–HCO3 exchange in Madin-Darby canine kidney cells. Role in intracellular pH regulation. J Biol Chem 262:4516–4520
86. Labotka RJ, Omachi A (1987) Erythrocyte anion transport of Phosphate analogs. J Biol Chem 262:305–311
87. Labotka RJ, Omachi A (1988) The pH dependence of red cell membrane transport of titratable anions studied by NMR spectroscopy. J Biol Chem 263:1166–1173
88. Lambert A, Lowe AG (1978) Chloride/bicarbonate exchange in human erythrocytes. J Physiol (Lond) 275:51–63

89. Lepke S, Passow H (1976) Effects of incorporated trypsin on anion exchange and membrane proteins in human red blood cell ghosts. Biochim Biophys Acta 455:353–370
90. Lieberman DM, Reithmeier RAF (1988) Localization of the carboxyl terminus of band 3 to the cytoplasmic side of the erythrocyte membrane using antibodies raised against a synthetic peptide. J Biol Chem 263:10022–10028
91. Lindsey AE, Schneider K, Simmons DM, Baron R, Lee BS, Kopito RR (1990) Functional expression and subcellular localization of an anion exchanger cloned from choroid plexus. Proc Natl Acad Sci USA 87:5278–5282
92. Low PS, Westfall MA, Allen DP, Appell KC (1984) Characterization of the reversible conformational equilibrium of the cytoplasmic domain of the erythrocyte membrane band 3. J Biol Chem 259:13070–13076
93. Low PS, Willardson BM, Thevenin B, Kannan R, Mehler E, Geahlen RL, Harrison M (1990) The other functions of erythrocyte membrane band 3. In: Hamasaki N, Jennings ML (eds) Anion transport protein of the red cell membrane. Elsevier, Amsterdam
94. Luckner H (1939) Über die Geschwindigkeit des Austausches der Atemgase im Blut. Pfluger Arch Gesamte Physiol 241:753–778
95. Luckner H (1948) Die Temperaturabhangigkeit des Anionenaustausches roter Blutkorperchen. Pflugers Arch Gesamte Physiol 250:303–311
96. Lutz HU, Fasler S, Stammler P, Bussolino F, Arese P (1988) Naturally occurring anti-band 3 antibodies and complement in phagocytosis of oxidatively-stressed and in clearance of senescent red cells. Blood Cells 14:175–195
97. Lux SE, John KM, Kopito RR, Lodish HF (1989) Cloning and characterization of band 3, the human erythrocyte anion-exchange protein (AE1). Proc Natl Acad Sci USA 86:9089–9093
98. Milanick MA, Gunn RB (1982) Proton-sulfate co-transport. Mechanism of H^+ and sulfate addition to the chloride transporter of human red blood cells. J Gen Physiol 79:87–113
99. Milanick MA, Gunn RB (1984) Proton-sulfate cotransport: external proton activation of sulfate influx into human red blood cells. Am J Physiol 247:C247–C259
100. Milanick MA, Gunn RB (1986) Proton inhibition of chloride exchange: asynchrony of band 3 proton and anion transport sites? Am J Physiol 250:C955–C969
101. Mond R (1927) Die Umkehr der Anionenpermeabilität roter Blutkörperchen in eine elektive Durchlässigkeit fur Kationen. Pflugers Arch Gesamte Physiol 217:618
102. Mueller H, Lutz HU (1983) Binding of autologous IgG to human red blood cells before and after ATP-depletion. Selective exposure of binding sites (autoantigens) on spectrin-free vesicles. Biochim Biophys Acta 729:249–257
103. Mueller TJ, Morrison M (1977) Detection of a variant of protein 3, the major transmembrane protein of the human erythrocyte. J Biol Chem 252:6573–6576 (abstract)
104. Nanri H, Hamasaki N, Minakami S (1983) Affinity labeling of erythrocyte band 3 proteins with pyridoxal 5-phosphate involvement of the 35000 dalton fragment in anion transport. J Biol Chem 258:5985–5989
105. Nasse H (1878) Untersuchungen über den Austritt und Eintritt von Stoffen (Transudation und Diffusion) durch die Wand der Haargefässe. Pflugers Arch 16:604
106. Nigg EA, Cherry RJ (1980) Anchorage of a band 3 population at the erythrocyte cytoplasmic membrane surface. protein rotational diffusion measurements. Proc Natl Acad Sci USA 77:4702–4706
107. Nikinmaa M, Railo E (1987) Anion movements across lamprey (Lampetra fluviatilis) red cell membrane. Biochim Biophys Acta 899:134–136
108. Oikawa K, Lieberman DM, Reithmeier RAF (1985) Confirmation and stability of the anion transport protein of human erythrocyte membranes. Biochemistry 24:2843–2848
109. Palfrey HC, Rao MC (1983) Na/K/Cl co-transport and its regulation. J Exp Biol 106:43–54
110. Palumbo AP, Isobe M, Huebner K, Shane S, Rovera G, Demuth D, Curtis PJ, Ballantine M, Croce CM, Showe LC (1986) Chromosomal localization of a human band 3-like gene to region 7q35 to 7q36. Am J Hum Genet 39:307–316
111. Passow H (1969) Passive ion permeability of the erythrocyte membrane. Prog Biophys Mol Biol 19:425–467
112. Passow H (1986) Molecular aspects of band 3 protein-mediated anion transport across the red cell membrane. Rev Physiol Biochem Pharmacol 103:61–203
113. Restrepo D, Kozody DJ, Spinelli LJ, Knauf PA (1989) Cl-Cl exchange in promyelocytic HL-60 cells follows simultaneous rather than ping-pong kinetics. Am J Physiol 257:C520–C527
114. Salhany JM, Gaines KC (1981) Connections between cytoplasmic proteins and the erythrocyte membrane. Trends Biochem Sci 6:13–15

115. Salhany JM, Rauenbuehler PB (1983) Kinetics and mechanism of erythrocyte anion exchange. J Biol Chem 258:245–249
116. Salhany JM, Shaklai N (1979) Functional properties of human hemoglobin bound to the erythrocyte membrane. Biochemistry 18:893–899
117. Sap J, Munoz A, Damm K, Goldberg Y, Ghysdael J, Leutz A, Beug H, Vennstrom B (1986) The c-erb-A protein is a high-affinity receptor for thyroid hormone. Nature 324:635–640
118. Scarpa A, Cecchetto A, Azzone GF (1968) Permeability of erythrocytes to anion and the regulation of cell volume. Nature 219:529–531
119. Schmidt A (1987) Über die Kohlensäure in der Blutkörperchen. Ber k Sachs Ges Wiss (Math-physikal Cl I) 19:30
120. Schnell KF (1977) Anion transport across the red blood cell membrane mediated by dielectric pores. J Membr Biol 37:99–136
121. Schuster VL, Bonsib SM, Jennings ML (1986) Two types of collecting duct mitochondria-rich (intercalated) cells: lectin and band 3 cytochemistry. Am J Physiol 251:C347–C355
122. Shoemaker DG, Bender CA, Gunn RB (1987) Phosphate transport in human red cells. J Gen Physiol 90:37a–37a (abstract)
123. Shoemaker DG, Bender CA, Gunn RB (1988) Sodium-phosphate cotransport in human red blood cells. J Gen Physiol 92:449–474
124. Showe LC, Ballantine M, Huebner K (1987) Localization of the gene for the erythroid anion exchange protein, band 3 (EMPB3), to human chromosome 17. Genomics 1:71–76
125. Simons TJB (1986) The role of anion transport in the passive movement of lead across the human red cell membrane. J Physiol (Lond) 378:287–312
126. Srivastava SK, Beutler E (1969) Transport of oxidized glutathione from human erythrocytes. J Biol Chem 244:9–16
127. Steck TL (1978) The band 3 protein of the human red cell membrane: a review. J Supramol Struct 8:311–324
128. Stone DK, Seldin DW, Kokko JP, Jacobson HR (1983) Anion dependence of rabbit medullary collecting duct acidification. J Clin Invest 71:1505–1508
129. Strapazon E, Steck TL (1976) Binding of rabbit muscle aldolase to Band 3, the predominant polypeptide of the human erythrocyte membrane. Biochemistry 15:1421–1424
130. Swanson ML, Keast RK, Jennings ML, Pessin JE (1988) Heterogeneity in the human erythrocyte Band 3 anion-transporter revealed by Triton X-114 phase partitioning. Biochem J 255:229–234
131. Swick-Vogelaar N (1989) Structural and mechanistic motifs in membrane proteins: the three-dimensional modeling of rhodopsin, band 3 and nicotinic acetylcholine receptor. Thesis, California Institute of Technology, Pasadena
132. Tanner MJA, Martin PG, High S (1988) The complete amino acid sequence of the human erythrocyte membrane anion-transport protein deduced from the cDNA sequence. Biochem J 256:703–712
133. Torrubia JOA, Garay R (1989) Evidence for a major route for zinc uptake in human red blood cells: [Zn(HCO3)2Cl]-influx through the [Cl – /HCO3 –] anion exchanger. J Cell Physiol 138:316–322
134. Tosteson DC (1959) Halide transport in red blood cells. Acta Physiol Scand 46:19–46
135. Tsai I-H, Murthy SN, Steck TL (1981) Effect of red cell membrane binding on the catalytic activity of glyceraldehyde-3-phosphate dehydrogenase. J Biol Chem 257:1438–1442
136. Tsuji A, Kawasaki K, Ohnishi S-I (1988) Regulation of band 3 mobilities in erythrocyte ghost membranes by protein association and cytoskeletal meshwork. Biochemistry 27:7447–7452
137. Tsuji A, Ohnishi S (1986) Restriction of the lateral motion of band 3 in the erythrocyte membrane by the cytoskeletal network: dependence on spectrin association state. Biochemistry 25:6133–6139
138. Tyler JM, Reinhardt BN, Branton D (1980) Associations of erythrocyte membrane proteins: binding of purified bands 2.1 and 4.1 to spectrin. J Biol Chem 255:7034–7039
139. Wagner S, Vogel R, Lietzke R, Koob R, Drenckhahn D (1987) Immunochemical characterization of a band 3-like anion exchanger in collecting duct of human kidney. Am J Physiol 253:F213–F221
140. Wainwright SD, Tanner MJA, Martin GEM, Yendle JE, Holmes C (1989) Monoclonal antibodies to the membrane domain of the human erythrocyte anion transport protein: localization of the C-terminus of the protein to the cytoplasmic side of the red cell membrane and distribution of the protein in some human tissues. Biochem J 258:211–220

141. Walder JA, Chatterjee R, Steck TL, Low PS, Musso GF, Kaiser ET, Rogers PH, Arnone A (1984) The interaction of hemoglobin with the cytoplasmic domain of band 3 of the human erythrocyte membrane. J Biol Chem 259:10 238–10 246
142. Wallert MA (1989) pH regulation in adult cardiac myocytes. Thesis, Emory University, Atlanta
143. Warburg EJ (1922) Studies on carbonic acid compounds and hydrogen ion activities in blood and salt solutons. A contribution to the theory of the equation of Lawrence J. Henderson and K. A. Hasselbalch. Biochem J 16:153–340
144. Wenzl E, Machen TE (1989) Intracellular pH dependence of buffer capacity and anion exchange in the parietal cell. Am J Physiol 257:G741–G747
145. Wieth JO, Bjerrum PJ (1982) Titration of transport and modifier sites in the red cell anion transport system. J Gen Physiol 79:253–282
146. Wieth JO, Bjerrum PJ, Borders CL Jr (1982) Irreversible inactivation of red cell chloride exchange with phenylgloxal, an arginine specific reagent. J Gen Physiol 79:283–312
147. Wieth JO, Bjerrum PJ, Brahm J, Andersen OS (1982) The anion transport protein of the red cell membrane. A zipper mechanism of anion exchange. Tokai J Exp Clin Med 7 [Suppl]:91–101
148. Wieth JO, Brahm J (1985) Cellular anion transport. In: Seldin DW, Giebisch G (eds) The kidney: physiology and pathophysiology vol 1. Raven, New York, pp 49–89
149. Zaki L, Julien T (1985) Anion transport in red blood cells and arginin specific reagents. Interaction between the substrate binding site and the binding site of arginine specific reagents. Biochim Biophys Acta 818:325–332
150. Zenke M, Kahn P, Disela C, Vennstrom B, Leutz A, Keegan K, Hayman MJ, Choi H-R, Yew N, Engel JD, Beug H (1988) v-erbA specifically suppresses transcriptions of the avian erythrocyte anion transporter (band 3) gene. Cell 52:107–119
151. Zuntz N (1867) Über den Einfluss des Partialdrucks der Kohlensäure auf die Verteilung dieses Gases im Blute. Centr Med Wiss 529

Cellular Calcium Transport Systems *

J.V. Bonventre

Introduction

In the 1880s Sidney Ringer [260–263] demonstrated that calcium was required for contraction of the frog heart, adhesion of cells to each other, and development of fertilized eggs and tadpoles. After Ringer's experiments, many physiologists confirmed the importance of Ca^{2+} for maintenance of the morphologic and functional integrity of animal tissues and for the regulation of multiple cellular processes [49].

Processes such as excitation-contraction coupling [354], endocytosis and exocytosis [182], cell motility [361], fertilization and cell division [248], hormone-effector systems [27], growth factor effector systems [29, 219], and various aspects of intermediary metabolism [92, 135, 351] are modulated by Ca^{2+}. Ca^{2+} binds to phospholipids and other membrane components, proteins, and nucleic acids. Through these interactions multiple structural, locomotive, and metabolic characteristics of cells are regulated.

In this chapter I will review the various Ca^{2+} transport and homeostatic processes that are important for cells to maintain the steady state as well as respond to stimuli in a specific way. I will exclude a detailed discussion of active Ca^{2+} ATPases. There has been a great deal of work devoted to the molecular characterization of many types of Ca^{2+} ATPases. Since this falls beyond the scope of the chapter, the reader is referred to many excellent recent reviews of Ca^{2+} ATPases [53, 55, 89, 93, 100, 167].

While there is a considerable amount of information regarding Ca^{2+} homeostasis and transport in epithelial tissues a great deal of work has been done in other cells and tissues. Since the general paradigms are similar I will draw from work in both epithelial and non epithelial tissues in an attempt to provide an overview of this large and rapidly expanding topic.

* The author's laboratory has been supported by the following grants from the National Institutes of Health: DK 39773, DK 38452, DK 38165, NS 10828. The author is an Established Investigator of the American Heart Association.

Intracellular Calcium Homeostasis

A large Ca^{2+} concentration gradient is maintained between the cytosolic compartment of eukaryotic cells (approximately 10^{-7} *M*) and the external milieu, where the free Ca^{2+} concentration is in the millimolar range (10^{-3} *M*). This Ca^{2+} gradient is maintained across the cell plasma membrane because of efficient energy-dependent Ca^{2+} extrusion mechanisms and low Ca^{2+} permeability [10, 283]. Primary pathways for Ca^{2+} movement across the plasma membrane and between various intracellular organelles are depicted schematically in Fig.1.

Although the cytosolic free Ca^{2+} ($[Ca^{2+}]_i$) is in the submicromolar range, total cell Ca^{2+} is in the millimolar range. Ca^{2+} can be bound to cytosolic Ca^{2+}

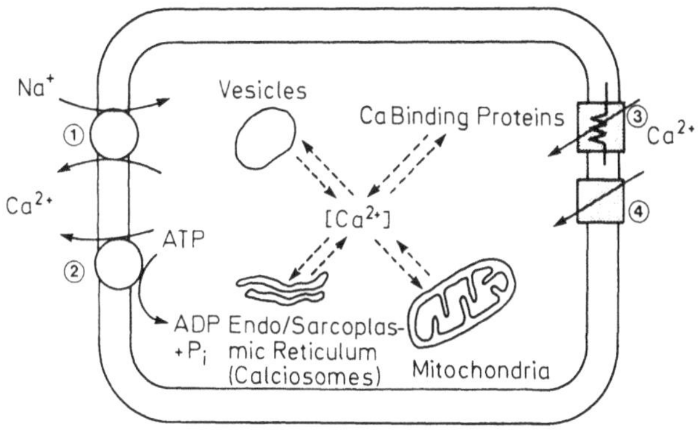

Fig. 1. Ca^{2+} transport processes. Summary of the primary transport processes involved with calcium homeostasis. Ca^{2+} can leave the cell via a Na^+/Ca^{2+} exchange process (*1*). Energy is provided for this exchange by the Na^+K^+ ATPase pump which establishes a Na^+ gradient across the plasma membrane. If the inwardly directed electrochemical gradient for Na^+ is greater than the electrochemical gradient for Ca^{2+} then the exchanger will function to move Ca^{2+} out of the cell. Under certain circumstances, in some cells, this exchanger may function in the reverse direction serving to move Ca^{2+} into the cell. A second mechanism for Ca^{2+} extrusion is the Ca^{2+} ATPase pump (*2*) which uses the energy of ATP. Ca^{2+} can enter the cell down its electrochemical concentration gradient via voltage-dependent (*3*) or voltage-independent (leak or receptor-operated) pathways (*4*).

Within the cell Ca^{2+} cycles in and out of the mitochondrial and endoplasmic reticulum compartments. There is some controversy about the composition of the nonmitochondrial storage sites. Specific stores, called calciosomes, have been proposed to be present. Ca^{2+} can also cycle in and out of nonendoplasmic reticular vesicular stores. There are also calcium-binding proteins in the cell which serve as Ca^{2+} storage depots but also act, along with Ca^{2+}, to transduce a change in intracellular Ca^{2+} concentration into a biochemical event altering cell function. These intracellular Ca^{2+} binding proteins may also be important for the regulation of transepithelial Ca^{2+} transport. $[Ca^{2+}]$, cytosolic free Ca^{2+} concentration.

The nucleus is not shown on this figure. Recent work in isolated nuclei suggests that the nuclear membrane represents a permeability barrier to calcium and that there is a nuclear membrane Ca^{2+}-ATPase distinct from that present in the plasma membrane and endoplasmic reticulum (235b). (From [28])

-binding proteins and cellular membranes or be sequestered within intracellular membrane-limited pools. It is likely that only a small amount of Ca^{2+} is bound to soluble Ca^{2+}-binding proteins. It is generally believed that these proteins are not primarily involved in the buffering of Ca^{2+} but rather are involved in the processing of the Ca^{2+} signal and, in epithelial cells, may be involved in the transcellular transport of Ca^{2+}. Such proteins include, but are not limited to, troponin C [98], calmodulin [71], the calpactins [88], and the intestinal Ca^{2+} binding protein [343, 344]. By contrast, negatively charged phospholipids can bind large amounts of Ca^{2+}. Thus cell membranes may constitute a major Ca^{2+}-binding substrate.

There are Ca^{2+}-transporting processes in the plasma membrane and in the intracellular membranes of the mitochondria and the endoplasmic and sarcoplasmic reticula. Less well defined are potential Ca^{2+}-transporting processes in the Golgi membranes [341], lysosomes [173], and nuclear membranes [344a, 348].

In the plasma membrane there is a Ca^{2+}-ATPase [283] that transports Ca^{2+} from inside to outside the cell. There is also a Na^+-Ca^{2+} exchanger [10] that can result in Ca^{2+} extrusion secondary to the energy harvested from extracellular Na^+ moving into the cell down its electrochemical gradient. It is generally believed that these processes act to maintain low levels of intracellular $[Ca^{2+}]$ although under certain conditions, in some cells, the Na^+/Ca^{2+} exchanger may function in a reverse configuration to drive Ca^{2+} into the cells [296]. There are also Ca^{2+} channels [321], which when open, allow Ca^{2+} to move into the cell. The channels may be dependent upon membrane potential (voltage-dependent) or voltage independent. In general less is known about the physiology and regulation of the voltage-independent channels. At the present time little is known about voltage-dependent Ca^{2+} channels in epithelial cells.

Mitochondria have a Ca^{2+} uniport system that transports Ca^{2+} into the mitochondria [195]. There is also a Na^+-Ca^{2+} exchanger that results in the Na^+-coupled release of Ca^{2+} from the mitochondria of excitable tissues [82]. Mitochondria isolated from nonexcitable tissues, such as liver and kidney, have a Na^+ independent Ca^{2+}-H^+ exchange system [82, 111].

The endoplasmic and sarcoplasmic reticulum have Ca^{2+}-ATPase uptake mechanisms distinct from that of the plasma membrane. There are also Ca^{2+}release processes that are incompletely characterized in the endoplasmic and sarcoplasmic reticula [66, 70].

In order to appreciate the physiological importance of these various Ca^{2+} transporting processes it is important to understand how each contributes to Ca^{2+} homeostasis under basal and activated states. The cell plasma membrane Ca^{2+} transporting processes are clearly critical to overall cellular Ca^{2+} balance since an imbalance between influx and efflux could rapidly lead to a Ca^{2+}depleted cell or a Ca^{2+} overloaded cell. The intracellular membrane transport properties are also critical, however, in the fine regulation of cytosolic and nuclear free Ca^{2+}, in the mediation of the Ca^{2+} response to various cell activators, and to the storage of Ca^{2+} under conditions where excess Ca^{2+} has been introduced into the cell.

Under normal conditions it is clear that the nonmitochondrial storage sites play a more important role in the buffering of Ca^{2+} than the mitochondria. The Michaelis-Menten kinetics constant (K_m) for Ca^{2+} for the mitochondrial Ca^{2+} uptake reaction is greater than $10 \mu M$, too high to be important for the submicromolar regulation of cytosolic free [Ca^{2+}]. We have shown that in kidney cells there is little Ca^{2+} in mitochondrial stores under physiologic conditions [70]. This has also been demonstrated by X-ray microanalysis in a number of tissues [193, 297, 298]. If the cells are Ca^{2+} loaded, however, then the mitochondria become important storage sites, representing a very high-capacity system for buffering Ca^{2+} [24].

Elevated levels of $[Ca^{2+}]_i$ may initiate a variety of cellular processes that are potentially detrimental to the cell [26, 28]. Under normal circumstances the cell, by maintaining low levels of $[Ca^{2+}]_i$ can avoid these effects; however, when the cell is stressed, $[Ca^{2+}]_i$ may increase into the range in which damaging processes are activated.

Each of the cellular Ca^{2+} transport processes, with the exception of the Ca^{2+} ATPases, will be examined in detail in subsequent sections of this chapter, as will the various roles of the mitochondrial and non mitochondrial Ca^{2+} storage depots.

Plasma Membrane Calcium Channels

A calcium channel is an ion channel with a high selectivity to calcium but which may pass other ions as well. High calcium selectivity is due to ion-channel as well as ion-ion interactions. Ca^{2+} ions effectively block flow of other ions through the channel by binding to a portion of the channel and allosterically altering the protein such that binding of other cations is altered [258]. Calcium movement through a pore is promoted by ion-ion repulsive interactions [145, 209]. The channels become permeable to Na^+ and other ions only when extracellular free Ca^{2+} concentration is very low ($< 10^{-6}$ M). The introduction of the patch-clamp technique, which permits single channel conductance characteristics to be determined, has contributed greatly to our understanding of Ca^{2+} channels.

Classically calcium channels have been subgrouped into the following primary categories: voltage dependent channels, agonist-activated channels and leak channels. This separation is somewhat artificial, however, since voltage dependent channels can be affected by agonist-receptor interactions. Furthermore voltage changes can have effects on receptor-operated channel activity as occurs with the *N*-methyl-D-aspartate (NMDA) form of the glutamate receptor in the brain. Agonist-activated channels can be further subdivided into two types: those which are opened directly through the receptor (receptor operated channels), and those opened indirectly through a diffusible second messenger (second-messenger-operated channel) [214]. These channels, when present in the plasma membrane can mediate movement of Ca^{2+} into or out of the cell. Furthermore Ca^{2+} release from intracellular stores can be mediated via Ca^{2+} channels in the membranes of intracellular organelles.

Voltage-Dependent Calcium Channels

Voltage-dependent calcium channels exist in heart, skeletal muscle, neurons, and some secretory cells [300]. There are at least three distinct types of calcium channels in neurons (L, N and T types [239, 321, 322]) and two in smooth muscle cells [14, 16, 202, 304] and heart cells [13, 237]. Channels are subtyped based upon: voltage; single channel kinetics and conductance; relative conductance of Ca^{2+} and Ba^{2+}; blockage with Ni^{2+} and Cd^{2+} and dihydropyridine sensitivity [322]. To date, potential activated Ca^{2+} channels have not been definitively demonstrated in epithelial cells [32]. Van Driessche and colleagues [334, 335] reported data suggesting the presence of Ca^{2+} channels in the apical membrane of the toad urinary bladder. When Ag^{+} was added to the mucosal side there was a marked increase in transepithelial current and conductance. La^{3+}, Cd^{2+}, verapamil and nicardipine, agents which inhibit Ca^{2+} channels in excitable tissue, inhibited the short circuit current. The rapidity of the response to Ag^{+}, as well as noise analysis, suggested that Ca^{2+} channels were present in the mucosal membrane. Ca^{2+} can be substituted by Ba^{2+} or Mg^{2+} and the short circuit current response is preserved.

Bouchelouche et al. [32] studied gall bladder epithelial cell Ca^{2+} entry characteristics using BAY K8644, an agent which enhances Ca^{2+} permeability through activation of preexisting potential-gated Ca^{2+} channels [146]. BAY K8644 increased $[Ca^{2+}]_i$, but two to three orders of magnitude higher concentrations were required when compared with the concentrations necessary to evoke similar increases in $[Ca^{2+}]_i$ in excitable cells. In addition the BAY K8644 effect was not blocked with verapamil and $[Ca^{2+}]_i$ did not increase in response to cellular depolarization. These results led the investigators to conclude that gall bladder epithelial cells do not have potential gated Ca^{2+} channels. There has been one preliminary recent report by Saunders and Isaacson [279] who studied calcium channel activity in the apical membranes of perfused isolated rabbit proximal straight tubules and cortical collecting ducts. Their data were consistent with a voltage-dependent Ca^{2+} channel of 12 ps, but the data are too preliminary to unequivocally establish the existence of these channels.

Much of the detailed characteristics of the various types of voltage dependent Ca^{2+} channels have been defined by Tsien's group in the chick sensory neuron. Hipocampal neurons also contain each of the three types while other neurons contain some but not each type.

T-Type Ca^{2+} Channels. These channels activate at very negative holding potentials, require only minimal depolarization for activation, carry very small Ba^{2+} currents and have a Ba^{2+} conductance that is roughly equal to that of Ca^{2+}. The T channel has been described as a low conductance (~ 8pS) and "low-threshold" channel [177]. These channels inactivate rapidly and are resistant to dihydropyridines [322].

N-type Ca^{2+} Channels. Like T-channels, N-channel currents are found at negative holding potentials, but unlike T channels the N channel currents require strong depolarizations to be seen. The conductance to Ba^{2+} is greater than that of Ca^{2+}. The N-channel inactivation rate is intermediate between that of the T and L-channels. The N-channel is dihydropyridine resistant and at present has only been identified in neurons [322].

L-Type Ca^{2+} Channels. These channels activate with strong depolarizations. They carry large Ba^{2+} currents and have a Ba^{2+} conductance that is greater than that of Ca^{2+}. These channels inactivate very slowly and are sensitive to dihydropyridine agonists and antagonists [322].

In smooth muscle cells a number of groups [14, 16, 202, 304] have identified two types of voltage dependent Ca^{2+} channels. These are similar to the T and L channels of neurons. One of the channel types (L) inactivates slowly while the other (T) rapidly inactivates. The L currents were blocked by dihydropyridine antagonists and stimulated by dihydropyridine agonists, such as BAY K8644. By contrast the T currents were not affected by the dihydropyridine agonists and antagonists. Channels with similar characteristics have been found in heart cells [236] and secretory cells [204]. T-channels have been found to be important for pacemaker depolarization in the heart [132] whereas the L-type dihydropyridine-sensitive Ca^{2+} channel is the predominant type of Ca^{2+} channel in the heart. Blockage of the cardiac L channel with dihydropyridines or phenylalkylamines decreases, and can completely suppress, cardiac contractility [113]. L-type channels are even much more abundant in skeletal muscle where they are present in large amounts in the transverse tubule membranes [115].

Structure of the Dihydropyridine-Sensitive Ca^{2+} Channel from Skeletal Muscle. The dihydropyridine-receptor has been purified from detergent-solubilized transverse tubule membranes from skeletal muscle [30, 31, 83, 85, 114, 196, 227, 233, 303, 330]. A schematic representation of the dihydropyridine sensitive channel is presented in Fig. 2. Various subunits of the Ca^{2+} channel have been purified and cloned. The channel is composed of at least four polypeptides which migrate with molecular weights of 175, 170, 52, and 32 kDa on non-reducing SDS-polyacylamide gels. These subunits have been designated α_1, α_2, β, and γ. The α_2 subunit is decreased in size under reducing conditions. It has been proposed by some investigators [48] that reduction of the α_2 subunit also liberates a γ subunit which has been observed by many labs. The α_1, α_2, β, and γ subunits appear to exist in a stoicheometry of 1:1:1:1; in the intact receptor. It

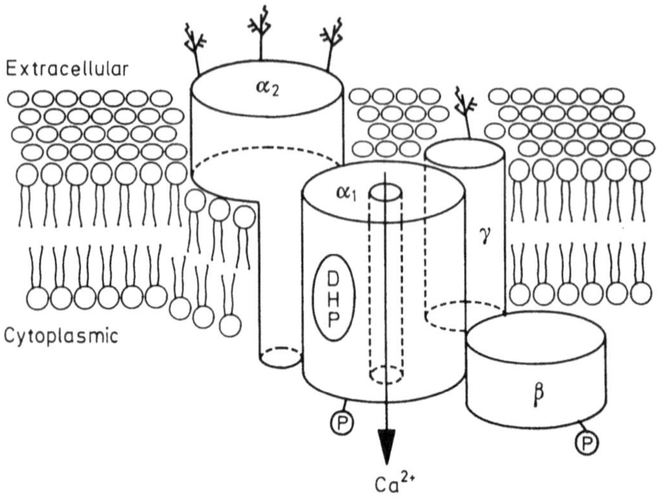

Fig. 2. A schematic representation of the dihydropyridine-sensitive channel. The symbols emanating from the α_2 and γ subunits represent glycosylation sites. *P*, Phosphorylation sites. The Ca^{2+} channel activity is believed to reside in the α_1 subunit. The α_1 subunit contains the binding site for the dihydropyridines. (From [48])

has been proposed [48] that the α_1 subunit is the major transmembrane spanning subunit which contains the binding sites for dihydropyridines, phenylalkylamines, and benzothiazepines, and forms the ion conducting pore [292, 330]. This subunit is a substrate at Ser^{687} for cAMP-dependent protein kinase [84, 153, 154, 159, 234, 267], Ca^{2+}-calmodulin dependent protein kinase [153], and protein kinase C [234]. Protein kinase C-induced phosphorylation of α_1 can be reversed by the Ca^{2+}-dependent phosphatase calcineurin [153].

The α_1 subunit has been cloned and has high sequence homology with the α subunit of the voltage sensitive Na^+ channel (316). Microinjection of the expression vector for the α_1 subunit cDNA into dysgenic muscle cells restores a dihydropyridine-sensitive Ca^{2+} current and excitation-contraction coupling [317]. The α_2 subunit also contains a transmembrane domain. This subunit is heavily glycosylated [124, 196, 312], does not bind to dihydropyridines, and is not a substrate for protein kinases [48]. The α_2 subunit has been cloned [350], and this subunit is present in dysgenic muscle cells [174].The β subunit is believed to be primarily cytosolic and is also a substrate for cAMP and cGMP dependent protein kinases [159, 234] and protein kinase C [234]. The β subunit has been cloned from rabbit skeletal muscle [273]. There are no typical membrane-spanning regions when the amino acid sequence is examined for local hydropathicity. The β subunit is tightly associated with the α_1 subunit and monoclonal antibodies against β have been found to activate the Ca^{2+} channel in a reconstitution system in lipid bilayers [340]. Thus the β unit may be important for the regulation of the Ca^{2+} channel. The γ subunit has a transmembrane configuration and is glycosylated. Antibodies against this subunit alter the activity of the Ca^{2+} channel [340] suggesting a role for this subunit in regulation of the channel.

Purification and Characterization of the Cardiac L-Channel. Hosey et al. have reported the purification and characterization of the L-channel from the chick heart [158]. The purified fractions were highly enriched in a 185-kDa peptide which copurified with dihydropyridine binding activity. Photoaffinity studies revealed the peptide to contain receptors for both dihydropyridines and phenylalkylamines. Antibodies used against the skeletal muscle dihydropyridine receptor peptide reacted poorly with the cardiac receptor suggesting that the heart and skeletal muscle L-channels were different.

Voltage-Sensitive Calcium Channels in Nonexcitable Cells. Chen and colleagues have found two types of voltage sensitive Ca^{2+} channels in 3T3 fibroblasts L and T types [69]. Cells that were transformed with retrovirus-mediated transfer of C-H-*ras*, EJ-*ras*, V-*src*, V-*fms*, or polyoma middle T, oncogenes coding for membrane proteins, lacked the T-type Ca^{2+} channel, whereas the L-type remained present. Cells transformed with oncogenes whose gene product was located primarily in the nucleus, polyoma large T, and SV40 large T, contained T-type and L-type channels although the T-type channels were at a lower density than in control cells.

Regulation of Voltage-Dependent Calcium Channels. As indicated above, cAMP-dependent protein kinases and protein kinase C have been implicated in Ca^{2+} channel regulation. These regulatory processes, involving phosphorylation, do not result in changes in single channel conductance but rather involve changes in the probabilities of channel opening or opening of channels previously unrecognized [322]. Chad et al. used patch clamp techniques to examine the effects of phosphorylation on the L-type Ca^{2+} channel in GH_3 rat pituitary tumor cells [62]. Ca^{2+} channel activity could be elicited by adding ATP and the catalytic subunit of the cAMP dependent protein kinase to the cytosolic face of the membrane patch. When a protein kinase inhibitor was added with the catalytic subunit and ATP there was no Ca^{2+} channel activity.

Neurotransmitters such as dopamine have been found to alter channel activity [153]. Norepinephrine, acetylcholine, dynorphin, adenosine, and neuropeptide Y alter channel activity [322]. cGMP-dependent protein kinase also modulates Ca^{2+}channel activity as reported in cardiac muscle and *Helix aspersa* [197, 243]

GTP-binding proteins may serve in the coupling of various agonists and Ca^{2+} channels [95, 142, 152]. G-proteins can be involved in the coupling to adenylate cyclase with increases in cAMP and activation of cAMP-dependent protein kinases which can alter channel activity. Alternatively G proteins may couple receptors to phosphatidylinositol turnover thus playing an important role in the agonist-induced increases in cellular diacylglycerol and activation of protein kinase C followed by channel activation.

A role for G-proteins in the regulation of voltage-dependent Ca^{2+} channel activity has been established using nonhydrolyzable GTP analogs: GTPγS and GTPβS, as well as pertussis toxin and cholera toxin. G-proteins have a heterotrimeric structure of α, β and γ subunits. In the various situations where G proteins serve as coupling elements in signal transduction, the hormone-receptor interaction is followed by a dissociation of the α subunit from the $\beta\gamma$ complex. The α subunit then binds GTP and, in this GTP bound configuration, carries out its cellular action. After GTP is hydrolyzed the α-GDP is inactive and ultimately the α subunit recombines with the $\beta\gamma$ subunits. When GTPγS is used experimentally the α subunit forms a stable complex with it and the action of α is potentiated. When GTPβS is used, by contrast, the α-GTPβS complex is inactive; thus the action of α subunit is inhibited. Pertussis toxin ADP-ribosylates the α subunit of the G_i and G_oG proteins [60]. As a result this toxin prevents functional coupling of agonists to these forms of G proteins.

Intracellular infusion of GTPγS markedly reduced Ca^{2+} currents in neurons [143, 144, 151] and mouse AtT-20 pituitary cells [198]. GTPβS prevented the agonist-induced inhibition of Ca^{2+} currents [143, 144, 151] in neurons. Pertussis toxin prevented agonist-induced inhibition of Ca^{2+} currents in neurons [146,151] and pituitary cells [198]. Intracellular infusion of G_i and G_o reconstituted the agonist induced inhibition of Ca^{2+} currents in pertussis toxin-treated neurons [143]. Intracellular infusion of the α subunit of G_o was sufficient to couple neuropeptide Y receptors to inhibition of Ca^{2+} currents in sensory neurons [103]. In the heart Yatani et al. [360] found that the α subunit of G_s stimulated Ca^{2+} channels. Activation of G-proteins with the nonhydrolyzable GTP analog GTPγS results in changes in neuronal Ca^{2+} channel responses to various Ca^{2+} channel blocking agents [286]. At high concentration GTPγS converted the blocking effects of these agents into agonist actions. In summary these observations suggest that G_s may be a stimulatory regulator of Ca^{2+} channels, while G_o and G_i may be inhibitory. It is likely, albeit not proven, that the effects of G proteins on Ca^{2+} channels are mediated by the α subunits of the G proteins.

It is unclear how the G proteins are coupled to channel activity. For example there may be a direct effect on the proteins that constitute the channel, altering their configuration. Alternatively the G protein may interact with protein kinases which are membrane associated and the alteration in channel activity may be secondary to an effect of these protein kinases.

There are many agonists which, upon cell stimulation, result in an increase in Ca^{2+} entry across the plasma membrane. However, with the large majority of these agonists the Ca^{2+} entry is not a direct result of coupling of the agonist-receptor binding process to a directly measurable Ca^{2+} entry. Rather there are intracellular intermediates which produce a secondary increase in plasma membrane Ca^{2+} channel activity. Thus Ca^{2+} released from intracellular stores due to inositol 1,4,5-trisphosphate (IP_3) generated from agonist-induced enhanced polyphosphoinositide turnover may alter plasma membrane Ca^{2+} channel activity. In aortic smooth muscle cells vasopressin and intracellularly injected IP_3 inhibits an L-type calcium channel [11]. Examples of second-messenger operated channels, especially as they relate to Ca^{2+} mobilization through polyphosphoinositide turnover, will be discussed at length in later sections of this chapter.

Leak Channels

By definition a "leak channel" is a Ca^{2+} channel which opens in the absence of membrane depolarization or ligand [16]. The presence of these leak channels were proposed by Cauvin et al. [61]. Kuno et al. [184] identified a voltage-independent channel in T-lymphocytes that passes Ca^{2+}. Benham and Tsien [16] identified a Ca^{2+} channel in vascular smooth muscle that opens without depolarization or ligands. The conductance of this channel was 16 pS.

Ligand-Gated Channels

Ligand-gated Ca^{2+} channels can be subtyped into receptor-operated channels and second-messenger-operated channels [322]. These channels may play the primary role in movement of Ca^{2+} across the plasma membranes of non-excitable cells, such as epithelial cells. The receptor operated channel is exemplified by the ATP induced Ca^{2+} channels in vascular smooth muscle [16], the thrombin-activated channel derived from platelets [333, 365] or the NMDA receptor in neurons, to name a few.

Benham and Tsien studied ATP activated Ca^{2+} channels in vascular smooth muscle cells [16]. They studied outside-out membrane patches with high concentrations of Ca^{2+} (110 m*M*) in the bath and very low free Ca^{2+} concentrations in the pipette. When the transmembrane voltage was maintained at highly negative levels to ensure quiescence of voltage-dependent channels, the addition of ATP to the outside of the patch resulted in a significant inward current although the single channel conductance was quite small (~ 5pS). The Ca^{2+} entry is insensitive to nifedipine or cadmium. The ATP-induced Ca^{2+} channel was similar to voltage dependent channels in only one characteristic; it transferred Na^+ ions more rapidly but Ca^{2+} appears more permeable than Na^+ when one compares reversal potentials [16].

A receptor-operated Ca^{2+} channel has been characterized in the thrombin-stimulated platelet by incorporating membrane vesicles from stimulated cells into lipid bilayers [366]. The channel shares some properties with potential-dependent Ca^{2+} channels including a high selectivity for divalent cations, a singe channel conductance of 10 pS and blockade by Ni^{2+}. However, this channel is not voltage-dependent and is not blocked by dihydropyridines. The fact that the thrombin-induced channels survive the fractionation procedure suggests that thrombin may result in a covalent modification of an existing membrane protein, or insertion of new channels into the plasma membrane [264]. Other examples of second-messenger operated channel include the Ca^{2+} pathway expressed in the endoplasmic reticulum in response to IP_3 [359] or perhaps a cell membrane Ca^{2+} channel opened by IP_3 [185]

Ca^{2+}-Dependent Channels

In many systems, including epithelial cells, stimuli have been found to activate Ca^{2+}-dependent channels by increasing cytosolic free [Ca^{2+}]. While these channels do not transport Ca^{2+} they are an important link between Ca^{2+} and vital membrane transport phenomena in epithelial and other cells. These channels are believed to be directly controlled by Ca^{2+} binding to the channel. Opening of Ca^{2+} dependent K^+ and Cl^- channels in epithelial cells is an important mechanism for cell volume regulation. In the acinar cells of the pancreas, secretagogues increase intracellular free [Ca^{2+}] and the increase in [Ca^{2+}], in turn, enhances the conductance of Ca^{2+}-dependent K^+ and Cl^- channels. This enhanced ion conductance is believed to play a role in the ion and fluid transport across the basolateral and luminal membranes [205, 247]

Ca^{2+}-dependent K^+ channels are found in the apical membrane of the cortical collecting duct [157] as well as cultured thick ascending limb cells [130], Madin-Darby canine kidney cells [186], and opossum kidney cells [324]. Calcium-dependent cation channels are present in the basolateral membrane of the cortical thick ascending limb [319] and the apical membrane of proximal convoluted tubule cells in primary culture [215]. In addition to Ca^{2+}, cAMP may play an important role in the activation of these channel conductances [131, 190]. Dual regulation by Ca^{2+} and cAMP may be important for the regulation of transepithelial fluid and electrolyte transport.

Sodium-Calcium Exchange

General Characteristics

An important mechanism for transcellular Ca^{2+} movement involves Na^+/Ca^{2+} exchange. In 1968 Reuter and Seitz [259] found that Ca^{2+} efflux from cardiac tissue could be enhanced by extracellular Na^+. They postulated a carrier which could combine with Na^+ or Ca^{2+} on the outer surface of the membrane and only Ca^{2+} on the inner surface with a resultant net outward transport of Ca^{2+} in exchange for inward transport of Na^+. The exchanger is reversible [41]. Baker et al. [9] found that the exchanger could be reversed with Ca^{2+} influx enhanced markedly when extracellular Na was replaced by LiCl or dextrose or by increased internal Na^+ concentration in the squid axon. They interpreted the data as evidence for a coupled inward movement of Ca^{2+} to outward movement of Na^+. The exchange is electrogenic, with one net positive charge moved into the cell with each transport cycle when the exchanger is working in its "normal" direction, that is, transporting Ca^{2+} out of the cell.

With the use of cardiac sarcolemmal vesicle preparations it was established that the exchanger had a lower affinity for Ca^{2+} than did the Ca^{2+} ATPase but had a greater transport capacity (approximately 30 times that of the Ca^{2+} ATPase) [58]. The V_{max} for the Na^+/Ca^{2+} exchanger was found to be 15 nmol $mg^{-1} s^{-1}$ while that of the ATP-dependent Ca^{2+} uptake process was 0.5 nmol $mg^{-1} s^{-1}$. The K_m for Ca^{2+} was 1.5 μ*M* for the Na^+/Ca^{2+} exchanger and 0.3 μ*M* for the Ca^{2+}ATPase. The exchanger was electrogenic with a probable stoichiometry of 3 Na^+ per Ca^{2+} ion [41, 256]. Reeves and Sutko [257] found that Ca^{2+} uptake into cardiac vesicles, driven by intravesicular Na^+, was competitively inhibited by extravesicular Na^+, with a K_i for Na^+ of 16 m*M*.

Na^+/Ca^{2+} exchangers are not the same in all cell types. In the rod outer segment it has been found that, when the exchanger is run in the "reverse" direction to drive Ca^{2+} into the cell, there is a strong dependency on external K^+ [61a]. The ability of external K^+ to perturb the equilibrium level of intracellular $[Ca^{2+}]$ has led to the conclusion that K^+ is cotransported with Ca^{2+} and the exchanger stoicheometry is actually 4 Na^+: 1 Ca^{2+}, 1 K^+. This has the advantage of harnessing the outward K^+ gradient to promote Ca^{2+} extrusion to very low levels of intracellular $[Ca^{2+}]$ when the exchanger is working in the "normal" Ca^{2+} extruding mode. Although K^+ ions activate the exchanger in mammalian cardiac muscle and in invertebrate neurons and muscle, K^+ is not transported by the exchanger.

Caroni and Carafoli [59] found that a kinase-mediated phosphorylation activates Na^+/Ca^{2+} exchange in heart sarcolemona. This activation was dependent upon Ca^{2+}, inhibited by anticalmodulin agents, and was not affected by cAMP. They suggested that the transporter was activated by a Ca^{2+} calmodulin-dependent, cAMP-independent protein kinase. The activation process increased the affinity of the exchanger (K_m decreased from approximately 10 μ*M* to 2μ*M*).

LeBlanc and Hume [189] have found that voltage dependent Na^+ entry into the cardiac cell can result in calcium release from ryanodine-sensitive internal stores. This Ca^{2+} release was dependent upon extracellular Ca^{2+}, suggesting that the Ca^{2+} which induced release of Ca^{2+} from intracellular stores entered the cell via the Na^{2+}/Ca^{2+} exchanger. The data indicates that Na^{2+}/Ca^{2+} exchange may be involved in physiological excitation-contraction coupling. To explain this effect, however, it is necessary to postulate a subplasma membrane compartment in which Na^+ concentration can increase significantly in response to the calculated amount of Na^+ that enters via the voltage-dependent Na^+ channel [191].

A Na^+/Ca^{2+} exchanger is also present in vascular smooth muscle, but its activity has generally been found to be much lower than that of cardiac cells

[129, 204, 206, 207, 225]. Greater activity was found in sarcolemmal vesicles from tracheal smooth muscle [204]; although the activity was still lower than in cardiac cells. The apparent V_{max} for the exchanger in the mesenteric artery was intermediate between values for uterine smooth muscle and tracheal smooth muscle. In summary, in nonepithelial tissue where a Na^{2+}/Ca^{2+} exchanger has been identified, there is controversy as to how important a role it plays in Ca^{2+} homeostasis. In the heart, where it likely plays an important role because of its high activity, it is believed to play a pivotal role in the physiological changes brought about by digitalis and related drugs. These agents, by inhibiting the Na^+,K^+ ATPase, alter the transmembrane Na^+ gradients and hence driving forces for Na^+/Ca^{2+} exchange. In vascular smooth muscle the Na^+/Ca^{2+} exchanger is generally considered to play a less significant role since the V_{max} of the exchanger is low and the K_m is above 2 μM (while cytosolic free Ca^{2+} concentration varies between 0.1 and 1.0 μM).

A Na^+/Ca^{2+} exchanger has been identified in a number of epithelial systems. Gmaj et al. [126] prepared vesicles from the renal brush border and from the basolateral membranes of rat cortical tubules. They found that the basolateral membrane vesicles contained a Na^+/Ca^{2+} exchange process. Na^+ ions had no effect upon the Ca^{2+} ATPase of this membrane. Jayakumar et al. [166] studied the Na^+ dependent Ca^{2+} uptake into rat cortical basolateral vesicles and found it to exhibit Michaelis-Menten kinetics with regard to Ca^{2+}. A $[Ca^{2+}]$ $K_{d0.5}$ of 8 μM was found. They found that the exchange was electrogenic and the activity of the exchanger could be modulated by parathyroid hormone. Talor and Arruda [315] solubilized the exchanger from the basolateral membranes of kidneys, partially purified it, and reconstituted it into liposomes. This permitted more extensive kinetic studies. They found the V_{max} to be 268 pmol Ca^{2+} mg^{-1} protein min^{-1} and the K_m for Ca^{2+} was 2.7 μM. Stoicheometry was $3Na^+$ for each Ca^{2+}. As in smooth muscle the renal Na^+/Ca^{2+} exchange V_{max} is much lower than that of cardiac tissue. By comparison the V_{max} of the Ca^{2+} ATPase is 50–80 nmol mg^{-1} protein min^{-1} with a K_m for Ca^{2+} of 0.1–0.2 μM [125].

It is possible that metabolic events such as phosphorylation may enhance the activity of Na^+/Ca^{2+} exchange as it does in the heart [59]. This has not been explored adequately in epithelial tissues. Gmaj and Murer [125] have estimated that a phosphorylation effect equivalent to that seen in the heart [59] could account for a change in Vmax to about 80 nmol mg^{-1} protein min^{-1} and K_m to 0.8 μM, which are likely to be upper limits of rate and affinity for the renal Na^{2+}/Ca^{2+} exchanger. However, these values will change with a reduction in the apparent affinity for Ca^{2+} if Na^+ competes with the Ca^{2+} binding site as it does in cardiac tissue [257]. Gmaj and Murer [125] assumed that intracellular Na^+ concentration was 17.5 mM in proximal tubule cells [362] and calculated a value for an effective K_m for Ca^{2+} of 2.6 μM. Since basal levels of cytosolic free calcium are approximately 100 nM, by using these assumptions one can conclude that the Ca^{2+} ATPase will be functioning much closer to its V_{max} than will the Na^+/Ca^{2+} exchanger under basal conditions.

The molecular properties of the Na^+/Ca^{2+} exchanger has been the topic of active investigation by many groups [4a]. Carafoli and colleagues have raised antibodies against a 33-kDa protein, which interact with this protein and partially inhibit Na^+/Ca^{2+} exchange in heart sarcolemma vesicles. Under nonreducing conditions these antibodies also react with proteins that migrate with apparent molecular weights of 70 and 140 kDa. This group has suggested that the monomeric apparent molecular weight of the exchanger is 33 kDa, and that intermolecular disulfide bridges associate monomers into dimeric and tetrameric forms [56].

By contrast, Hale et al. [133] concluded that the Na^+/Ca^{2+} exchanger is approximately 225 kDa in size or larger. This conclusion was based upon irradiation inactivation-target sizing analysis and gel permeation high performance liquid chromatography. For the latter studies, cardiac sarcolemma vesicles were solubilized in detergent and fractionated by size. Fractions were then reconstituted into proteoliposomes and assayed for Na^+/Ca^{2+} exchange activity. Activity was localized to the fraction including a protein with apparent molecular weight of 234 kDa, and when transferred to nitrocellulose, this protein bound ^{45}Ca in a specific manner.

Cook and Knaupp [80] purified a glycoprotein with apparent molecular weight of 220 kDa to over 90% homogeneity from bovine retinal rod outer segments. This protein was estimated to constitute approximately 1% of total membrane protein and exhibited $Na^+–Ca^{2+}$ exchange properties when reconstituted into proteoliposomes.

Nicoll et al. [235a] recently identified a complementary DNA clone encoding a canine cardiac sarcolemmal Na^+/Ca^{2+} exchange protein. Antibodies raised to the deduced amino acid sequence react with sarcolemmal proteins of 70, 120, and 160 kDa. The open reading frame encodes a protein of 970 amino acids with a predicted molecular size of 108 kDa. Since the antibodies react with a 160 kDa protein it is possible that the mature protein is heavily glycosylated. There are six potential N-glycosylation sites on the exchanger. When compared with other known protein sequences the exchanger shows little sequence homology. There is a short span of 23 amino acids with 48% homology to the a subunit of the Na^+ K^+ ATPase. The hydropathy plots of the exchanger show similarities to those of the Na^+, K^+ ATPase and Ca^{2+} ATPases of the plasma membrane and sarcoplasmic reticulum.

Physiological Importance of Na^+/Ca^{2+} Exchange in the Epithelial Cell

Ullrich et al. [327] found that calculated rates of "active" Ca^{2+} transport across the proximal tubule were reduced if they experimentally reduced peritubular Na^+ concentration or treated the tubules with ouabain. Lee et al. [192] measured cytosolic free calcium ion activity in *Necturus* proximal tubule epithelial cells, using Ca^{2+} sensitive electrodes. When extracellular sodium concentration was decreased there was an increase in intracellular cytosolic

Ca^{2+} activity. They argued that Na^+/Ca^{2+} exchange played an important role in the maintenance of low cytosolic Ca^{2+} activity in the proximal tubule cell. Lorenzen et al. [203] found that ouabain increased both intracellular free Ca^{2+} concentration ($[Ca^{2+}]_i$) and cytosolic Na^+ activity. This was taken as further evidence that a basolateral Na^+/Ca^{2+} exchange process was of physiological importance.

In LLC-PK_1 cultured kidney epithelial LLC-PK_1 cells we found that replacement of extracellular Na^+ with choline, or ouabain exposure, resulted in an increase in cytosolic free Ca^{2+} concentration [24] (Figs. 3, 4). $[Ca^{2+}]_i$ was monitored in these studies with Quin 2. These observations can be interpreted as evidence for Na^+/Ca^{2+} exchange since choline would not replace Na^+ on the exchanger and an increase in intracellular Na^+, brought about as a result of ouabain, would decrease the electrochemical potential for Na^+ entry and thus decrease Ca^{2+} extrusion from the cell. The net effect on $[Ca^{2+}]_i$ was small, however, even after 30 min of exposure to choline substitution or ouabain. In perfused proximal tubules of *Necturus* similar manuevers were found to have greater effects on Ca^{2+} activity, as measured with microelectrodes. Low peritubular $[Na^+]$ resulted in an increase in Ca^{2+} activity from 82 ± 7 to 585 ± 106 and ouabain increased Ca^{2+} activity to 835 ± 185 [118].

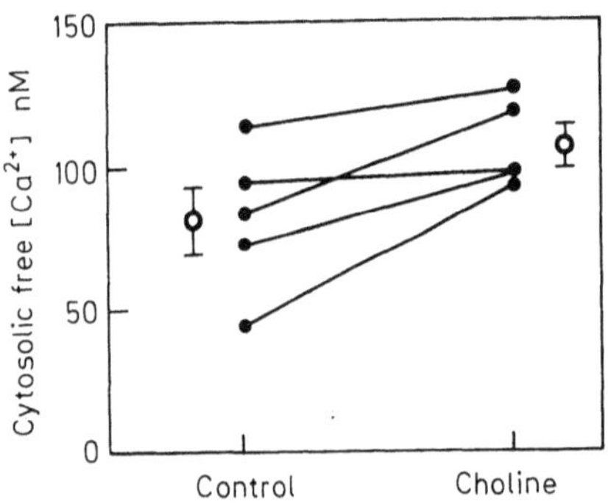

Fig. 3. Cytosolic free $[Ca^{2+}]$ ($[Ca^{2+}]_i$) determined after 30 min of incubation of LLC-PK_1 cells in a buffer containing 148 m*M* $[Na^+]$ (*control*) or containing 28 m*M* $[Na^+]$ and 120 m*M* choline (*choline*) as primary cations. In each of five paired experiments $[Ca^{2+}]_i$ was slightly higher in the choline group ($p < .05$). (From [24])

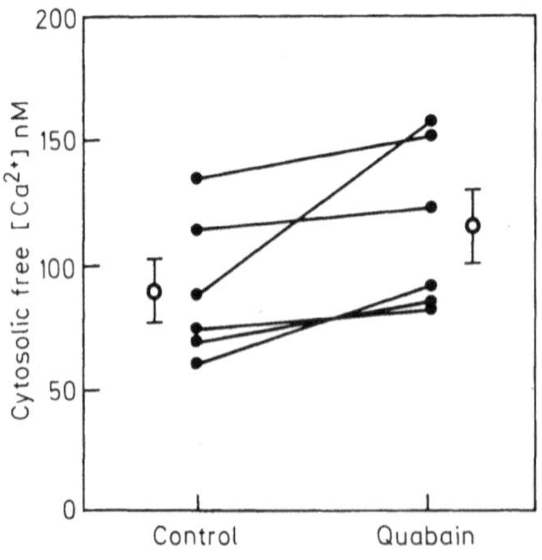

Fig. 4. Cytosolic free $[Ca^{2+}]$ ($[Ca^{2+}]_i$) determined after 30 min exposure of LLC-PK_1 cells to control buffer or buffer containing ouabain (10^{-4} *M*). In each of six paired experiments $[Ca^{2+}]_i$ was greater in the ouabain treated cells ($p < .05$). (From [24])

Snowdowne and Borle [296] proposed that the exchanger can work in either the Ca^{2+} extrusion mode or Ca^{2+} entry mode in monkey kidney cells depending upon the relative levels of $\Delta\mu$ Na^+ versus $\Delta\mu$ Ca^{2+}. These investigators found that the Na^+/Ca^{2+} exchanger was inactive in the basal state in these cells. A reduction in $\Delta\mu$ Na^+ resulted in activation of the exchanger in a reverse mode Ca^{2+} out-to-in, Na^+ in-to-out). Dominguez and colleagues [96] found that a reduction in extracellular $[Na^+]$ caused an increase in $[Ca^{2+}]_i$, an increase in ^{45}Ca influx and a decrease in ^{45}Ca efflux. They also found that ouabain increased $[Ca^{2+}]_i$. The increases in $[Ca^{2+}]_i$ that they found were similar to those found in our laboratory. When a prior reduction in extracellular $[Na^+]$ was combined with ouabain exposure the increase in $[Ca^{2+}]_i$ was blunted suggesting that it was the ouabain-induced increase in intracellular $[Na^+]$ that accounted for the increase in $[Ca^{2+}]_i$ due to a resultant decrease in the transmembrane $\Delta\mu$ Na^+.

Role of Na^+/Ca^{2+} Exchange in Transepithelial Transport

The function of the Na^+/Ca^{2+} exchange, in transepithelial transport remains unclear. Grinstein and Erlig [128] found that decreased serosal Na^+ was associated with increased ^{45}Ca uptake by frog skin epithelial cells. This reduction in serosal Na^+ concentration also resulted in decreased Na^+ transport by the frog skin. These investigators postulated that the reduction in serosal Na^+ concentration led to a reversal of Na^+/Ca^{2+} exchange to promote Ca^{2+} uptake and the increased intracellular Ca^{2+} led to decreased transepithelial Na^+ transport. Low peritubular Na^+ concentration also decreases Na^+ reabsorption in isolated perfused proximal [117] and collecting tubules [119] of the rabbit. In cell-attached patch clamp studies on apical membranes of cortical collecting tubule cells, it has been shown that ionomycin causes a marked inhibition of Na^+ channel activity [241]. The mean number of open channels decreased to 11% of the control value 10 min after addition of ionomycin. In excised inside-out patches, Ca^{2+} had no effect on the Na^+ channel activity. Thus it was concluded that an increase in $[Ca^{2+}]_i$ inhibits apical membrane Na^+ conductance by an indirect mechanism. In the small intestine Na^+/Ca^{2+} exchange was proposed to play a role in Ca^{2+} extrusion from the basolateral membrane [231]. It is believed however, that this process is only of little importance in the transepithelial transport of Ca^{2+} across the rat intestine [21]

Mitochondrial Calcium Transport

Calcium Uptake Pathway

Slater and Cleland [293] in 1953 observed that mitochondria could take up Ca^{2+} from the medium. Since they found that Ca^{2+} uptake could occur at 0°C they concluded that the uptake was passive. It was found in the early 1960s that Ca^{2+} uptake could be coupled to oxidative phosphorylation [337] or driven by ATP hydrolysis in the absence of coupled respiration [91]. Vasington and Murphy [338] found that kidney mitochondria could take up approximately 2 μmol Ca^{2+} per mg of protein, an amount approximately 500–1000 times greater than the amount of Ca^{2+} mitochondria normally contain in vivo. Vasington and Murphy also found that, although inhibitors of mitochondrial respiration, such as antimycin A and cyanide, inhibit Ca^{2+} uptake, uptake of Ca^{2+} was not associated with phosphorylation, indicating that Ca^{2+} uptake was an alternative to oxidative phosphorylation.

Rossi and Lehninger [269] identified differences between Ca^{2+} transport characteristics when mitochondria were exposed to low (100–500 μ*M*) concentrations of medium Ca^{2+} from characteristics under conditions of high Ca^{2+} concentrations of the medium (2–4 m*M*). When low concentrations of Ca^{2+} are employed, Ca^{2+} uptake is stoichiometric with electron transport without an associated uptake of phosphate. Maximal uptake of Ca^{2+} in the absence of phosphate was approximately 100 nmol/mg mitochondrial protein. By contrast phosphate uptake accompanies Ca^{2+} uptake when the ambient Ca^{2+} concentration is high. When low concentrations of Ca^{2+} are employed, phosphate accumulates with Ca^{2+} only when there is ATP and Mg^{2+} present in the medium. Massive amounts of Ca^{2+} can be taken up into mitochondria in the presence of phosphate and ATP. Under these conditions calcium phosphate precipitates within the mitochondria [42]. Whereas with modest amounts of Ca^{2+} loading there is no apparent damage to the mitochondria structurally or functionally, large amounts of Ca^{2+} uptake result in mitochondrial swelling [194], uncoupling of oxidative phosphorylation [67, 338], and release of free fatty acids [355] which are themselves uncoupling agents. The ATP (or ADP) dependence for uptake seems to be related to the fact that adenine nucleotides precipitate together with Ca^{2+} and phosphate, perhaps "priming" the precipitate within the mitochondria or stabilizing it [50]. The perturbations in the mitochondrial membrane permeabilities depend, not only on the matrix calcium, but also on the conformation state of the ADP/ATP carrier (195a).

Scarpa and Azzone in 1970 [280] studied the uptake of Ca^{2+} into liver mitochondria in the absence of metabolism. They observed Ca^{2+} uptake when the mitochondria were treated with the K^+ ionophore, valinomycin. The characteristics of the uptake were similar to the Ca^{2+} translocation observed under aerobic conditions. The uptake rate was of the same order of magnitude observed under aerobic conditions and there was a strong inhibition by La^{3+} The data were interpreted according to the Mitchell chemiosmotic hypothesis [220] by which either metabolism or K^+ efflux enhances transmembrane potential differences and the latter drives Ca^{2+} uptake. Selwyn et al. [288] suspended mitochondria in isosmotic solutions of calcium salts containing metabolic inhibitors and found that rapid swelling, resulting from Ca^{2+} accumulation, occurred when the anion was permeant. These observations indicated that the uptake of Ca^{2+} was an electrophoretic process. In order to further characterize the Ca^{2+} uptake process, Rottenberg and Scarpa [271] compared the distribution of Rb^+ and Ca^{2+} across liver mitochondrial membranes in the presence of ionophores. Acetate was used as an anion in an attempt to prevent intramitochondrial Ca^{2+} precipitation. In the presence of valinomycin the $^{86}Rb^+$ distribution was used to calculate the mitochondrial membrane potential using the Nernst equation:

$$E = \frac{RT}{nF} \ln \frac{[Rb^+]_i}{[Rb^+]_0}$$

Where n = net change transfer per molecule transported. The Ca^{2+} distribution ratio was determined using ^{45}Ca. There was a very good correlation between the calculated membrane potential using ^{86}Rb and the log of the $^{45}Ca^{2+}$ distribution ratio ($[^{45}Ca^{2+}]_i$):

$$\log \frac{[Ca^{2+}]_i}{[Ca^{2+}]_0} = 2 \log \frac{[Rb^+]_i}{[Rb^+]_0}$$

This relationship led to the conclusion that the net charge transfer associated with Ca^{2+} uptake is 2.

Crompton and Heid [80] analyzed this question in heart mitchondria using a different approach which avoids the potential criticism of the above technique in which intramitochondrial Ca^{2+} activity is not directly measured. Crompton and Heid used the relationship that at equilibrium:

$$\Delta E = \frac{RT}{F} \ln \frac{[Rb^+]_i}{[Rb^+]_0} = \frac{RT}{nF} \ln \frac{[Ca^{2+}]_i}{[Ca^{2+}]_0}$$

to derive the following expression:

$$\log [Ca^{2+}]_e = -n \log \frac{[RB^+]_i}{[Rb^+]_0} + \log [Ca^{2+}]_i$$

They then performed experiments to change equilibrium values of $[Ca^{2+}]_e$ and ΔE by varying values of $[Ca^{2+}]_e$ and concentrations of the uncoupler, carbonylcyanide-*m*-chlorophenylbydrazone so that the maximum amounts of Ca^{2+} taken up by the mitochondria were the same. Using this technique they confirmed the results of Rottenberg and Scarpa [271] finding a net charge transfer of 2 per Ca^{2+} transported. When the kinetics of the Ca^{2+} uptake process were examined it was found that the K_m of the influx process was approximately 10 μM [51, 82]. Under conditions designed to maximize the rate of Ca^{2+} uptake, in the absence of added Mg^{2+}, Vercesi et al. [339] found the V_{max} for Ca^{2+} uptake to be approximately 0.5 μmol of Ca^{2+} per minute per milligram of heart mitochondrial protein under optimal conditions.

Inhibitors of Mitochondrial Ca^{2+} Uptake

Mela [212] found that lanthanum was a potent inhibitor of Ca^{2+} uptake into liver mitochondria. She suggested that La^{3+} bound to phospholipids and occupied a carrier for Ca^{2+} translocation. The concentration of lanthanides needed for complete inhibition of Ca^{2+} uptake was less than 0.1 nmol/mg protein [213]. Mela proposed this number to be the maximal number of Ca^{2+} carriers in the mitochondrial membrane. Other rare earth cations such as Nd^{3+} and Tm^{3+} also inhibit Ca^{2+} uptake. With time La^{3+} is accumulated by mitochondria and the Ca^{2+} induced stimulation of respiration returns [255]. By contrast, the mucopolysaccharide and glycoprotein stain, ruthenium red, inhibits Ca^{2+} uptake [223] and remains effective over a long period of time [255]. While the inhibition of Ca^{2+} uptake by La^{3+} is affected by the concentration of Ca^{2+} due to competitive inhibition, the inhibition by ruthenium red is not, indicating that the latter inhibits noncompetitively [255]. The concentration of ruthenium red-binding sites was found to be 0.08 ± 0.01 nmol/mg protein.

Mg^{2+} is a potent inhibitor of Ca^{2+} uptake in heart mitochondria [299]. The effect upon liver mitochondria was less [3, 165]. This inhibition of Ca^{2+} uptake occurs at Mg^{2+} concentrations which approximate those present in the normal cell suggesting that Mg^{2+} may play an important regulatory role in mitochondrial Ca^{2+} uptake in vivo.

Mitochondrial Ca^{2+} Release Mechanisms

Na^{+}-Dependent Ca^{2+} Release. Ca^{2+} is continuously cycling across the inner mitochondrial membrane. This cycling is due to the presence of Ca^{2+} efflux pathways which are distinct from the Ca^{2+} uptake pathways we have discussed. Ca^{2+} is extruded from mitochondria by Na^{+}-dependent and Na^{+}-independent mechanisms. In the presence of ruthenium red, Ca^{2+} is released from heart [299], liver [251] and kidney [70] mitochondria. This provided early evidence for an efflux pathway independent of the ruthenium red uptake pathway. In 1974 Carafoli et al. [56] showed that Ca^{2+} efflux from heart mitochondria could be induced by increasing the Na^{+} concentration of the incubation medium, within the physiological range ($\approx$ 5 m*M*) when ruthenium red was present to block reuptake of Ca^{2+} into the mitochondria. This exchange is electroneutral ($2Na^{+}$ for $1Ca^{2+}$) [2]. The exchanger is particularly active in heart and nervous tissue mitochondria and much less active in liver and lung [138]. Mitochondria from the kidney medulla have a more active Na^{+}-dependent extrusion process than mitochondria from the renal cortex [138]. In heart, brain, and kidney medulla mitochondria the maximal rate of egress is low (0.01–0.02 μmol/mg mitochondrial protein per minute [56, 138]). The maximal rate of Ca^{2+} uptake by the uniporter exceeds the maximal rate of egress by at least one order of magnitude.

The affinity of the exchanger for Ca^{2+} has been measured at the external aspect of the mitochondrial membrane and has been found to be approximately 10 μ*M* [79]. The affinity for Ca^{2+} on the matrix face of the membrane is estimated from studies where matrix Ca^{2+} concentration is approximated by indirect methods. Using a null point titration method or using a Ca^{2+} ionophore, in the presence of ruthenium red, it is possible to establish relationships between total mitochondrial Ca^{2+} and free matrix Ca^{2+}. Using these techniques values for the K_m for Ca^{2+} on the matrix side of the membrane have been found to be 3–6 μM [76, 136]. Na^+ competes with Ca^{2+} for binding to the exterior surface of the mitochondrial inner membrane [140]. Diltiazem inhibits this exchanger [329].

Na^+-Independent Calcium Release. Mitochondria also can release Ca^{2+} in the absence of extramitochondrial Na^+ when ruthenium red inhibits the uniporter. This pathway is more important in kidney and liver mitochondria [81] than it is in heart mitochondria [79, 140]. In general, mitochondria that exhibit high rates of Na^+-independent Ca^{2+} extrusion are those which have low activities of Na^+/Ca^{2+} exchange. This extrusion process may involve Ca^{2+}/H^+ exchange [112] but is poorly characterized. There is even some question about whether a carrier exists at all [78]. No inhibitors of Na^+ independent Ca^{2+} transport have been found.

Implications of Mitochondrial Ca^{2+} Transport Properties

Under normal physiological conditions the Ca^{2+} uptake pathway is inhibited at normal cellular Mg^{2+} concentration. This together with the relatively slow Na^+/Ca^{2+} exchange process, which is operating at approximately one-half V_{max}, results in Ca^{2+} cycling across the mitochondrial membrane in a manner that is not likely to contribute in an important way to cytosolic Ca^{2+} homeostasis. This does not, however, imply that this sophisticated machinery is without an important function. It has been proposed that a primary function of the mitochondrial calcium transport processes is to allow intramitochondrial matrix $[Ca^{2+}]$ to follow that of the cytosol in order to modulate oxidative metabolism [210]. There are three intramitochondrial dehydrogenases (pyruvate dehydrogenase, NAD^+-isocitrate dehydrogenase, and α-keto glutarate dehydrogenase) whose activity is altered by changing concentrations of calcium over the 0.1–10 μM range. Ca^{2+} decreases the K_m of both NAD^+-isocitrate and 2-oxoglutarate dehydrogenases for their respective substrates and activates pyruvate dehydrogenase phosphate phosphatase, thus increasing the amount of active, dephosphorylated form of the enzyme. All three dehydrogenases have been found to be activated when extramitochondrial $[Ca^{2+}]$ is increased over the physiological range observed with hormone activation [210a]. Increases in cytosolic $[Ca^{2+}]$ under physiological or pathophysiological states would result in increased mitochondrial matrix $[Ca^{2+}]$, enhanced dehydrogenase activity, and enhanced ATP generation. Wan et al. [342a] have utilized fura-2 to directly measure matrix free $[Ca^{2+}]$ in isolated rat heart mitochondria, and correlated the matrix free $[Ca^{2+}]$ with metabolic flux in the citric acid cycle. They argue that under physiological conditions of substrate availability that α-ketoglutarate dehydrogenase may be a more physiologically relevant target of

Ca^{2+} regulation than pyruvate dehydrogenase. Stimulation of mitochondrial dehydrogenases by Ca^{2+} may contribute significantly to the rate of generation of ATP by resulting in enhanced supply of substrates to the respiratory chain. This may explain the stimulated oxidative phosphorylation observed in response to enhanced ATP demand under conditions of stimulated secretion or contraction when there are no changes in the ATP/ADP ratio in the cell [210a].

Intracellular Nonmitochondrial Calcium Transport Processes

Nonmitochondrial Ca^{2+} Storage Compartments

As discussed previously, the role of intracellular organelles in the regulation of intracellular free Ca^{2+} has been studied in intact or permeabilized cells, as well as fractions of cells, such as isolated mitochondria and microsomes. In many studies cells are permeabilized with saponin (digitonin). Saponins form insoluble complexes with cholesterol which is concentrated in the cell plasma membrane. At low concentrations saponin does not affect intracellular organelles [111]. In such studies it has generally been found that at least two intracellular compartments are involved in an important way in the regulation of cytosolic $[Ca^{2+}]$: the mitochondrial compartment which can lower extramitochondrial $[Ca^{2+}]$ to the micromolar range and an ATP-dependent, nonmitochondrial compartment which can lower $[Ca^{2+}]$ to approximately 100–200 μ*M*, a value close to that of the resting levels of cytosolic free Ca^{2+} in epithelial [70], as well as nonepithelial cells [250].

In the absence of Mg^{2+} and ATP when Ca^{2+} is added to permeabilized LLC-PK_1 cells it is rapidly taken up into an intracellular compartment and cytosolic Ca^{2+} (in equilibrium with bath free $[Ca^{2+}]$) is maintained at approximately basal levels of 100 n*M*. This Ca^{2+} uptake is inhibited by ruthenium red and the accumulated Ca^{2+} is released by FCCP. Thus, under these Mg^{2+}, ATP depleted conditions the mitochondria were effective Ca^{2+} buffering organelles. By contrast, under more physiological conditions, in the presence of ATP and Mg^{2+}, additions of Ca^{2+} results in a biphasic response (Figs. 5, 6). A slow, low capacity uptake process maintains Ca^{2+} buffered at the 50–300 n*M* concentration range. Vanadate, which inhibits Ca^{2+} ATPases, inhibits this uptake but ruthenium red does not. After this storage site, which we interpreted to be the endoplasmic reticulum, was saturated, the faster, higher capacity, ruthenium red-sensitive mitochondrial compartment then became dominant but buffered $[Ca^{2+}]_i$ to a higher steady state value, in the 600–1500 n*M* range [70].

In permeabilized enterocytes van Corven et al. [333] found similar results. Using oxalate to enhance Ca^{2+} uptake into a compartment which they called endoplasmic reticulum they found a fourfold stimulation of Ca^{2+} uptake when free $[Ca^{2+}]$ was 100 n*M* indicating that Ca^{2+} uptake into this compartment is

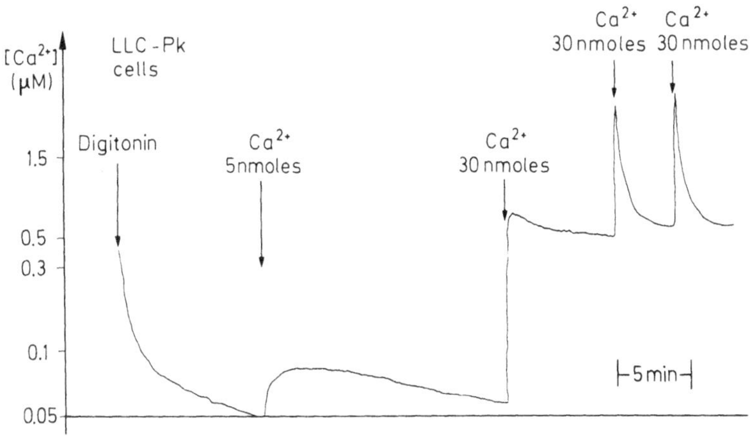

Fig. 5. Ca^{2+} regulation by nonmitochondrial and mitochondrial compartments in LLC-PK_1 cells. Bath $[Ca^{2+}]$ is monitored with a Ca^{2+} selective electrode. When digitonin is added, Ca^{2+} is taken up into the cells and bath $[Ca^{2+}]$ decreases to approximately 50 nM in this experiment. When small amounts of Ca^{2+} are added, Ca^{2+} is taken up into a pool which effectively returns bath $[Ca^{2+}]$ to a level close to baseline. When larger amounts of Ca^{2+} are added, however, this pool is saturated and a new steady state $[Ca^{2+}]$ is established at approximately 600 nM. This pool, with its high Ca^{2+} "set point," can take up large amounts of added Ca^{2+} while maintaining its ability to buffer $[Ca^{2+}]$. [From (70)]

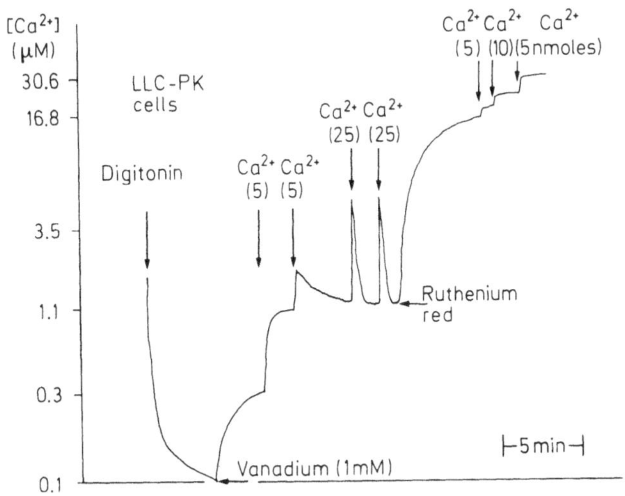

Fig. 6. Effects of vanadate and ruthenium red on Ca^{2+} regulation by permeabilized LLC-PK_1 cells. Vanadate (1 mM) inhibits Ca^{2+} uptake into the storage site that is otherwise able to buffer $[Ca^{2+}]$ at very low levels ($\approx$ 100 nM in this experiment). When vanadate is added there is release from this site due to unmasking of a continuous Ca^{2+} cycling in and out of this organelle(s), with vanadate inhibition of the entry pathway. After subsequent addition of Ca^{2+} this storage site cannot buffer the added Ca^{2+}. However, when enough Ca^{2+} is added to obtain levels of approximately 1.1 μM, then uptake of added Ca^{2+} is observed to take place. Thus vanadate does not inhibit uptake into this compartment which has a higher "set point" of $[Ca^{2+}]$. This compartment is mitochondrial as reflected by release of Ca^{2+} in response to Ruthenium red and inability to take up additional added Ca^{2+}. (From [70])

important at Ca^{2+} concentrations prevailing under normal physiological conditions. Endoplasmic reticulum Ca^{2+} uptake is much less important when $[Ca^{2+}]$ is 1.0 μ*M*. Vanadate inhibited uptake at 100 n*M* but had little effect at 1.0 μ*M* $[Ca^{2+}]$. Mitochondrial inhibitors had no effect at 100 n*M* but inhibited Ca^{2+} uptake completely at 1.0 μ*M* $[Ca^{2+}]$. The authors conclude, as we had previously in the LLC-PK_1 cell, that the endoplasmic reticulum of this epithelial cell has a high affinity, low capacity Ca^{2+} transport system with a K*m* for Ca^{2+} uptake of 100 n*M* Ca^{2+}.

Ca^{2+} Induced Ca^{2+} Release

The sarcoplasmic reticulum of cardiac and striatal muscle is a Ca^{2+} storage site critical to the control of excitation contraction coupling. Ca^{2+} release from this storage site is elicited by the Ca^{2+} entry into the cell which occurs in response to cell membrane depolarization. This Ca^{2+}-induced Ca^{2+} release is likely mediated by Ca^{2+} channels which have been characterized by incorporating sarcoplasmic reticulum vesicles into plasma phospholipid bilayers and studying single-channel conductance by voltage clamp techniques [103, 349]. Cardiac channels obtained from the dog are activated by Ca^{2+} additions to the cytoplasmic side in the presence of ATP. Channels opened rarely at a $[Ca^{2+}]$ of 1 μ*M* and with increasing frequency as $[Ca^{2+}]$ varied from 10 μ*M to* 1 *mM*. Channel opening does not depend upon hydrolysis of ATP as the nonhydrolyzable analog, AMP-PCP, was just as effective. Mg^{2+}, at millimolar concentrations, inhibited Ca^{2+}-induced Ca^{2+} release. Functionally, the Ca^{2+} release channel of the sarcoplasmic reticulum is believed to be the ryanodine receptor, a 450-kDa protein [158a, 159].

IP_3-Induced Release of Ca^{2+} from Intracellular Nonmitochondrial Storage Sites

The ATP-dependent nonmitochondrial storage site is the compartment which responds to IP_3 with release of Ca^{2+}. Ca^{2+} uptake and IP_3-induced release can be found in the microsomal fraction in a number of different cell types [169, 301]. Initially it was believed that this compartment was the endoplasmic reticulum. However, the microsomal fraction contains the Golgi and plasma membrane, and likely other organelles, in addition to the endoplasmic reticulum. Krause and Lew [178] explored this issue in more detail in the human neutrophil. Using differential centrifugation to separate various compartments of the microsomal fraction, and using biochemical markers to identify the fractions, they were unable to find Ca^{2+} pumping activity in the endoplasmic reticulum. They concluded that nonendoplasmic reticulum organelles, such as endosomes, Golgi or other unidentified structures, play important roles in Ca^{2+} homeostasis in the neutrophil. Van Corven et al. [333] used gradient centrifugation to produce endoplasmic reticulum, Golgi and basolateral membranes from the duodenal epithelial cell. All three populations were reported to have a high affinity Ca^{2+}-ATPase with a K_m for Ca^{2+} in the submicromolar (0.3–0.5 μM) range. They also found that the Ca^{2+} ATPase in each fraction was inhibited by calmodulin inhibitors. Lew's group coined the term "calciosome" which they believe represents a novel IP_3-responsive organelle [342]. With recent advancements in the characterization of the IP_3 receptor it has been possible to develop antibodies and use immunohistochemical methods to localize the IP_3 binding site. Using polyclonal antiserum from rabbits immunized with the purified IP_3 receptor, Ross et al. [268] localized the IP_3 receptor, in cerebellar Purkinje neurons, to the endoplasmic reticulum (ER), including portions of the rough ER, smooth ER, subplasma membrane cisternae and nuclear membranes, but not to the mitochondria or cell membrane. By contrast, the monoclonal antibodies raised against the IP_3 receptor by Furuichi et al. [121] stain the plasma membrane as well as the ER. This is particularly of interest given the question of whether IP_3 acts on a Ca^{2+} channel in the plasma membrane ([185], see below). In addition these groups disagree with regard to nuclear localization since Furuichi et al. found the staining of the nuclear membrane to be only weakly positive.

Ehrlich and Watras [101] demonstrated that IP_3 opened channels in plasma lipid bilayers into which vesicles from endoplasmic reticulum of aortic smooth muscle were incorporated. The single channel conductance was approximately 10 pS. Ca^{2+} channel opening depends upon the binding of at least three molecules of IP_3 [217]. The channels open within 4 s and the opening is not dependent upon the presence of ATP [217].

Ca^{2+} release from the IP_3 -sensitive pool is dependent upon the presence of permeant cations in the fluid surrounding this compartment. K^+, Na^+, Rb^+ and Cs^+ are all approximately equally effective in supporting IP_3-mediated Ca^{2+} release. Li^+ does not support IP_3-induced release [168]. The presence of cations does not alter IP_3 binding to its receptor. This dependency on cations is most likely due to the ability of cations to replace the extruded Ca^{2+} preventing the development of large potentials associated with Ca^{2+} extrusion. IP_3 also releases $^{54}Mn^{2+}$ and $^{85}Sr^{2+}$ from nonmitochondrial storage sites [168].

An important feature of the IP_3-mediated Ca^{2+} release is its property of not desensitizing (Fig. 7). Repeated additions of IP_3 result in equivalent release of Ca^{2+} from nonmitochondrial stores.

Evidence for Heterogeneity of Nonmitochondrial Ca^{2+} Storage Sites

In 1986 Gill and colleague [123] found that GTP markedly enhanced $^{45}Ca^{2+}$ release from saponin-permeabilized NIE-115 neuroblastoma cells, previously loaded with ^{45}Ca in a buffer containing 100 nM Ca^{2+} with 1 mM ATP. The half-maximal activation of release was at approximately 0.8 μM GTP. The GTP-induced release (at 10 μM GTP) was as rapid as that observed with A23187. The total amount of Ca^{2+} released was 65%–75% of the ionophore-releasable Ca^{2+}. Other triphosphates did not mimic the effect of GTP. Non-

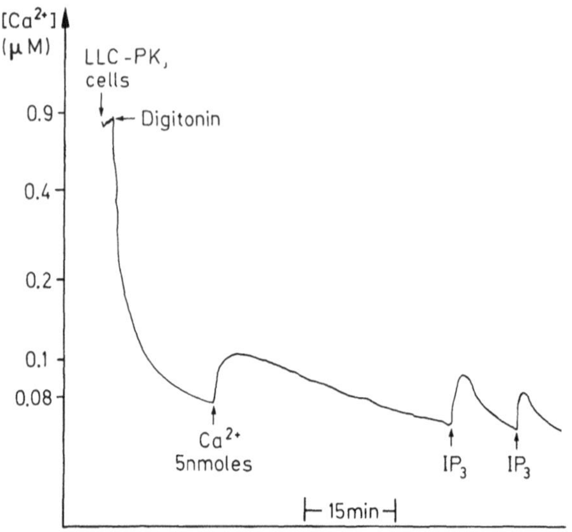

Fig. 7. Mobilization of cellular Ca^{2+} stores by IP_3. Cells were rendered permeable with digitonin. $[Ca^{2+}]$ of the bath, as measured with a Ca^{2+} selective electrode, was buffered to approximately 80 μM. When IP_3 was added to the cells, there was a release of Ca^{2+}, reflected as an increase in bath $[Ca^{2+}]$. Ca^{2+} then was taken back into cellular storage sites. Repeat addition of IP_3 resulted in Ca^{2+} release thus demonstrating the short term effect of the IP_3 and the lack of tachyphylaxis. (From [70])

hydrolyzable analogs of GTP, GppNHp, or GTPγS, had no effect on Ca^{2+} release. GTP also released Ca^{2+} from NIE-115 cell microsomes [326]. GTP-mediated Ca^{2+} release from permeabilized cells was enhanced by polyethylene glycol [PEG] although Ca^{2+} release was also present in the absence of PEG. By contrast GTP-induced release from microsomes was completely dependent upon the presence of PEG.

It appears that the GTP-mediated Ca^{2+} release process is very different from the release process mediated by IP_3 [74]. PEG does not modify IP_3-induced Ca^{2+} release. GTPγS blocks GTP-induced Ca^{2+} release but has no effect upon IP_3-induced release. Reducing temperature to 4°C eliminates the effect of GTP but reduces that of IP_3 by only 20%. When free Ca^{2+} concentration is increased from 0.1 to 10 μ*M*, in the presence of ruthenium red to block mitochondrial Ca^{2+} uptake, IP_3-induced release is completely inhibited whereas GTP-induced release remains at approximately 50% of ionophore-releasable Ca^{2+}. The effect of GTP on Ca^{2+} movements was reversed in the presence of oxalate, resulting in uptake and accumulation of Ca^{2+} rather than release [228]. This uptake process demonstrated the same guanine nucleotide specificity and sensitivity and Gill and colleagues suggested that both uptake and release are activated by the same GTP regulatory process. Since both processes were enhanced in the presence of PEG, and since PEG favors membrane-membrane interactions it was proposed that GTP-induced a translocation of Ca^{2+} from one pool to another which differed with regard to oxalate permeability. In a subsequent study this same laboratory found that Ca^{2+} entry into the IP_3-releasable Ca^{2+} pool is mediated by a GTP-regulatory mechanism [229]. This conclusion was reached as a result of the following set of observations:

1. IP_3 releases Ca^{2+} from an oxalate sensitive compartment in both NIE-115 cells as well as DDTIMF-2 smooth muscle cells.
2. In the absence of oxalate, GTP, like IP_3, inhibited Ca^{2+} accumulation into permeabilized cells. IP_3 inhibited accumulation by 30% whereas GTP inhibited it by 60%. Presumably this was due to enhanced release with GTP and IP_3.
3. In the presence of oxalate, GTP continued to inhibit Ca^{2+} accumulation initially (≈ 2 min). This was followed by a sustained increase in the rate of accumulation. This second phase was blocked by IP_3. Thus IP_3 released Ca^{2+} from the same pool into which GTP promoted accumulation.
4. The effects of IP_3 and GTP on Ca^{2+} accumulation were not additive. Thus the larger GTP-releaseable pool may contain a smaller IP_3-sensitive pool. The authors proposed a model to explain these results (Fig. 8). Two pools are proposed which can accumulate Ca^{2+} via an ATP-dependent process. One compartment is permeable to oxalate and the other is not. GTP mediates transfer of Ca^{2+} between the compartments and releases Ca^{2+} from the oxalate- permeable compartment. This release process and the GTP-dependent translocation of Ca^{2+} between compartments is aided by close association between membrane compartments which is potentiated by PEG.

Thevenod and colleagues [285, 320] studied Ca^{2+} uptake and release in permeabilized pancreatic acinar cells and functionally identified two non-mitochondrial Ca^{2+} storage pools, an IP_3-sensitive and an IP_3-insensitive pool. Ca^{2+} uptake into the IP_3-sensitive pool was proposed to be a MgATP-dependent process in which Ca^{2+} ions are exchanged for H^+ ions. This pool was proposed to have a H^+ ATPase pump. The other Ca^{2+} pool was insensitive to IP_3 and accumulated Ca^{2+} via a vanadate (10^{-4} *M*) sensitive Ca^{2+} ATPase.

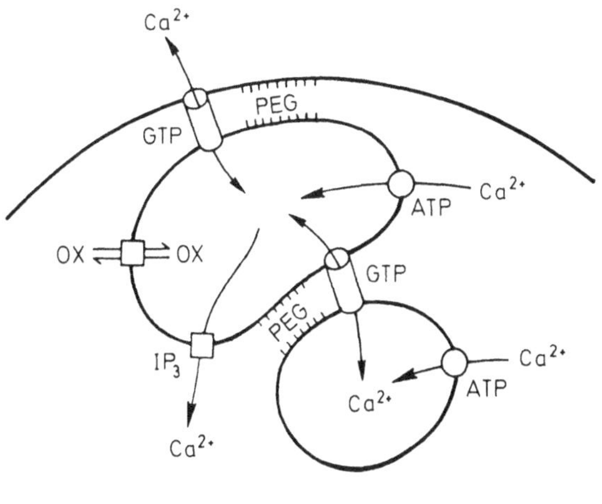

Fig. 8. Model proposed by Gill and colleagues of Ca^{2+} release from intracellular nonmitochondrial stores. Two pools are proposed, one of which is oxalate (*OX*) and IP_3 sensitive. GTP regulates the transfer of Ca^{2+} from the non-IP_3-responsive compartment to the IP_3-responsive one and mediates release of Ca^{2+} from the IP_3-sensitive compartment. Ca^{2+} is taken up into both pools by an ATP-dependent process. Polyethylene glycol (*PEG*) potentiates the close association between the compartments and the plasma membrane. (From [229])

Thevenod et al. proposed that the IP_3 sensitive pool filled rapidly at levels of Ca^{2+} (approximately 1 μM) achieved during cell stimulation, whereas the IP_3 insensitive pool is filled more rapidly at basal physiological levels of Ca^{2+} concentration (100 nM). Both pools are sensitive to high concentrations (1 nM) of vanadate. Whether cross-talk between the pools is mediated by GTP-dependent mechanisms or IP_4 was not assessed in this study. Berridge and Irvine [19] have summarized various models of how IP_3 might act to induce increased Ca^{2+} entry into cells and increased cytosolic free [Ca^{2+}] (Fig. 9).

Inhibitors of IP_3-induced Ca^{2+} Release

Heparin. IP_3-induced release of Ca^{2+} is very sensitive to heparin. The half-maximal inhibition of Ca^{2+} release to supramaximal concentrations of IP_3 (10 μM) is 0.6 μg/ml heparin. In microsomes the sensitivity was even greater with half-maximal inhibition at approximately 0.2 μg/ml. Heparin binds reversibly and can be displaced by IP_3. Heparin can displace IP_3 and stop Ca^{2+} release even after prolonged activation of release with IP_3. These data support the view [309, 359] that the IP_3 receptor is a heparin sensitive binding site.

Other Inhibitors. In brain membranes, where the IP_3-binding protein is present in high abundance [359], IP_3 binding is inhibited by supraphysiological levels of calcium concentration. This may serve in a feedback control system to limit the

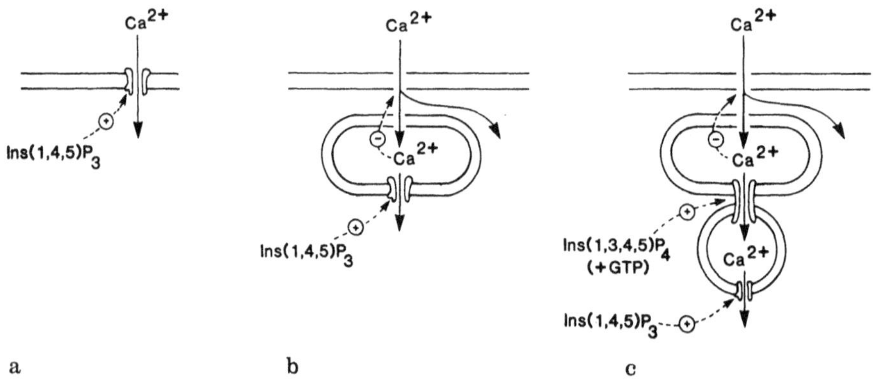

Fig. 9. Summary of suggested mechanisms by which inositol 1,4,5-triphosphate [Ins (1,4,5)P_3] can alter Ca^{2+} entry into cells. *a*, Ins (1,4,5)P_3 may act directly on a membrane channel to allow more Ca^{2+} to move through the channel. Ins (1,4,5)P_3 may act upon a membranous compartment closely associated with the plasma membrane. *b*, Upon emptying of Ca^{2+} from this compartment the compartment is refilled by direct entry of Ca^{2+} from the extracellular environment. This is the "capacitance" model of Putney. *c*, It is possible that Ins (1,4,5)P_3 mediates release from a compartment with Ins (1,3,4,5)P_3 and GTP potentiating movement of Ca^{2+} from a second compartment closely associated with the cell membrane. Upon transfer of Ca^{2+} between compartments the second compartment is refilled by Ca^{2+} entry from the extracellular space. (From [19])

rise in cytosolic free [Ca^{2+}] in response to agonists. IP_3 binding is increased by increasing pH, which occurs under many physiological conditions in which agonists result in the activation of protein kinase C and subsequent increase in Na^+/H^+ exchange across the cell plasma membrane.

Seiler et al. [287] examined the effects of various agents upon IP_3-induced Ca^{2+} release from platelet membrane vesicles. The slow Ca^{2+} channel blockers, nifedipine, verapamil, bepridil, and diltiazem, at concentrations as high as 50 μ*M* had no effect upon IP_3-induced Ca^{2+} release. Cinnarizine and flunarizine did inhibit release with an IC_{50} below 10^{-6} *M*. Other agents, which have been characterized as "intracellular Ca^{2+} antagonists", including ryanodine, 8-(*N*, *N*-dietylamino) octyl-3,4,5-trimethoxybenzoate (TMB-8), dantroleine, trifluoperazine, and chlorpromazine, at concentrations of 50 μ*M*, had no effect upon Ca^{2+} release. The local anesthetics, benzocaine and lidocaine, inhibited IP_3-induced Ca^{2+} release with IC_{50} values of 5 and 50 μ*M*, respectively. It is very possible, therefore, that the reduction in cytosolic [Ca^{2+}] transients observed with agents such as TMB-8 may be the result of other actions of these drugs in the Ca^{2+} signaling cascade of the cell.

Permeant anions, at concentrations above 40 m*M*, inhibited IP_3-mediated Ca^{2+} release. Cl^-, Br^-, I^-, and SO_4^- were equally effective as inhibitors [168]. The mechanisms for this inhibition are not known. Since DIDS does not alter IP_3-induced Ca^{2+} release, it is unlikely that an anion exchanger is involved in the release process. Also furosemide does not inhibit release indicating that a cation-anion cotransporter, similar to that present in the thick ascending limb of the nephron, is not involved [168].

The IP_3 Receptor

Worley et al. in 1987 identified high levels of specific binding sites for IP_3 in brain membranes [358]. In a subsequent paper [359] they characterized the binding further and examined regulation of binding by Ca^{2+} and pH. The binding sites were selective for IP_3 with a K_d of approximately 40 n*M*. Other inositol phosphates such as IP_4 and IP_2 exhibited much lower affinity for binding ($K_i \approx 10$ μ*M*). Binding increased sharply with increasing pH. Ca^{2+} inhibited binding with an IC_{50} of approximately 300 n*M*. This group subsequently found, however, that Ca^{2+} concentrations as high as 1.5 m*M* did not inhibit IP_3 binding to the purified receptor [307]. The receptor was purified using the fact that heparin bound to it [309]. The purified receptor has a molecular weight of 260 kDa. The receptor is a glycoprotein and is a substrate for cAMP-dependent protein kinase [308, 309]. After reconstitution of the receptor into lipid vesicles IP_3 stimulated release of Ca^{2+} from these reconstituted vesicles indicating that the purified IP_3 binding protein is the physiological receptor mediating IP_3- induced Ca^{2+} release [108]. The binding properties for IP_3 in the reconstituted vesicles were essentially the same as those of the purified unreconstituted protein. While IP_3 and inositol (1, 3, 4, 5)tetrakisphosphate (IP_4) at 10 μM and 20 μM, respectively, inhibited IP_3 binding, they did not induce Ca^{2+} release. These data revealed that a single protein mediates binding of IP_3 and stimulation of calcium flux. Furuichi et al. [121] have cloned the IP_3 receptor. They have found that the clone hybridized in cerebellar sections. When transfected into NG108-15, a mouse neuroblastoma/rat glioma hybrid cell line, these cells expressed binding sites for IP_3. The predicted protein structure was very similar to that of the ryanodine receptor [314], which mediates Ca^{2+} release from the sarcoplasmic reticulum of skeletal muscle.

Chadwick et al. [63] purified the IP_3 receptor from smooth muscle to near homogeneity. The smooth muscle receptor has a lower K_d for binding for IP_3 than does the brain receptor. The smooth muscle receptor is similar to the ryanodine receptor with regard to size and fourfold symmetry but appears to be different in important ways. Ryanodine does not bind to smooth muscle microsomes. The apparent size by migration on SDS/PAGE is smaller than the size predicted by migration of the ryanodine receptor.

Phosphatidylinositol Phospholipases C, and Ca^{2+} Transport

A number of hormones, growth factors and cytokines are believed to exert their effects upon cells by increasing cytosolic free Ca^{2+} due to activation of phospholipase C, enhanced turnover of membrane polyphosphoinositides, and subsequent release of intracellular Ca^{2+} stores and/or enhanced entry of Ca^{2+} from the extracellular milieu. Therefore, in order to understand cellular Ca^{2+} homeostasis it is important to understand how phospholipase C is activated, what the products of activation are, and how these products act to alter cellular Ca^{2+} transport.

Hokin and Hokin in 1953 reported that secretagogues enhanced the turnover of phospholipids in the pancreas [148]. Subsequently these lipids were identified as phosphatidylinositol and phosphatidic acid. Michell in 1975 proposed that inositol lipid hydrolysis, yielding lipophilic 1,2-diacylglycerol and a water-soluble inositol phosphate, occurs as a result of receptor activation and this process results in a rise in cytosolic free Ca^{2+} concentration [218]. It then became clear that agonists stimulated the turnover, primarily, of a less abundant derivative of phosphatidylinositol, phosphatidylinositol 4,5-bisphosphate [1, 110, 172]. When IP_3, the water soluble product of phospholipase C-mediated breakdown of phosphatidylinositol 4,5-bisphosphate, was found to release Ca^{2+} from intracellular stores of epithelial and nonepithelial cells [18, 70, 302], the link between receptor-polyphosphoinositol interaction and changes incytosolic free Ca^{2+} concentration was established. There are, however, a number of issues which remain unclear: Is the breakdown of polyphosphoinositides, as described above, responsible for the increases in cytosolic free $[Ca^{2+}]$ observed with physiological concentrations of agonists or does the increase in $[Ca^{2+}]_i$ precede the activation of phospholipase C? What are the roles of inositol phosphates in oscillations in $[Ca^{2+}]_i$? Do the products of metabolism of polyphosphoinositols play a direct role in Ca^{2+} transport across the cell membrane? Are inositol phosphates involved in the spatial heterogeneity of $[Ca^{2+}]_i$ in cells? Does diacylglycerol, the other product of phospholipase C-mediated breakdown of membrane phosphoinositides, modulated the Ca^{2+} signal or responsiveness of cellular processes to the Ca^{2+} signal?

Polyphosphoinositide Turnover and the Increase in Cytosolic Free $[Ca^{2+}]$

Most of the published data on hormone, cytokine, or growth factor-induced activation of phospholipase C and increase in $[Ca^{2+}]_i$ has led to the conclusion that the increase in $[Ca^{2+}]_i$ comes about due to the receptor-coupled activation of phospholipase C with subsequent generation of IP_3, which in turn releases Ca^{2+} from intracellular non-mitochondrial stores (18, 19, 29, 162). There is some controversy about this, however (104).

In order for polyphosphoinositide turnover to be implicated as the primary step in a cascade of cellular events leading to a rise in cytosolic $[Ca^{2+}]$ it is necessary to demonstrate that the enzyme mediating the conversion of phosphatidylinositol bisphosphate to diacylglycerol and IP_3, phospholipase C, can be activated at basal levels of cytosolic $[Ca^{2+}]$. In addition it is necessary to demonstrate that increased levels of IP_3 occur prior to a change in cytosolic free $[Ca^{2+}]$. Phosphatidylinositol 4, 5 bisphosphate is cleaved in vivo by phospholipase C at very low levels of $[Ca^{2+}]$ [352]. By contrast, the cleavage of phosphatidylinositol requires Ca^{2+} with a K_m of approximately 1 μM. Exton [104] has argued that many agonists, at lower physiological concentrations, increase cytosolic free $[Ca^{2+}]$ without increasing IP_3 levels or releasing intracellular Ca^{2+}.

Rapid kinetic measurements of changes in cytosolic $[Ca^{2+}]$ have been made using stopped flow fluorimetry. In human platelets, Sage and Rink [276] found that the mean time delay observed, before a detectable rise in $[Ca^{2+}]_i$ was measured in response to stimulation with thrombin, was 250 ± 10 ms. Thrombin activates polyphosphoinositide turnover in the platelet. In the absence of extracellular Ca^{2+} the delay was 300 ± 20 ms. By contrast, ADP, which does not enhance polyphosphoinsolitide turnover, increases $[Ca^{2+}]_i$ with a lag time of only 20 ± 10 ms, a time which was lengthened to 210 ± 10 ms in the absence of extracellular Ca^{2+}. These data demonstrate: (a) measurements of inositol phosphate turnover with less that 1 s kinetics will be necessary to appreciate fully the role of phospholipase C and IP_3 in the $[Ca^{2+}]_i$ transients; and (b) agonists can release Ca^{2+} from intracellular stores via processes that are independent of inositol phosphates. In carbachol-treated rat parotid acinar cells there was a dose-dependent lag of 70–220 ms before an increase in $[Ca^{2+}]_i$ could be measured. This lag was equivalent in the presence or absence of extracellular Ca^{2+} [216].

The Role of Inositol Phosphates in the Oscillations in $[Ca^{2+}]_i$

It has become recognized recently that cytosolic free $[Ca^{2+}]$ undergoes an oscillatory behavior in many excitable and nonexcitable cells [5, 86, 127, 132a,

149, 164, 182, 200, 221, 249, 289, 325, 353, 357]. Periodicity of the oscillations has generally been reported to be 5–120 s. The frequency of the oscillations has been found to be dependent upon agonist concentrations in some cells [164, 249, 357] while dependent of agonist concentration in others [127, 363].

These Ca^{2+} oscillators have been measured using aequorin, fura-2 or a Ca^{2+} electrode. The mechanisms responsible for these oscillations are not clear at present, nor are the implications of the oscillations. The oscillations persist in some cells after removing extracellular Ca^{2+} [249] whereas they cease in other cells under these conditions [164]. Prentki et al. [249] found that the response of insulinoma cells to carbamylcholine was very heterogenous. However, a single cell demonstrated a response which was reproducible on repeated exposure to the agonist. The amplitude, frequency, shape of the transient, as well as the latency in onset of response after agonist was added, were characteristic for a particular cell and constituted a "Ca^{2+} finger print." We have recently reported oscillations of $[Ca^{2+}]_i$ in mesangial cells with an oscillatory frequency dependent upon vasopressin concentration and temperature. Furthermore we established that the frequency of oscillation was directly dependent upon basal intracellular $[Ca^{2+}]$ (132a).

An attractive hypothesis for the oscillations is that they are the result of oscillatory patterns in cellular IP_3 in single cells. Unfortunately, levels of IP_3 in single cells cannot be measured at the present time. It is not known whether concentrations of IP_3 also oscillate. Another possibility is that IP_3 levels may remain constant but the response to IP_3 may vary temporally resulting in an oscillatory pattern. $[Ca^{2+}]$ oscillations are observed when IP_3 is injected into cells [17]. It is known, for example, that IP_3 binding to brain membranes is inhibited by peak levels of $[Ca^{2+}]_i$ achieved [359]. In pancreatic cells injection of the nonmetabolizable phosphorothioate derivative of IP_3 evoked $[Ca^{2+}]$ oscillations identical to those produced by acetylcholine . Since elevated levels of $[Ca^{2+}]$ inhibit IP_3 binding to its receptor [359], it may be that when peak levels of $[Ca^{2+}]$ are achieved, IP_3 no longer binds, Ca^{2+} is taken back up into IP_3-sensitive stores, and $[Ca^{2+}]_i$ falls. Another possibility that has been proposed [162, 363] emphasizes the role of Ca^{2+} in catalyzing the conversion of IP_3 to IP_4 [20]. As $[Ca^{2+}]_i$ increases, IP_3 will be converted to IP_4 which is ineffective in releasing Ca^{2+} from internal stores. With the associated reduction in IP_3 $[Ca^{2+}]_i$ will fall. At the same time the increased levels of IP_4 will potentially lead to enhanced uptake of Ca^{2+} into the IP_3 sensitive pool [147], thus priming the system for the next increase in IP_3 which results from the lower $[Ca^{2+}]_i$ and hence lower IP_3-kinase activity. Another model relies on the assumption that, as the IP_3 sensitive pool empties, it primes another pool which then releases Ca^{2+} via a Ca^{2+}-induced Ca^{2+} release process [17]. We will discuss these interactions further in subsequent sections of this chapter where the interactions between inositol phosphates and Ca^{2+} are discussed in more depth. A potential implication of oscillatory behavior is possible control of Ca^{2+}-mediated cell processes by frequency modulation rather than amplitude modulation.

Effect of IP_3 and IP_4 on Plasma Membrane Ca^{2+} Permeability

There are two phases to the Ca^{2+} response observed with agonists that activate phospholipase C. The first phase is transient and is characterized by an acute large increment in cytosolic free Ca^{2+} concentration. Subsequently, the Ca^{2+} concentration does not return to normal, however. There is a sustained elevation in intracellular Ca^{2+} concentrations, which usually can be prevented by removing extracellular Ca^{2+}. The first, early, response is generally believed to be related to the IP_3-mediated Ca^{2+} release from intracellular stores. The sustained elevation in $[Ca^{2+}]_i$ is believed to be related to enhanced Ca^{2+} entry across the plasma membrane. As indicated previously, plasma membrane Ca^{2+} channels may be opened as a direct result of agonist interaction with the cell or may open secondary to an intracellular intermediate. There has been considerable controversy regarding possible roles for IP_3 and IP_4 in the opening of these Ca^{2+} channels and the role of these Ca^{2+} channels in the sustained response to the agonist. Kuno and Gardner [185] were the first investigators to report a direct effect of IP_3 on plasma membrane Ca^{2+} channels using patch clamp techniques. They reported an IP_3-activated conductance in excised patches of jurket EG-1 human T-cell lymphocyte plasma membrane. In another study IP_3 appeared to increase the permeability of platelet plasma membrane vesicles to Ca^{2+} [366].

In other studies, however, direct exposure of plasma membranes vesicles of adipocytes [90], liver cells [87], or neurons [326] to IP_3 did not result in changes in Ca^{2+} permeability. Others have proposed that IP_3 is involved in enhancing Ca^{2+} entry into cells but not by a direct action on the cell plasma membrane channel. Putney [252], based upon experiments performed with the rat parotid gland, proposed that Ca^{2+} entry into the cell occurred secondary to release of Ca^{2+} from an IP_3 sensitive intracellular endoplasmic reticulum pool which is closely associated with the plasma membrane. According to this "capacitative" model, Ca^{2+} enters the endoplasmic reticulum directly due to close association of the plasma membrane and membrane of the IP_3-sensitive pools. Emptying of the IP_3-sensitive pool tiggers Ca^{2+} entry across the plasma membrane.

When IP_3 is injected into sea urchin eggs [161] and lacrimal acinar cells [226] investigators have found a maximal physiological response on Ca^{2+} entry only when Ins (1, 3, 4, 5)IP_4 is also coinjected. In the latter study on lacrimal acinar cells, using intracellular perfusion techniques, IP_3 or IP_4 alone had no effect, whereas together they evoked a sustained increase in Ca^{2+}-activated K^+ current which was dependent on external Ca^{2+}. In other experiments a cooperative effect of IP_4 was not observed either in patch clamp studies [244] or in isolated plasma membrane vesicles [90]. It has been suggested that IP_4 regulates the transfer of Ca^{2+} between intracellular pools [163] (Fig. 9). If IP_4 regulates transfer of Ca^{2+} between a membrane associated endoplasmic reticulum pool and an IP_3-sensitive pool [19], then various apparently disparent experimental results could be reconciled by considering that plasma membrane vesicles or patches may or may not have one or both of these "intracellular compartments" attached to them.

Spatial Heterogeneity of $[Ca^{2+}]_i$ and Inositol Phosphate Generation

There is spatial heterogeneity of the $[Ca^{2+}]_i$ changes in response to agonists. For example, in the adrenal chromaffin cell, agents, such as nicotine and high K^+, which would be expected to raise Ca^{2+} by increasing Ca^{2+} entry across the plasma membrane, resulted in initial increases in $[Ca^{2+}]_i$ just beneath the plasma membrane. This was followed by a much larger increase in $[Ca^{2+}]_i$ emanating from the internal aspects of the cells [240], perhaps due to Ca^{2+} induced Ca^{2+} release from internal stores [105] or Ca^{2+} induced IP_3 production [99]. In a number of cell types a Ca^{2+} wave has been identified to move across the cell in response to an agonist. This response was first observed in the egg upon fertilization [102] but also has been observed in other cells including cardiac myocytes and hepatocytes [275]. Agonists, such as bradykinin, angiotensin II, and muscarinic agonists, which stimulate polyphosphoinositide turnover, result in $[Ca^{2+}]_i$ increases which emanate from the internal aspects of the cell.

Effects of Diacylglycerol and Protein Kinase C on Ca^{2+} Transport

When a phospholipase C-linked hormone reacts with its receptor and polyphosphoinoside cleavage occurs diacylglycerol is generated in addition to IP_3. The IP_3-induced increase in $[Ca^{2+}]_i$, together with the increase in membrane diacylglycerol, results in the translocation of Ca^{2+}-insensitive protein kinase C to the plasma membrane where it becomes Ca^{2+}-sensitive due to the association of the enzyme with plasma membrane lipids. This activated protein kinase C then phosphorylates a number of proteins, some of which are directly related to Ca^{2+} transport. Park and Rasmussen [242] found that stimulation with angiotensin II resulted in distinct early (1 min) and late (30 min) patterns of phosphorylated proteins. The early set of proteins could be phosphorylated, in the absence of angiotensin II, by A23187, whereas the later phosphorylation pattern could be induced by A23187 and phorbol esters, two manuevers which would be expected to activate protein kinase C. The sustained phase of enhanced Ca^{2+} entry across the plasma membrane was found to be important for this late pattern of phosphorylation seen with angiotensin since it was prevented when angiotensin treated cells were cotreated with the Ca^{2+}-channel blocking agent nitrendipine, an agent which blocks the angiotensin II-induced sustained increase in Ca^{2+} influx [175]. When $[Ca^{2+}]_i$ initially increases in the cell, as a result of increased levels of IP_3, there is an activation of a number of calmodulin-dependent enzymes, including the plasma membrane Ca^{2+}-ATPase [282, 295]. The pump activity is also directly stimulated by protein kinase C activation [295]. Thus, during the latter phases of cell activation when Ca^{2+} influx is enhanced there is also enhanced Ca^{2+} efflux. During this time there is thus enhanced cycling of Ca^{2+} across the plasma membrane with enhanced efflux counterbalancing enhanced influx. Rasmussen and colleagues [4, 253] have proposed that it is this enhanced "cycling" that controls the rate and characteristics of protein phosphorylation that regulate the sustained physiological response to an agonist.

Calcium Binding Proteins

Calcium binding proteins are important for the transduction of Ca^{2+} signals within the cell. In addition they may be important for the physical translocation of Ca^{2+} within cells of transporting epithelia.

The EF Hand Family of Ca^{2+} Modulated Proteins

The "EF hand" (Fig. 10) was derived from the crystal structure of parvalbumin [179, 180, 222], a low molecular weight, acidic, calcium binding protein, found in muscle. Parvalbumin consists of six α-helices (A–F). The EF hand consists of two α-helices, E and F, which are bound together by a connecting calcium-binding loop. Functional EF hands appear to always occur in pairs and this motif has been identified in many [246] high affinity Ca^{2+} binding proteins, including calmodulin, troponin C, calbindin, calpain, the S100 family, and the intestinal Ca^{2+} binding protein. There are other important Ca^{2+} binding proteins which do not contain the motif, including protein kinase C, phospholipase A_2 and the annexins (46).

Calmodulin

Calmodulin is a 16.5-kDa protein found in most vertebrate tissues, which binds 4 mol Ca^{2+} per mole of protein [71]. The crystal structure has been recently

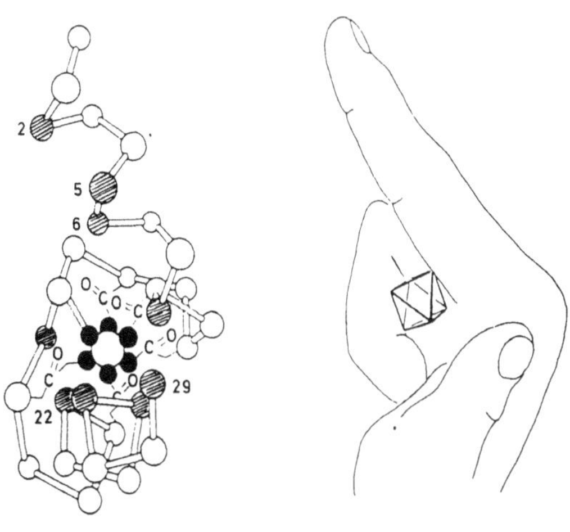

Fig. 10. A model demonstrating Ca^{2+} binding to the "EF hand". The first nine residues (helix E) symbolized by the index finger of the hand. The curved middle finger symbolized the amino acid residues which form a Ca^{2+} binding site. Residues 22 through 29, symbolized by the thumb, form the F helix. (From [246])

published [8]. There are four EF hand domains related in pairs. There are little differences between calmodulin present in organisms which diverged in evolution millions of years ago [246]. Calmodulin, in a Ca^{2+}-dependent fashion, activates cAMP phosphodiestease, adenylate cyclase, phosphorylase b kinase, phospholamban, and the Ca^{2+} ATPase of the cell membrane [7]. Calcium-calmodulin activates adenylate cyclase in the pancreatic islet cell [331], and this may play an important role in insulin secretion. There is no evidence for the presence of calmodulin in mitochondria.

Ca^{2+} and calmodulin have been implicated in the control of proliferation and differentiation [37, 44]. In general, however, it has been difficult to dissect out the role of Ca^{2+} itself from other early events occurring with mitogen activation, such as phospholipase turnover, increased Na^+/H^+ exchange, activation of protein kinase C or tyrosine kinases. Ca^{2+} has been implicated to be important in the regulation of the progression of cells from G_1 to S phase [141, 323, 345]. Studies in the lymphocyte have suggested an obligatory role for Ca^{2+} uptake for mitogen stimulated T-cell proliferation [199]. Intracellular injection of calcium stimulates mammalian oocytes to divide [120]. Furthermore it has been noted that transient increases in Ca^{2+} are noted during nuclear membrane breakdown, chromatin condensation [245], and the onset of anaphase [170]. Ca^{2+} had been related to the expression of protooncogenes [225a]. Regulation of intranuclear [Ca^{2+}] may play an important role in gene regulation. We have found that there is a Ca^{2+} dependent enhancement in mRNA levels of the immediate-early response gene, Egr-1(24a), in renal mesangial cells in culture. Nuclear [Ca^{2+}] increases when cells are activated by growth factors [360a, 344a]. Concentration gradients have been reported between cytoplasm and nucleus [344a, 360a, 348a]. Isolated rat liver nuclei contain an ATP-and calmodulin-dependent Ca^{2+} uptake process [235b]. The inositol 1,4,5-trisphosphate receptor has been localized to the nuclear membrane [268]. In summary, the nucleus may represent a distinct Ca^{2+} cellular compartment with its membrane representing a Ca^{2+} permeability barrier and possessing distinct Ca^{2+} transport properties and the ability to respond to cellular messengers. Control of nuclear [Ca^{2+}] may be important for cell proliferation and differentiation. In NB_2 lymphoma cells, TMB-8, which decreases the peak elevation of Ca^{2+} observed with many agonists [252], inhibited c-*myc* expression and DNA synthesis [232]. There is an important difference between the Ca^{2+} sensitivity of cell proliferation in normal and neoplastic cells [38]. Neoplastic cells lose their dependence on calcium for proliferation and mitosis. Cells transformed by avian sarcoma virus B77 [40] or Rous sarcoma virus [39] lose Ca^{2+} dependency.

There is a good deal of evidence that calmodulin is important for cell proliferation. Calmodulin has been localized to the mitotic spindle [211,364]. Many studies have demonstrated inhibition of cell proliferation with "calmodulin antagonists," such as trifluoperazine, chlorpromazine, or the napthalene-sulfonamides [64, 65, 278]. These studies can be criticized, however, since these antagonists are not specific for calmodulin. Nevertheless, there is considerable additional evidence to implicate calmodulin in proliferation [254].

Calmodulin has been shown to be important for phosphorylation of some nuclear proteins [203a] and progression of the cell cycle [64, 65, 278].

Recently it has been reported (292a) that the adenosine 3′,5′-monophosphate (cAMP) response element-binding protein functions as a Ca^{2+}-dependent transcription factor and serves as a substrate for Ca^{2+}-calmodulin dependent protein kinases, thus providing a link between changes in $[Ca^{2+}]_i$ and changes in gene transcription.

Intestinal Ca^{2+} Binding Protein

One of the primary actions of vitamin D is to enhance intestinal calcium absorption. While the mechanism of this enhanced absorption is not clear, one of the effects of vitamin D is to enhance the synthesis of a calcium binding protein via the active metabolic product 1,25-dihydroxyvitamin D_3 [$1,25(OH)_2D_3$]. Wasserman and Taylor [344] noted that vitamin D administration to rachitic chicks resulted in the intestinal synthesis of a Ca^{2+} binding protein. The avian intestinal Ca^{2+} binding protein is a heat-stable, acidic protein of approximately 28 kDa. It has four high-affinity calcium binding sites (six EF-hand domains with two unable to bind Ca^{2+}). The protein found in the avian intestine is also found in the avian kidney and shell gland and all vertebrate tissues examined, except striated muscle and mammalian intestine [22]. This protein has also been called calbindin. In the mammalian intestine a distantly related 9-kDa EF hand protein replaces calbindin [22]. This protein binds two calcium ions per molecule. Its crystal structure has been delineated [311]. It is found throughout the small intestine with the highest concentration found in the duodenum [6]. Kretsinger and colleagues [181] proposed that the 9-kDa protein acts to carry Ca^{2+} across the intestinal epithelial cell. Feher [106, 107] has provided evidence for such a facilitated diffusion process, and Bonner and colleagues [43] found a direct relationship between duodenal Ca^{2+} transport and the amount of Ca^{2+} binding protein. In the rat kidney the Ca^{2+} binding protein, calbindin, is found only in the distal nephron with highest levels found in the distal convoluted tubule [318], the site of high active Ca^{2+} reabsorption.

Calbindin has been cloned [356]. Rat kidney contains three RNA species [1.9, 2.8, and 3.2 kb pairs] for calbindin. Each is dependent upon $1,25(OH)_2D_3$ for expression. Increased levels of mRNA are observed 2 h after a single injection of $1,25(OH)_2D_3$ and levels peak at 12 h [155, 336]. Borke et al. [47] found that intercalated cells of the distal nephron did not stain with a polyclonal antibody directed against the 28-kDa Ca^{2+} binding protein, while principal cells stained.

S100 Proteins

S100 proteins refers to a family of dimeric proteins. The subunits are of two types: α and β and the hetero- and homodimeric forms have different tissue distribution. Two Ca^{2+} ions bind each monomer but the affinity is lower than for many other EF hand proteins (K_d 60–200). Several proteins in this family vary with the state of growth and differentiation of the cell, and some bind to microtubule-related proteins [246]. The role of these proteins in Ca^{2+} transport remains ill-defined.

Mechanisms of Transepithelial Calcium Transport

Calcium can move across an epithelial layer by moving between the cells (e.g., by passive facilitated or non-facilitated diffusion or solvent drag) or through the cells. If calcium moves through the cytoplasm of the cell there must be tight coupling of apical calcium entry with basolateral calcium extrusion so that transepithelial calcium flux can occur without significant increases in cytosolic free [Ca^{2+}]. Another possible pathway for transcellular calcium movement that would "protect" the various cytoplasmic compartments against increases in [Ca^{2+}] would involve endocytosis with Ca^{2+} packaged in endocytic vesicles.

Intestine

Ca^{2+} transport involves both a transcellular, vitamin D regulated, saturable process and a non-saturable process. In the intestine the saturable process consists of facilitated calcium entry at the luminal membrane, intracellular diffusion, and a Ca^{2+}-ATPase mediated Ca^{2+} extrusion step on the basolateral side of the cell [122]. The intracellular diffusion step appears to be rate limiting. As described above the intracellular Ca^{2+} diffusion step may be facilitated by the intestinal Ca^{2+} binding protein. This saturable transport process predominates in the proximal intestine. In addition to this saturable process there is a nonsaturable process present throughout the intestine. Bonner et al. [21] showed that 16% of luminal Ca^{2+} is absorbed per hour from the lumen of the duodenum of a vitamin D deficient animal. When the saturable Ca^{2+} transport was blocked with trifluoperazine, absorption of calcium from rightside-out duodenal, jejunal or ileal sacs was linear with time and was virtually identical to that of absorption of phenol [23]. The rate of Ca^{2+} absorption was similar to the previously determined nonsaturable rate of Ca^{2+} transfer from in situ loop experiments by the same group [21]. This non-saturable Ca^{2+} transport may occur via a paracellular route. Nellans has proposed that the same paracellular pathway for Ca^{2+} may also account for net loss of Ca^{2+} due to paracellular Ca^{2+} secretion in the absence of sufficient dietary Ca^{2+} [235].

Kidney

The concentration of Ca^{2+} in glomerular filtrate is approximately equal to that of free (nonbound) Ca^{2+} in the plasma [137, 188]. Approximately 98% of filtered calcium is reabsorbed by the nephron.

Proximal Tubule. Approximately 60% of filtered calcium is reabsorbed in the proximal tubule. Lassiter et al. [187] showed that Ca^{2+} concentration along the proximal tubule did not change from that in the glomerular filtrate. In many studies the tubular fluid to glomerular filtrate [Ca^{2+}] ratio is 1.0–1.2 [310]. Although the contribution of passive processes to transepithelial Ca^{2+} transport in this nephron segment is not clear, Ullrich and colleagues estimated that only 20% of proximal convoluted tubule transepitheial calcium transport is due to active transport [327]. Although Ca^{2+} can entry the cell via passive processes down its electrochemical concentration gradient its extrusion involves an Ca^{2+} ATPase or Na^+–Ca^{2+} exchance process dependent upon a transmembrane Na^+ gradient established by a Na^+K^+ ATPase as described previously. Passive movement can occur due to passive diffusion down a Ca^{2+} concentration gradient from lumen to peritubular space or due to solvent drag associated with the large amount of water movement in this nephron segment. A significant amount of fluid reabsorption is believed to move via a paracellular route [45].

Murayama et al. [230] examined bidirectional fluxes of Ca^{2+} and concluded that the proximal tubule is highly permeable to Ca^{2+}. In the S_2 segment of the proximal tubule, where the bicarbonate concentration is lower and the chloride concentration higher than in the S_1 segment, the transepithelial potential becomes lumen positive [205]. Therefore there is an electrochemical driving force for passive Ca^{2+} reabsorption in addition to solvent drag in this segment. Sacks and Bourdeau [254] have found that ionic diffusion can account for all the calcium absorption in the isolated S_2 segment from the rabbit kidney. These data, however, do not exclude the possibility of a small component of active Ca^{2+} transport undetectable by their techniques. By contrast Rouse and colleagues [272] reported that, in the pars recta approximately 75% of transepithelial calcium transport is active and this transport is not sensitive to ouabain suggesting that it does not involve Na^+/Ca^{2+} exchange [272]. From their report, however, it is not clear whether the segments studied were from the S_2 or S_3 segment of the proximal tubule.

Loop of Henle. Both the isolated rabbit thin descending limbs and the thin ascending limbs are relatively impermeable to Ca^{2+} [266, 272]. The thick ascending limb reabsorbs approximately 20% of filtered calcium. Ca^{2+} transport in the medullary thick ascending limb is passive and can be explained by the transepithelial potential difference. There is controversy regarding the movement of Ca^{2+} in the cortical thick ascending limb. Rocha et al. [266], Imai [160], and Suki et al. [305] concluded that there was evidence for active Ca^{2+} transport since the transepithelial Ca^{2+} flux could not be explained by the transepithelial voltage. Boudeau and Burg [33] found that Ca^{2+} flux in the isolated rabbit thick ascending limb was voltage dependent. They concluded that the movement of Ca^{2+} was passive finding no evidence for active transport. They further suggested that the movement of Ca^{2+} was by single file diffusion. They subsequently found that parathyroid hormone (PTH) increased Ca^{2+} transepithelial movement by a mechanism proposed to be mediated by adenylate cyclase. Bourdeau et al. [305] found complete inhibition of the Ca^{2+} flux when transepithelial voltage was reduced to 0 with furosemide, whereas Imai [160] and Suki et al. [305] found little change in Ca^{2+} flux with furosemide. Dietl and Oberleithner [94] studied Ca^{2+} transport in the diluting segment of the frog (*Rana pipiens*) kidney. Their results indicate that Ca^{2+} reabsorption is passive, driven by the transepithelial potential, which in turn, is dependent upon NaCl reabsorption. They suggest that this passive movement occurs via a paracellular route.

Distal Convoluted Tubule. Approximately 10%–15% of filtered calcium is reabsorbed in the distal convoluted tubule. Lassiter et al. [187] showed that Ca^{2+} was absorbed between the early and late distal tubule. Costanzo and Windhager later confirmed these studies and found that Ca^{2+} reabsorption was proportional to Na^{+} reabsorption. The intraluminal Ca^{2+} concentration was always lower than that of plasma. This, together with the fact that the distal convoluted tubule transepithelial voltage is lumen negative and Ca^{2+} permeability was negligible, as measured in backflux studies, led to the conclusion that the transport of Ca^{2+} is active in this segment. This is consistent with the presence of the highest amount of Ca^{2+} ATPase in this nephron segment [97] and the large amounts of the vitamin D dependent Ca^{2+} binding protein in this segment as discussed previously.

Collecting Duct. The cortical collecting duct plays a relatively minor role in overall kidney tubule transport of Ca^{2+}. The electrochemical potential gradient favors secretion and in fact there may be a small secretory Ca^{2+} leak, which is however, very limited by the low passive Ca^{2+} permeability [36]. In two studies [290, 291] no active Ca^{2+} transport was found in isolated "light" cortical segments. Shareghi and Stoner [290] found Ca^{2+} transport in the earliest part of the collecting tubule or "granular" segment (connecting tubule) only. Holt and Lechene [150] found net Ca^{2+} absorption which could be inhibited by vasopressin. Using microcatheterization, Bengele et al. [15] demonstrated Ca^{2+} absorption in the inner medullary segments of the collecting duct. The (fluid/plasma) Ca/inulin ratio decreased from 2.2 at the beginning of the inner medullary collecting duct to 0.84 in the urine. Ca^{2+} ATPase is present in relatively high amounts in the cortical collecting tubule but there are very low amounts in the medullary collecting tubule [97].

Hormonal Involvement in Ca^{2+} Transport

Vitamin D_3. As discussed above $1,25(OH)_2D_3$ plays an important role in intestinal Ca^{2+} reabsorption but the mechanisms by which this is accomplished are not completely understood. Rizzoli and colleagues [265] found, in thyroparathyroidectomized rats, that $1,25(OH)_2D_3$ increased plasma $[Ca^{2+}]$ but also increased urinary calcium excretion. In humans urinary Ca^{2+} increases within 7 h of administration of $1,25(OH)_2D_3$, without any direct inhibitory affect upon

PTH production [201]. These effects are due to the enhanced transport of calcium across the intestinal epithelium. Vitamin D and its metabolites enhances overall renal calcium reabsorption in the rat (22). The primary action of vitamin D metabolites is on distal convoluted tubular reabsorption of calcium, where the receptors for 1,25-$(OH)_2$-D_3 are localized. The effect of 1,25-$(OH)_2$-D_3 is likely facilitated by the 28 kD calcium binding protein, calbindin.

Parathyroid Hormone. Both proximal and distal renal tubules have receptors for PTH. In the proximal tubule PTH decreases solute (including calcium) and water absorption. With increasing plasma levels of PTH distal tubular reabsorption of Ca^{2+} increases. PTH may enhance transepithelial Ca^{2+} transport via its effect to enhance levels of 1,25$(OH)_2D_3$. PTH, however, also has a direct effect on the tubule. It increases Ca^{2+} transport by the isolated cortical thick ascending limb but not the medullary thick ascending limb of the rabbit [305]. cAMP has been implicated in the action of PTH [249] since the hormone increases cAMP levels in proximal tubules, cortical thick ascending limb and distal convoluted tubules [224]. It has been suggested that the effect of PTH is not related to active transport [33]; however, Friedman [116] has reported that PTH and cAMP analogs increased active calcium transport in this segment isolated from the mouse.

Calcitonin. Calcitonin enhances Ca^{2+} transport in the medullary thick ascending limb of the rabbit but has no effect on Ca^{2+} transport in the cortical thick ascending limb [306]. This can be correlated with a much greater calcitonin-induced stimulation of adenylate cyclase in the medullary versus cortical thick limbs [68, 224]. 8-Bromo cAMP induces enhanced Ca^{2+} transport in both cortical and medullary thick limbs. Thus it was proposed that, like PTH, the effects of calcitonin may be due to changes in cell cAMP levels [306]. The increase in net calcium transport in the medullary thick ascending limb with calcitonin and cAMP occurs without a change in voltage and it has been suggested by Suki and Rouse [306] that calcitonin either uncovers an active Ca^{2+} transport process normally operating at low levels, or induces rectification of Ca^{2+} movement across the cell membrane, allowing greater Ca^{2+} efflux from the cell relative to Ca^{2+} influx.

Conclusions

In this review I have attempted to provide an overview of passive calcium transport processes. I have had to be selective and, in doing, have excluded references to many important works in this extensive field. Much more is known about these processes of calcium transport in nonepithelial tissue than in epithelial structures. Many paradigms developed in cardiac, skeletal and smooth muscle can be observed in epithelial cells but it is also clear that there are likely to be fundamental differences between excitable cells and nonexcitable cells especially with regard to Ca^{2+} channel characteristics.

I hope that I have conveyed some of the excitment surrounding many of the unresolved issues in cellular calcium metabolism. While we have learned a great deal there remains a great deal to be learned.

Acknowledgement. The author wishes to thank Dr. Joseph Cheung for reading the manuscript and offering many helpful suggestions.

References

1. Abdel-Latif AA, Akhtar R, Hawthorne JN (1977) Acetylcholine increases the breakdown of triphosphoinositide of rabbit iris muscle prelabelled with [^{32}P] phosphate. Biochem J 162:61–73
2. Affolter H, Carafoli E (1980) The Ca^{2+}-Na^{+} antiporter of heart mitochondria operates electroneutrally. Biochem Biophys Res Commun 95:193–196
3. Akerman KEO, Wikstrom MKF, Saris N-E (1977) Effect of inhibitors on the sigmoidicity of the calcium ion transport kinetics in rat liver mitochondria. Biochim Biophys Acta 464:287–294
4. Alkon DL, Rasmussen H (1988) A spatial-temporal model of cell activation. Science 239:998–1005

4a. Allen TJA, Noble D, Reuter H (eds) (1989) Sodium-calcium exchange. Oxford University Press, Oxford

5. Ambler SK, Poenie M, Tsien RY, Taylor P (1988) Agonist-stimulated oscillations and cycling of intracellular free calcium in individual cultured muscle cells. J Biol Chem 263:1952–1959
6. Arnold BM, Kuttner M, Willis DM, Hitchman AJW, Harrison JE, Murray TM (1975) Can J Physiol Pharmacol 53:1135
7. Ausiello DA, Skorecki KL, Verkman AS, Bonventre JV (1987) Vasopressin signaling in kidney cells. Kidney Int 31:521–529
8. Babu YS, Bugg CE, Cook WJ (1988) Structure of calmodulin refined at 2.2 Å resolution. J Mol Biol 204:191–204
9. Baker PF, Blaustein MP, Hodgkin AL, Steinhardt RA (1969) The influence of calcium on sodium efflux in squid axons. J Physiol (Lond) 200:431–458
10. Baker PF, Baker PF, Blaustein MP Hodgkin AL, Steinhardt RA (1969) The influence of calcium on sodium efflux in squid axons. J Physiol (Lond) 200:431–458
11. Barhanin J, Borsotto M, Coppola T, Fosset M, Hosey MM, Mourre C, Pauron D, Qar J, Romey G, Schmid A, Vandaele S, Van Rentergheim C, Lazdunski M (1989) Biochemistry, molecular pharmacology, and functional control of Ca^{2+} channels. Ann NY Acad Sci 360:15–26
12. Beam KG, Knudson CM (1988) Calcium currents in embryonic and neonatal mammalian skeletal muscle. J Gen Physiol 91:781–798
13. Bean BP (1985) Two hands of calcium channels in canine atrial cells. Differences in kinetics, selectivity and pharmacology. J Gen Physiol 86:1–30
14. Bean BP, Sturek M, Puga A, Hermsmeyer K (1986) Calcium channels in muscle cells isolated from rat: modulation by dihydropyridine drugs. Circ Res 59:229–235
15. Bengele HH, Alexander EA, Lechene CP (1980) Calcium and magnesium transport along the inner medullary collecting duct of the rat. Am J Physiol 239 (Renal Fluid Electrolyte Physiol) 8:F24–F29
16. Benham CD, Tsien RW (1986) Calcium-permeable channels in vascular smooth muscle: voltage-activated, receptor-operated and leak channels. In: Mandel LJ, Eaton DC (eds) Cell calcium and the control of membrane transport. Rockefeller University Press, New York, pp 45–64

17. Berridge MJ, Cobbold PH, Cuthbertson KSR (1988) Spatial and temporal aspects of cell signalling. Philos Trans R Soc Lond [Biol] 320:325–343
18. Berridge MJ (1987) Inositol trisphosphate and diacylglycerol: two interacting second messengers. Annu Rev Biochem 56:159–193
19. Berridge MJ, Irvine RF (1988) Inositol phosphates and cell signalling. Nature 341:197–205
20. Biden TJ, Wollheim CB (1986) Ca^{2+} regulates the inositol tris/tetrakisphosphate pathway in intact and broken preparations of insulin-secreting RIN m5F cells. J Biol Chem 261:11 931–11 934
21. Bonner F, Pansu D, Stein WD (1986) An analysis of calcium transport across the rat intestine. Am J Physiol 250 (Gastrointest Liver Physiol 13):G561–G569
22. Bonner F (1989) Renal calcium transport: mechanisms and regulation—an overview. Am J Physiol 257 (Renal Fluid Electrolyte Physiol) 26:F707–F711
23. Bonner F, Spence K (1988) Non-saturable Ca transport in the rat intestine is via the paracellular pathway. In: Bonner F, Peterlik M (eds) Cellular calcium and phosphate transport in health and disease. Liss, New York, pp 277–283
24. Bonventre JV, Cheung JY (1986) Cytosolic free calcium concentration in cultured renal epithelial cells. Am J Physiol 250:F329–F338
24a. Bonventre JV, Sullivan S, Sukhatme VP, Ouellette AJ (1989) Vasopressin increases the expression of the early growth response gene, Egr-l in cultured mesangial cells. Kidney Int 35:169
25. Bonventre JV, Swidler M (1988) Calcium dependency of prostaglandin E_2 production in rat glomerular mesangial cells. Evidence that protein kinase C modulates the Ca^{2+}-dependent activation of phospholipase A_2. J Clin Invest 82:168–176
26. Bonventre JV (1988) Mediators of ischemic renal injury. Annu Rev Med 39:531–544
27. Bonventre JV, Skorecki KL, Kreisberg JI, Cheung JY (1986) Vasopressin increases cytosolic free calcium concentration in glomerular mesangial cells. Am J Physiol 251 (Renal Fluid Electrolyte Physiol) 20:F94–F102
28. Bonventre JV, Malis CD, Cheung JY (1990) Role of calcium in ischemic injury. Mechanisms and consequences of tissue injury. In: Zelenock GB (ed) Clinical ischemic syndromes. Mosby, St Louis, pp 227–242
29. Bonventre JV, Weber P, Gronich JH (1988) PAF and PDGF increase cytosolic $[Ca^{2+}]$ and phospholipase activity in mesangial cells. Am J Physiol 254:F87–F94
30. Borsotto M, Barhanin J, Fosset M, Lazdunski M (1985) The 1,4-dihydropyridine receptor associated with the skeletal muscle voltage-dependent Ca^{2+} channel. J Biol Chem 260:14 255–14 263
31. Borsotto M, Barhanin J, Norman RI, Lazdunski M (1984) Purification of the dihydropyridine receptor of the voltage-dependent Ca^{2+} channel from skeletal muscle transverse tubules using (+) ^{3}H PN 200-110. Biochem Biophys Res Commun 122:1357–1366
32. Bouchelouche PN, Hainau B, Frederiksen O (1989) Effect of BAY K 8644 on cytosolic free calcium in isolated rabbit gall-bladder epithelial cells. Cell Calcium 10:37–46
33. Bourdeau JE, Burg MB (1979) Voltage dependence of calcium transport in the thick ascending limb of Henle's loop. Am J Physiol 236 (Renal Fluid Electrolyte Physiol) 5:F357–F364
34. Bourdeau JE, Burg MB (1980) Effect of PTH on calcium transport across the cortical thick ascending limb of Henle's loop. Am J Physiol 239(Renal Fluid Electrolyte Physiol) 8:F121–F126
35. Bourdeau JE, Buss SL, Vurek GG (1982) Inhibition of calcium absorption in the cortical thick ascending limb of Henle's loop by furosemide. J Pharmacol Exp Ther 221:815–819
36. Bourdeau JE, Hellstrom-Stein RJ (1982) Voltage-dependent calcium movement across the cortical collecting duct. Am J Physiol 242(Renal Fluid Electrolyte Physiol) 11:F285–F292
37. Boynton AL (1988) Calcium and epithelial cell proliferation. Miner Electrolyte Metab 14:86–94
38. Boynton AL, Whitfield (1976) Different calcium requirements for proliferation of conditionally and unconditionally tumorigenic mouse cells. Proc Natl Acad Sci USA 73:1651–1654
39. Boynton AL, Whitfield JF (1978) Calcium requirement for the proliferation of cells infected with a temperature-sensitive mutant of Rous sarcoma virus. Cancer Res 38:1237–1240
40. Boynton AL, Whitfield JF, Isaacs RJ, Tremblay RG (1977) Different extracellular calcium requirements for proliferation of non-neoplastic, preneoplastic and neoplastic mouse cells. Cancer Res 37:2657–2661
41. Bridge JHB, Bassingthwaighte JB (1983) Uphill sodium transport driven by an inward calcium gradient in heart muscle. Science 219:178–180

42. Brierley GP, Slautterback DB (1964) Studies on ion transport. IV. An electron microscope study of the accumulation of Ca^{2+} and inorganic phosphate by heart mitochondria. Biochim Biophys Acta 82:183–186
43. Bronner F, Pansu D, Stein WD (1986) An analysis of intestinal calcium transport across the rat intestine. Am J Physiol 250(Gastrointest Liver Physiol) 13:G561–G569
44. Brown WF (1984) The basic connection. Nature 312:312–313
45. Burg MB (1986) Renal handling of sodium, chloride, water, amino acids, and glucose. In: Brenner BM, Rector FC Jr (eds) The kidney. Saunders, Philadelphia, pp 145–175
46. Burgoyne RD, Geisow MJ (1989) The annexin family of calcium-binding proteins. Cell Calcium 10:1–10
47. Burke JL, Caride A, Verma AK, Penniston JT, Kumar R (1989) Plasma membrane calcium pump and 28-kDa calcium binding protein in cells of rat kidney distal tubules. Am J Physiol 257(Renal Fluid Electrolyte Physiol) 26:F842–F849
48. Campbell KP, Leung AT, Sharp AH (1988) The biochemistry and molecular biology of the dihydropyridine-sensitive calcium channel. Trends Neuro Sci 11:425–430
49. Campbell AK (1983) Intracellular calcium: its universal role as regulator. Wiley, New York
50. Carafoli E (1982) The transport of calcium across the inner membrane of mitochondria. In: Carafoli E (ed) Membrane transport of calcium. Academic, New York, pp 109–139
51. Carafoli E (1987) Intracellular calcium homeostasis. Annu Rev Biochem 56:395–433
52. Carafoli E (1988) Intracellular calcium regulation, with special attention to the role of the plasma membrane calcium pump. J Cardiovasc Pharmacol 12[Suppl 3]:S77–S84
53. Carafoli E (1989) Calcium pump of the plasma membrane. Physiol Rev 71:129–153
54. Carafoli E, Longoni S (1987) The plasma membrane in the control of the signalling function of calcium. Soc Gen Physiol Ser 42:21–29
55. Carafoli E, Niggli V, Malmstrom K, Caroni P (1980) Calmodulin in natural and reconstituted calcium transporting systems. Ann NY Acad Sci 356:258–266
56. Carafoli E, Tiozzo R, Lugli G, Crovetti F, Kratzing C (1974) The release of calcium from heart mitochondria by sodium. J Mol Cell Cardiol 6:361–371
57. Carbone E, Lux HD (1984) A low voltage-activated, fully inactivating Ca channel in vertebrate sensory neurons. Nature 310:501–502
58. Caroni P, Carafoli E (1981) The Ca^{2+}-pumping ATPase of heart sarcolemma. Characterization, calmodulin dependence, and partial purification. J Biol Chem 256:3263–3270
59. Caroni P and Carafoli E (1983) The regulation of the Na^{+}-Ca^{2+} exchanger of heart sarcolemma. Eur J Biochem 132:451–460
60. Casey PJ, Gilman AG (1988) G protein involvement in receptor-effector coupling. J Biol Chem 263:2577–2580
61. Cauvin C, Loutzenhizer R, Van Breemen C (1983) Mechanisms of antagonists induced vasodilation. Annu Rev Toxicol 23:373–396
61a. Ceretlo L, Lagnado L, Perry RJ, Robinson DW, McNaughton PA (1989) Extrusion of calcium from rod outer segments is driven by both sodium and potassium gradients. Nature 337:740–743
62. Chad J, Kalman D, Armstrong D (1987) The role of cAMP-dependent phosphorylation in the maintenance and modulation of voltage-activated calcium channels. In: Mandel LJ, Eaton DC (eds) Cell calcium and the control of membrane transport. Rockefeller University Press, New York, pp 167–186
63. Chadwick CC, Saito A, Fleischer S (1990) Isolation and characterization of the inositol trisphosphate receptor from smooth muscle. Proc Natl Acad Sci USA 87:2132–2136
64. Chafouleas JG, Bolton We, Hidaka H, Boyd AE III, Means AR (1982) Calmodulin and the cell cycle: involvement in regulation of cell-cycle progression. Cell 28:41–50
65. Chafouleas JG, Legace L, Bolton WE, Boyd AE III, Means AR (1984) Changes in calmodulin and its mRNA accompany reentry of quiescent (G0) cells into the cell cycle. Cell 36:73–81
66. Chamberlain BK, Volpe P, Fleisher S (1984) Calcium-induced calcium release from purified cardiac sarcoplasmic reticulum vesicles: general characteristics. J Biol Chem 259:7540–7546
67. Chance B (Editor) (1963) Energy-linked function of mitochondria. Academic, New York
68. Charbardes D, Imbert-Teboul M, Montegut M, Clique A, Morel F (1976) Distribution of calcitonin-sensitive adenylate cyclase activity along the rabbit kidney tubule. Proc Natl Acad Sci USA 73:3608–3612
69. Chen C, Corbley MJ, Roberts TM, Hess P (1989) Transformation changes the functional expression of calcium channels in fibroblasts. Ann NY Acad Sci 560:118–123
70. Cheung JY, Constantine JM, Bonventre JV (1986) Regulation of cytosolic free calcium

concentration in cultured renal epithelial cells. Am J Physiol 251:F690–F701
71. Cheung WY (1980) Calmodulin plays a pivotal role in cellular regulation. Science 207:19–27
72. Cheung WY (1980) Calmodulin plays a pivotal role in cellular regulation. Science 207:19–27
73. Choi DW (1988) Calcium-mediated neurotoxicity: relationship to specific channel types and role in ischemic damage. Trends Neurosci 11:465–469
74. Chueh S-H, Gill DL (1986) Inositol 1,4,5-trisphosphate and guanine nucleotides activate calcium release from endoplasmic reticulum via distinct mechanisms. J Biol Chem 261:13 883–13 886
75. Cognard C, Lazdunski M, Romey G (1986) Different types of Ca^{2+} channels in mammalian skeletal muscle cells in culture. Proc Natl Acad Sci USA 83:517–521
76. Coll KE, Joseph SK, Corkey BE, Williamson JR (1982) Determination of the matrix free Ca^{2+} concentration and kinetics of Ca^{2+} efflux in liver and heart mitochondria. J Biol Chem 257:8696–8704
77. Cook NJ, Knaupp UB (1988) Solubilization, purification, and reconstitution of the sodium-calcium exchanger from bovine retinal rod outer segments. J Biol Chem 263:11 382–11 388
78. Crompton M (1985) The regulation of mitochondrial calcium transport in heart. Curr Top Membr Transp 25:231–276
79. Crompton M, Capano M, Carafoli E (1976) The sodium-induced efflux of calcium from heart mitochondria. A possible mechanism for the regulation of mitochondrial calcium. Eur J Biochem 69:453–462
80. Crompton M and Heid I (1978) The cycling of calcium, sodium, and protons across the inner membrane of cardiac mitochondria. Eur J Biochem 91:599–608
81. Crompton M, Moser R, Ludi H, Carafoli E (1978) The interrelations between the transport of sodium and calcium in mitochondria of various mammalian tissues. Eur J Biochem 82:25–31
82. Crompton M, Sigel E, Salzmann M, Carafoli E (1976) A kinetic study of the energy-linked influx of Ca^{2+} into heart mitochondria. Eur J Biochem 69:429–434
83. Curtis BM, Catterall WA (1984) Purification of the calcium antagonist receptor of the voltage-sensitive calcium channel from skeletal muscle transverse tubules. Biochemistry 23:2113–2118
84. Curtis BM, Catterall WA (1985) Phosphorylation of the calcium antagonist receptor of the voltage-sensitive calcium channel by cAMP-dependent protein kinase. Proc Natl Acad Sci USA 82:2528–2532
85. Curtis BM, Catterall WA (1985) Reconstitution of the voltage-sensitive calcium channel purified from skeletal muscle transverse tubules. Biochemistry 25:3077–3083
86. Cuthbertson KSR, Cobbold PH (1985) Phorbol ester and sperm activate mouse occytes by inducing sustained oscillations in cell Ca^{2+}. Nature 316:541–542
87. Dargemont C, Hilly M, Claret M, Mauger J-P (1988) Characterization of Ca^{2+} fluxes in rat liver plasma-membrane vesicles. Biochem J 256:117–124
88. Davidson FF, Dennis EA, Powell M, Glenney JR Jr (1987) Inhibition of phospholipase A_2 by "lipocortins" and calpactins. J Biol Chem 262:1698–1705
89. Dean WL (1989) Structure, function and subcellular localization of a human platelet Ca^{2+}-ATPase. Cell Calcium 10:289–297
90. Delfert DM, Hill S, Pershadsingh HA, Sherman WR, MacDonald JM (1986) Myoinositol 1,4,5-trisphosphate mobilizes Ca^{2+} from isolated adipocyte endoplasmic reticulum but not from plasma membranes. Biochem J 236:37–44
91. Deluca HF, Engstrom G (1961) Calcium uptake by rat kidney mitochondria. Proc Natl Acad Sci USA 47:1744–1750
92. Denton RM, McCormack JC (1985) Ca^{2+} transport by mammalian mitochondria and its role in hormone action. Am J Physiol 249:E543–E554
93. Dhalla NS, Zhao D (1988) Cell membrane Ca^{2+}/Mg^{2+} ATPase. Prog Biophys Mol Biol 52:1–37
94. Dietl P, Oberleithner H (1988) Paracellular Ca^{2+} transport in diluting segment of frog kidney. In: Bonner F, Peterik M (eds) Cellular calcium and phosphate transport in health and disease. Liss, New York, pp 285–292
95. Dolphin AC, Scott RH (1987) Calcium channel currents and their inhibition by (−)-baclofen in rat sensory neurones: modulation by guanine nucleotides. J Physiol (Lond) 386:1–17
96. Dominguez JH, Rothrock JK, Macias WL, Price J (1989) Na^{+} electrochemical gradient and Na^{+}–Ca^{2+} exchange in rat proximal tubule. Am J Physiol 257 (Renal Fluid Electrolyte Physiol 26):F531–F538
97. Doucet A, Katz AI (1982) High affinity Ca-Mg-ATPase along the rabbit nephron. Am J Physiol 242 (Renal Fluid Electrolyte Physiol 11):F346–F352

98. Ebashi S, Endo M (1968) Calcium ion and muscle contraction. Prog Biophys Mol Biol 18:123
99. Eberhard DA, Holz RW (1987) Cholinergic stimulation of inositol phosphate formation in bovine adrenal chromaffin cells: distinct nicotinic and muscarinic mechanisms. J Neurochem 49:1634–1643
100. Eggermont JA, Vrolix M, Wuytack F, Raeymaekers L, Casteels R (1988) The (Ca^{2+}–Mg^{2+})-ATPases of the plasma membrane and of the endoplasmic reticulum in smooth muscle cells and their regulation. J Cardiovasc Pharmacol 12 [Suppl 5]:S51–S55
101. Ehrlich BE, Watras J (1988) Inositol 1,4,5-trisphosphate activates a channel from smooth muscle sarcoplasmic reticulum. Nature 336:583–586
102. Eisen A, Kiehart DP, Wieland SJ, Reynolds GT (1984) Temporal sequence and spatial distribution of early events of fertilization in single sea urchin eggs. J Cell Biol 99:1647–1654
103. Ewald DA, Sternweis P, Miller RJ (1988) Guanine nucleotide-binding protein G_o-induced coupling of neuropeptide Y receptors to Ca^{2+} channels in sensory neurons. Proc Natl Acad Sci USA 85:3633–3637
104. Exton JH (1988) Mechanisms of action of calcium-mobilizing agonists: some variations on a young theme. FASEB J 2:2670–2676
105. Fabiato A, Fabiato F (1977) Calcium release from the sarcoplasmic reticulum. Circ Res 40:119–129
106. Feher JJ (1983) Facilitated calcium diffusion by intestinal calcium-binding protein. Am J Physiol 244 (Cell Physiol 13):C303–C307
107. Feher JJ (1984) Measurement of facilitated calcium diffusion by soluble calcium-binding protein. Biochim Biophys Acta 773:91–98
108. Ferris CD, Huganir RL, Supattapone S, Snyder SH (1989) Purified inositol 1,4,5-trisphosphate receptor mediates calcium flux in reconstituted lipid vesicles. Nature 342:87–89
109. Fill M, Ma J, Knudson CM, Imagawa T, Campbell KP, Coronado R (1989) Role of the ryanodine receptor of skeletal muscle in excitation-contraction coupling. Ann NY Acad Sci 560:155–162
110. Fisher SK, Agranoff BW (1981) Enhancement of the muscarinic synaptosomal phospholipid labeling effect by the ionophore A23187. J Neurochem 37:968–981
111. Fiskum G (1985) Intracellular levels and distribution of Ca^{2+} in digitonin-permeabilized cells. Cell Calcium 6:25–37
112. Fiskum G, Lehninger AL (1979) Regulated release of Ca^{2+} from respiring mitochondria by $Ca^{2+}/2H^+$ antiport. J Biol Chem 254:6236–6239
113. Fleckenstein A (1983) History of calcium antagonists. Circ Res 52 [Suppl I]:I3–I16
114. Flockerzi V, Oeken HJ, Hofmann F (1986) Purification of a functional receptor for calcium-channel blockers from rabbit skeletal-muscle microsomes. Eur J Biochem 161:217–224
115. Fosset M, Jaimovich E, Delpont E, Lazdunski (1983) [^{3}H] Nitrendipine receptors in skeletal muscle. J Biol Chem 258:6086–6092
116. Friedman PA (1988) Basal and hormone-activated calcium absorption in mouse renal thick ascending limbs. Am J Physiol 254 (Renal Fluid Electrolyte Physiol 23):F62–F70
117. Friedman PA, Figueriredo JF, Maack T, Windhager EE (1981) Sodium-calcium interactions in the renal proximal convoluted tubule of the rabbit. Am J Physiol 240:F558–F568
118. Frindt G, Lee CO, Yang JM, Windhager EE (1988) Potential role of cytoplasmic calcium ions in the regulation of sodium transport in renal tubules. Miner Electrolyte Metab 14:40–47
119. Frindt G, Windhager EE (1981) Effect of quinidine, low peritubular [Na] or [Ca] on Na transport in isolated perfused rabbit collecting tubules. Fed Proc 42:305
120. Fulton BP, Whittingham DG (1978) Activation of mammalian oocytes by intracellular injection of calcium. Nature 273:149–151
121. Furuichi T, Yoshikawa S, Miyawaki A, Wada K, Maeda N, Mikoshiba K (1989) Primary structure and functional expression of the inositol 1,4,5-trisphosphate-binding protein P_{400}. Nature 342:32–38
122. Ghijsen WEJM, Delong MD, van Os CH (1982) ATP-dependent calcium transport and its correlation with Ca^{2+} ATPase activity in basolateral plasma membranes of rat duodenum. Biochim Biophys Acta 689:327–336
123. Gill DL, Ueda T, Chueh S-H, Noel MW (1986) Ca^{2+} release from endoplasmic reticulum is mediated by a guanine nucleotide regulatory mechanism. Nature 320:461–464
124. Glossmann H and Ferry DR (1983) Solubilization and partial purification of putative calcium channels labelled with ^{3}H-nimodipine. Naunyn Schmiedbergs Arch Pharmacol 323:279–291
125. Gmaj P. Murer H (1988) Calcium transport mechanisms in epithelial cell membranes. Miner Electrolyte Metab 14:22–30

126. Gmaj P, Murer H, Kinne R (1979) Calcium ion transport across plasma membranes isolated from rat kidney cortex. Biochem J 178:549–557
127. Gray PTA (1988) Oscillations of free cytosolic calcium evoked by cholinergic and catecholaminergic agonists in rat parotid acinar cells. J Physiol (Lond) 406:35–53
128. Grinstein S, Erlig D (1978) Intracellular calcium and the regulation of sodium transport in the frog skin. Proc R Soc Lond 202:353–360
129. Grover AK, Kwan CY, Daniel EE (1981) Na-Ca exchange in rat myometrium membrane vesicles highly enriched in plasma membranes. Am J Physiol 240(Cell Physiol 9):C175–C182
130. Guggino SE, Guggino WB, Green N, Sacktor B (1987) Ca^{2+}-activated K^{+} channels in cultured medullary thick ascending limb cells. Am J Physiol 252:C121–C127
131. Guggino SE, Suarez-Isla BA, Guggino WB, Sacktor B (1985) Forskolin and antidiuretic homone stimulate a Ca^{2+}-activated K^{+} channel in cultured kidney cells. Am J Physiol 249:F448–F455
132. Hagiwara N, Irisawa H, Kameyama M (1988) Contribution of two types of calcium currents to the pacemaker potentials of rabbit sino-atrial node cells. J Physiol (Lond) 395:233–253
132a. Hajjar RJ, Bonventre JV (1991) Oscillations of intracellular calcium induced by vasopressin in individual fura-2-loaded mesangial cells. Frequency dependence on basal calcium concentration, agonist concentration, and temperature. J Biol Chem 266: 21589–21594
133. Hale CC, Kleiboeker SB, Carlton CG, Rovetto MJ, Jung C, Kim HD (1988) Evidence for high molecular weight Na-Ca exchange in cardiac sarcolemmal vesicles. J Membr Biol 106:211–218
134. Hammill OP, Marty A, Neher E, Sakmann B, Sigworth FJ (1981) Improved patch-clamp technique for high-resolution current recording from cells and cell-free membrane patches. Pflugers Arch 391:85–100
135. Hansford RG (1987) Relation between cytosolic free Ca^{2+} concentration and the control of pyruvate dehydrogenase in isolated cardiac myocytes. Biochem J 241:145–151
136. Hansford RG, Castro F (1981) Effect of micromolar concentrations of free calcium ions on the reduction of heart mitochondrial NAD(P) by 2-oxoglutarate. Biochem J 198:525–533
137. Harris CA, Baer PG, Chirito E, Dirks JH (1974) Composition of mammalian glomerular filtrate. Am J Physiol 227:972–976
138. Haworth RA, Hunter DR, Befkoff HA (1980) Na^{+} releases Ca^{2+} from liver, kidney and lung mitochondria. FEBS Lett 110:216–218
140. Hayat LH, Crompton M (1982) Evidence for the existence of regulatory sites for Ca^{2+} on the Na^{+}/Ca^{2+} carrier of cardiac mitochondria. Biochem J 202:509–518
141. Hazelton B, Mitchell B, Tupper JT (1979) Calcium, magnesium, and growth control in the WI-38 human fibroblast cell. J Cell Biol 83:487–498
142. Hescheler J, Rosenthal W, Trautwein W, Shultz G (1987) The GTP-binding proteins, G_o, regulates neuronal calcium channels. Nature 325:445–447
143. Hescheler J, Rosenthal W, Trautwein W, Schultz G (1987) The TRP-binding protein, G_o, regulates neuronal calcium channels. Nature (Lond) 325:445–447
144. Hescheler J, Rosenthal W, Wulfern M, Tang M, Yajima M, Trautwein W, Schultz G (1988) Involvement of the guanine nueleotide-biding protein, N_o, in the inhibitory regulation of neuronal calcium channels. Adv Second Messenger Phosphorprotein Phosphorylation Res 14:173–187
145. Hess P, Tsien RW (1984) Mechanism of ion permeation through calcium channels. Nature 309:453–456
146. Hess P, Lansman JB, Tsien RW (1984) Different modes of Ca channel grating behavior favoured by dihydropyridine Ca agonists and antagonists. Nature 311:538–544
147. Hill TD, Dean NM, Boynton AL (1988) Inositol 1,3,4,5-tetrakisphosphate induces Ca^{2+} sequestration in rat liver cells. Science 242:1176–1178
148. Hokin MR, Hokin LE (1953) Enzyme secretion and incorporation of P^{32} into phospholipides of pancreas slices. J Biol Chem 203:967–977
149. Holl RW, Thorner MO, Mandell GL, Sullivan JA, Sinha YN, Leony DA (1988) Spontaneous oscillations of intracellular calcium and growth hormone secretion. J Biol Chem 263:9682–9685
150. Holt WF, Lechene C (1981) ADH-PGE_2 interactions in cortical collecting tubule. II. Inhibition of Ca and P reabsorption. Am J Physiol 241 (Renal Fluid Electrolyte Physiol 10):F461–F467
151. Holz GG IV, Rane SG, Dunlap K, (1986) TRP-binding proteins mediated transmitter inhibition of voltage-dependent calcium channels. Nature (Lond) 319:670–672
152. Holz GG, Rane SC, Dunlap K (1986) GTP-binding proteins mediate transmitter inhibition of voltage-dependent calcium channels. Nature 319:670–672

153. Hosey MM, Bosortto M, Lazdunski M (1986) Phosphorylation and dephosphorylation of sensitive voltage-dependent Ca^{2+} channel in skeletal muscle membranes by cAMP- and Ca^{2+}-dependent processes. Proc Natl Acad Sci USA 83:3733–3737
154. Hosey MM, Bosortto M, Lazdunski M (1986) Photoaffinity labelling and phosphorylation of a 165 kilodalton peptide associated with dihydropyridine and phenylalkylamine-sensitive calcium channels. Biochem Biophys Res Commun 147:1137–1145
156. Huang YC, Christakos S (1988) Modulation of rat calbindin-D_{28} gene expression by 1,25-dihydroxyvitamin D_3 and dietary alteration. Mol Endocrinol 2:928–935
157. Hunter M, Lopes AG, Boulpaep EL, Giebisch GH (1984) Single channel recordings of calcium-activated potassium channels in the apical membrane of rabbit cortical collecting tubules. Proc Natl Acad Sci USA 81:4237–4239
158. Hosey MM, Chang FC, O'Callahan CM, Ptasienski J (1989) L-type calcium channels in cardiac and skeletal muscle. Ann NY Acad Sci 560:27–38
158a. Imagawa T, Smith JS, Coronado R, Campbell KP (1989) Purified ryanodine receptor from skeletal muscle sarcoplasmic reticulum is the Ca^{2+} permeable pore of the calcium release channel. J Biol Chem 262:16636–16643
159. Imagawa T, Leung AT and Campbell KP (1987) Phosphorylation of the 1,4-dihydropyridine receptor of the voltage-dependent Ca^{2+} channel by an intrinsic kinase in isolated triads from rabbit skeletal muscle. J Biol Chem 262:8333–8339
160. Imai M (1978) Calcium transport across the rabbit thick ascending limb of Henle's loop perfused in vitro. Pflügers Arch 374:255–263
161. Irvine RF, Moor RM (1986) Micro-injection of inositol 1,3,4,5-tetrakis-phosphate activates sea urchin eggs by a mechanism dependent on external Ca^{2+}. Biochem J 240:917–920
162. Irvine RF (1989) Functions of inositol phosphates. In: Michell RH, Drummond AH, Downes CP (eds) Inositol lipids in cell signalling. Academic, London, pp 135–161
163. Irvine RF, Moor RM, Pollock WK, Smith PM, Wreggeth KA (1988) Inositol phosphates: proliferation, metabolism and function. Philos Trans R Soc Lond [Biol] 320:281–298
164. Jacob R, Merritt JE, Hallam TJ, Rink TJ (1988) Repetitive spikes in cytoplasmic calcium evoked by histamine in human endothelial cells. Nature 335:40–45
165. Jacobus WE, Tiozzo R, Lugli G, Lehninger AL, Carafoli E (1975) Aspects of energy-linked calcium accumulation by rat heart mitochondria. J Biol Chem 250:7863–7870
166. Jayakumar A, Cheng L, Liang CT, Sacktor B (1984) Sodium gradient-dependent calcium uptake in renal basolateral membrane vesicles. J Biol Chem 259:10827–10833
167. Jencks WP (1989) How does the calcium pump pump calcium? J Biol Chem 264:18855–18858
168. Joseph S, Williamson JR (1988) Characteristics of inositol trisphosphate-mediated Ca^{2+} release from permeabilized hepatocytes. J Biol Chem 261:14658–14664
169. Joseph SJ, Williams RJ, Corkey BE, Matschinsky FM, Williamson JR (1984) The effect of inositol triphosphate on Ca^{2+} fluxes in insulin-secreting tumor cells. J Biol Chem 259:12952–12955
170. Keith CH, Ratan R, Maxfield FR, Bajer A, Shelanski ML (1985) Local cytoplasmic calcium gradients in living mitotic cells. Nature 316:848–850
171. Kirchberger MA, Antonetz T (1982) Calmodulin-mediated regulation of calcium transport and (Ca^{2+} + Mg^{2+})-activated ATPase activity in isolated cardiac sarcoplasmic reticulum. J Biol Chem 257:5685–5691
172. Kirk CJ, Creba JA, Downes CP, Michell RH (1981) Hormone-stimulated metabolism of inositol lipids and its relationship to hepatic receptor function. Biochem Soc Trans 9:377–379
173. Klempner M (1985) An adenosine trisphosphate-dependent calcium uptake pump in human neutrophil lysosomes. J Clin Invest 76:303–310
174. Knudson CM, Chaudhari N, Sharp AH, Powell JA, Beam KG, Campbell KP (1989) Specific absence of the α_1 subunit of the dihydropyridine receptor in mice with muscular dysgenesis. J Biol Chem 264:1345–1348
175. Kojima I, Kojima K, Rasmussen H (1985) Role of calcium fluxes in the sustained phase of angiotensin II-mediated aldosterone secretion from adrenal glomerulosa cells. J Biol Chem 260:9177–9184
176. Kort AA, Capogrossi MC, Lakatta EG (1985) Frequency, amplitude, and propagation velocity of spontaneous Ca^{2+}-dependent contractile waves in intact adult rat cardiac muscle and isolated myocytes. Circ Res 57:844–855
177. Kostyuk P, Akaike N, Osipchuk YU, Savchenko A, Shuba YA (1989) Gating and permeation of different types of Ca Channels. Ann NY Acad Sci 560:63–79

178. Krause K-H, Lew PD (1987) Subcellular distribution of Ca^{2+} pumping sites in human neutrophils. J Clin Invest 80:107–116
179. Kretsinger RH (1980) Crystallographic studies of calmodulin and homologs. Ann NY Acad Sci 356:14–19
180. Kretsinger RH (1980) Structure and evolution of calcium-modulated proteins. CRC Crit Rev Biochem 8:119–174
181. Kretsinger RH, Mann JE, Simmons JG (1982) Model of the facilitated diffusion of calcium by the intestinal calcium binding protein. In: Norman AW, Schaefer K, Herrath DV, Gringoleit H-G (eds) Vitamin D chemica, Biochemical and Clinical Endocrinology of Calcium Metabolism. De-Gruyter, New York, pp 233–246
182. Kruskal BA, Maxfield FR (1987) Cytosolic free calcium increases before and oscillate during frustrated phagocytosis in macrophages. J Cell Biol 105:2685–2693
184. Kuno M, Goronzy J, Weyand CM, Gardner P (1986) Single-channel and whole-cell recordings of mitogen-regulated inward currents in human cloned helper T lymphocytes. Nature 323:269–273
185. Kuno M and Gardner P (1987) Ion channels activated by inositol 1,4,5-trisphosphate in plasma membrane of human T-lymphocytes. Nature 326:301–304
186. Lang F, Friedrich F, Paulmichl M, Schobersberger W, Jungwith A, Ritter M, Steidl M, Weiss H, Woll E, Tshernko E, Paulmichl R, Hallbrucker C (1990) Ion Channels in Madin-Darby canine kidney cells. Renal Physiol Biochem 13:82–93
187. Lassiter WE, Gottschalk CW, Mylle M (1963) Micropuncture study of renal tubular reabsorption of calcium in normal rodents. Am J Physiol 204:771–775
188. Le Grimellec C, Poujeol P, de Rouffignac C (1975) ^{3}H-inulin and electrolyte concentrations in Bowman's capsule in rat kidney. Comparison with artificial ultrafiltration. Pflugers Arch 354:117–131
189. Leblanc N, Humes JR (1990) Sodium current-induced release of calcium from cardiac sarcoplasmic reticulum. Science 248:372–376
190. Lechleiter JD, Dartt DA, Brehm P (1988) Vasoactive intestinal peptide activities Ca^{2+}-dependent K^+ channels through a cAMP pathway in mouse lacrimal glands. Neuron 1:227–235
191. Lederer WJ, Niggli E, Hadley RW (1990) Sodium-calcium exchange in excitable cells: fuzzy space. Science 248:283
192. Lee CO, Taylor A, Windhager EE (1980) Cytosolic calcium ion activity in epithelial cells of necturus kidney. Nature 287:859–861
193. LeFurgey A, Ingram P, Mandel LJ (1986) Heterogeneity of calcium compartmentation: Electron probe analysis of renal tubules. J Membr Biol 94:191–196
194. Lehninger AL (1962) Water uptake and extrusion by mitochondria in relation to oxidative phosphorylation. Physiol Rev 42:467–517
195. Lehninger AL (1970) Mitochondria and calcium ion transport. Biochem J 119:129–138
195a. Lê-Quôc D and Lê-Quôc K (1989) Relationships between the NAD(P) redox state, fatty acid oxidation, and inner membrane permeability in rat liver mitochondria. Arch. Biochem. Biophys. 273:466–478
196. Leung AT, Imagawa T, Campbell KP (1987) Structural characterization of the 1,4-dihydropyridine receptor of the voltage-dependent Ca^{2+} channel from rabbit skeletal muscle. J Biol Chem 262:7943–7946
197. Levi RC, Aloatti G, Fischmeister R (1989) Cyclic GMP regulates the Ca-channel current in guinea pig ventricular myocytes. Pflügers Arch 413:685–687
198. Lewis DL, Wieght FF, Luini A (1986) A guanine nucleotide-binding protein mediates the inhibition of voltage-dependent calcium current by somatostatin in a pituitary cell line. Proc Natl Acad Sci USA 83:9035–9039
199. Lichtman AH, Segel GB, Lichtman MA (1983) The role of calcium in lymphocyte proliferation. Blood 61:413–422
200. Lipscombe D, Madison DV, Poenie M, Reuter H, Tsien RY, Tsien RW (1988) Spatial distribution of calcium channels and cytosolic calcium transients in growth cones and cell bodies of sympathetic neurons. Proc Natl Acad Soc USA 85:2398–2402
201. Llach F, Coburn JW, Brickman AW, Kurokawa K, Norman AW, Canterbury JM, Reiss E (1977) Acute actions of 1,25-dihydroxyvitamin D_3 in normal man: effect on calcium and parathyroid status. J Clin Endocrinol Metab 44:1054–1060
202. Loirand G, Pacaud P, Mironneau C, Mironneau J (1986) Evidence for two distinct calcium

channels in rat vascular smooth muscle cells in short-term primary culture. Pflügers Arch 407:566–568

203. Lorenzen M, Lee CO, Windhager EE (1984) Cytosolic Ca^{2+} and Na^{+} activities in perfused proximal tubules of *Necturus* kidney. Am J Physiol 247(Renal Fluid Electrolyte Physiol 16):F93–F102

203a. Maizels E, Jungman RA (1983). Ca^{2+}-calmodulin-dependent phosphorylation of soluble and nuclear proteins in the rat ovary Endocrinology 112:1895–1902

204. Marchetti C and Brown AM (1988) Protein kinase activator-oleoyl-2-acetyl-sn-glycerol inhibits two types of calcium currents in GH_3 cells. Am J Physiol 254:C206–C210

205. Marty A, Tan YP, Trautmann A (1984) Three types of calcium-dependent channel in rat lacrimal glands. J Physiol (Lond) 357:293–325

206. Matlib MA, Schwartz A and Yamori Y (1985) A Na^{+}–Ca^{2+} exchange process in isolated sarcolemmal membranes of mesenteric arteries from WKY and SHR rats. Am J Physiol 249 (Cell Physiol 18):C166–C172

207. Matlib MA, Reeves JP (1987) Solubilization and reconstitution of the sarcolemmal Na^{+}–Ca^{2+} exchange system of vascular smooth muscle. Biochim Biophys Acta 904:145–148

208. Mayer ML, Westbrook GL (1987) The physiology of excitatory amino acids in the vertebrate central nervous system. Prog Neurobiol 28:197–276

209. McCleskey EW, Almers W (1985) The Ca channel in skeletal muscle is a large pore. Proc Natl Acad Sci USA 82:7149–7153

210. McCormick JG, Denton RM (1986) Ca^{2+} as a second messenger within mitochondria. Trends Biochem Sci 11:258–262

210a. McCormick JG, Denton RM (1990) The role of mitochondrial Ca^{2+} transport and matrix Ca^{2+} in signal transduction in mammalian tissues Biochim Biophys Acta 1018:287–291

211. Means AR, Dedman JR (1980) Calmodulin-an intracellular calcium receptor. Nature 285:73–77

212. Mela L (1968) Interactions of La^{3+} and local anesthetic drugs with mitochondrial Ca^{2+} and Mn^{2+} uptake. Arch Biochim Biophys 123:286–293

213. Mela L (1969) Inhibition and activation of calcium transport in mitochondria. Effect of lanthanides and local anesthetic drugs. Biochemistry 8:2481–2485

214. Meldolesi J, Pozzan T (1987) Pathways of Ca^{2+} influx at the plasma membrane: voltage-, receptor-, and second messenger-operated channels. Exp Cell Res 171:271–283

215. Merot J, Bidet M, Gachot B, LeMaout S, Tauc M, Poujeol P (1988) Patch clamp study on primary culture of isolated proximal convoluted tubules. Pflügers Arch 413:51–61

216. Merritt JE, Rink TJ (1987) Rapid increases in cytosolic free calcium in response to muscarinic stimulation of rat acinar cells. J Biol Chem 267:4958–4960

217. Meyer T, Holowka D, Stryer L (1988) Highly cooperative opening of calcium channels by inositol 1,4,5-trisphosphate. Science 240:653–656

218. Michell RH (1975) Inositol phospholipids and cell surface receptor function. Biochim Biophys Acta 415:81–147

219. Miller BA, Scaduto RC Jr., Tillotson DL, Botti JJ, Cheung JY (1988) Erythropoietin stimulates a rise in intracellular free calcium concentration in single early human erythroid precursors. J Clin Invest 82:309–315

220. Mitchell P (1961) Coupling of phosphorylation to electron and hydrogen transfer by a chemiosmotic type of mechanism. Nature 191:144–148

221. Miyazaki S-I, Hashimoto N, Yoshimoto Y, Kishimoto T, Igusa Y, Hiramoto Y (1986) Temporal and spatial dynamics of the periodic increase in intracellular free calcium at fertilization of golden hamster eggs. Dev Biol 118:259–267

222. Moews PC, Kretsinger RH (1975) Preliminary crystallographic data for a calcium binding protein for bovine intestine. J Mol Biol 97:661–664

223. Moore CL (1971) Specific inhibition of mitochondrial Ca^{2+} transport by ruthenium red. Biochem Biophys Res Commun 42:298–305

224. Morel F, Doucet A (1986) Hormonal control of kidney functions at the cell level. Physiol Rev 66:377–468

225. Morel N and Godfraind T (1982) Na-Ca exchange in heart and smooth muscle microsomes. Arch Int Pharmacodyn Ther 258:319–321

225a. Morgan JI, Curran C (1986). Role of ion fluxes in the control of c-fos expression. Nature 322:552–555

226. Morris AP, Gallacher DV, Irvine RF, Peterson OH (1987) Synergism of inositol trisphosphate and tetrakisphosphate in activating Ca^{2+}-dependent K^{+} channels. Nature 330:653–655

227. Morton ME, Froehner SC (1987) Monoclonal antibody identifies a 200-kDa subunit of the dihydropyridine-sensitive calcium channel. J Biol Chem 262:11 904–11 907
228. Mullaney JM, Chueh S-H, Ghos, TK, Gill DL (1987) Intracellular calcium uptake activated by GTP. J Biol Chem 262:13 865–13 872
229. Mullaney JM, Yu M, Ghosh TK, Gill DL (1988) Calcium entry into the inositol 1,4,5-trisphosphate-releasable calcium pool is mediated by a GTP-regulatory mechanism. Proc Natl Acad Sci USA 85:2499–2503
230. Murayama Y, Morel F, LeGrimellec C (1972) Phosphate, calcium and magnesium transfer in proximal tubules and loops of Henle, as measured by single nephron microperfusion experiments in the rat. Pflügers Arch 333:1–16
231. Murer H, Hildmann B (1981) Transcellular transport of calcium and inorganic phosphate in the small intestine epithelium. Am J Physiol 240(Gastrointest Liver Physiol 3):G409–G416
232. Murphy PR, Dimattiea GE, Frisen HG (1988) Role of calcium in prolactin-stimulated *c-myc* gene expression and mitogenesis in NB2 lymphoma cells. Endocrinology 122:2476–2485
233. Nakayama N, Kirley TL, Vaghy PL, Schwartz A (1987) Purification of a putative Ca^{2+} channel protein from rabbit-skeletal muscle. J Biol Chem 262:6572–6576
234. Nastainczyk W, Rohrkasten A, Sieber M, Rudolph C, Schächtele C, Marmè D, Hofman F (1987) Phosphorylation of the purified receptor for calcium channel blockers by cAMP kinase and protein kinase C. Eur J Biochem 169:137–142
235. Nellans HN (1988) Contributions of cellular and paracellular pathways to transepithelial intestinal calcium transport. In: Cellular calcium and phosphate transport in health and disease. Liss, New York, pp 269–276
235a. Nicoll DA, Longoni S, Philipson KD (1990) Molecular cloning and functional expression of the cardiac sarcolemmal Na^+–Ca^{2+} exchanger. Science 250:562–565
235b. Nicotera P, McConkey DJ, Jones DP, Orrenius S (1989) ATP stimulates Ca^{2+} uptake and increases the free Ca^{2+} concentration in isolated rat liver nuclei. Proc Natl Acad Sci USA 86:453–457.
236. Nilius B, Hess P, Lansmann JB, Tsien RW (1985) A novel type of cardiac calcium in ventricular cells. Nature 316:443–446
237. Nilus B, Hess P, Lansman JB, Tsien RW (1985) A novel type of cardiac calcium channel in ventricular cells. Nature 316:443–446
238. Nishizuka Y (1984) The role of protein kinase C in cell surface signal transduction and tumour production. Nature 308:693–698
239. Nowyeky MC, Fox AP, Tsien RW (1985) Three types of neuronal calcium channel with different calcium agonist sensitivity. Nature 316:440–443
240. O'Sullivan AJ, Cheek TR, Moreton RB, Berridge MJ, Burgoyne RD (1989) Localization and heterogeneity of agonist-induced changes in cytosolic calcium concentrations in single bovine adrenal chromaffin cells from video imaging of fura-2. EMBO J 8:401–411
241. Palmer LG, Frindt G (1986) Epithelial Na channels; characterization using the patch clamp technique. Fed Proc 45:2708–2712
242. Park S, Rasmussen H (1986) Carbachol-induced protein phosphorylation changes in bovine tracheal smooth muscle. J Biol Chem 261:15 734–15 739
243. Paupardin-Tritsch D, Hammond C, Gershenfeld HM, Nairn AC and Greengard P (1986) cGMP-dependent kinase enhances Ca^{2+} current and potentiates the serotonin-induced Ca^{2+} current increase in snail neurones. Nature 323:812–814
244. Penner R, Matthews G, Neher E (1988) Regulation of calcium influx by second messengers in rat mast cells. Nature 334:499–504
245. Peonie M, Alderton J, Tsien RY, Steinhardt RA (1985) Changes of free calcium levels with stages of the cell division cycle. Nature 315:147–149
246. Persechini A, Noncrief ND, Kretsinger RH (1989) The EF-hand family of calcium modulated proteins. Trends Neursci 12:462–467
247. Petersen OH, Maryuama Y (1984) Calcium-activated potassium channels and their role in secretion. Nature 307:693–696
248. Poenie M, Alderton J, Steinhardt R, Tsien R (1985) Calcium rises abruptly and briefly throughout the cell at the onset of anaphase. Science 233:886–889
249. Prentki M, Glennon MC, Thomas AP, Morris RL, Matschinksy FM, Corkey BE (1988) Cell-specific patterns of oscillating free Ca^{2+} in carbomylcholine-stimulated insulinoma cells. J Biol Chem 263:11 044–11 047
250. Prentki M, Wollheim CB, Lew PD (1984) Ca^{2+} homeostasis in permeabilized human neutrophils. J Biol Chem 259:13 777–13 782

251. Puskin JS, Gunter TE, Gunter KK, Russell PR (1976) Evidence for more than one Ca^{2+} transport mechanism in mitochondria. Biochemistry 15:3834–3842
252. Putney JW Jr (1986) A model for receptor-regulated calcium entry. Cell Calcium 7:1–12
253. Rasmussen H (1989) The cycling of calcium as an intracellular messenger. Sci Am October: 66–73
254. Rasmussen CD, Means AR (1989) Calmodulin, cell growth and gene expression. Trends Neurosci 12:433–438
255. Reed KC, Bygrave FL (1974) The inhibition of mitochondrial calcium transport by lanthanides and Ruthenium Red. Biochem J 140:143–155
256. Reeves JP, Hale CC (1984) The stoicheometry of the cardiac sodium-calcium exchange system. J Biol Chem 259:7733–7739
257. Reeves JP, Sutko JL (1983) Competitive interactions of sodium and calcium with the sodium-calcium exchange system of cardiac sarcolemmal vesicles. J Biol Chem 258:3178–3182
258. Reuter H, Porzig H (1988) Calcium channels. Diversity and complexity. Nature 336:113–114
259. Reuter H, Seitz N (1968) The dependence of calcium efflux from cardiac muscle on temperature and external ion composition. J Physiol (Lond) 195:451–470
260. Ringer S (1883) A further contribution regarding the influence of different constituents of the blood on the contractions of the heart. J Physiol 4:29–42
261. Ringer S (1890) Concerning experiments to test the influence of lime, sodium, and potassium salts on the development of ova and growth of tadpoles. J Physiol 11:79–84
262. Ringer S, Sainsbury H (1894) The action of potassium, sodium and calcium salts of Tubiflex rivulorum. J Physiol 16:1–9
263. Ringer S, Sainsburg H (1883) An investigation regarding the action of strontium and barium salts compared with the action of lime on the ventricle of the frog's heart. Practitioner 31:81–93
264. Rink TJ (1988) A real receptor-operated calcium channel? Nature 334:649–650
265. Rizzoli R, Fleisch H, Bonjour JP (1977) Effect of thyroparatrhyroidectomy on calcium metabolism in the rat: role of 1,25 dihydroxyvitamin D_3. Am J Physiol 233:E160–E164
266. Rocha AS, Magaldi JB, Kokko JP (1977) Calcium and phosphate transport in isolated segments of rabbit Henle's loop. J Clin Invest 59:975–983
267. Rohrkasten A, Meyer HE, Nastainczyk W, Sieber M, Hofmann F (1988) cAMP-dependent protein kinase rapidly phosphorylates serine-687 of the skeletal muscle receptor for calcium channel blockers. J Biol Chem 263:15 325–15 329
268. Ross CA, Meldolesi J, Milner TA, Satoh T, Supattopone S, Synder SH (1989) Inositol 1,4,5-trisphosphate receptor localized to endoplasmic reticulum in cerebellar Purkinje neurons. Nature 339:468–470
269. Rossi CS, Lehninger AL (1964) Stoichiometry of respiratory stimulation, accumulation of Ca^{2+} and phosphate and oxidative phosphorylation in rat liver mitochondria. J Biol Chem 239:3971–3980
270. Rossi CS, Vasington FD, Carafoli E (1973) The effect of ruthenium red on the uptake and release of Ca^{2+} by mitochondria. Biochem Biophys Res Commun 50:846–852
271. Rottenberg H, Scarpa A (1974) Calcium uptake and membrane potential in mitochondria. Biochemistry 13:4811–4817
272. Rouse D, Ng RCK, Suki WN (1980) Calcium transport in the pars recta and thin descending limb of Henle of the rabbit, perfused in vitro. J Clin Invest 65:37–42
273. Ruth P, Rohrkasten A, Biel M, Bosse E, Regulla S, Meyer HE, Flockerzi V, Hofmann F (1989) Primary structure of the β subunit of the DHP-sensitive calcium channel from skeletal muscle. Science 245:1115–1118
274. Sacks P, Bourdeau JE (1989) Ca^{2+} absorption in the pars recta of cortical S_2 rabbit proximal tubules: role of diffusion. Am J Physiol 257 (Renal Fluid Electrolyte Physiol 26):F262–F267
275. Saez JC, Connor JA. Spray DC, Bennett MVL (1989) Hepatocyte gap junctions are permeable to the second messenger, inositol 1,4,5-trisphosphate and to calcium ions. Proc Natl Acad Sci USA 86:2708–2712
276. Sage SO, Rink TJ (1986) Kinetic differences between thrombin-induced and ADP-induced calcium influx and release from internal stores in fura-2-loaded human platelets. Biochem Biophys Res Commun 136:1124–1129
277. Sanchez J, Stefani E (1978) Inward calcium current in twitch muscle fibres of the frog. J Physiol (Lond) 283:197–209
278. Sasaki Y, Hidaka H (1982) Calmodulin and cell proliferation. Biochem Biophys Res Commun 104:451–456

279. Saunders JCJ, Isaacson LC (1989) Non-selective cation and Ba permeable apical channels in rabbit renal tubules. International Congress on Physiology, Helsinki
280. Scarpa A, Azzone GF (1970) The mechanism of ion translocation in mitochondria. 4. Coupling of K^+ efflux with Ca^{2+} uptake. Eur J Biochem 12:328–335
281. Schafer JA, Barfuss DW (1982) The study of pars recta function by the perfusion of isolated tubule segments. Kidney Int 22:434–448
282. Scharff O, Foder B, Skibsted U (1983) Hysteretic activation of the Ca^{2+} pump revealed by calcium transients in human red cells. Biochim Biophys Acta 730:295–305
283. Schatzman JH (1966) ATP-dependent Ca^{2+}-extrusion from human red cells. Experientia 22:364–365
284. Schatzman HJ (1966) ATP-dependent Ca^{2+}-extrusion from human red cells. Experientia 22:364–365
285. Schulz I, Thevenod F, Dehlinger-Kremer (1989) Modulation of intracellular free Ca^{2+} concentration by IP_3-sensitive and IP_3-insensitive nonmitochondrial Ca^{2+} pools. Cell Calcium 10:325–336
286. Scott RH, Dolphin AC (1987) Activation of a G protein promotes agonist response to calcium channel ligands. Nature 330:760–762
287. Seiler SM, Arnold AJ, Stanton HC (1987) Inhibitors of inositol trisphosphate-induced Ca^{2+} release from isolated platelet membrane vesicles. Biochem Pharmacol 36:3331–3337
288. Selwyn MJ, Dawson AP, Dunnett SJ (1970) Calcium transport in mitochondria. FEBS Lett 10:1–5
289. Shangold GA, Murphy SN, Miller RJ (1988) Gonadotrophin-releasing hormone-inducing Ca^{2+} transients in single identified gonadotropes require both intracellular Ca^{2+} mobilization and Ca^{2+} influx. Proc Natl Acad Sci USA 85:6566–6570
290. Shareghi GR, Stoner LC (1978) Calcium transport across segments of the rabbit distal nephron in vitro. Am J Physiol 235:F367–F375
291. Shareghi GR, Agus ZS (1982) Phosphate transport in the light segment of the rabbit coritcal collecting duct. Am J Physiol 242 (Renal Fluid Electrolyte Physiol 11):F379–F384
292. Sharp AH, Imagawa T, Leung AT, Campbell KP (1987) Identification and characterization of the dihydropyridine-binding subunit of the skeletal muscle dihydropyridine receptor. J Biol Chem 262:12 309–12 315
292a. Sheng M, Thompson MA, Greenberg ME (1991) CREB: A Ca^{2+}-regulated transcription factor phosphorylated by calmodulin-dependent kinases. Science 252:1427–1430
292b. Shull GE, Greeb J (1988) Molecular cloning of two isoforms of the plasma membrane Ca^{2+}-transporting ATPase from rat brain. Structural and functional domains exhibit similarity to Na^+,K^+ and other cation transport ATPases. J Biol Chem 263:8646–8657
293. Slater EC, and Cleland KW (1953) The effect of calcium on the respiratory and phosphorylative activites of heart-muscle sarcosomes. Biochem J 55:566–580
294. Slaughter RS, Welton AF and Morgan DW (1987) Sodium-calcium exchange in sarcolemmal vesicles from tracheal smooth muscle. Biochim Biophys Acta 904:92–104
295. Smallwood JI, Gugi B, Rasmussen H (1988) Regulation of erythrocyte Ca^{2+} pump activity by protein kinase C. J. Biol Chem 263:2195–2202
296. Snowdowne KW and Borle AB (1985) Effects of sodium on cytosolic ionized calcium. Na^+–Ca^{2+} exchange as a major influx pathway in kidney cells. J Biol Chem 260:14 998–15 007
297. Somlyo AP, Urbanics R, Vadasz G, Kovach AGB and Somlyo AV (1985) Mitochondrial calcium and cellular electrolytes in brain cortex frozen in situ: electron probe analysis. Biochem Biophys Res Commun 132:1071–1078
298. Somlyo AP, Bond M, Somlyo AV (1985) Calcium content of mitochondria and endoplasmic reticulum in liver frozen rapidly in vivo. Nature 314:622–625
299. Sordahl (1974) Effects of magnesium, Ruthenium Red and the antibiotic ionophore A-23187 on initial rates of calcium uptake and release by heart mitochondria. Arch Biochem Biophys 167:104–115
300. Stanfield PR (1981) Voltage-dependent calcium channels of excitable membranes. Br Med Bull 42:359–367
301. Streb H, Bayerdoffer E, Haase W, Irvine RF, Schulz I (1984) Effect of inositol-1,4,5-trisphosphate on isolated fractions of rat pancreas. J Membr Biol 81:241–253
302. Streb H, Irvine RF, Berridge MJ, Schulz I (1983) Release of Ca^{2+} from a nonmitochondrial intracellular store in pancreatic acinar cells by inositol-1,4,5-triphosphate. Nature 306:67–68
303. Striessnig J, Moosburger K, Goll A, Ferry DR, Glossmann H (1986) Stereoselective photo-

affinity labelling of the purified 1,4-dihydropyridine receptor of the voltage-dependent calcium channel. Eur J Biochem 161:603–609

304. Sturek M, Hermsmeyer K (1986) Calcium and sodium channels in spontaneously contracting vascular muscle cells. Science 233:475–478
305. Suki WN, Rouse D, Ng RCK, Kokko JP (1980) Calcium transport in the thick ascending limb of Henle. Heterogeneity of function in the medullary and cortical segments. J Clin Invest 66:1004–1009
306. Suki WN, Rouse D (1981) Hormonal regulation of calcium transport in thick ascending limb renal tubules. Am J Physiol (Renal Fluid Elect Physiol 10):F171–F174
307. Supattapone S, Worley PF, Baraban JM, Snyder SH (1988) Solubilization, purification, and characterization of an inositol trisphosphate receptor. J Biol Chem 263:1530–1534
308. Supattapone S, Danoff SK, Theibert A, Joseph SK, Steiner J, Snyder SH (1988) Cyclic AMP-dependent phosphorylation of a brain inositol trisphosphate receptor decreases its release of calcium. Proc Natl Acad Sci USA 85:8747–8750
309. Supattapone S, Worley PF, Baraban JM, Snyder S (1988) Solubilization, purification, and characterization of an inositol trisphosphate receptor. J Biol Chem 263:1530–1534
310. Sutton RAL and Dirks JH (1986) Calcium and mangesium: renal handling and disorders of metabolism. In: Brenner BM, Rector FC Jr (eds) The Kidney. Saunders, Philadelphia, pp 551–618
311. Szebenyi DME, Moffat K (1987) Determination of the three-dimensional structure of vitamin D-dependent calcium binding proteins from bovine intestine. Methods Enzymol 139:585–610
312. Takahashi M, Seagar MJ, Jones JF, Reber BFX, Catterall WC (1987) Subunit structure of dihydropyridine-sensitive calcium channels from skeletal muscle. Proc Natl Acad Sci USA 84:5478–5482
313. Takai Y, Kishimoto A, Iwusa Y, Kawakara Y, Mori T, Nishizuka (1979) Calcium-dependent activation of a multi-functional protein kinase by phospholipids. J Biol Chem 254:3692–3695
314. Takeshima H, Nishimura S, Matsumoto T, Ishida H, Kangawa K, Minamino N, Matsuo H, Ueda M, Hanaoka M, Hirose T, Numa S (1989) Primary structure and expression from complementary DNA of skeletal muscle ryanodine receptor. Nature 339:439–445
315. Talor Z, Arruda JAL (1985) Partial purification and reconstitution of renal basolateral Na^+/Ca^{2+} exchanger into liposomes. J Biol Chem 260:15473–15476
316. Tanabe T, Takeshima H, Mikami A, Flockerzi V, Takahashi H, Kangawa K, Kojima M, Matsuo H, Hirose T, Numa S (1987) Primary structure of receptor for calcium channel blockers from skeletal muscle. Nature 328:313–318
317. Tanabe T, Beam KG, Powell JA, Numa S (1988) Restoration of excitation-contraction coupling and slow calcium current in dysgenic muscle by dihydropyridine receptor complementary DNA. Nature 336:134–139
318. Taylor AN, Gleason We, Lankford GL (1984) Immunocytochemical localization of rat intestinal vitamin D-dependent calcium-binding protein. J Histochem Cytochem 32:153–158
319. Teulon J, Paulais M, Boutheir M (1987) A Ca^{2+}-activated cation-selective channel in the basolateral membrane of the cortical thick ascending limb of Henle's loop of the mouse. Biochim Biophys Acta 905:125–132
320. Thevenod F, Deblinger-Kremer M, Kemmer TP, Christian A-L, Potter BVL, Schulz I (1989) Characterization of inositol 1,4,5-trisphosphate-sensitive (IsCap) and-insensitive (IisCap) non-mitochondrial Ca^{2+} pools in rat pancreatic acinar cells. J Membr Biol 109:173–186
321. Tsien RW (1983) Calcium channels in excitable cell membranes. Annu Rev Physiol 45:341–358
322. Tsien RW, Lipscombe D, Madison DV, Bley KR, Fux AP (1988) Multiple types of neuronal calcium channels and their selective modulation.Trends Neurosci 11:431–437
323. Tupper JT, Kaufman L, Bodine PV (1980) Related effects of calcium and serum on the G1 phase of the human W138 fibroblast. J Cell Physiol 104:97–103
324. Ubl J, Murer H, Kolb HA (1988) Hypotonic shock evokes opening of Ca^{2+}-activated K channels. Pflügers Arch 412:551–553
325. Ueda S, Oiki S, Okada Y (1986) Oscillations of cytoplasmic concentrations of Ca^{2+} and K^+ in fused L cells. J Membr Biol 91:65–72
326. Ueda T, Chueh S-H, Noel MW, Gill DL (1986) Influence of inositol 1,4,5-trisphosphate and guanine nucleotides on intracellular calcium release within NIE-115 neuronal cell line. J Biol Chem 261:3184–3192
327. Ullrich KJ, Rumrich G, Kloss S (1976) Acute Ca^{2+} reabsorption in the proximal tubule of the rat kidney. Dependence on sodium-and buffer transport. Pflugers Arch 364:223–228

328. Ullrich KJ, Rumrich G, Kloss S (1976) Active Ca^{2+} reabsorption in the proximal tubule of the rat kidney. Pflugers Arch 364:223–228
329. Vaghy PL, Johnson JD, Matlib MA, Wang T, Schwartz A (1982) Selective inhibition of Na^{+}-induced Ca^{2+} release from heart mitochondria by diltiazem and certain other Ca^{2+} antagonist drugs. J Biol Chem 257:6000–6002
330. Vaghy P L, Striessnig J, Miwa K, Knaus HG, Itagaki K, McKenna E, Glossmann H, Schwartz A (1987) Identification of a novel 1,4-dihydropyridine-and phenylalkylamine-binding polypeptide in calcium channel preparations. J Biol Chem 267:14337–14342
331. Valverde I, Vandermeers A, Anjaneyulu R, Malaisse WJ (1979) Calmodulin activation of adenylate cyclase in pancreatic islets. Science 206:225–227
332. Van Breemen C, Cauvin C, Yamamoto H, Schaver AZ (1987) Vascular smooth muscle calcium channels. J Cardiovasc Pharmacol 10 [Suppl 10]:510–515
333. Van Corven EJJM, Timmermans JAH, Mircheff AK, van Os CH (1988) Characterization of intracellular Ca^{2+}-stores in rat duodenal epithelium. In: Bonner F, Peterik M (eds) Cellular calcium and phosphate transport. Liss, New York, pp 127–132
334. Van Driessche W (1987) Ca^{2+} channels in the apical membrane of the toad urinary bladder. Pflugers Arch 410:243–249
335. Van Driessche W, Van Keer H, Witters K (1987) Activation of a mono- and divalent cation-selective pathway in the apical membrane of the toad urinary bladder. Arch Int Physiol Biochim 95:P18
336. Varghese S, Lee S, Huang YC, Christakos S (1988) Analysis of rat vitamin D-dependent calbindin-D_{28K} gene expression. J Biol Chem 263:9776–9784
337. Vasington FD, Murphy JV (1961) Active binding of calcium by mitochondria. Fed Proc 20, 146
338. Vashington FD, Murphy JV (1962) Ca^{2+} uptake by rat kidney mitochondria and its dependence on respiration and phosphorylation. J Biol Chem 237:2670–2677
339. Vercesi A, Reynafarje B, Lehninger AL (1978) Stoicheometry of H^{+} ejection and H^{+} uptake coupled to electron transport in rat heart mitochondria. J Biol Chem 253:6379–6385
340. Vilven J, Leung AT, Imagawa T, Sharp AH, Campbell KP, Coronádo R (1988) Interaction of calcium channels of skeletal muscle with monoclonal antibodies specific for its dihydropyridine receptor. Biophys J 53:556a
341. Virk SS, Kirk CJ, Shears SB (1985) Ca^{2+} transport and Ca^{2+} dependent ATP hydrolysis by Golgi vesicles from lactating rat mammary glands. Biochem J 226:741–748
342. Volpe P, Krause K-H, Hashimoto S, Zorzato F, Pozzan T, Meldolesi J, Lew DP (1988) "Calciosome," a cytoplasmic organelle: the inositol 1,4,5-trisphosphate-sensitive Ca^{2+} store of nonmuscle cells? Proc Natl Acad Sci USA 85:1091–1095
342a. Wan B, LaNoue KF, Cheung JY, Scaduto RC Jr (1989) Regulation of citric acid cycle by calcium. J Biol Chem 264:13430–13439
343. Wasserman RH, Fullmer CS (1982) In: Cheung WY (ed) Calcium and cell function, (vol 2). Academic, New York, pp 175–216
344. Wasserman RH, Taylor AN (1966) Vitamin D_3-Induced calcium-binding protein in chick intestinal mucosa. Science 152:791–793
344a. Waybill MM, Yelamarty RV, Zhang Y, Scaduto RC Jr, LaNoue KF, Hsu C, Smith BC, Tillotson DL, Yu FTS, Cheung JY (1991) Nuclear calcium gradients in cultured rat hepatocytes. Am. J. Physiol. (Endocrinol. and Metab). 261: E49–E57
345. Whitfield JF, MacManus JP, Rixon RH, Boynton AL, Youdale T, Swierenga S (1976) The positive control of cell proliferation by the interplay on calcium ions and cyclic nucleotides. A review. In Vitro 12:1–18
346. Wibo M, Morel N, Godfraind T (1981) Differentiation of Ca^{2+}-pumps linked to plasma membrane and endoplasmic reticulum in the microsomal fraction from intestinal smooth muscle. Biochim Biophys Acta 649:651–660
347. Wideman RF Jr (1987) Renal regulation of avian calcium and phosphorus metabolism. J Nutr 117:808–8115
348. Williams DA, Fogarty KE, Tsien RY, Fay FS (1985) Calcium gradients in single smooth muscle cells revealed by the digital imaging microscope using fura-2. Nature 318:558–561
348a. Williams DA, Becker PL, Fay FS (1987) Regional changes in calcium underlying contraction of single smooth muscle cells. Science 235:1644–1648
349. Williams AJ, Ashley RH (1989) Reconstitution of cardiac sarcoplasmic reticulum calcium channels. Ann NY Acad Sci 560:163–173
350. Williams ME, Ways NR, Brenner R, Sharp AH, Leung AT, Campbell KP, McKenna E, Koch

WJ, Hui A, Schwartz A, Harpoid MM (1988) Sequence and expression of mRNAs encoding the α_1 and α_2 subunits of a DHP-sensitive calcium channel. Science 241:1661–1664

351. Williamson Jr, Cooper RH, Hoek JB (1981) Role of calcium in the hormonal regulation of liver metabolism. Biochim Biophys Acta 639:243–293
352. Wilson DB, Bross TE, Hofmann SL, Majerus PW (1984) Hydrolysis of polyphosphoinositides by purified sheep seminal vesicle phospholipase C enzymes. J Biol Chem 259:11 718–11 724
353. Wilson HA, Greenblatt D, Poenie M, Finkelman FD, Tsien RY (1987) Crosslinkage of a B lymphocyte surface immunoglobulin by anti-Ig or antigen induces prolonged oscillation of intracellular ionized calcium. J Exp Med 166:601–606
354. Winegrad S (1979) Electromechanical coupling in the heart muscle. In: Berne RM, Sperelakis N (eds) The heart. American Physiological Society, Bethesda (Handbook of physiology, vol 1)
355. Wojtczak L, Lehninger AL (1961) Formation and disappearance of an endogeneous uncoupling factor during swelling and contraction of mitochondria. Biochim Biophys Acta 51:442–456
356. Wood TL, Kobayashi Y, Frantz G, Varghese S, Christakos S, Tobin AJ (1988) Molecular cloning of mammalian 28,000 M_r vitamin D-dependent calcium binding protein (calbindin-D_{28K}): expression of calbinding-D_{28K} RNAs in rodent brain and kidney. DNA 7:585–593
357. Woods NM, Cuthbertson KSR, Cobbold PH (1986) Repetitive transient rises in cytoplasmic free calcium in hormone-stimulated hepatocytes, Nature 319:600–602
358. Worley PF, Baraban JM, Colvin JS and Snyder SH (1987) Inositol trisphosphate receptor localization in brain: variable stoichiometry with protein kinase C. Nature 325:159–161
359. Worley PF, Baraban JM, Supattapone S, Wilson VS, Snyder SH (1987) Characterization of inositol trisphosphate receptor binding in brain:regulation by pH and calcium. J Biol Chem 262:12 132–12 136
360. Yatani A, Codina J, Imoto Y, Reeves JP, Birnbaumer L, Brown AM (1987) A G protein directly regulates mammalian cardiac calcium channels. Science 238:1288–1292

360a. Yelamarty RV, Miller RV, Scaduto RC Jr, Yu FTS, Tillotson DL, Cheung JY (1990) Three-dimensional intracellular calcium gradients in single human burst-forming units-erythroid-derived erythroblasts induced by erythropoietin. J Clin Invest 85:1799–1809

361. Yin HL, Stossel TP (1979) Calcium control of actin network structure by gelsolin. In: Cheung WY (ed) Calcium and cell function, vol 2. Academic, Orlando
362. Yoshitomi K, Fromter E (1985) How big is the electrochemical potential difference of Na^+ across rat renal proximal tubular cell membranes in vivo? Pflügers Arch (Suppl 1) S121–S126
363. Yule DI, Gallacher DV (1988) Oscillations of cystolic calcium in single pancreatic acinar cells stimulated by acetylcholine. FEBS Lett 239:358–362
364. Zavortink M, Welsh MJ, McIntosh JR (1983) The distribution of calmodulin in living mitotic cells. Exp Cell Res 149:375–385
365. Zschauer A, van Breemen C, Buhler FR, Nelson MT (1988) Calcium channels in thrombin-activated human platelet membrane. Nature 334:703–705
366. Zschauer A, Van Breemen C, Buhler FR, Nelson MT (1988) Calcium channels in thrombin-activated human platelet membrane. Nature 334:703–705

Membrane Mechanisms in Volume Regulation in Vertebrate Cells and Epithelia

E.K. Hoffmann and H.H. Ussing

Introduction

The cell membranes of virtually all animal cells are highly permeable to water, and hence the cell volume is determined by the cellular content of osmotically active solutes and by the osmolality of the extracellular fluid. It should be noted, however, that in epithelial cells only the membrane on the serosal side is highly water permeable; the mucosal side membrane can be quite tight, as in the epithelial cells of late distal tubules and collecting ducts of the kidney in the absence of antidiuretic hormone.

The cells contain impermeant anionic macromolecules, and the colloid-osmotic swelling resulting from entry of diffusible ions and water (see [299] for a complete treatment of the Gibbs–Donnan equilibrium) is constantly counteracted by ion pumps which extrude ions at a rate equal to that of the dissipative "leak" entry. This "pump-leak steady state" concept was developed in detail by Leaf [202], Ussing [307], and Tosteson and Hoffman [304], but it was first formulated by Krogh in his Croonian Lecture [177], where he wrote:

> When cations inside the erythrocytes exert normal osmotic pressure and are free to move in the whole of the corpuscle water, when, further, the concentrations of the single ions differ more from the corresponding concentrations outside than can be explained by a Donnan effect, and when finally the erythrocyte membrane is definitely, even if slightly, permeable to these ions, the concentration differences can be produced and maintained only by some active transport: we are face to face with a steady state and not with any equilibrium.

This is formulated even more clearly in a memorandum by Krogh in 1945: "We must look upon the cell as being normally in a *steady state* in which the passive loss of potassium to the surrounding medium is exactly made good by the active potassium uptake."

Although essentially correct, the main shortcoming of this description is that passive ion leaks and active ion pump rates have turned out to be less static than originally suggested. Over the last 10–15 years evidence has accumulated to demonstrate that a constant cell volume is maintained not only by adjustment of ion pump rates but also to a large extent by dynamic and controlled changes in a variety of leak pathways. The passive ion leaks play an active role in cell volume control.

Besides maintaining a constant cell volume under "resting" conditions, most cells investigated are also capable of counteracting volume perturbations by volume recovery processes. Swollen cells tend to reduce their volume toward initial values by loss of KCl and concomitant loss of cell water (regulatory volume decrease, RVD), and shrunken cells often show the capacity to increase their volume toward initial values by net uptake of KCl and concomitant uptake of cell water (regulatory volume increase, RVI).

The ion transport systems involved in volume recovery and the maintenance of a constant cell volume are often quiescent or operate only at a low level in the "resting" cell but are activated by perturbations of the cell volume or by various external stimuli.

The ion transport systems involved in volume regulation and their activation have been studied in vitro mainly by observing the volume recovery processes following osmotic perturbations of the cell volume. In anisotonic media animal cells initially behave as nearly perfect osmometers by osmotic cell swelling in hypotonic media or by cell shrinkage in hypertonic media, but subsequently the cells regulate their volume toward initial values by a RVD or RVI response. (In some cell types the RVI response is observed only following pretreatment of the cells in hypotonic media.)

Under in vitro conditions, the ion transport systems involved in the RVD or RVI response are usually activated not by changes in the osmolality of the extracellular fluid, which is normally kept essentially constant by body fluid homeostasis (see below), but rather by other challenges of cell volume regulation, such as changes in the cell content of osmotically active solutes and various stimuli affecting the activity of the various specific ion transport systems involved in cell volume control (e.g. hormones, growth factors and other external stimuli, the cell membrane potential, and cytoplasmic second messengers). It should thus be stressed that most of the experiments discussed in this chapter are performed in order to reveal membrane mechanisms rather than to imitate likely biological conditions.

The intense studies of cell volume regulation in recent years have shown that a wide variety of cells and tissues share volume regulatory capacities but have disclosed a surprising diversity between different cell types in the nature of the ion transport systems involved.

Biological Role of Volume Regulation

The biological origin of RVD probably dates back to the very first appearance of cells. Without such a "safety valve" cells would suffer cytolysis when exposed to sudden drop in salinity of the medium. In fact, regulatory volume decrease is well known from microorganisms (for references see Chamberlin and Strange [40] and Schultz [288]). The mechanism may still play a similar role for amphibia whose skin is damaged when attacked by a predator. Terrestrial vertebrates are not exposed to this risk, and for most of their body cells the composition of surrounding body fluids is so well regulated that their capacity for rapid volume responses is never exploited. Nevertheless, it appears that mechanisms which limit cell swelling or actually restore cell volume when the cell is exposed to a hypotonic environment have been conserved from the earliest forms of bacterial life, through invertebrate predecessors, and are general characteristics of vertebrate cells today. On the other hand, mechanisms that respond to cell shrinkage in response to exposure to a hypertonic environment appear to have been only sporadically conserved. From a teleological point of view, shrinkage of cells that lack walls is certainly a lesser threat to survival than cell lysis.

For the majority of body cells in a terrestrial vertebrate the volume regulation is a steady state process: a slow steady uptake of Cl^- and of K^+ and a leakage of one or both of these ions which balance the uptake. Some cells, however, are exposed to different and varying osmolalities. Transporting epithelia and secreting or absorbing organs of epithelial nature frequently encounter media which differ from a plasma ultrafiltrate with respect to both osmolarity and composition. These cells typically face a plasma ultrafiltrate at one boundary (basolateral) and a solution of varying composition at the other (apical). For such cells rapid adaptation of the transport pathways in both apical and basolateral membranes is necessary in order to escape exorbitant swelling or shrinkage.

Intestinal epithelial cells and presumably blood cells traversing intestinal capillaries are exposed to hypotonicity after excessive water intake. Moreover, kidney medullary cells and blood cells traversing the capillaries of kidney medulla are exposed to high osmolalities, in particular during antidiuresis, where the osmolarity in this region of the mammalian kidney can reach levels higher than 1200 mosmol [162, 211]. Under pathophysiological conditions disturbances of body fluid homeostasis can be encountered, and plasma fluid osmolalities ranging between about 220 and 350 mosmol/kg H_2O have been observed. Under these extreme conditions the cells of the body would—in the absence of volume regulating mechanisms—swell and shrink by about 30% and 20%, respectively [217, 219]. Under pathophysiological conditions the anisotonic state usually develops over a matter of hours or days. Lohr and Grantham [215] have recently introduced a novel approach to studying the volume regulatory behavior of cells in vitro, under conditions approaching pathophysiological states. They demonstrated [215] that rabbit kidney proximal tubule epithelial cells have the capability to keep their cell volume constant (within $\pm$ 3% or less) with exquisite sensitivity and precision over a wide range of varying osmolalities, provided that the external osmolality is gradually reduced or increased at a rate not exceeding a few milliosmoles per minute. This capability for cell volume maintenance in anisotonic media is referred to as isovolumetric cell volume regulation [215].

Cell volumes can also be influenced by factors not immediately related to their external osmotic environment [218]. Thus, for example, cells in the mammalian brain are known to swell under the influence of a variety of nonosmotic stimuli (for references see [321]). Cells in the kidney swell during metabolic blockade [266, 209] or during ischemia [80]; their volumes can also be influenced by the activity of the Na^+ pump and by extracellular osmotic pressure [210].

An increase in the cellular content of osmotically active organic solutes and ions may also represent a challenge to cellular volume regulation and activate membrane transport processes involved in the RVD response. This is presumably the case in epithelial cells in relation to Na^+-coupled transport of sugars and amino acids [191, 289], and in liver cells in relation to Na^+-coupled uptake and accumulation of amino acids [172, 173]. As discussed by Kristensen [172], metabolic events may probably also lead to a rapid increase in the cellular content of organic solutes and hence activate a RVD response. This could conceivably be the case in hepatic glycogenolysis and in adipocyte lipolysis. The metabolic events in exercising skeletal muscle could also lead to accumulation of intracellular organic solutes and activation of volume regulatory responses [278].

Schultz [287] has proposed a hypothesis for the parallelism between basolateral membrane K^+ conductance and Na^+-coupled transport of sugars and amino acids (Fig. 1). The hypothesis is as follows. (a) An increase in the rate of Na^+ entry across the apical membrane results in an increase in intracellular osmolyte content; this is double true for Na^+ entry processes coupled to the entry of other solutes such as sugars, amino acids, and Cl^-. (b) This results in cell

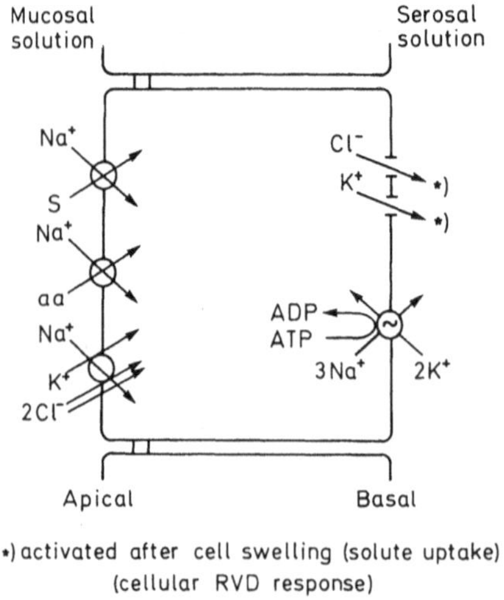

Fig. 1. Diagram of a solute- and Na^+-absorbing epithelium. The figure shows three of the possible entry paths for Na^+ at the apical membrane; sodium may enter the cell in a coupled reaction with a sugar, an amino acid or with chloride and possibly K^+. In all cases the sodium uptake leads to an increase in cellular content of osmolytes, resulting in cell swelling. This in turn leads to opening of chloride and potassium channels at the basal membrane (cellular RVD response)

swelling, which activates stretch-activated channels in the basolateral membranes. (c) These stretch-activated channels may be K^+ and/or Cl^- channels; and/or Ca^{2+} channels that result in an increase in intracellular Ca^{2+} and the resulting activation of Ca^{2+}-dependent K^+ and/or Cl^- channels. Alternatively, as suggested by Wong and Chase [326], an increase in intracellular Ca^{2+} could promote exocytotic processes that bring about the insertion of K^+ and/or Cl^- channels from cytoplasmic sources.

Another, but by no means mutually exclusive, explanation for the pump-leak parallelism stems from the recent finding that the K^+ leak pathways in the basolateral membranes of frog [184] and *Necturus* renal proximal tubules [276, 156], *Amphiuma* renal collecting tubule [148] and *Necturus* small intestine [49] exhibit voltage-dependent properties; that is, the K^+ conductance decreases when the basolateral membrane is depolarized and increases with hyperpolarization of that barrier. Inasmuch as the Na^+/K^+ pump is rheogenic (or electrogenic), an increase in pump rate brings about a hyperpolarization of the electrical potential difference across the basolateral membrane, which in turn promotes an increase in the K^+ conductance of that barrier. However, an increase in the K^+ conductance of that barrier, mediated by a separate set of volume-activated K^+ channels may bring about an even larger hyperpolarization which, in turn, could activate the voltage-sensitive K^+ channels.

Finally, cell volume regulation may conceivably play an essential role in cell growth and proliferation which include a balanced increase in cell content of organic solutes and ions, in cell volume, and in cell membrane surface area.

In some cases cell volume regulatory processes are an integrated part of the function of an epithelium; as an example we shall consider the secreting epithelia.

The Role of Cellular Volume Regulation in Secreting Epithelia

The term volume regulation may lead to the misconception that cells have one preferred volume. This is, however, not the case. The "preferred" volume depends on the functional state of the cell.

As an example let us consider salivary gland cells. In the resting state, the apical membrane is virtually tight to all ions but possesses dormant Cl^- channels. This is strongly suggested for the exocrine acinar cells [263] and studied in great detail in the shark rectal gland [98]. The basal membrane in gland cells possesses the usual Na^+/K^+-ATPase and a cotransport system for ($Na^+,K^+,2Cl^-$) [263, 158, 98]. Furthermore, it has K^+ channels, which, however, are mostly closed in exocrine acinar cells [263]. In the resting state the cells therefore take up Na^+,K^+, and Cl^-, and, being able to expell excess Na^+ and replace $3Na^+$ with $2K^+$, they ultimately attain a steady state where the near equilibrium is determined by the stoichiometry of the cotransporter. This means a high concentration of chloride and a large volume. Now, when the glands are stimulated and secretion is initiated (in part by an increase in cellular Ca^{2+}),

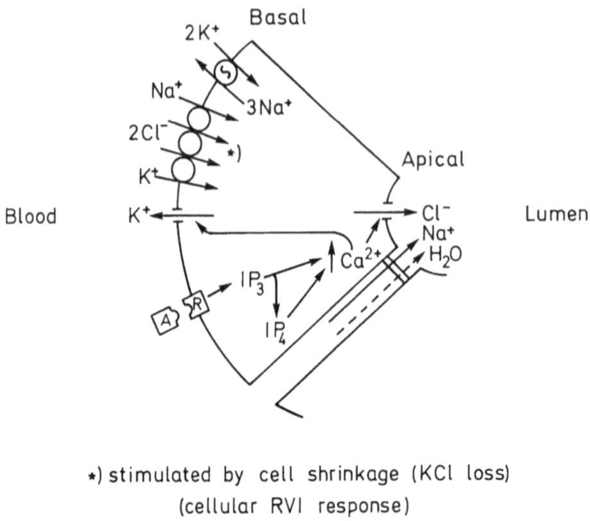

Fig. 2. Activation of a gland cell. When the agonist (*A*) reacts with the receptor (*R*) inositol triphosphate (IP_3) is released. This, in turn, releases Ca^{2+} from an intracellular pool, and Ca^{2+} then activates apical chloride and basolateral potassium channels. In the resting cell, the chloride and potassium concentrations are so high that the Na^+,K^+,$2Cl^-$ cotransporter of the basolateral membrane is virtually standing still (i.e., the forward and backward rates are about equal). The opening of the Cl^- and K^+ channels leads to rapid loss of KCl (and concomitant cell shrinkage), and the Na^+,K^+,$2Cl^-$ cotransporter is activated. The apical exit of Cl^- and the basolateral exit of K^+ creates an electric potential drop between blood and lumen which drives Na^+ (and water) through the cation-selective junction between the gland cells. The Na^+/K^+ pump maintains the ionic gradients, so that the secretion of Na^+ and Cl^- can continue as long as the stimulation lasts. The model thus accounts for the electrolyte secretion. However, the formation of an isotonic secretion requires recycling of Na^+ via apical channels (see Figs. 3, 4). (Adapted from [263])

there is a simultaneous opening of the basal K^+ channels [264] and the apical Cl^- channels (Fig. 2). Thus, KCl is rapidly lost, and the cells shrink. This leads to activation of the cotransport system, but even though there is a rapid uptake of Na^+, K^+, and Cl^-, the cells stay shrunk as long as the stimulation is maintained. A steady state is reestablished but with lower cellular Cl^- and K^+ levels, a slightly increased Na^+ level, and lower cell volume. In the new steady state the net process is a secondary active, outward transport of chloride at the expense of a strongly increased pumping rate of the Na^+/K^+ pump and an increased metabolic rate. This secondarily active chloride transport in turn creates the potential difference between blood and lumen, which drives the counterions to Cl^- (mostly Na^+) through the junction between the gland cells.

The Role of Cellular Volume Regulation in Isotonic Water Transport

It is well known that most primary gland secretions are isotonic or nearly isotonic with blood plasma. It has been suggested [317] that the processes of cell volume regulation are involved in the regulatory processes leading to isotonicity. As a background for discussing this problem it is necessary to realize that there is a problem.

Traditionally, isotonic transport through composite biological structures is discussed in terms of osmotic differences en route and reflection coefficients at the membranes involved. Unfortunately, however, the parameters must be measured at more or less inaccessible places. However, an alternative description of the interaction of solutes and water is possible in terms of variables which can be measured or calculated accurately. By a reformulation of Fick's law it is easily seen that the water/solute interactions can be expressed in terms of (a) forward and backward solute fluxes (b) free diffusion coefficients of the solutes, and (c) length of the transport path. An analysis in terms of this principle shows that isotonic water transport through a cell is impossible, even if the cell membranes were infinitely permeable to water [317]. The reason is that the friction of water against the stagnant solutes in the cell (for instance, proteins) forces the solvent (water) to lag behind the driving (transported) solute.

Paracellular transport offers more favorable conditions for isotonic transport because there are no cell membranes to cross, and there need not be any stagnant solutes. In glands, however, the gland cells are connected by cation-selective junctions. Thus, through the whole length of the paracellular pathway chloride and other anions are stagnant and by frictional interactions delay water relative to the driving species, Na^+. Nevertheless, the paracellular pathway is still much more favorable for water transport than that through the cells, and isotonic water transport via the gland is possible if part of the dominating cation (Na^+), after having passed the junction from blood to acinus, is reabsorbed by the gland cells and returned to the blood side (Fig. 3). In this way sodium can pass the junction and exert its frictional interaction with water more than once.

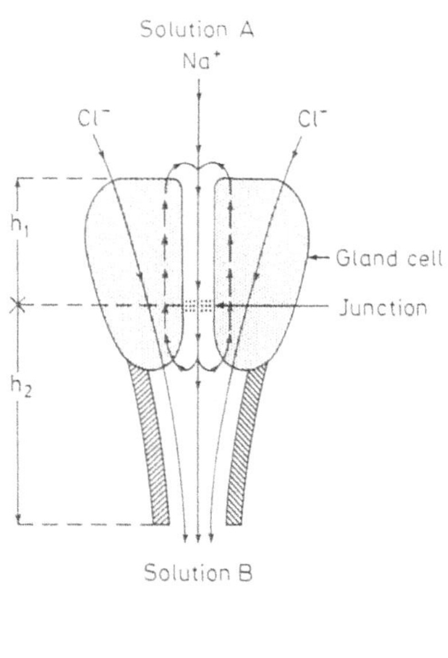

Fig. 3. Schematic model of skin gland. Only two gland cells are shown. The two cells are connected by a cation-selective junction, designated as "leaky seal". *Full lines*, the pathways for Na^+ and Cl^-; *dashed loops*, the recycling paths for Na^+. h_1, Distance from the inside solution to the junction; h_2, distance from the junction to the outside bath

Experimental evidence for this recycling of Na^+ [317] strongly supports this hypothesis. Furthermore, it is experimentally possible to vary the relative amounts of Na^+ and Cl^- which pass through the glands. It can then be shown that water secretion is strongly positively correlated to the net sodium efflux but not correlated to the net chloride flux [241].

The recycling theory thus is fairly well substantiated, but how do the gland cells "know" how much Na^+ should be recycled? The hypothesis has been advanced that the glands themselves are sensors for the osmotic pressure of the secretion. Let us assume that the apical Cl^- and Na^+ channels react to volume changes in the same way as those of Ehrlich ascites tumor cells [136, 137, 142]: cell swelling increases chloride permeability and reduces Na^+ permeability, whereas shrinkage has the opposite effects. Let us now assume that the secretion is becoming too hypotonic: the cells swell and more chloride is secreted whereas less Na^+ is recycled (Fig. 4, upper panel). Since the chloride secretion determines the total electrolyte secretion, whereas the Na^+ recycling determines the amount of water secreted, the secretion becomes more concentrated and the cells shrink somewhat. However, if the secretion now beomes too concentrated, the cell shrinks and the recycling of Na^+ is increased relative to the chloride secretion (Fig. 4, lower panel), more water can follow, and the hypertonicity is corrected. Similar feedback regulations probably are at work in all transporting epithelia.

The assumption that the apical absorption of Na^+ goes via Na^+ channels in the gland cells is supported by the finding of such channels in the secreting epithelium of canine airways [26]. Recently Dissing and Nauntofte [53] have suggested that cell shrinkage in isolated parotid acini stimulates the Na^+ uptake via Na^+ channels in the apical membrane.

Principal Ion Transport Systems for Na^+,K^+,Cl^-, and H^+ Involved in Volume Regulation

Single Cells

The principal modes by which ions cross cell membranes are electrodiffusion, primary active transport (ATP-fuelled ion pumps), electroneutral ion exchange, and electroneutral cotransport. The ion transport systems reported to be involved in volume regulation in various single cell types are summarized in Fig. 5.

Electrodiffusion must respond to the electrochemical potential of the ion at the two sides of the membrane. Such ion electrodiffusion, illustrated by the K^+,Cl^- and Na^+ channels in Fig. 5, tends to dissipate the ion gradients created by secondary and primary active transport processes and hence may be unfavourable for the economy of cell metabolism. In many cells it has been found

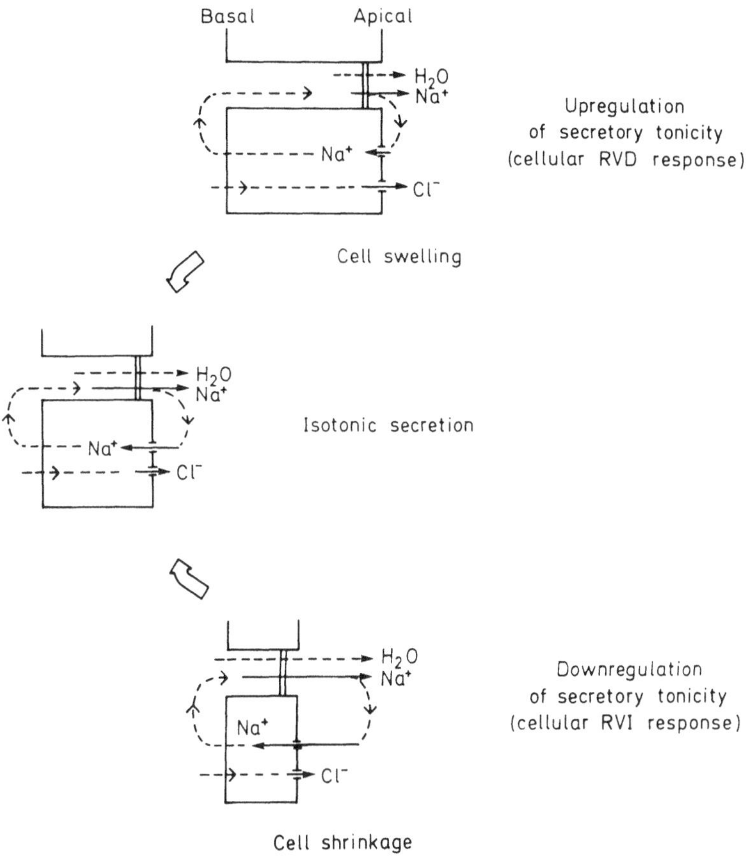

Fig. 4. Feedback regulation of the tonicity of gland secretion. It is assumed that the apical membrane of the gland cells is much more permeable to water than the basal membrane, so that the cell volume is determined by the osmolarity of the secretion. *Middle panel* (*isotonic secretion*) shows the cell volume when the secretion is isotonic. If it attains a lower tonicity (*upper panel*), the cells will swell. This elicits a RVD response (increased chloride transport and decreased Na^+ recycling), i.e., increased tonicity of the secretion. Conversely, if the tonicity of the secretion should become too high (*lower panel*), the cells shrink, and we see a RVI response, with reduced chloride transport and an increased recycling of Na^+, i.e., reduced electrolyte secretion and increased water transport, the net result being a reduced tonicity of the secretion

that the ion leak by electrodiffusion is in fact very small under normal steady-state conditions. For muscle and nerve cells, secreting glands and cells in "tight" epithelia, this is of course only true in the resting state. In many cell types ion channels may be activated by perturbation of cell volume and play important roles during volume regulation.

The active transport of sodium and potassium via the Na^+,K^+ pump is of crucial importance in maintaining the inwardly directed electrochemical Na^+ gradient across the cell membrane, which, as discussed below, represents the

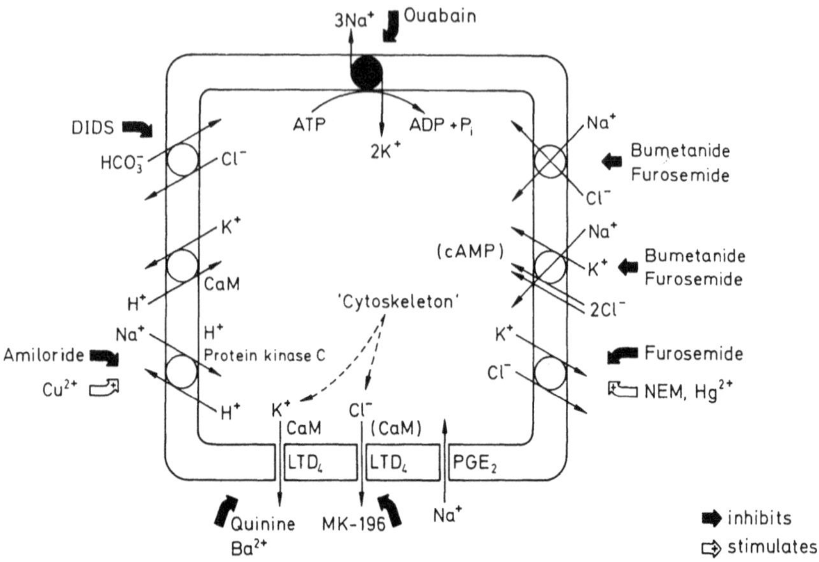

Fig. 5. Principal ion transport systems for Na^+,K^+,Cl^-, and H^+ in the cell membrane involved in volume and pH regulation in single cells. Several of the ion transport systems are quiescent in the resting cell. The molecular mechanisms proposed to be involved in the activation process are indicated on the figure. *Arrows*, blocking or stimulating effects of various substances on the different transport systems. For details see text. (From [143])

driving force for the Na^+-coupled cotransport systems ("secondary active" transport systems).

Electroneutral ion exchange, involving a one-for-one exchange of ions, is illustrated by the Cl^-/HCO_3^-, K^+/H^+, and Na^+/H^+ exchange systems in Fig. 5. In this case, ion influx and efflux are tightly coupled, so that the conductive ion permeability is much lower than that measured by isotope self-exchange or by heteroexchange experiments in which one ion is exchanged for another across the cell membrane. Such systems play a volume-regulatory role only in certain cell types (see below).

Electroneutral cotransport of anions and cations has attracted considerable interest in recent years and seems to play important roles in volume regulation.

Cotransport is often identified by the dependence of cation flow on anion concentration, and viceversa. Such experiments, however, only provide evidence for anion-activated cation flow or cation-activated anion flow. The demonstration of true, tightly coupled cotransport requires evidence that the flows of anions and cations are mutually dependent, usually with a defined stoichiometry, and that the gradient of one solute can drive transport of its cosolute against its gradient. Such transport systems have a high degree of chemical specificity toward the transported ions; thus, the Na^+,Cl^- and Na^+,K^+,$2Cl^-$ cotransport systems transport chloride, but not nitrate, thiocyanate, or sulfate. These cotransport systems (see Fig. 5) are able to accumulate chloride ions in the cell against an electrochemical gradient. The uphill Cl^- transport is energized by the downhill movement of sodium ions and is possible only in the presence of an inwardly directed gradient for sodium, and it thus depends on subsequent removal of sodium ions from the cell by the action of the Na^+,K^+-pump.

Ion transport systems proposed to be activated during RVD in various cell types involve conductive K^+ and Cl^- channels (see "Volume-Activated Cl^- and K^+ Conductance Pathways"), K^+,Cl^- cotransport ("Relationship Between Volume Regulation and pH Regulation"), and functionally coupled K^+/H^+ and Cl^-/HCO_3^- exchange ("K^+,Cl^- Cotransport").

RVI has been proposed to involve Na^+,Cl^- and Na^+,K^+,2Cl^- cotransport (see "Volume-Activated Cotransport") and functionally coupled Na^+/H^+ and Cl^-/HCO_3^- exchange ("Structure of the Na^+,K^+,Cl^- Cotransport").

Epithelia

As an example we use the frog skin epithelium, as one of the best described epithelial models. Figure 6 shows the essential features of the functional organization of the frog skin epithelium.

Control of Cell Volume at the Steady State

Donnan Equilibrium

The membrane is conceived of as being permeable to the solvent and to some, but not all, ions. The impermeable ions can be small ions or macromolecular polyelectrolytes. The equilibrium state in this situation is called the Donnan equilibrium. It is characteristic of the Donnan distribution that at equilibrium there is between the two compartments: (a) an uneven (asymmetrical) distribution of the diffusible ions, (b) an electrical potential difference (the Donnan potential), and (c) an osmotic pressure difference. A treatment of the Donnan distribution may be found in Sten-Knudsen [299] (the first issue in the present series).

The Gibbs–Donnan rule states that the products of diffusible anions and cations on both sides of the membrane are equal at equilibrium. A state of electroneutrality must exist in each compartment, and as a consequence a positive osmotic pressure difference (π) develops between inside and outside:

$$\pi = RT[C^i_{Na} + C^i_K + C^i_{Cl} + C^i_A - 2C^o_{Cl}] > RT[C^i_A] \tag{1}$$

where C indicates the concentration and A the impermeable anion. R is the gas constant and T the absolute temperature. Water enters the cell and tends to level out the difference in osmotic pressure, but at the same time the ions redistribute

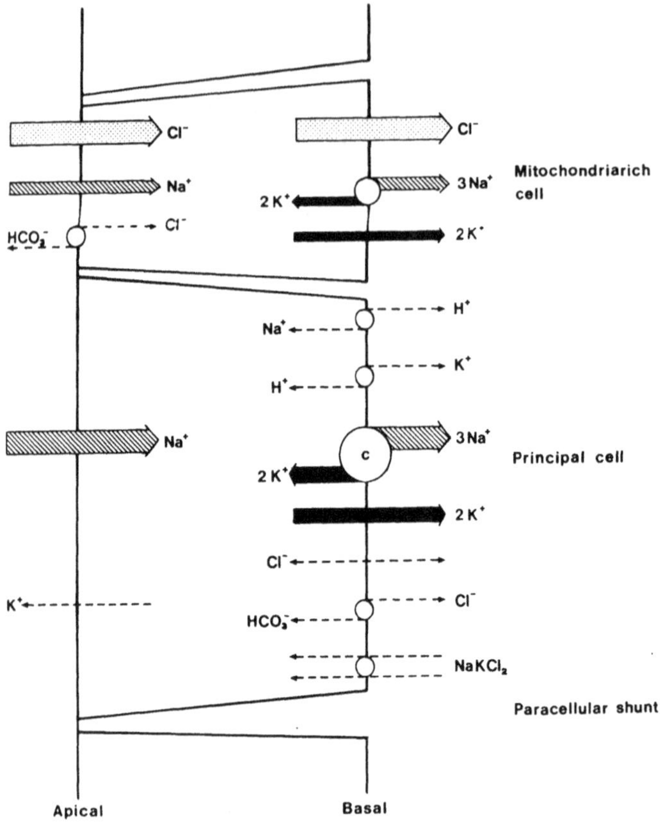

Fig. 6. A model drawing of the functional unit of a tight, salt-absorbing epithelium, based mainly on studies of anuran skins. A unit comprises the most abundant, syncytium-forming principal cells, the mitochondria-rich cells, (MR cells), and a paracellular shunt. The permeability of the latter is small compared to the cellular pathways, and it has a small preference for cations. Both types of cells contain the functional components of the original two-membrane hypothesis: A Na^+ channel (amiloride-sensitive) in the apical (mucosal) membrane and an Na^+/K^+ exchange pump together with a passive K^+ permeability in the basal (serosal) membrane. The major anion pathway is located to the MR cells that have a large Cl^- permeability in the basal membrane and a voltage-sensitive Cl^- permeability in the apical membrane. Active secretion of K^+ is possible through K^+ channels in the apical membrane of the principal cells in combination with the serosal Na^+/K^+ exchange pump. The basal membranes of the principal cells further contain an $Na^+/K^+/2Cl^-$ cotransport system and a Cl^- channel. The former is under normal conditions almost quiescent but may be activated by a drop in cellular Cl^- concentration. The mucosal anion exchanger of the MR cell is probably responsible for uptake of chloride from dilute solutions, and it may be related to the Cl^- channel on the molecular level. The amiloride-sensitive Na^+/H^+ exchanger and the K^+/H^+ exchanger reveal their presence only under special experimental circumstances

according to the Gibbs–Donnan rule. This continues until all the diffusible ions are inside or until the cells burst (colloid osmotic lysis). This means that animal cells must, even in isotonic solutions, somehow prevent passive distribution of permeant ions across the cell membrane.

Pump and Leak: Volume Maintenance

There is general agreement that control of cellular volume reflects a balance between passive and active ion movements across the cellular membrane, with the colloid pressure of intracellular macromolecules being offset by the extrusion of sodium ions from the cells. This "pump and leak" concept was developed about 25 years ago by Leaf [202], Ussing [307], and Tosteson and Hoffman [304]. The necessary equations for describing such a system can be found in Tosteson and Hoffman [304] and shall not be given here.

Intracellular Cl^- and the net charge of intracellular impermeable macromolecules are two of the factors determining the volume of a cell. This is supported by the following arguments. The extracellular fluid contains different ions and solutes with a total concentration (C_t^o). The intracellular fluid contains $Na^+(C_{Na}^i)$, $K^+(C_K^i)$, and $Cl^-(C_{Cl}^i)$ uncharged solutes $X(C_x^i)$ and other charged solutes $A^z(C_A^i)$ with mean valency z (a negative number, which, for example, red blood cells is between 3 and 4 at physiological pH). These factors, influencing intracellular osmolarity and the cell concentration of impermeant (fixed) charges $(-z \cdot C_A^i)$ are represented in the following equations, which state that the osmotic pressure must be the same on each side of the membrane, and that the difference between the total number of negative charges and the total number of positive charges is negligible (electroneutrality):

$$C_{Na}^i + C_K^i + C_{Cl}^i + C_x^i + C_A^i = C_t^o \text{ equal osmolality} \tag{1}$$

$$C_{Na}^i + C_K^i - C_{Cl}^i + zC_A^i = 0 \text{ electroneutrality} \tag{2}$$

By subtraction:

$$2C_{Cl}^i + C_x^i + C_A^i - zC_A^i = C_t^o$$

and by multiplication with the cell volume (V) (or strictly speaking cell water content):

$$2 \cdot C_{Cl}^i \cdot V + C_x^i \cdot V + C_A^i \cdot V - zC_A^i \cdot V = C_t^o \cdot V$$

or

$$2M_{Cl}^i + M_x^i + M_A^i - zM_A^i = C_t^o \cdot V$$

hence

$$V = \frac{2M_{Cl}^i + M_x^i + (1 - z)M_A^i}{C_t^o} \tag{3}$$

where M_x is the cell content of uncharged solutes; M_A, the cell content of charged solutes other than Na^+, K^+, and Cl^- (average valency: z); and M_{Cl}^i the cell content of chloride.

This equation shows the parameters that influence cell volume (V). For instance, cells shrink upon increase in osmolarity (C_t^o) and swell upon increase in cellular content (M^i_x) of uncharged solutes (e.g., amino acids or sorbitol) or of chloride (M^i_{Cl}). Intracellular Cl^- content is regulated by the rates of cotransport and uncoupled fluxes as described in more detail by Geck [87]. The role of these Cl^- movements in cell volume regulation is described in the following sections. Alterations in M^i_x (e.g., variation in cellular content of amino acids or sorbitol) may also be used for regulation of cell volume (see "Organic Osmolytes").

Alterations in M^i_A or in intracellular pH (which controls the mean valency, z) are also expected to affect cell volume. A decrease in cellular pH results in a decrease in the negative charge of A^z (less negative z) and thus in a decrease in cellular content of fixed charges. Equation (3) implies that provided two cells have equal chloride contents (M^i_{Cl}) the one with the lowest content of cellular, fixed charges has the smallest cell volume.

Alterations in cellular fixed charges (zM^i_A) are, however, in a cell with a functional anion exchanger compensated by corresponding alterations in intracellular Cl^- content (M^i_{Cl}). In most cells the cellular Na^+ and K^+ contents remain essentially constant after a change in cellular pH, and a fast one-for-one electroneutral exchange of Cl^- for HCO_3^- is taking place:

From equation (1) and (2) follows by addition:

$$2(C^i_{Na} + C^i_K) + C^i_x + (1 + z)C^i_A = C^o_t$$

by multiplication with the cell volume (V)

$$2(M^i_{Na} + M^i_K) + M^i_x + (1 + z)M^i_A = C^o_t \cdot V$$

and hence

$$V = \frac{2(M^i_{Na} + M^i_K) + M^i_x + (1 + z)M^i_A}{C^o_t} \tag{4}$$

The distribution ratio (cell to extracellular medium) for the exchangeable anions can be derived from Eqs. (2) and (4):

$$r = \frac{2(M^i_{Na} + M^i_K + zM^i_M)}{2(M^i_{Na} + M^i_K) + (1 + z)M^i_M} \tag{5}$$

With z negative the anion distribution ratio (r) is less than 1. When pH_i and z change, the anion distribution ratio (r) and the cell volume both change. Thus, *as pH_i decreases* and the negative charge (z) on the impermeant anions decreases (z becomes less negative), the cell content of negative, fixed charges decreases, and *the anion distribution ratio as well as the cell volume increases* (provided that M^i_{Na} and M^i_K are kept constant). For further discussion see Hladky and Rink [134a].

Regulatory Volume Decrease

In hypotonic media vertebrate cells initially swell by osmotic water equilibration, but subsequently regulate their volume (RVD) by a net loss of KCl and concomitant loss of cell water (see reviews by Kregenow [168], Cala [33], Hoffmann [137, 138], Grinstein et al. [108, 114], Lauf [195], Siebens [291], Larson and Spring [190], Hoffmann and Simonsen [143], Lang et al. [186], and Geck [87]).

In various cell types several ion transport mechanisms have been proposed to be activated during RVD, and the principal types are shown in Fig. 7. Also, organic osmolytes have been implicated in anisosmotic cell volume regulation. (This is further described under "Organic Osmolytes.")

Separate, conductive K^+ and Cl^- transport pathways have been reported to be activated during RVD in human lymphocytes [41, 99, 100, 108, 114, 279–280], in Ehrlich ascites tumor cells [136, 138, 142], in human platelets [212], and recently in human fibroblasts [272a]. A volume-sensitive conductive Cl^- transport pathway has been reported in the basolateral membrane of frog skin epithelial cells [309, 176]. A conductive K^+ transport pathway activated by cell swelling has been reported in the basolateral membrane of frog and toad

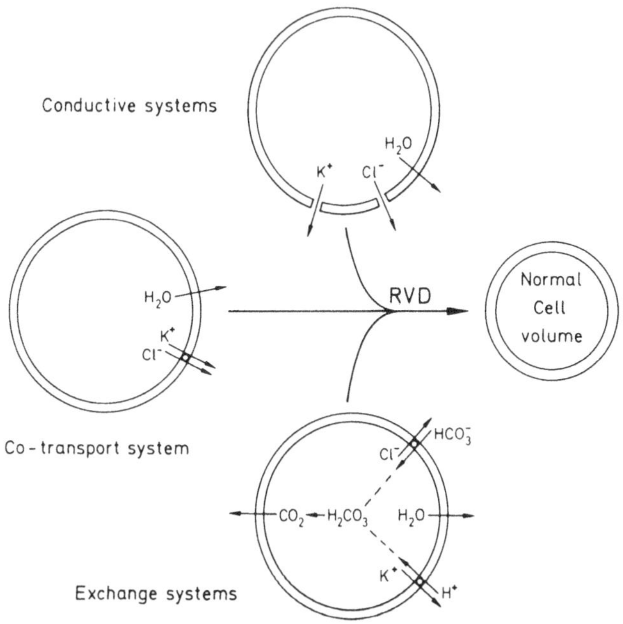

Fig. 7. Principal types of ion fluxes activated during regulatory volume decrease. (a) Conductive K^+ flux functionally coupled to conductive Cl^- flux. (b) Electroneutral K^+,Cl^- cotransport. (c) Electroneutral K^+/H^+ exchange functionally coupled to Cl^-/HCO_3^- exchange, giving KCl loss without change in pH_i. (From [143])

urinary bladder epithelial cells [50, 51], in mouse and rabbit kidney epithelial cells [296, 323, 70, 159, 160; 272], in enterocytes [191, 289], and in liver cells [172, 173].

An electroneutral, functionally coupled exchange of K^+/H^+ and Cl^-/HCO_3^- is involved in the RVD response in *Amphiuma* red cells [32, 35, 36, 168, 291] and frog skin epithelium [316].

An electroneutral K^+,Cl^- cotransport involved in RVD has been reported in red cells from several species, including duck [224, 168, 223], low-potassium (LK) sheep [58, 64, 65, 194], and fish [192, 27, 25]. In human red cells a volume-sensitive K^+,Cl^- cotransport is latent, but can be stimulated by high hydrostatic pressure [64, 65, 124]. A Cl^--dependent K^+ transport can also be induced by *N*-ethylmaleimide (NEM) in LK sheep and LK goat red cells [199, 194], and in human red cells [63, 64, 65, 194, 196, 231], or by Ca^{2+} depletion in LK sheep red cells [197, 195]. The effect of NEM and Ca^{2+} depletion are not additive, and the NEM-stimulated as well as the basal K^+,Cl^- cotransport is inhibited by Ca^{2+} plus ionophore A23187. It is proposed that cell swelling, NEM and Ca^{2+} depletion all affect the same transport molecule [195]. K^+,Cl^- cotransport can also be induced by NEM or by Ca^{2+} depletion in Ehrlich ascites tumor cells [163].

Dog red cells lack the Na^+,K^+-pump and have a high Na^+ content. The presence of a volume-sensitive K^+,Cl^- cotransport system can be demonstrated in cells where the ion content has been manipulated [253]. In physiological high-Na^+ cells, however, the RVD response is mediated by Na^+/Ca^{2+} exchange, coupled with the ATP-dependent Ca^{2+} pump [249, 252, 259, 261]. In the unperturbed cell under physiological conditions these systems serve to extrude Na^+ which leaks into the cells and maintain normal cell volume in much the same way that the Na^+,K^+ pump does in most cells. It has recently been reported that the Na^+/Ca^{2+} exchange can be turned on by formation of a critical disulfide bridge [259].

In *Necturus* gall bladder epithelial cells, an activation of both K^+,Cl^- cotransport and K^+ conductance in the basolateral membrane during RVD has been suggested [189, 190]. In Ehrlich cells at low pH and after Ca^{2+} depletion the combined activation during RVD of K^+, Cl^- cotransport and of conductive K^+ and Cl^- transport has also been reported [163].

Volume-Activated Cl^- and K^+ Conductance Pathways

In early papers on RVD in isolated cells, only a volume-induced increase in K^+ permeability, which was presumed to be conductive, was considered, whereas the Cl^- conductance was simply assumed to be initially high [21, 134, 271]. A swelling-induced increase in the conductive Cl^- permeability of the basolateral membrane of frog skin epithelial cells was suggested already in 1961 by MacRobbie and Ussing ([220]; see also [176, 309, 310]). Activation by cell swelling of separate, conductive fluxes of both K^+ and Cl^- in single cells was first proposed for Ehrlich ascites tumor cells [136]. K^+ and Cl^- tracer flux measurements showed an increase in permeabilities during volume regulation, and since the Cl^- conductance of the unperturbed cell is low ([131, 295]; see also [146]), it was inferred that the increase in tracer flux, as illustrated in Fig. 8, reflects an increase in the Cl^- conductance during RVD. As discussed below, this conclusion was confirmed in later studies of RVD in Ehrlich cells [142, 144]. Activation of ion channels in hypotonically swollen Ehrlich cells has recently been directly demonstrated in studies using the patch-clamp technique [150, 45], and the K^+ and Cl^- channels activated by cell swelling have been characterized in single-channel studies [45]. Activation of conductive K^+ and Cl^- fluxes during RVD has also been reported in lymphocytes and described in detail [99, 100, 114, 279, 280, 281]. To illustrate the principles we shall use the Ehrlich cell and the frog skin epithelial cell as examples.

Ehrlich Ascites Tumor Cells

The Cl^- conductance is estimated at 0.6 μS/cm² in Ehrlich cells at 37°C, similar to the value reported for human red blood cells [20]. It should be noted that, whereas in the human red cell the conductive permeabilities to Na^+ and K^+ are about two orders of magnitude lower than the conductive permeability to Cl^-, in the Ehrlich cell the conductive Na^+ and K^+ permeabilities are higher than that to Cl^- [116, 146]. The membrane conductances in Ehrlich cells are estimated at 10.4 μS/cm² for K^+, 3.0 μS/cm² for Na^+, and 0.6 μS/cm² for Cl^- [183].

When Ehrlich cells are suspended in hypotonic medium, a RVD response occurs which decreases the Na^+ permeability but greatly increases both the K^+

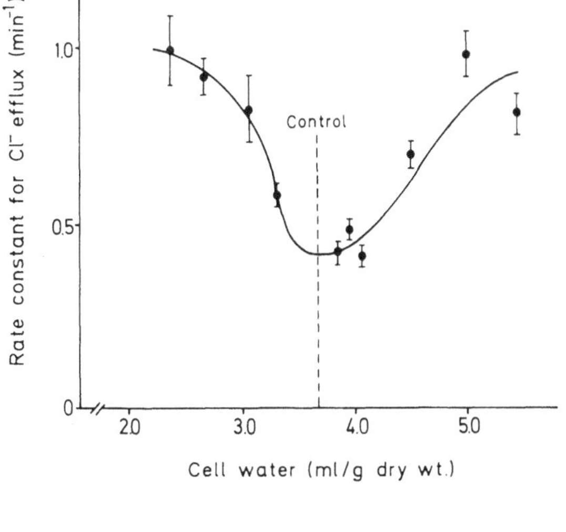

Fig. 8. Volume dependence of the rate constant for Cl^- efflux in Ehrlich ascites tumor cells measured as unidirectional ^{36}Cl efflux at steady state. The cells were preincubated with ^{36}Cl in a medium containing 75 m*M* NaCl with the addition of varying amounts of sucrose. *Broken line* (control): The cell water content measured for cells suspended in standard incubation medium in parallel experiments (3.66 ± 0.39 ml/g dry weight, mean ± SEM, $n = 6$). It should be noted that the unidirectional Cl^- flux at isosmotic cell volume (control) is dominated by electroneutral Cl^-/Cl^- exchange (see [146]). The increase in Cl^- flux upon cell swelling reflects an activation of conductive Cl^- channels, whereas the increase seen in shrunken cells reflects an activation of Na^+,Cl^- cotransport. For details see text. (Redrawn from [136])

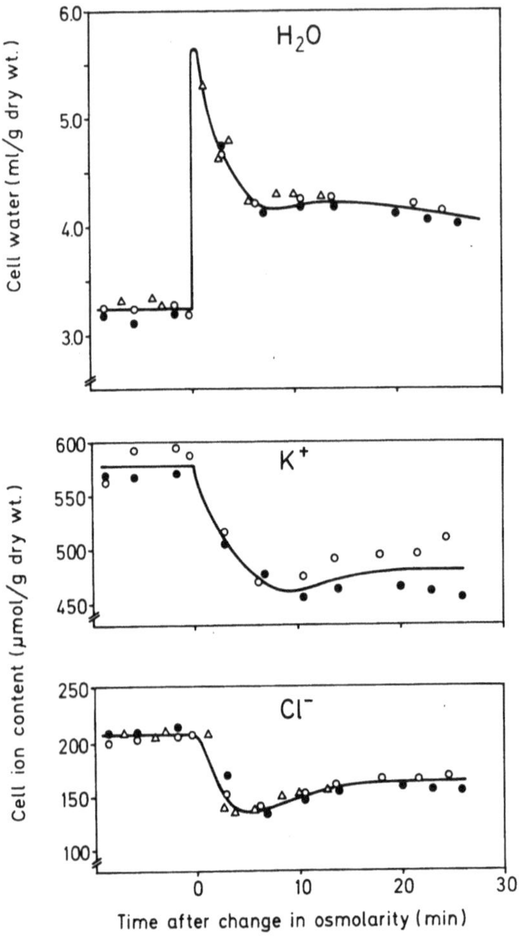

Fig. 9. Regulatory volume decrease in Ehrlich ascites tumor cells after reduction of external osmolarity. The cells were preincubated in 300 mosmol of chloride medium for 40–50 min, and at time zero resuspended in 150 mosmol chloride medium. Cell water, K^+ and Cl^- content were followed with time. Measurements from three independent experiments are indicated (○, ●, △). (From [144])

and Cl^- permeability of the cell membrane and causes a net efflux of K^+ and Cl^- with concomitant loss of cell water to restore the cell volume [134, 136]. The KCl loss during RVD is illustrated in Fig. 9.

The consequences of a loss of K^+ and Cl^- via separate, conductive transport pathways have been analyzed in detail by Lew and Bookchin [204] in terms of an integrated mathematical cell model. For a cell with a Cl^-/HCO_3^- exchanger in its membrane, the K^+ loss can be predicted to be partially balanced by HCO_3^- loss, producing a proton uptake via the Jacobs-Stewart cycle [151] with resulting intracellular acidification, and hence the K^+ loss exceeds the Cl^- loss to an extent which depends upon the cellular pH buffering capacity [79, 204].

In the case of the Ehrlich ascites tumor cell, the K^+ loss during RVD exceeds the Cl^- loss by a factor about 1.6 (Fig. 9; see [134]). Moreover, a cytoplasmic acidification during RVD can be demonstrated in nominally HCO_3^--free medium using an intracellular pH-sensitive fluorescent probe [213]. The relationship between volume regulation and pH regulation is further discussed below.

The chloride permeability (P_{Cl}) is increased to a larger extent than that of K^+, and it is concluded that the K^+ permeability limits the rate of volume change during RVD since volume changes are more rapid when valinomycin [144] or gramicidin [142] are used to provide a parallel pathway for K^+, as illustrated in Fig. 10. The depolarization (about 20 mV) observed during RVD in measurements using conventional microelectrodes [185] or a fluorescent probe [183] is consistent with a larger increase in the conductive Cl^- permeability than in the K^+ permeability. The increase in the Cl^- conductance is estimated at 68-fold when cells are transferred to a medium with half the osmolarity of the standard medium, whereas the concomitant increase in the K^+ conductance is only twofold [183]. It may be noted that the depolarization will increase the driving force for K^+ efflux, and hence account for at least part of the net K^+ loss [185].

Activation and Inactivation of K^+ and Cl^- Transport Pathways: Time-Dependence and Role of Ca^{2+}. The volume change in the presence of gramicidin has been used to monitor the time course of the changes in P_{Cl}. As shown in Fig. 11, P_{Cl} increases abruptly when the cells are swollen, but the change in P_{Cl} is transient, with inactivation within about 10 min. The initial increase in P_{Cl} is, as mentioned above, about 60-fold. In ^{42}K flux experiments the K^+ permeability was found to be about twofold increased 50 min after the hypotonic exposure [134] similar to the initial increase [183], indicating a much slower inactivation of the K^+ channel.

Separate, conductive pathways for K^+ and Cl^- can also be activated in isotonic cells. Addition of the Ca^{2+} ionophore A23187 in isotonic medium

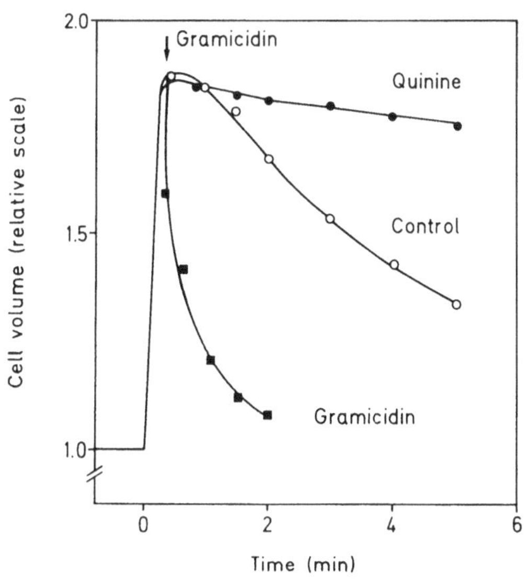

Fig. 10. Effect of quinine and gramicidin on regulatory volume decrease in Ehrlich ascites tumor cells in Na^+-free choline medium. Ehrlich cells were preincubated at 3%–8% cytocrit in standard incubation medium containing Ca^{2+} (1 m*M*). At time zero a sample of the cell suspension was diluted 500- to 1500-fold (final cell density about 70000 cells/ml) in hypotonic (150 mosmol) choline medium containing Ca^{2+} (0.5 m*M*), and the cell volume was followed with time using a Coulter counter. The cell volume (*ordinate*) is given relative to the initial cell volume, measured by dilution of a parallel sample of the cell suspension in standard incubation medium (mean value of two experiments). In one group quinine (1 m*M*) was present in the hypotonic choline medium (○). In the other group gramicidin (0.5 μ*M*) was added at the time indicated by the *arrow* (■). The experiment illustrated was performed at pH 8.2 but similar results were obtained in experiments at pH 7.4. (From [142])

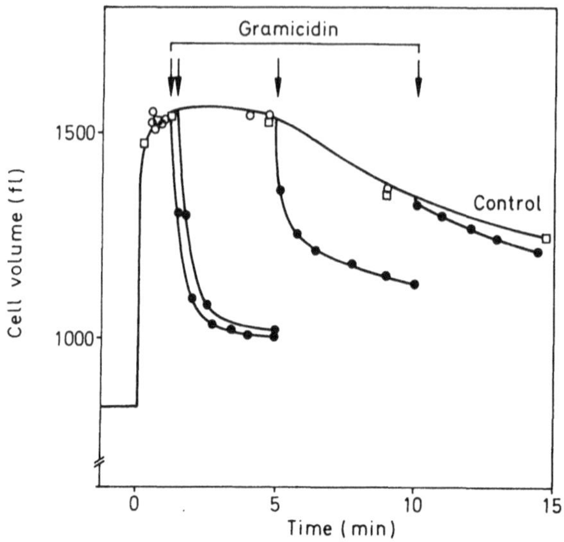

Fig. 11. Transient increase of Cl^- permeability of Ehrlich ascites tumor cells after hypotonic cell swelling in choline medium containing quinine. Experimental protocol as in Fig. 5. Quinine (1 m*M*) was added to the hypotonic choline medium in order to block the K^+ channels (control; *open symbols*). At the times indicated by the *arrows* 0.5 μ*M* gramicidin (○) was added to impose a high cation permeability. The figure is representative of two independent experiments. (From [142])

induces a fast net loss of KCl both in Ca^{2+}-containing and in Ca^{2+}-free media [145]. The net loss of KCl induced by A23187 shows several parallels to the KCl loss induced during RVD, and it has been demonstrated that A23187 induces a substantial increase of both the conductive K^+ and the conductive Cl^- permeability [142]. The K^+ and Cl^- conductances in A23187-treated Ehrlich cells are estimated at 134 and 30 μS/cm², respectively. Figure 12 shows the activation of K^+ channels after addition of A23187 in studies using the patch-clamp technique [45].

The finding that Ca^{2+} is somehow involved in the activation of the Cl^- transport pathway appears to be at variance with the findings in human lymphocytes where the increase in Cl^- net permeability induced by A23187 plus Ca^{2+} is significant [100], but not appreciable compared to that induced by cell swelling [280, 281]. A Ca^{2+}-activated Cl^- transport pathway has recently been reported in several cell types, for example, *Xenopus* oocytes [17], tracheal epithelial cells [81], neurones [245] kidney epithelial (MDCK) cells [28], and in the luminal membrane of salivary gland epithelial cells [237, 238]. In Ehrlich cells the Ca^{2+} activation of the Cl^- channels is found to be indirect (see below).

The A23187-induced increase in P_{Cl} in Ca^{2+}-free media (which is presumably mediated by release of Ca^{2+} from internal stores) is transient like the activation during the volume response, whereas the activation of P_{Cl} is persistent in Ca^{2+}-containing media (Fig. 13). In the latter case the Ca^{2+} pump is probably swamped by the A23187-induced Ca^{2+} leak across the cell membrane, although the Ca^{2+} pump in Ehrlich cells has been reported to be extremely powerful [47, 161]. In Ca^{2+}-free media, A23187-induced release of Ca^{2+} from internal stores [10] and a subsequent hysteretic activation of the Ca^{2+} pump [283] are likely to produce a transient increase in cytosolic Ca^{2+}. These findings suggest that it may be a transient increase in free cytosolic Ca^{2+} which accounts for the transient activation of the Cl^- transport pathway during RVD.

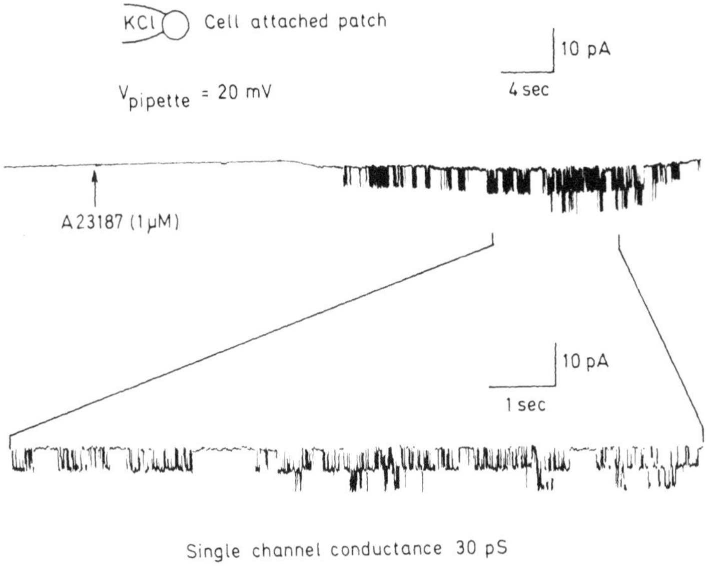

Fig. 12. Patch current from a K^+ channel in a cell-attached patch with KCl in the electrode. At the time indicated by the *arrow* 1 μ*M* of the Ca^{2+} ionophore A23187 is added to the bath solution. Inward K^+ currents are shown as negative (*downward*) deflections. *Bottom panel*, details on an expanded time scale. (Result from [45])

Selectivity of the Anion Transport Pathway. The net permeability of the cell membrane is substantially higher to nitrate and thiocyanate than to Cl^-, as seen by comparing the rate of cell shrinkage induced by valinomycin in these media [142, 163]. The volume recovery during RVD is unaffected by substitution of nitrate or thiocyanate for Cl^- [142, 144] also in the presence of gramicidin, added in order to ensure a high K^+ permeability (Fig. 14). The volume-induced anion transport pathway in human lymphocytes was also found to be rather unselective [99]. Following ionophore A23187-induced activation of the anion transport pathway the rate of cell shrinkage in the presence of valinomycin is substantial both in Cl^-, Br^-, NO_3^-, and SCN^- medium, and the absolute increase in the rate of cell shrinkage is of similar magnitude in the four media, indicating the activation by Ca^{2+} of a rather unselective anion transport pathway, in sharp contrast to the Cl^--dependent cotransport systems [142].

Inhibitors. The conductive chloride transport pathway has a different sensitivity to inhibitors than the inorganic-anion exchange system and the chloride-dependent cotransport system in Ehrlich cells [142, 144]. The volume recovery during RVD and the ionophore A23187 plus Ca^{2+}-induced increase in ^{36}Cl influx were both rather insensitive to DIDS, which blocks the inorganic-anion exchange system, and to bumetanide, which blocks the Cl^--dependent cotransport system. The Cl^- transport pathway is also activated during RVD in the presence of quinine which blocks the K^+ transport pathway [142]. The Cl^-

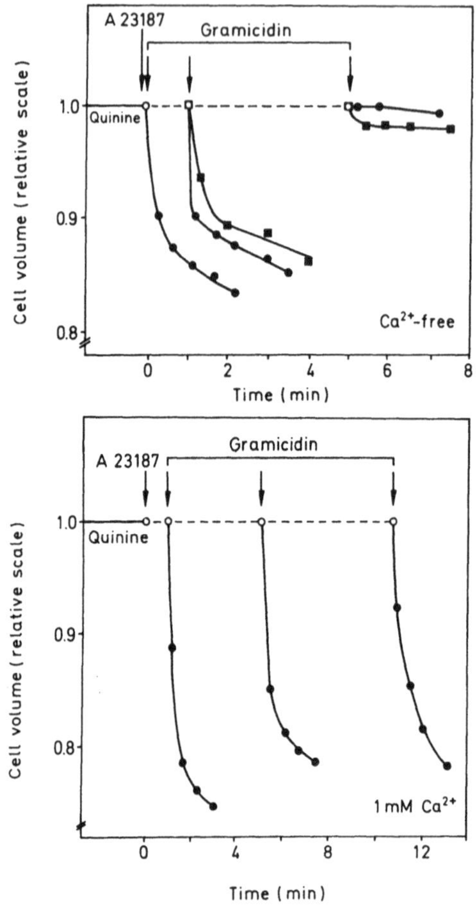

Fig. 13. Time dependence of Cl^- permeability following addition of ionophore A23187 in Ca^{2+}-free and Ca^{2+}-containing choline medium. The media contained quinine to block the Ca^{2+}-dependent K^+ channels. Ionophore A23187 (2 μ*M*) and gramicidin (0.5 μ*M*) were added at the times indicated by the *arrows*; gramicidin was added to impose a high cation permeability. Experimental protocol as in Fig. 5. The choline medium was nominally Ca^{2+} free, containing 0.5 m*M* EGTA (*upper panel*), or contained 1 m*M* Ca^{2+} (*lower panel*). In both media 0.8 m*M* choline was replaced by Na^+ in order to provide an equilibrium potential for Na^+ near that for K^+. The experiment is representative of two and three experiments in Ca^{2+}- free and Ca^{2+}-containing medium, respectively. The cell volume is given relative to the value measured before addition of A23187 and gramicidin (*open symbols*). The curves shown in the upper panel are compiled from two experiments marked individually. (From [142])

transport pathway is strongly inhibited by diphenyl-amine-2-carboxylate (DPC) and indacrinone (MK-196) [1], drugs known to inhibit Cl^- channels in Cl^--transporting epithelia [54] and the potential-gated Cl^- conductance of toad skin [61]. This is seen in Fig. 15, which shows that MK-196 also inhibits RVD in the presence of gramicidin, added in order to ensure a high K^+ permeability. In mouse proximal tubule epithelial cells and in Madin-Darby canine kidney (MDCK) cells, the anion channel activated by cell swelling was not blocked by the chloride channel blocker anthracene-9-COOH [185].

Role of Calmodulin. A number of drugs reported to inactivate the Ca^{2+}-binding protein calmodulin has been demonstrated to inhibit the KCl loss induced by cell swelling or by A23187 in lymphocytes [108] and in Ehrlich cells [142, 144]. The inhibition of RVD by pimozide, a diphenylbutyl piperidine anti-calmodulin drug, is illustrated in Fig. 16.

Pimozide blocks RVD also in the presence of anions with a high conductive permeability (nitrate or thiocyanate; see Fig. 16, right panel) and also when a

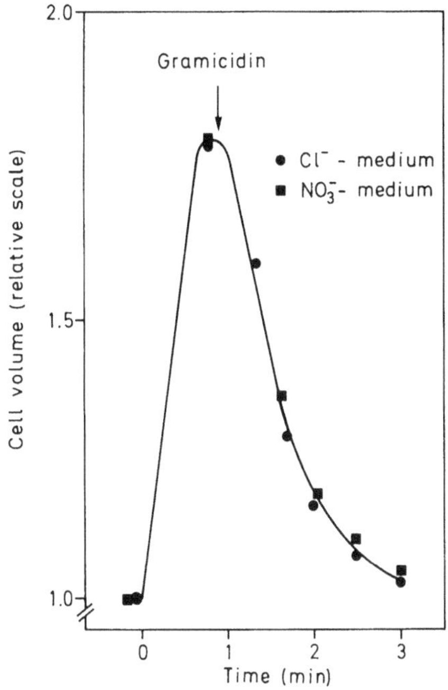

Fig. 14. Selectivity of the swelling-induced anion transport parthway. Ehrlich ascites cells were preincubated at 4% cytocrit in Cl^- medium (standard incubation medium) or NO_3^- medium. At time zero the cell suspensions were diluted 600-fold in hypotonic (150 mosmol) choline chloride medium and *N*-methyl-D-glucamine nitrate medium, respectively, and the cell volume was followed with time. Gramicidin (0.5 μ*M*) was added at $t = 0.9$ min (*arrow*). The cell volume is given relative to the initial cell volume, measured by dilution of a parallel sample of the cell suspension in the corresponding isotonic incubation medium (unpublished results)

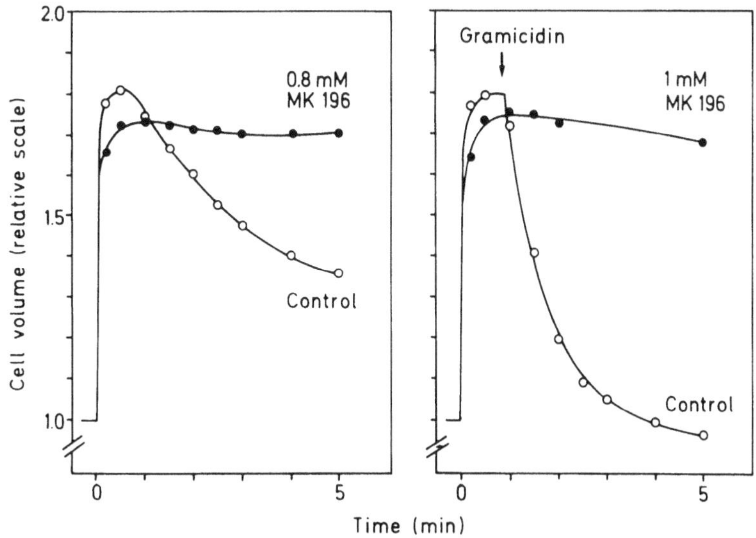

Fig. 15. Effect of indacrinone (MK-196) on RVD in Ehrlich cells after hypotonic cell swelling in Na^+-free choline medium. Cells were prepared and preincubated as described in the legend to Fig. 4. MK-196 was added at the time of the hypotonic exposure (time zero). Gramicidin (0.5 μ*M*) was added at $t = 0.9$ min as indicated by the *arrow*. (Based on data from [1])

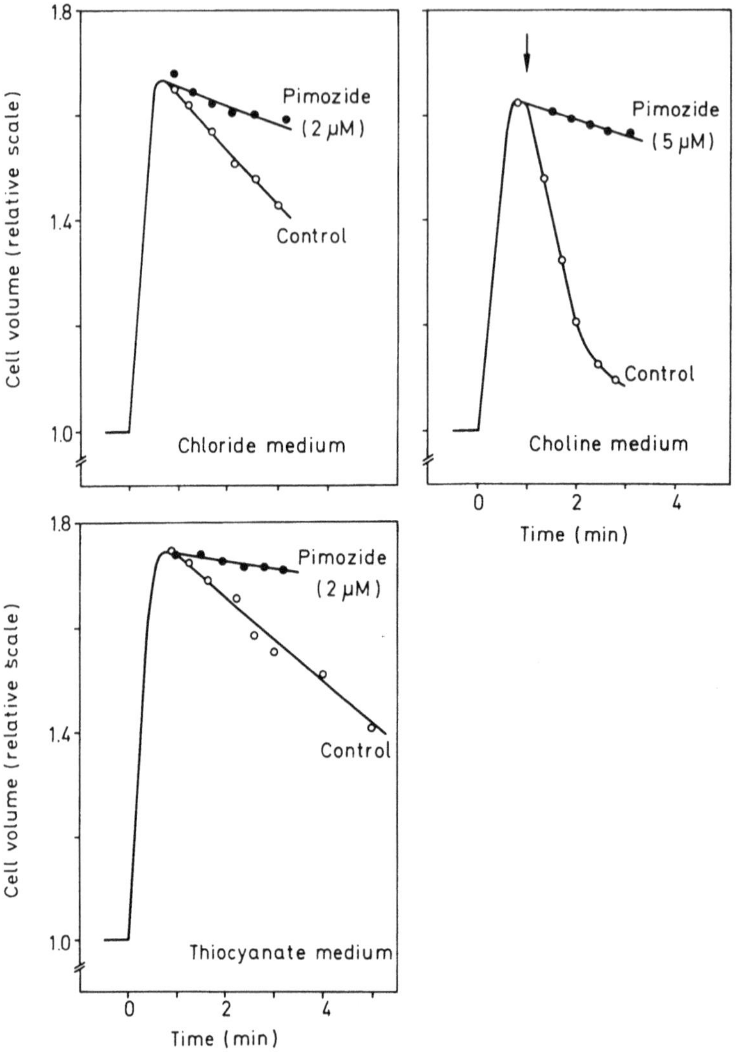

Fig. 16. Inhibition by pimozide of the increase in Cl^- and K^+ net permeabilities induced by cell swelling. Parallel groups of cells were preincubated at 4% cytocrit for 25 min in Cl^- medium (standard incubation medium), choline medium, or thiocyanate medium (*left*, *middle*, *right panels*, respectively) with a shift of the medium once after 15-min preincubation. The media were buffered at pH 8.2 and contained 0.15 mM Ca^{2+} and 0.15 mM Mg^{2+}. At time zero the cell suspensions were diluted 1000-fold in the corresponding media with the tonicity reduced to 150 mosmol, and the cell volume followed with time. Pimozide was added to the experimental groups (*closed symbols*) in the concentrations indicated. In the experiment in choline medium (*middle panel*) gramicidin (0.5 μM) was added at $t = 1$ min to contol and experimental groups in order to impose a high K^+ permeability. The cell volume is given relative to the initial cell volume, measured by dilution of a parallel sample of the cell suspension in the corresponding isotonic incubation medium. Similar results were obtained in nitrate medium (data not illustrated). (From [142])

high K^+ net permeability is ensured by addition of gramicidin (Fig. 16, middle panel). These findings demonstrate that the volume-activated K^+ and Cl^- transport pathways are both inhibited by pimozide. A similar inhibition by pimozide was observed in the case of ionophore A23187 plus Ca^{2+}-induced cell shrinkage. The inhibition by pimozide of the A23187-activated K^+ transport pathway is, moreover, directly demonstrated by ^{42}K flux experiments [142]. Although pimozide is not selective as an anti-calmodulin drug, the findings suggest that calmodulin is involved in the activation of both the K^+ and Cl^- transport pathway in Ehrlich cells. Calmodulin is found in the cytoplasm of Ehrlich cells and the cell membranes bind calmodulin in a Ca^{2+}-dependent manner [2]. Taken together, the above findings support a role for calmodulin in the Ca^{2+} activation of the channels. Evidence for the involvement of calmodulin in the operation of Ca^{2+}-activated K^+ channels in mouse fibroblasts has been presented [244].

Properties of the Volume- or Ca^{2+}-Activated K^+ Channel. Data obtained using cell-attached patches indicate that the same K^+ channel is activated by cell swelling and by addition of Ca^{2+} plus ionophore A23187. The K^+ channel is also observed in inside-out patches. This K^+ channel is a Ca^{2+}-activated inward rectifier channel with 40 pS conductance for inward currents and 15 pS for outward currents [45], similar to the small, Ca^{2+}-activated K^+ channel in red cells [115] and HeLa cells [282]. A comparison of the A23187-induced K^+ conductance estimated from ^{42}K tracer flux measurements at high external K^+ and from net K^+ flux measurements suggests single-file behavior of the Ca^{2+}-activated K^+ channel in Ehrlich cells [142]. Single-file diffusion has been demonstrated in K^+ channels in giant axons [19, 135], in frog striated muscle [149], in human red cells [319], and for the K^+ channel in the baso-lateral membrane of frog skin epithelial cells [68] as well as in the apical membrane [69]. From the estimated single-channel conductance the number of volume- or Ca^{2+}-activated K^+ channels in the Ehrlich cell can be estimated at 20×10^6 cm^{-2}, or about 250 per cell 45. It is tempting to suggest that the volume-induced conductive K^+ transport seen in many cells is always mediated by this type of Ca^{2+}-activated channels; this may, however, be an oversimplification. In human red blood cell "inside-out" vesicles exposed to hypotonicity a high K^+ permeability was demonstrated which showed a large Ca^{2+}-independent component [270]. In frog red blood cells two classes of K^+ channels were demonstrated in patch clamp studies. In osmotically swollen cells an increased K^+ permeability was demonstrated which was mediated predominantly by a small conductance channel which did not display the same strong Ca^{2+} sensitivity as the larger conductance channel [125]. In the acinar cell upon cholinergic stimulation the increased cytosolic Ca^{2+} activates opening of maxi K^+ channels [264].

Functional Reconstitution of the Isolated Volume-regulatory K^+ Channel. The Ca^{2+}/calmodulin complex is involved in the activation of the K^+ channel during RVD (see above). We have isolated calmodulin-binding proteins from cholate-solubilized membranes of Ehrlich cells by affinity chromatography, using calmodulin as ligand. Proteins retarded by the affinity material were eluted by a Ca^{2+}-free buffer, containing 5 m*M* ETDA. A crude membrane protein fraction and the proteins isolated by the calmodulin gel were reconstituted into phospholipid vesicles, and the K^+ transport activity of these vesicles was assayed by measurement of $^{86}Rb^+$ uptake. The reconstituted systems were in both cases found to perform a K^+ transport, and this transport was inhibited by 5 m*M* Ba^{2+}, similar to the volume-activated K^+ channel in intact cells (Fig. 17). The $^{86}Rb^+$ uptake rate (nmol mg^{-1} protein min^{-1}) after reconstitution of the proteins isolated by the calmodulin gel was about ten times higher than the $^{86}Rb^+$ uptake rate after reconstitution of the crude membrane preparation [154]. A large part of the transport protein has presumably been inactivated during the isolation procedure, which would explain the relatively low increase in specific activity.

Properties of the Volume-Activated Cl^- Channel. The volume-activated Cl^- channel is a small channel with a channel conductance of 2–5 pS. The channel is seen in both cell-attached and inside-out patches in a state of low activity. In cell-attached patches, it is fully activated by addition of the Ca^{2+} ionophore A23187 and by hypotonic cell swelling. Figure 18 shows typical records of currents obtained with the patch pipette attached to the cell after a hypotonic chock. The current-voltage relationship intersects the zero current level at about 30 mV, which fits well with the difference between the membrane potential (− 60 mV) and the Nernst potential for Cl^- (− 28 mV), resulting in zero driving force for Cl^- exit across the patch. The bottom record (Fig. 18, D) shows that the activation of the Cl^- channel after hypotonic exposure is delayed about 30–50 s. The channel cannot be activated by Ca^{2+} in an inside-out patch and it is not stretch activated.

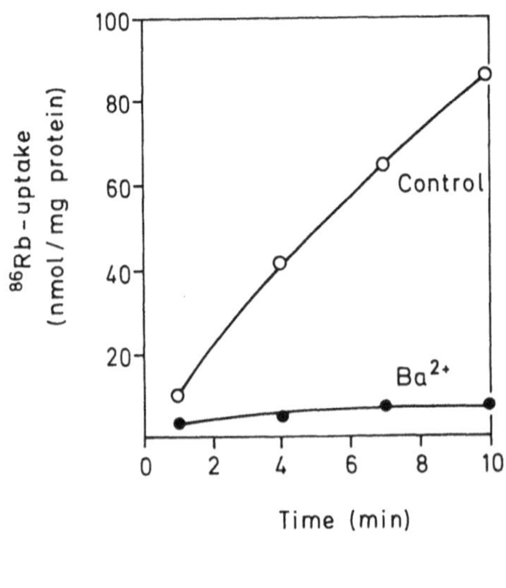

Fig. 17. K^+ transport activity (using ^{86}Rb as tracer) in phospholipid vesicles reconstituted with solubilized membrane proteins that are Ca^{2+}-dependently retained by calmodulin affinity material. (Result from [154])

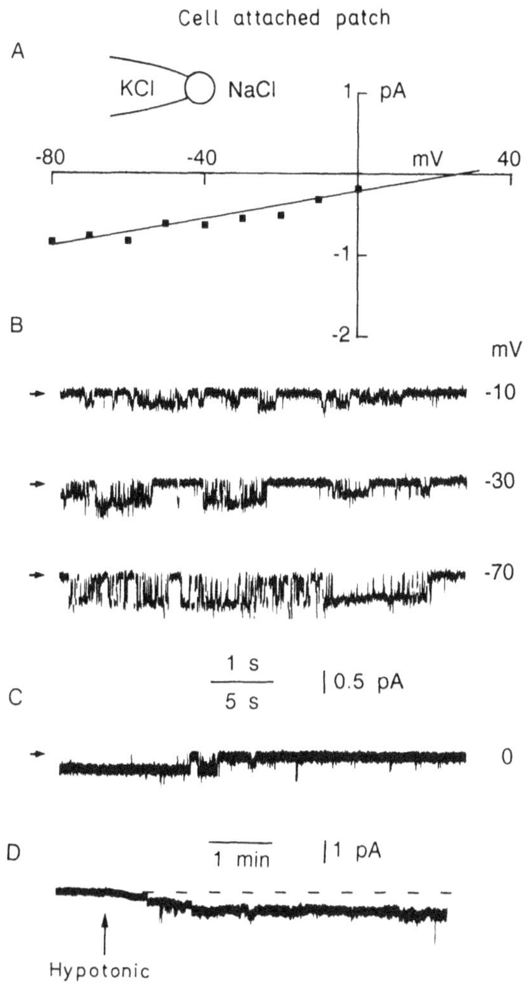

Fig. 18. Activation of a small Cl^- channel in hypotonically swollen Ehrlich cells. Current recordings for a chloride channel from an attached patch with KCl in the electrode. *A*, Single-channel current versus bath potential for the Cl^- channel (i/v curve). *B* and *C*, Details with two different time resolutions. *D*, Activation after cell swelling. At the time indicated by the *arrow* the fluid bathing the cell is being changed from a normal Ringer's solution to a solution with half the tonicity. (Results from [45])

Properties of the Basolateral Volume Regulatory Anion Channels in Frog Skin.

Some of the properties of the basolateral anion channels [313, 316] are illustrated in Fig. 19. (With respect to the theoretical basis for the experiments, see "Cotransport of $Na^+ + K^+ + 2Cl$ Responsible for High Cl^- Concentration in Principal Cells of the Frog Skin.") The outside medium is 1/20 ordinary Ringer throughout. The inside medium initially is full ordinary Ringer. The potential difference is 88 mV across the skin, which is close to the expected basolateral potential difference. The value is very close to the K^+ diffusion potential, in agreement with the assumption that the anion channels are mostly closed. At the first arrow, marked SCN/2, the inside medium is replaced by thiocyanate Ringer, diluted to half. This leads to cell swelling, which again induces an opening of anion channels as evidenced by a drastic decrease of the

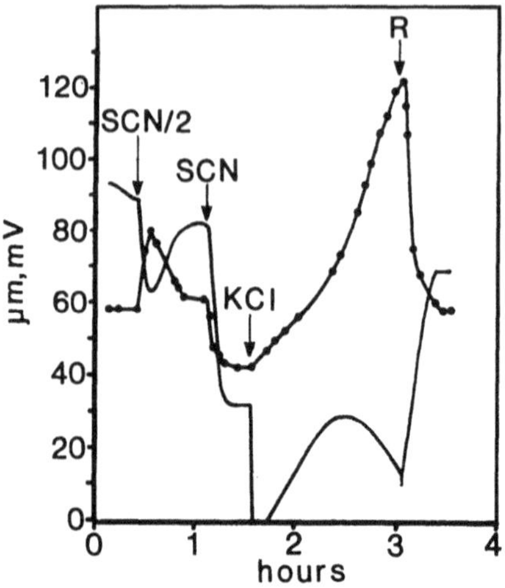

Fig. 19. Basolateral chloride channels open even when the epithelium is shrunken due to chloride depletion depolarization of the basolateral membrane, brought about by replacing Na^+ by K^+, so that an exorbitant, but reversible, swelling is brought about. Outside medium 1/20 Ringer throughout. Inside medium originally Ringer, followed at *first arrow* by half thiocyanate Ringer and at *second arrow* by full thiocyanate Ringer. At the *third arrow* (*KCl*) the inside medium is made full KCl Ringer (all Na^+ is replaced by K^+). At the *fourth arrow* (*R*) the inside bath is again made ordinary Ringer. ([312])

PD. As the cells shrink while losing K^+ and Cl^-, the PD increases again, because the dominating anion (now thiocyanate) approaches Donnan distribution across the basolateral membrane. At the second arrow, marked SCN, full thiocyanate Ringer replaces the dilute medium. The epithelium shrinks immediately, and the PD drops drastically. This must mean that both K^+ channels and anion channels close since both K^+ and SCN diffusion potentials should remain the same when cellular and inside bath concentrations are doubled. At the third arrow, marked KCl, the medium is changed to a Ringer where all Na^+ has been replaced by K^+. This results in a violent swelling of the epithelium, which in time may triple its volume. The swelling can be obtained with K^+ Ringer having thiocyanate, nitrate, iodide, or any other small anion replacing chloride [310]. It is seen that the high K^+ concentration leads to a violent drop in PD which actually may reverse sign. The depolarization probably activates both K^+ channels and anion channels (we shall return to this below). The PD first increases pari passu with the cell swelling but then starts falling again as both K^+ and Cl^- approach Donnan distribution between the cytoplasm and the high-K^+ medium. At the fourth arrow, marked R, the inside medium is made normal Ringer. At this stage evidently both K^+ channels and anion channels must be wide open, so that both K^+ and Cl^- are lost extremely rapidly when the inside bath is made ordinary Ringer. The loss of electrolytes and water stops abruptly when the thickness of the epithelium has attained its original value as measured at the beginning of the experiment. At the same time the PD has increased steeply from 10 mV at the beginning of the ordinary Ringer period to 70 mV at the time that the stationary situation has been established. This pattern of events is the same whether the excessive swelling has been obtained with KCl Ringer as in this experiment or with high-K^+ Ringer having nitrate or thiocyanate as the dominating anion. It appears then that the anion channels of the basolateral membrane are wide open at 10 mV and closed at 70 mV.

Depolarization of the basolateral membranes can also be brought about by closure of K^+ channels. In the experiment shown in Fig. 20 the inside bath is originally Ringer. At the first arrow this solution is made 2 mmol with respect to Ba^{2+}. Thus a sort of cascade effect is initiated: As more chloride channels open, the PD across the basolateral membrane approaches the chloride equilibrium potential which is somewhere between 20 and 30 mV. Following the steep drop, PD shows a slow increase. This has to do with a normally silent K^+ pathway, which possibly is a calcium-gated K^+ channel activated by Ca^{2+} entering the cells from the inside bath via a potential-gated Ca^{2+} channel (cf. [314]).

At the second arrow the inside bathing solution is diluted to half. The epithelium swells, but this leads to increased permeability to both K^+ and Cl^-, and the cells lose KCl and volume. As both K^+ and Cl^- of the epithelium approach Donnan equilibrium with the inside medium, the PD increases, but this apparently leads to closure of the Cl^- channels. There is a sharp break in the rate of cell shrinkage, and there is simultaneously a burst of spike like depolarizations. These irregular oscilations may be due to the fact that different groups of cells reach the conditions for chloride channel closing a different times.

At the third arrow the inside bath is made full Ringer with 2 m*M* Ba^{2+}. It is seen that the cells regain the volume lost, despite the presence of Ba^{2+}, which evidently does not inhibit the cotransport. However, the PD remains low, about 30 mV, until, at the fourth arrow, the inside medium is made Ba^{2+}-free ordinary Ringer.

Since the anion channels appear to be permeable to most small anions, one might have thought that the potential responses would be the same if chloride were replaced by nitrate. However, this is not the case. Figure 21 shows such an experiment. The inside solution initially is ordinary chloride Ringer. At the first

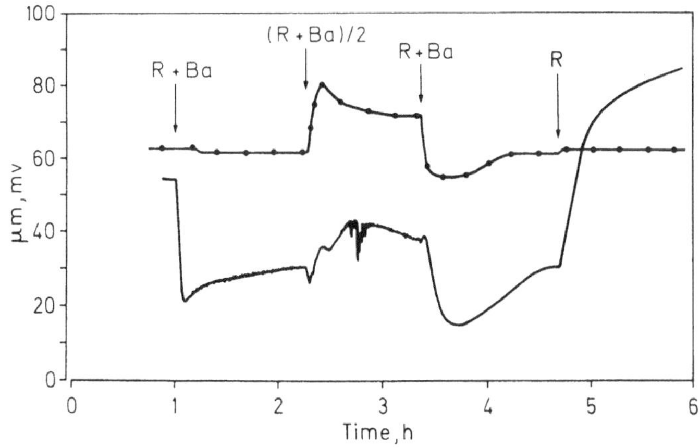

Fig. 20. The effect of making the chloride Ringer, bathing the inside of the skin, 2 m*M* with respect to Ba^{2+}. Ordinates and abscissa as in Fig. 19. Outside bath 1/20 nitrate Ringer throughout. Inside bath initially Ringer. The sequence of changes of inside bath shown by *arrows*. (From [313])

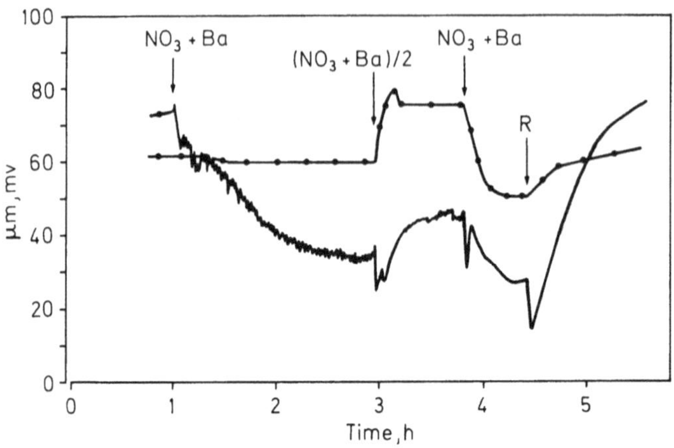

Fig. 21. The effects of making the nitrate Ringer, bathing the inside of the skin, 2 m*M* with respect to Ba^{2+}. Ordinates and abscissa as in Fig. 19. Outside bath 1/20 nitrate Ringer throughout. Changes of inside bath indicated by *arrows* [313])

arrow this solution is replaced by nitrate Ringer, made 2 m*M* with respect to Ba^{2+}. The PD immediately drops about 10 mV, but then the slope declines, so that the drop to the anion equilibrium potential lasts about 2 h.

The epithelial volume has not changed significantly during these 2 h, and thus the total electrolyte content has not changed significantly. The following explanation of the observations suggests itself: the introduction of Ba^{2+} in the inside medium brings about inhibition of K^+ channels, and the PD starts dropping. As in the foregoing experiment, this results in gating of anion channels. However, since nitrate (like thiocyanate) diffuses faster than chloride in the channels, the cell membrane polarizes near the open channels, and this leads to rapid closure of the channels in question. Thus the drop in potential is delayed. Thiocyanate has the same effect. This ion is known to diffuse at a rate about twice that of chloride in the apical anion channels of the mitochondria-rich (MR) cells [174].

Relationship Between Volume Regulation and pH Regulation

Most of the processes involved in cell volume regulation are also involved in the regulation of cellular pH, and a few basal interactions between volume and pH regulation must be discussed here.

In the case of the Ehrlich ascites tumor cell, the K^+ loss during RVD exceeds the Cl^- loss by a factor of about 1.6 (see above). Moreover, a cytoplasmic acidification during RVD, which is to be expected for a cell with a fast Cl^-/HCO_3^- exchanger, has been demonstrated using an intracellular pH-sensitive fluorescent probe (Fig. 22). To investigate the assumption that the source of the acidification is actually recycling of extracellular Cl^- for cellular

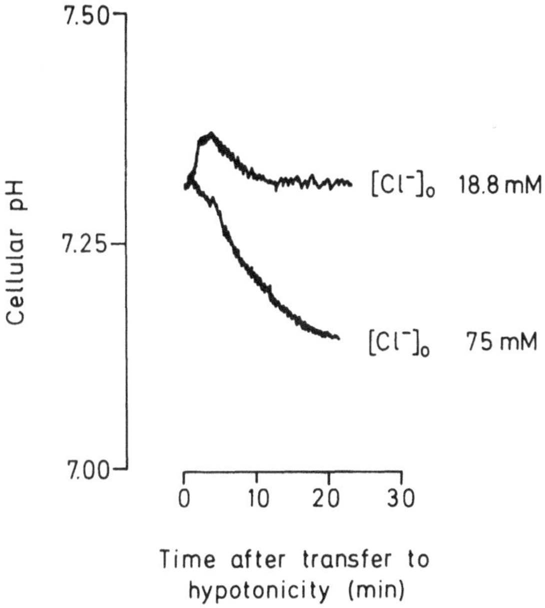

Fig. 22. Cytoplasmic acidification during RVD in Ehrlich cells. Effect of Cl^- substitution on the shifts in intracellular pH as a function of time after transfer to hypotonic conditions. At time zero BCECF-loaded cells were transferred to a hypotonic (150 mosmol) standard medium (75 m*M* Cl^-) or a hypotonic (150 mosmol) medium where 3/4 of the NaCl was substituted by Na-gluconate in equimolar amounts (18.75 m*M* Cl^-). (From [213])

HCO_3^- via the anion exchanger [79], the volume change experiment is performed at varying external Cl^- concentrations. Figure 22 shows that, at an extracellular Cl^- concentration in the hypotonic solution of 19 m*M*, which is well below the initial cytoplasmic Cl^- concentration immediately after the cell swelling, acidification was no longer seen, showing that an inward chemical gradient for chloride is required [213]. The acidification results in an activation of the Na^+/H^+ exchanger (the intracellular pH "set point" for the activation is estimated to be around 7.0), leading to an incomplete RVD. In support of this conclusion, amiloride and Na^+-free medium, known to limit the Na^+/H^+ exchange, indeed enhance the RVD response. Intracellular acidification and activation of Na^+/H^+ exchange may be a general response of cells undergoing RVD. Similar results have previously been demonstrated for blood platelets [212], thus possibly indicating a general pattern in cells undergoing RVD. As described below for the frog skin some cells are apparently able to "close down" the anion exchanger during RVD when the Cl^- channels open.

Three mechanisms dominate the pH regulation of most cells, namely the Cl^-/HCO_3^- exchanger, the H^+/Na^+ exchanger, and the Na^+-dependent Cl^-/HCO_3^- exchanger. In most cells at rest the pH is close to the one predicted for the Cl^-/HCO_3^- exchanger, and the two other systems seem to be almost quiescent. However, as soon as the cell pH drops significantly, the Na^+/H^+ exchanger and the Na^+-dependent Cl^-/HCO_3^- exchanger are activated. Clearly, this would lead to a paradoxical situation, where one powerful exchange system would try to lock the pH at one value, whereas two other systems would attempt to establish another equilibrium. The net result would be a volume change instead of a regulation of pH. Under extreme experimental conditions such paradoxical volume responses are actually seen (Fig. 43), but under more biological conditions the cells seem to solve the problem by closing up the Cl^-/HCO_3^- system when the others are activated. If we take the frog skin epithelium as an example, there is good evidence for the assumption that the principal cells possess a powerful Cl^-/HCO_3^- exchanger, which must define the cellular pH. If, however, the basolateral membrane is depolarized by addition of Ba^{2+} to the inside bath, we find that the exchange of nitrate in the bath against cellular chloride will take hours (Fig. 21) [313].

When, finally, equilibration between extracellular and intracellular nitrate and chloride does take place, it seems to do so via anion channels which are normally closed, but which open as the basolateral membrane gradually depolarizes. In other words, it looks as if depolarization leads to disappearance of the anion exchanger and appearance of anion channels. Such a relationship has been demonstrated beyond doubt in the apical membrane of the MR cells of the amphibian skin [175]. Tentatively, these authors propose that the anion exchanger and the anion channel are two modes of the same protein, and that the transition between the two modes is potential-dependent.

K^+, Cl^- Cotransport

As the best described examples of K^+, Cl^- cotransport involved in volume regulation, we shall look at duck red cells and sheep red cells.

Duck Red Cells

The study of volume regulation by duck red blood cells was pioneered by Kregenow [164, 167], who showed that following osmotic swelling the duck red cell regulates its volume back to control levels (RVD) in about 90 min. This RVD results from activation of a transport system which mediates an efflux of K^+ and Cl^- from the cells driven by the K^+ concentration gradient [164]. This K^+ flux is dependent on the presence of Cl^- (or Br^-), and a component of the Cl^- flux parallels the increased K^+ flux [169], suggesting cotransport of one Cl^- with one K^+ (see also review by Kregenow [168]). Electroneutrality of the transport process has been established [223]. The K^+ flux increases linearly with increasing Cl^- concentration. The KCl flux is inhibited by furosemide [168] and also by bumetanide, but only at high concentrations [222]. The inhibition by bumetanide is not affected be external Cl^- [222], but is enhanced by extracellular K^+ [216]. The stilbene derivatives, DIDS or SITS, which have little effect on the $Na^+,K^+,2Cl^-$ cotransport system (see "Structure of the Na^+,K^+,Cl^- cotransporter"), have, however, been demonstrated to inhibit the K^+,Cl^- cotransport system. The stilbene derivative, 4,4′-dinitrostilbene-2,2-disulfonic acid (DNDS), has, however, no effect on the K^+,Cl^- cotransport system, and the inhibition by SITS and DIDS is reversible, in contrast to the covalent inhibition of the inorganic-anion exchanger [216]. It may be noted that the inhibition of volume-induced K^+,Cl^- cotransport by inhibitors of the inorganic-anion exchange system has also been reported in fish red cells [24].

Experiments by Haas and McManus [122] suggest that the RVD response in swollen cells may be absent under in vivo conditions because the swelling-induced activation of the K^+,Cl^- cotransport system is overridden by the effect of endogenous catecholamines which activate the Na^+,K^+,$2Cl^-$ cotransport system operating during RVI (see "Volume-Activated Cotransport"), and because the K^+,Cl^- cotransport system and the Na^+,K^+,$2Cl^-$ cotransport system seem to represent different modes of operation of a common transport system in duck red cells [122, 222].

Sheep Red Cells

Cell swelling also increases the K^+ permeability of LK sheep red cells, while Na^+ permeability is unaffected [58]. This volume-induced response is absent or only weak in high-potassium (HK) sheep red cells [63], except in reticulocytes [193]. The volume-sensitive K^+ flux in LK red cells is Cl^- dependent (Br^- substitutes for Cl^-, and the K^+ flux is even stimulated about twofold when Cl^- is replaced by Br^-) [58]. A very high furosemide or bumetanide concentration (about 2 m*M*) is required for 50% inhibition [63]. The K^+ flux in LK sheep red cells is inhibited by anti-L_1, a specific component of an antiserum against LK sheep red cells [297]. The K^+ flux must involve a small number of transport sites, since there are only about 850 anti-L binding sites per cell [297], or else the effect of anti-L_1 must be highly cooperative or mediated through a second messenger. The number of cotransport sites may even be an order of magnitude lower than the number of antigen sites [297]. Compression at high hydrostatic pressure also induces a large Cl^--dependent K^+ flux in LK sheep red cells which shows many of the characteristics of the hypotonically induced flux [64, 65].

The sulfhydryl-reactive reagent NEM greatly stimulates a Cl^--dependent, K^+-selective flux in LK sheep red cells [18, 63, 199, 214]. There are several similarities between the NEM-stimulated and the volume-stimulated Cl^--dependent K^+ flux [194]. On the other hand, neither response blocks the other: NEM-treated cells are still volume sensitive [63], and cells with different volumes show about the same response to NEM [194]. The simplest and most likely hypothesis is that the K^+ transport system itself is the same in both cases, but the triggering mechanism is different. For further discussion on the relationship between volume- and NEM-stimulated K^+, Cl^- flux in LK sheep red cells, see the review by Lauf [195].

K^+/H^+ Exchange Functionally Coupled to Cl^-/HCO_3^- Exchange

Amphiuma Red Blood Cells

The *Amphiuma* red cell RVD response involves a net loss of K^+ and Cl^- [32, 168], which has been reported to be mediated via a K^+/H^+ exchange mechanism functionally coupled to the Cl^-/HCO_3^- exchange system [32–37]. The net proton influx via the K^+/H^+ exchange mechanism tends to decrease intracellular pH. Consequently, intracellular HCO_3^- ions react with protons forming CO_2 and water. the decrease in intracellular HCO_3^- generates a disequilibrium for the anion exchanger, leading to a net influx of HCO_3^- and a net efflux of Cl^-. CO_2 then diffuses out of the cell and is hydrated to HCO_3^- and H^+ in the extracellular compartment. This notion is based on the findings that (a) the membrane potential (measured with microelectrodes) does not change during RVD, (b) the volume-regulatory K^+ loss is largely unaffected by valinomycin-induced changes in membrane potential, and (c) the valinomycin-induced hyperpolarization is similar with and without volume-induced activation of the K^+ flux, indicating that the membrane conductance remains low during RVD [33, 34]. In addition (d), during RVD in the presence of ionophore A23187, a thermodynamic analysis indicates that the driving force for net K^+ flux is the difference between the chemical potential difference for K^+ across the cell membrane and the chemical potential difference for H^+, as opposed to the electrochemical potential difference for K^+ [36]. The K^+/H^+ exchange in *Amphiuma* red cells can also be activated by NEM [3].

The above analysis of the relevant driving force for net K^+ flux was made with cells exposed to ionophore A23187 plus Ca^{2+} in order to produce a membrane hyperpolarization with a resulting deviation of the H^+ distribution across the cell membrane from electrochemical equilibrium. Under these conditions a dissipative H^+ transport could be demonstrated to drive a conservative K^+ transport.

It should be noted, as discussed by Haas et al. [123] and Cala [33], that in red cells with a rapid Cl^-/HCO_3^- exchange, the driving force for electroneutral K^+/H^+ exchange is equal to the driving force for electroneutral K^+, Cl^- cotransport, and hence a thermodynamic distinction between these K^+ transport modes is not possible. Cl^- dependence of cation transport has often been taken to indicate the involvement of cotransport systems. The demonstration of Cl^- dependence may be inconclusive, however, since anion replacements may exert their primary effect upon reactions involved in the activation of inducible transport pathways [3]. In the case of the *Amphiuma* red cell, the strongest argument against a K^+,Cl^- cotransport is that DIDS-treated, hypotonically swollen cells lose K^+ without a concomitant Cl^- loss. Under these conditions the net K^+ loss via K^+/H^+ exchange is balanced by Na^+ gain via Na^+/H^+ exchange [36].

In the *Amphiuma* red cell the K^+ loss during RVD exceeds the Cl^- loss [33, 34] and is moreover accompanied by cellular acidification [32]. As discussed by Lew and Bookchin [204] in their analysis of an integrated mathematical cell model, these findings are predicted also in the case of KCl loss via separate, conductive K^+ and Cl^- transport pathways. To account for a K^+ loss exceeding the Cl^- loss about fivefold, as observed in the case of the *Amphiuma* red cell would, however, require an unexpectedly large pH buffering capacity of these cells [204].

As in the Ehrlich cell and the lymphocyte system, a role for intracellular Ca^{2+} in controlling the RVD response has been proposed in *Amphiuma* red cells

[33, 34]. The ionophore A23187 plus Ca^{2+}-induced K^+ loss shows several similarities to the volume-induced K^+ loss, and is also reported to be largely electroneutral by virtue of coupling with H^+ [33, 34]. The Ca^{2+}-induced conductive K^+ loss was found to contribute only a minor part of the total K^+ loss. It seems as if an increase in cytosolic-free Ca^{2+} is involved in activation of the K^+/H^+ exchange by osmotically swollen *Amphiuma* red blood cells [38]. It is suggested that K^+/H^+ exchange and Na^+/H^+ exchange may represent different transport modes of the same membrane component(s) [37].

Evidence for a K^+/H^+ Exchanger in the Basolateral Membrane of the Frog Skin

When the cytoplasm is acidified by increase of the CO_2 tension, the Na^+/H^+ exchanger is activated (see below; Fig. 23). If, however, this latter process is inhibited by addition to the inside solution of amiloride, a different response is seen. As usual, the inside bathing solution initially is ordinary Ringer. If this solution is replaced by half nitrate Ringer, we see initial swelling and then a drop in volume which is the result of the loss of virtually all cellular chloride, while nitrate is not taken up at all. Then the solution is made 10^{-3} *M* with respect to amiloride and acidified with 90% O_2 + 10% CO_2. This leads to shrinkage of the epithelium, so that it ends up with a volume smaller than the initial one,

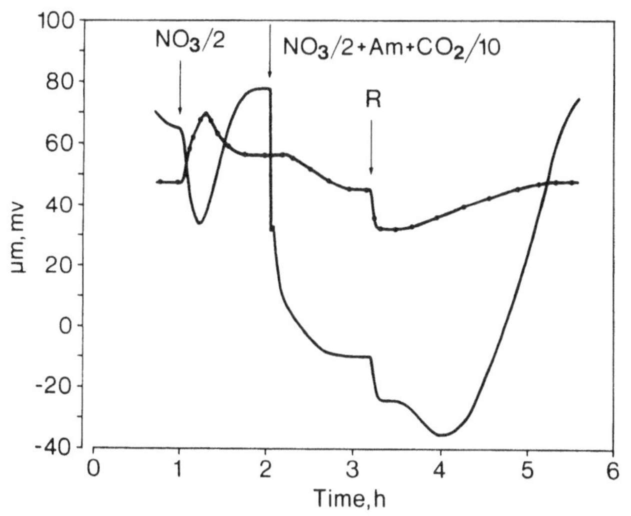

Fig. 23. Paradoxical shrinkage of anion-depleted cells when exposed to CO_2 acidification after inhibition of the basolateral Na^+/H^+ exchanger. Ordinates and abscissa as in Fig. 19. Outside medium 1/20 nitrate Ringer throughout. Inside bath initially Ringer. At the *first arrow* change to half-strength nitrate Ringer. At the *second arrow* this medium is made 2 m*M* with respect to amiloride and, at the same time, both inside and outside media are being equilibrated with 10% CO_2 and 90% O_2. At the *third arrow* the inside bath is made ordinary Ringer, and both this medium and the outside medium are equilibrated with atmospheric air

despite the fact that the cytoplasm of the cells is half Ringer isotonic. The only possibility left, especially since the K^+ channels are closed, is that loss of K^+ is balanced by an uptake of H^+ which titrates negative charges of the nondiffusible anions (proteins, phosphate esters, etc.). In other words, there must be a K^+/H^+-exchanger. If the acidification is discontinued, the cells recover (Fig. 23).

Organic Osmolytes

Organic osmolytes have been implicated in anisosmotic cell volume regulation of animal, plant, bacterial, fungal, and protist cells exposed to anisosmotic environments. Although the accumulation or loss of any organic solute would serve to restore the cell volume, only a restricted number of organic osmolytes are used in biological systems. The general classes of organic osmolytes are sugars, polyols, amino acids, methylamines, and urea. Polyols such as sorbitol and inositol are accumulated in renal medullary cells during hypertonic stress [236, 30]. Taurine is involved in volume regulation in several cell types (see below). The quaternary amine betaine is found in mammalian renal cells [12, 236, 328]. High concentrations of trimethylamine oxide are found in several vertebrates such as hagfish, skates, rays, chimaeras, and coelocanths [329, 16]. Another quaternary amine, glycerophosphoryl-choline, is found in mammalian renal medulla [12, 328]. Urea is an important intracellular solute in some vertebrates, most notably elasmobranchs, coelocanths, amphibia and mammals (renal medulla) [12, 328]. A role for this compound in RVD has recently been postulated [16]. The intracellular concentration of organic osmolytes can be modulated either through changes of their metabolism or through modification of the inward or outward fluxes across the cell membrane.

Many organisms and cell types such as bacteria, invertebrates, algae, and certain mammalian cells utilize both inorganic ions and organic compounds for volume maintenance. Small hypoosmotic shocks are normally compensated for by the loss of intracellular Na^+, K^+, and Cl^-, whereas large hypoosmotic stresses result in the efflux of both inorganic ions and organic osmolytes. Thus there seems to be different "set points" for controlling organic and inorganic efflux pathways. Moreover, it is a general phenomenon that readily available inorganic ions mediate short-term volume regulation, but that long-term volume maintenance depends on the accumulation of more compatible organic solutes. These phenomena have mainly been observed in some euryhaline invertebrates [265] but are likely to occur in many other organisms and cell types, including cells of the mammalian renal medualla. Renal medullary cells can be exposed to relatively rapid and large hyperosmotic stresses, but the accumulation of organic solutes is in many cases rather slow (see e.g., [306). RVI could be similarly slow in these cell types or it could initially be mediated by inorganic ion transport pathways. Studies are needed in a variety of cell types to determine the relative contribution made by organic and inorganic osmolytes to volume regulation, the relative rates at which these two classes of osmolytes are lost or accumulated during volume stress, and the cellular signals which regulate the intracellular concentration of these solutes (see [40] for further discussion).

Transport of Amino Acids and Taurine

In some vertebrate cell types a net loss of free amino acids and related compounds during RVD contribute to the volume recovery, as is the case for volume regulation in all invertebrate phyla studied so far. The dominating part of the net loss of osmotically active nonelectrolytes is accounted for by taurine (β-amino ethane sulfonic acid), glycine, alanine, and other small, nonessential amino acids.

A net loss of taurine during RVD is found in red cells, heart muscle, and brain of the little skate [157], in red cells from the euryhaline fish species flounder and eel [74, 83, 84], in fish myocardium [320], in Ehrlich ascites tumor cells [139], and in mammalian heart [303]. In trout red cells, on the other hand, amino acids make no significant contribution to the volume-induced net loss of osmotically active solutes [25].

The use of amino acids for RVD requires that their concentration within the cells are high. The taurine concentration of teleosts red cells are high, 25–45 mmol/l cell water [74]. This is also the case in a mammalian cell like the Ehrlich ascites tumour cell, where the taurine concentration is measured at 53 mmol/l cell water. In Ehrlich cells taurine is the dominant organic osmolyte. Figure 24 shows that the cellular concentration gradients of taurine and glycine is diminished in diluted media. Table 1, furthermore, shows that the amino acids are released from the cells into the medium during the regulatory decrease. This indicates that the decrease in the intracellular concentration caused by transfer to hypotonic conditions is due to a change in amino acid and taurine transport parameters [14].

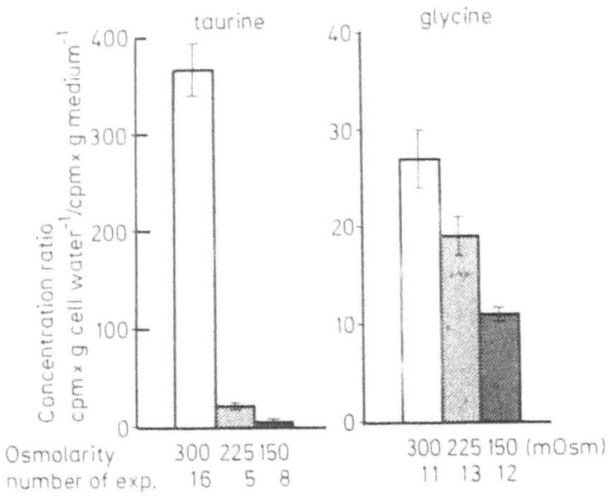

Fig. 24. Taurine and glycine gradients in Ehrlich ascites cells suspended in standard medium (300 mosmol) or hypotonic solutions (225 mOsm; 150 mOsm). The gradients are measured as the concentration ratio cell water to external medium for [^{35}S]-taurine (2.5 μM; 0.2 μCi/ml) and [^{14}C]-glycine (9.5 μM; 1.0 μCi/ml) and 30–110 min incubation. Values are mean ± SE of mean. (From [141])

Table 1. Volume-dependent amino acid transport in Ehrlich ascites tumor cells

	Amino acid concentration at steady state			
	mmol/l cell water		mmol/l medium	
	300 mOsm ($n = 6$)	150 mOsm ($n = 6$)	300 mOsm ($n = 3$)	150 mOsm ($n = 3$)
Taurine	53.0 ± 2.8	6.9 ± 0.2	0.07 ± 0.05	1.04 ± 0.3
Aspartic acid	1.8 ± 0.2	0.6 ± 0.06	0.02 ± 0.003	0.09 ± 0.02
Glutamic acid	1.2 ± 0.09	0.3 ± 0.02	0.01 ± 0.002	0.04 ± 0.005
Proline	2.1 ± 0.03	1.2 ± 0.2	0.03 ± 0.007	0.04 ± 0.01 (NS)
Glycine	9.1 ± 0.5	6.2 ± 0.9	0.17 ± 0.02	0.45 ± 0.10
Alanine	1.3 ± 0.3	0.4 ± 0.08	< 0.01	< 0.01(NS)

Cellular and extracellular concentrations of amino acids and taurine in Ehrlich ascites cells were suspended in standard medium (300 mOsm) or hypotonic solution (150 mOsm). The samples were taken 40 min after transfer to the experimental solution, at which time the cells had reached a new steady state. Values are mean ± SE with the number of independent experiments indicated. The cellular and extracellular values at 150 mOsm were significantly different from those at 300 mOsm ($p < 0.05$, Student's t test) unless otherwise indicated (NS) (from [141]). In three paired experiments (included in this table) the losses of cellular taurine, aspartic acid, and glutamic acid were estimated at 124, 5.0, 3.4 μmol/g cell dry weight, respectively, and the concomitant increases in extracellular content were 91, 4.9, 2.4 μmol/g cell dry weight, respectively.

The net loss of taurine during RVD in Ehrlich cells and in flounder and eel red cells is caused mainly by an increased efflux [74, 83, 141], presumably by stimulation of a Na^+-independent transport pathway. A reduction of the active taurine uptake, however, contributes to the RVD response. Under physiological conditions active uptake in Ehrlich cells occurs via a pH-dependent, Na^+- and Cl^--dependent system which has a high affinity for β-amino acids but a low transport capacity [178, 179]. It is proposed that taurine transport via this system takes place as an electroneutral $2Na^+$, Cl^-, taurine-carrier complex. Taurine efflux, in contrast, is predominantly mediated via a simple, pH-sensitive leak pathway, which is neither Na^+- nor Cl^--dependent. During the regulatory volume decrease following swelling in hypotonic medium the taurine uptake is reduced while the taurine efflux increases significantly. A concomitant membrane depolarization and a volume-induced increase in leukotriene synthesis are proposed as possible mechanisms involved in the increase in taurine efflux during RVD [181a]. Ca^{2+}/calmodulin seems to be involved in the transduction of the signal generated by the cell swelling [179].

In skate red cells the swelling-induced release of taurine can be mimicked by the Ca^{2+}-ionophore A23187 [203]. Moreover, hypotonic stress stimulates the inositol phosphate cycle, consistent with the notion that an increase in cytosolic, free Ca^{2+} mediates the volume regulatory amino acid release [221]. Calmodulin is probably not involved in the response to hypotonicity, since anti-calmodulin drugs failed to inhibit the taurine efflux [203].

In cultured human fibroblasts a number of transport systems for amino acids have been extensively characterized. One of these systems, the concentrative, Na^+-dependent system A, is modulated by a variety of environmental stimuli, including hormones and the availability of substrate amino acids. These operational features render system A a putative target for mechanisms aimed at regulating the intracellular concentration of organic osmolytes. System A is activated after cell shrinkage and inhibited after cell swelling. It is suggested that de novo protein synthesis may be involved in the activation because no activation occurs in the presence of the protein synthesis inhibitor cycloheximide. A slow time course for the activation also indicates involvement of the genome [86a].

For reviews on the volume-regulatory role of nonelectrolytes see [82, 92–94, 96, 200].

Activation Mechanisms

The mechanisms by which cells sense osmotically induced changes in turgor pressure, volume, and/or intracellular composition and transduce those signals into a regulatory response is an important and exciting area of research. A recent study by Lohr and Grantham [215] demonstrated the exquisite sensitivity of volume sensors and/or osmosensors in rabbit proximal tubules. These investigators demonstrated that proximal tubule cells could sense alterations in cell volume/intracellular composition of less than 3% and/or transmembrane osmotic gradients in the order of a few milliosmoles.

In Ehrlich cells it has been shown that the stimulation of the Cl^- channel activity during RVD is "triggered" by an average increase in cell volume that does not exceed 5% [150]. The actual mechanism of activation of the RVD response is, however, still only partly understood, and several factors have been assigned a regulatory function, e.g., Ca^{2+}, calmodulin, cAMP, eicosanoids, polyphosphoinositide metabolism, protein kinases, and the microfilament network. Factors, which have been proposed to play a role in the RVD response in Ehrlich cells, are shown in Fig. 25 in a schematic form.

There is evidence that Ca^{2+} and calmodulin play a role in the RVD in *Amphiuma* red cells, human lymphocytes, and Ehrlich cells (for references, see above). In epithelial cells from *Necturus* gall bladder [78], frog urinary bladder [51], and in cultured human epithelial cells [128], there is also evidence that RVD depends on calcium, and calmodulin seems to be involved in the RVD response in *Necturus* gall bladder epithelial cells [78]. In the *Amphiuma* red cell, Ca^{2+} seems to modulate the K^+/H^+ exchange [34], whereas there is convincing evidence that Ca^{2+} is involved in the activation of a conductive K^+ transport pathway in lymphocytes [108, 114, 279, 280], in Ehrlich cells [142, 144, 183], and in intestinal epithelial cells [128]. Patch clamp studies have shown, as discussed above that the same inward rectifier potassium channel is activated in Ehrlich cells after cell swelling and after addition of the Ca^{2+} ionophore

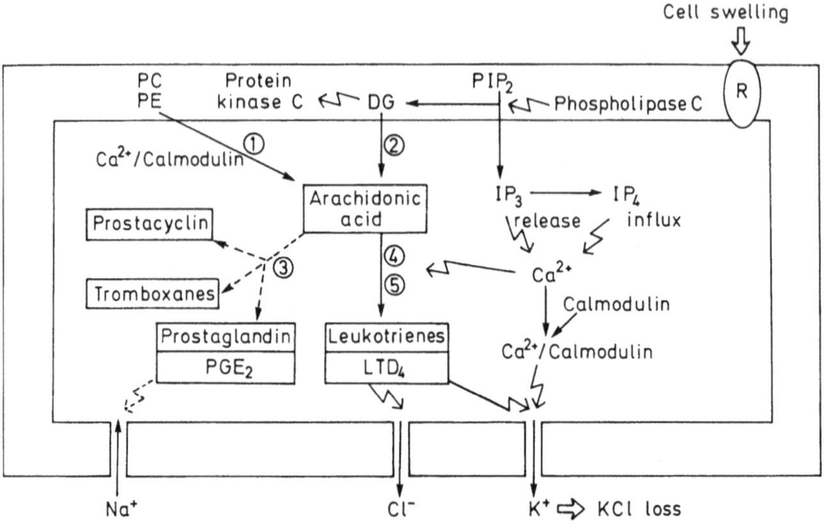

Fig. 25. Activation mechanisms during RVD in Ehrlich ascites tumor cells. *PE*, Phosphatidylethanolamine; *PC*, phosphatidylcholine; *DG*, diacylglycerol; *PIP_2* Phosphatidylinositol 4,5-bisphosphate; *IP_3*, inositol 1,4,5-trisphosphate; *IP_4*, inositol 1,3,4,5-tetrakisphosphate; *R*, volume-sensitive "receptor." Enzymes: *1*, phospholipase A_2; *2*, diacylglycerol lipase; *3*, cyclooxygenase; *4*, 5-lipoxygenase; *5*, dehydrase. (From [145])

A23187. Figure 26 shows the Ca^{2+} dependence of gating for the inward rectifier K^+ channel from Ehrlich cells in inside-out patches bathed in KCl on both sides and at a membrane potential at -40 mV. It can be seen that activation of the K^+ channel takes place after an increase in $[Ca^{2+}]_i$ from $10^{-7}\,M$ to $10^{-6}\,M$, which is in the physiological range, supporting the hypothesis that the volume-induced K^+ transport is mediated by the Ca^{2+}-activated K^+ channels.

A Cl^- transport pathway in Ehrlich cells is also activated by an increase in intracellular Ca^{2+} and inhibited by anti-calmodulin drugs [142, 144], whereas the Cl^- transport pathway in lymphocytes [108, 280] and in human epithelial cells [128] seems to be essentially Ca^{2+} independent. The Cl^- channel cannot be activated by Ca^{2+} in inside-out patches from Ehrlich cells [45]. The activation of Cl^- channels in intact Ehrlich cells by increase in intracellular Ca^{2+} is probably an indirect effect, caused by a stimulated synthesis of leukotrienes (see below).

A number of studies on epithelial cells have demonstrated that extracellular Ca^{2+} is a prerequisite for RVD [51, 326, 128], and an increased cellular ^{45}Ca uptake during RVD has been directly demonstrated [326]. Ca^{2+} entry across the cell membrane is, however, not necessary to activate RVD in Ehrlich cells [134, 144] or in lymphocytes [108, 114]. In these cell types release of Ca^{2+} from internal stores has been proposed to be involved in the RVD response. An increase in cytosolic-free Ca^{2+} induced by cell swelling has been demonstrated in toad bladder epithelial cells using the intracellular Ca^{2+} indicator Quin-2

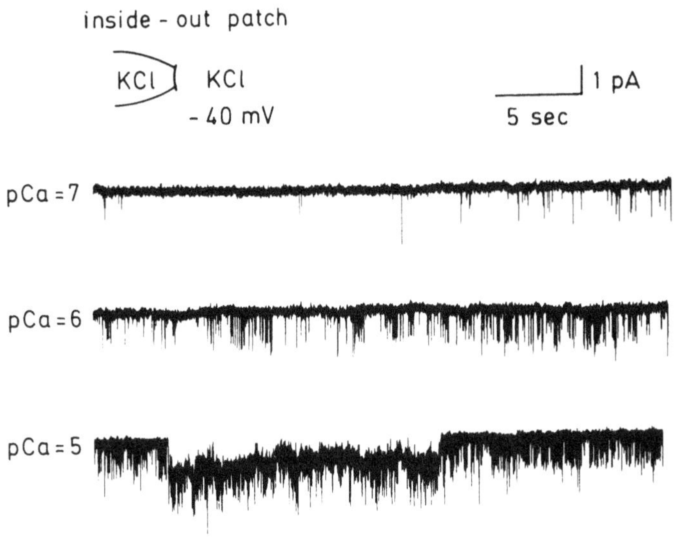

Fig. 26. Patch current from the inward rectifier K^+ channel in an inside-out excised patch in symmetrical KCl solutions with a bath potential of -40 mV. The inward rectifier K^+ channel is activated by internal Ca^{2+} ions (i.e. Ca^{2+} ions in the bath). At 10^{-7} *M* there is very little channel activity but the channel is activated at 10^{-6} *M* Ca^{2+}. (Result from [45])

[326]. In *Amphiuma* red blood cells, Cala et al. [38] have presented evidence of an increase in cellular free Ca^{2+} during RVD using the calcium-sensitive extracellular dye arsenazo III. Moreover, Hazama and Okada [128] have reported a sizeable increase in cytosolic, free Ca^{2+} during RVD in cultured intestinal epithelial cells. No change in cytosolic, free Ca^{2+} concentration could, however, be detected in human lymphocytes during RVD [269]. A modulation of the Ca^{2+} sensitivity of the transport pathways has been suggested to play a role during RVD in Ehrlich ascites cells [144]. For a recent review of the role of Ca^{2+} in cell volume regulation see [266].

The link between cell swelling and the putative transient increase in the intracellular level of free Ca^{2+}, by Ca^{2+} entry across the cell membrane or by Ca^{2+} release from intracellular stores, is unknown. Depletion of cell membrane phosphatidyl-inositol 4,5-bisphosphate (PIP_2) and release of inositol 1,4,5-trisphosphate (IP_3) have been demonstrated to play a key role in receptor-mediated Ca^{2+} mobilization [23, 290]. A subsequent activation of Ca^{2+} entry across the cell membrane has been proposed to be induced by metabolic conversion of inositol trisphosphate into inositol 1,3,4,5-tetrakisphosphate (IP_4) [23]. Christensen et al. [46] found that the RVD response in Ehrlich cells is associated with a rapid decrease in membrane PIP_2 and a concomitant, substantial increase in cytoplasmic IP_3. This is seen in Fig. 27. In addition, a slightly delayed increase was found in IP_4 and in IP_3. These results suggest that IP_3 and perhaps IP_4 may play a role as second messengers in the Ca^{2+} mobilization during the RVD response in Ehrlich cells.

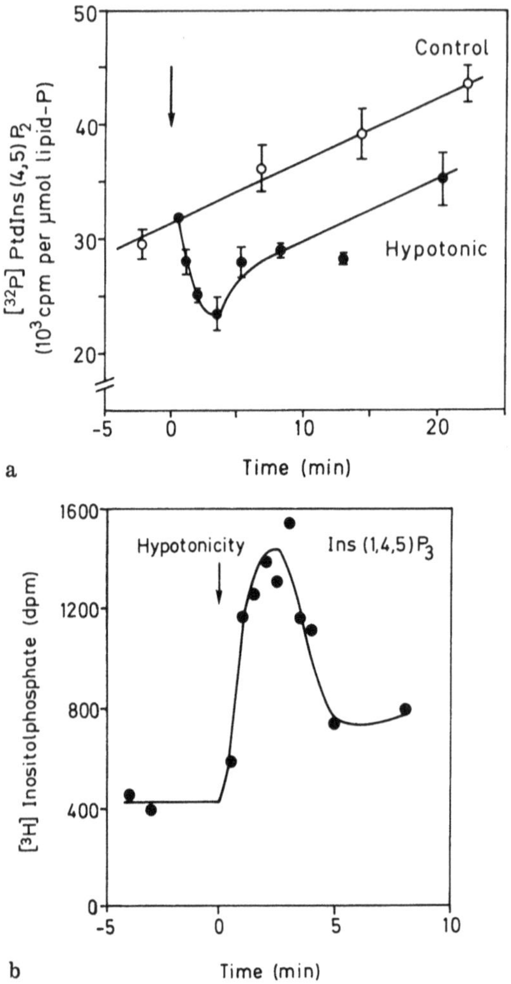

Fig. 27A,B. Time course of cellular [^{32}P]phosphatidylinositol 4,5-bisphosphate [^{32}P]-*PtdIns*(4,5)P_2) and [^{3}H]-inositol 1,4,5-triphosphate-([^{3}H]*Ins*(1,4,5)P_3) in hypotonically swollen Ehrlich cells. **A** The cells were preloaded with ^{32}P for 60 min, and time zero (*arrow*) diluted with 1 vol P_i-free hypotonic (50 mOsm) Ringer's solution to give a final osmolority of 175 mOsm. The total cellular lipids were extracted with chloroform-methanol-HCl, and their water-soluble deacylation products separated on Dowex anion exchange columns. Total cell lipid P was analyzed as a measure of cell mass. *Error bars*, indicate $\pm$ SEM for tri- or quadruplicate samples. **B** The cells were preloaded with [^{3}H]-myoinositol for 2 h and after addition of 10 mM LiCl diluted with hypotonic Ringer. Samples were precipitated with trichloroacetic acid (5%) and analyzed by high-performance liquid chromatogaphy on a SAX-10 column, eluted by an ammonium formate gradient. (**A** from [46]; **B**, reproduced from [140].)

The hypothesis that the inositol phosphate cycle is involved in the RVD response implies that agonists known from other systems to induce receptor-mediated activation of phosphoinositidase C with depletion of PIP_2 and release of IP_3 should also induce a volume response which mimics the RVD response seen after cell swelling. We have found that thrombin (0.5 IU/ml) and bradykinin (0.5 μM) both induce IP_3 release and a transient increase in free cytosolic Ca^{2+}, as measured using the fluorescent probe Fura-2 and the "ratio" method based on dual wavelength excitation (Fig. 28). Both agonists induce cell shrinkage when added to isotonic cells and, moreover, accelerate the RVD response when added at the time of maximal cell swelling in hypotonic medium [294]. These findings support the notion that the inositol phosphate signal system is involved in the RVD response in Ehrlich cells [46, 145, 294].

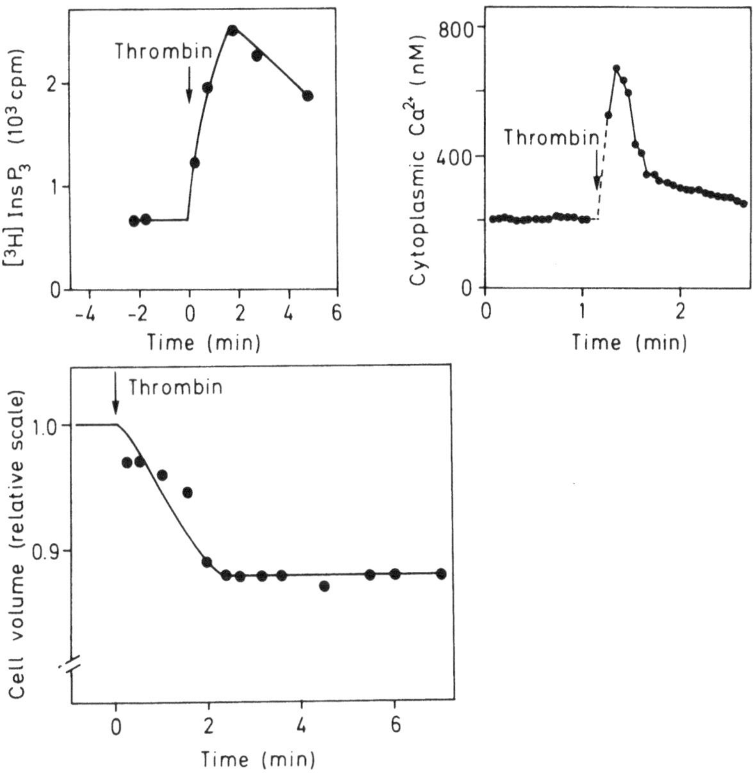

Fig. 28. Effect of thrombin (1 IU/ml) on free inositol triphosphate (*$InsP_3$*), cytoplasmic Ca^{2+} and on cell volume in Ehrlich cells. Cell volume (*lower panel*) was followed with time using a Coulter counter. Bumetamide (30 μM) was added to inhibit a RVI response. Free cytoplasmic Ca^{2+} (*upper right panel*) was measured using the fluorescent probe Fura-2 and the "ratio method" based on dual wavelength excitation. $InsP_3$ (*upper left panel*) was measured on cells pre-labeled with [^{3}H]myoinositol. The inositol phosphates were separated by Dowex anion exchange chromatography. (Results from [294])

It is unknown whether the diacylglycerol (DAG) also released from PIP_2 plays an additional role in the response, for example a protein kinase C-mediated modulation of the Ca^{2+} sensitivity of the transport pathways. A role for protein kinase C as activator of volume-sensitive K^+/H^+ exchange in *Amphiuma* red blood cells has been proposed by Cala [37] based on a stimulation of K^+/H^+ exchange by phorbol esters.

The DAG released by inositol phospholipid breakdown is a source of arachidonic acid [22]. In addition, calcium and calmodulin play an essential role in the release of arachidonic acid from membrane phospholipids [318, 327] and in the conversion of arachidonic acid into leukotrienes via activation of the lipoxygenase pathway [72]. It has recently been demonstrated [182] that Ehrlich ascites cells possess the ability to synthesize and release the arachidonic acid metabolites, prostaglandins and leukotrienes, and that addition of arachi donic acid stimulates the production of both compounds. During RVD the

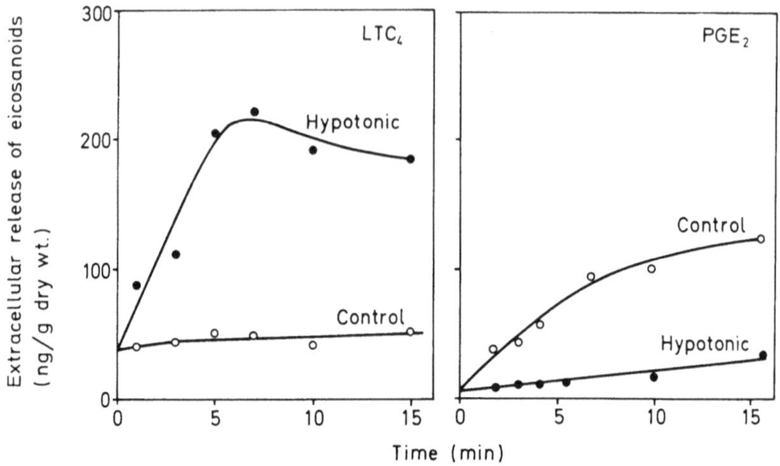

Fig. 29. Leukotriene (*LTC_4*) and *PGE_2* synthesis in Ehrlich cells suspended in standard incubation medium and in hypotonic incubation medium. Ehrlich cells preincubated in standard incubation medium were gently spun down and resuspended in either standard or hypotonic (150 mosmol) incubation medium, and the cytocrit adjusted to 6%. Release of *LTC_4* and *PGE_2* was followed with time by serially isolating cell-free medium by centrifugation and measuring the *LTC_4* and *PGE_2* concentrations in the supernatant by radioimmunoassay. The extracellular eicosanoid release (ng/g cell dry wt.) was calculated from the measured concentration in the medium (ng/ml) and the cell density of the cell suspension (g dry wt./ml). The figure is representative of four experiments. (Reproduced from [182])

synthesis of leukotrienes is stimulated while, concomitantly, prostaglandin synthesis is reduced (Fig. 29). Addition of LTD_4 accelerates RVD (which is rate-limited by the K^+ conductance), while addition of inhibitors of leukotriene synthesis blocks the volume response, even when a high K^+ conductance has been ensured by the presence of gramicidin (Fig. 30). It has been proposed that the activation of the K^+ and Cl^- transport pathways after hypotonic swelling involves an increase in leukotriene synthesis [182], and moreover that the effect of Ca^{2+} and calmodulin on the activation of the Cl^- transport pathway is indirect and mediated via stimulation of leukotriene synthesis [180, 181]. This was based on the observation that the inhibition of the RVD response seen in the presence of the anti-calmodulin drug pimozide can be lifted by addition of LTD_4 as seen in Fig. 31.

Addition of prostaglandin E_2 (PGE_2) inhibits the RVD response in hypotonic medium containing Na^+ but not in sodium-free hypotonic media, suggesting that PGE_2 increases the passive Na^+ permeability in Ehrlich cells. The reduced prostaglandin synthesis during RVD could thus account for the concomitant reduction in the passive permeability for Na^+ previously reported by Hoffmann [136]. Moreover, the increased synthesis of PGE_2 following addition of arachidonic acid could well explain the stimulating effect of arachidonic acid on the Na^+ conductance, and hence the observed inhibition of RVD by arachidonic acid [180].

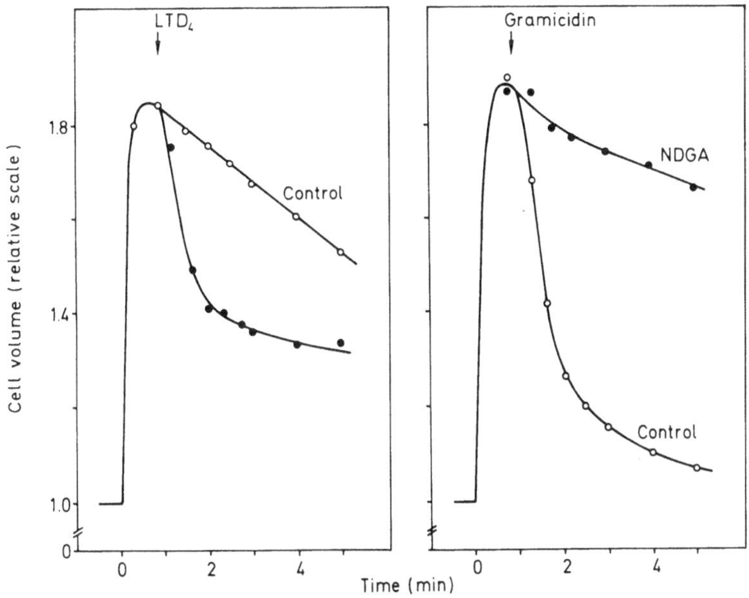

Fig. 30. Effect of added leukotriene (*LTD_4*) and of the leukotriene synthesis inhibitor nordihydroguaiaretic acid (*NDGA*) on regulatory volume decrease in Ehrlich cells after swelling in hypotonic standard incubation medium. Cells were preincubated in standard incubation medium for 40 min or more, and at time zero diluted 500-fold with hypotonic (150 mOsm) incubation medium. *Left panel*, *LTD_4* (60 n*M*) was added at the time of maximum swelling (*arrow*). The cell volume was followed with time using a Coulter counter and given relative to the initial cell volume in standard incubation medium. The curves are representative of three independent experiments. Leukotrienes *LTB_4*, *LTC_4*, and *LTE_4* (300 n*M*) had only marginal effect on RVD. *Right panel*, the cells were preincubated for 3 min *NDGA* and exposed to 50 μ*M* *NDGA* during hypotonic challenge (choline chloride, 150 mOsm). Gramicidin (0.5 μ*M*) was added at the time of maximum swelling (*arrow*) to control and experimental groups in order to ensure a high K^+ permeability. The curves are representative of three independent experiments. (Data from [182])

In conclusion, the proposed model for activation of the RVD response in Ehrlich cells is as follows (Fig. 25). Cell swelling activates (by an unknown mechanism) phosphoinositidase C with resulting depletion of cell membrane PIP_2 and release of IP_3 and DAG. The increase in IP_3 concentration stimulates the release of Ca^{2+} from intracellular stores and the formation of the Ca^{2+}-calmodulin complex which in turn activates lipoxygenase, converting arachidonic acid (released from DAG) to leukotriene D_4. The leukotriene then activates either directly or indirectly both Cl^- and K^+ conductances. Whether Ca^{2+}-calmodulin is involved in the decreased production of prostaglandins by the down-regulation of cyclooxygenase (which converts arachidonic acid into prostaglandins) is unknown. The end result of this sequence of events is the loss of KCl and also the inhibition of Na^+ uptake mechanism(s). Thus one can state that the Ehrlich ascites cells during the RVD response have provided optimal conditions for net loss of cell osmolytes, by the concomitant stimulation of osmolyte loss and inhibition of osmolyte gain.

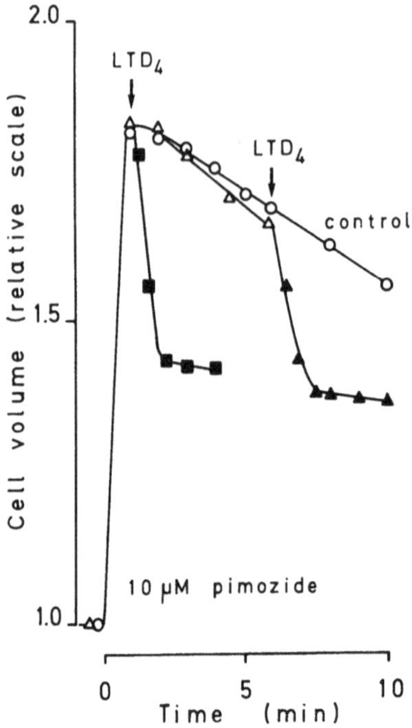

Fig. 31. Leukotriene (LTD_4) induced cell shrinkage in hypotonically swollen Ehrlich ascites cells in the presence of pimozide. Cells were prepared and exposed to hypotonicity as described in the legend to Fig. 20 with pimozide (10 µ*M*) being added within the first seconds after reduction in osmolarity in order to inactivate the Ca^{2+}-binding protein calmodulin (see Fig. 11). LTD_4 (100 n*M*; ■) was added as indicated by *arrows*. Cell volume was followed with time using a Coulter counter and given relative to the initial cell volume in isotonic media. The curves are representative of five independent experiments. (From [181])

The Cytoskeleton. In isolated membrane vescicles from lymphocytes (rabbit thymocytes) the K^+ permeability increase in response to either cytoplasmic Ca^{2+} or hypotonic swelling was absent, suggesting that an intact cytoarchitecture may be required for the RVD response [107]. Cytochalasin B, which is known to disrupt microfilaments, inhibits RVD in rat liver slices [319], in *Necturus* gallbladder epithelial cells [78], in Ehrlich cells [140], and in various cultured mammalian cells [48], suggesting that an intact microfilament network is also a prerequisite for a normal RVD response. It is proposed that actin filament-mediated fusion of cytoplasmic vesicles with the cell membrane could result in insertion of new ion channels [78, 208]. A possible role of microtubules and the microfilament network during volume regulation has also been suggested in cultured rat pheochromocytoma cells [52]. Mills and Skiest [229] and Mills and Lubin [228] have analyzed the role of the cytoskeleton in cell volume regulation using MDCK cells. They propose that the state of organization of F-actin, which can be influenced by changes in the levels of cAMP, could have an effect on membrane elements that play a role in volume control processes. Changes in the organization of F-actin have recently been demonstrated in a cultured mammalian cell [48]. The possible role of the cytoskeleton in the mechanism of cell volume control has been reviewed by Mills [227].

Three possible mechanisms by which microfilaments can regulate ion channels are (a) by the insertion of channels from a cytoplasmic store into the membrane, (b) by acting as the mechanical linkage by which membrane stretch is transmitted through the cytoskeletal element and directly causing channel activation (see below), and (c) by activating a second messenger cascade, the products of which either activate or modulate membrane ion channels. At present, it is not known which of the above mechanisms is subserved by the cytoskeleton in the regulation of ion channels.

Stretch-Activated (SA) Channels and Their Function During Volume Regulation

The nature of the sensory mechanisms detecting the volume changes during RVD is, as yet, totally unknown. Over the last couple of years, however, ion channels activated by membrane stretch, and presumably attached to cytoskeletal strands, have been detected in various vertebrate cells, for example chick skeletal muscle [117], mammalian endothelial cells [187], *Necturus* renal proximal tubule cells [274], choroid plexus from salamander [44], opossum kidney cell line [305], and human fibroblast cell line [300] (for review see [273, 232]). On this basis it is tempting to speculate that stretch-activated (SA) ion channels, probably linked with components of the cytoskeleton, could be involved in cell volume regulation. The membrane SA ion channels are in general nonspecific, cation-selective channels which are permeable to Ca^{2+}, at least in the case of endothelial cells and choroid plexus, and may allow sufficient Ca^{2+} to enter to serve a second messenger function [44, 187].

Christensen [44] showed (a) that amphibian choroid plexus cells contain SA channels (i.e., channels that could be activated by suction) which could bring Ca^{2+} into the cell, and (b) that Ca^{2+}-activated K^+ channels are activated after cell swelling. He thus concluded that SA channels were activated by the swelling-induced increase in membrane tension, and that the increase in cytoplasmic Ca^{2+} was caused by influx of Ca^{2+} through these channels. Swelling-induced activation of the SA channel could, however, not be demonstrated in choroid plexus cells. Recently, however, single channel reports have appeared showing such swelling-induced activation of SA cation channels (SACat):

1. Opossum kidney cells were exposed to 60% of normal osmolarity, while the channel activity in a cell-attached patch was monitored [305]. In a given patch, an identical number of channels (up to six per patch) could be activated either by swelling or by direct mechanical stress (i.e., suction). The characteristics of the swelling- and suction-activated channels were identical; a nonselective channel of small conductance.
2. Falke and Misler [71], using 65%–79% hypotonic shock on neuroblastoma cells, observed that SACat channels were activated with a time course similar to cell swelling and that, as expected if the stimulus is mechanical, activity subsided during the regulatory volume decrease phase.
3. Ehrlich ascites tumour cells were exposed to 50% of normal osmolarity and the channel activity in a cell-attached patch was monitored. A typical result is

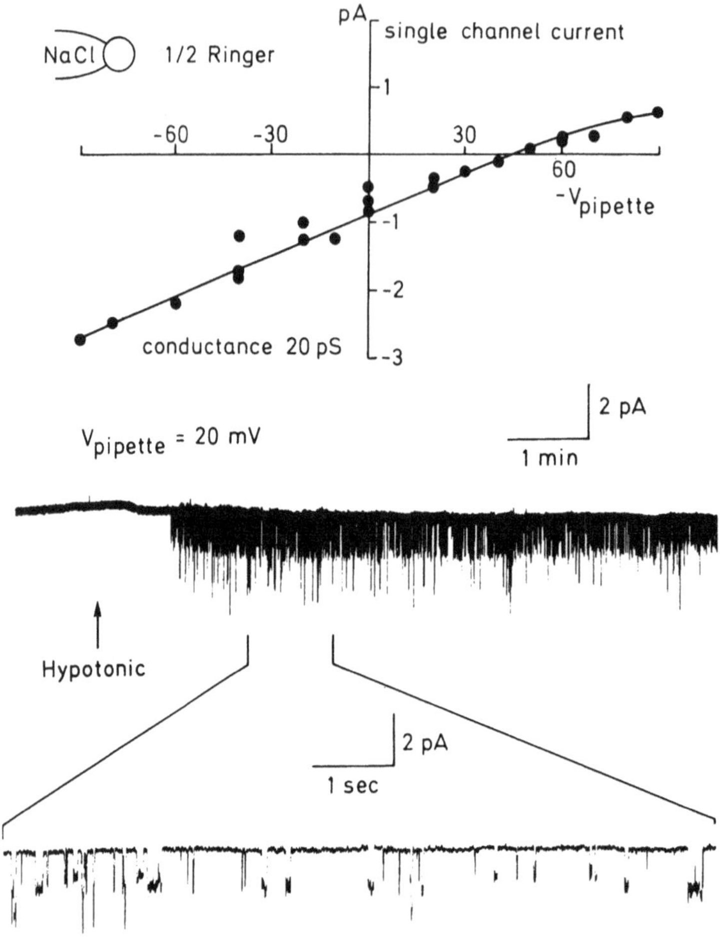

Fig. 32. Current recordings for a nonselective cation channel from a cell-attached patch with NaCl in the electrode and 20 mV pipette voltage (hyperpolarization of the patch). At the time indicated by the *arrow* the fluid bathing the cell is being changed from normal Ringer's solution to a solution with half the tonicity. *Upper panel*, single-channel current versus bath potential for the cation channel (i/v curve). *Bottom panel*, details with a higher time resolution. Inward current through the SA channels is shown as negative deflections. (Results from [45])

seen in Fig. 32. After the onset of osmotically induced cell swelling nonselective cation channels became activated. Channels with identical characteristics could be activated by suction, thus identifying the channels as SACat channels. It is interesting to note that the channel activation does not occur instantaneously but only after 0.5–1 min. At that time the cells are at least 25% swollen. Disregarding membrane infoldings, Sackin [276] estimated for a spherical cell that membrane stretch induced by cell volume increase as small as 1% is sufficient for SA channel activation. Cells are, however, not osmometric balloons. They can expand membrane folds [136, 145] when subjected to hypotonic swelling. Volume-activated channels have previously been demonstrated in Ehrlich cells after a volume increase of only 5% [150]. The delay 0.5–1 min in SA channel activation (see above) may perhaps reflect the involvement of other factors than simply the degree of swelling. It is a crucial point in evaluation of the significance of the SA channels in cell volume regulation that they are evoked after only a small cell swelling. The use of unphysiologically large swelling optimizes the chance of seeing changes in SA channel activity but leaves open the question of whether small-scale swelling of cells would actually activate SA channels.

The SA channels described so far are mainly nonselective cation channels. This property seems to contradict an involvement of SA channels as pathway for the K^+ and Cl^- efflux. The volume-activated K^+ and Cl^- channels described above for Ehrlich cells *cannot* be activated by mechanical stress (suction) [45]. Therefore, it seems likely that the function of the former SA channels is correlated with their observed Ca^{2+} permeability [44, 305]. A resulting increase in $[Ca^{2+}]_i$ could be part of the volume-activated signal system. Moreover, a significant cell depolarization during RVD has been observed for several cell systems and may play a role in K^+ channel activation [143]; such depolarization is enhanced by the activation of nonselective stretch-activated cation channels.

Considering that membrane tension can selectively affect specific channels while having no effect on others, it is conceivable that classes of other membrane proteins, for example, enzymes, may also be sensitive to membrane tension. The missing link between cell swelling and increased PIP_2 break down (see Fig. 25) may thus be related to membrane tension, resulting in some changes in the membrane protein-cytoskeleton interaction.

Volume Regulation: Regulatory Volume Increase

Following shrinkage in hypertonic saline, RVI can be observed in some tissues and cell types (e.g., *Amphiuma* red blood cells [32] and *Necturus* gall bladder epithelial cells [190]. Other cells, such as Ehrlich ascites tumor cells, simply shrink as osmometers with no sign of volume recovery (Fig. 33, left panel). The same kind of behavior is shown by lymphocytes [144], frog skin epithelial cells [309], and frog urinary bladder epithelial cells [51]. Nearly all cells, however, show a volume recovery when hypotonic pretreatment—resulting in a net loss of KCl (the RVD response)—is followed by restoration of tonicity. After the initial osmotic shrinkage the cells recover their volume with an associated KCl uptake (Fig. 33, right panel). It is not clear why some cells show an RVI response only when the "RVI-after-RVD protocol" is applied. This is called by Lewis and Donaldson [205] "pseudo-regulatory volume increase;" it may reflect a requirement for the reduction of a critical intracellular solute, perhaps Cl^-.

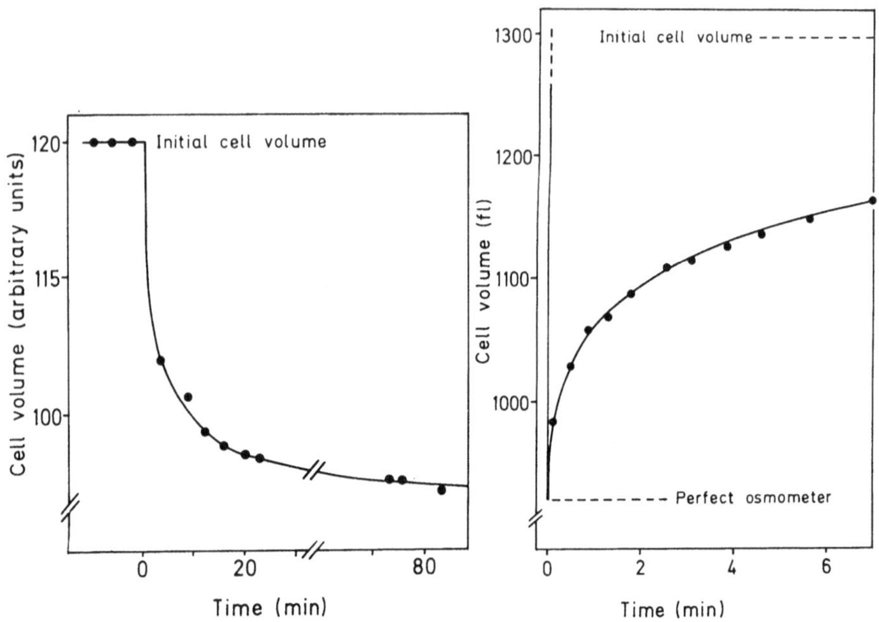

Fig. 33. *Left panel*, cell volume in Ehrlich ascites tumor cells after an increase in external osmolarity. At time zero sucrose was added to increase osmolarity from 300 to 400 mOsm, and the cell volume was followed with time using a Coulter counter (redrawn from [134]). *Right panel*, regulatory volume increase in Ehrlich ascites tumor cells after increase in external osmolarity. The cells were pretreated by exposure to low external osmolarity (150 mOsm) for 10 min. At time zero a sample of the cell suspension was diluted 1000 times with isotonic (300 mOsm) medium, and the cell volume was followed with time using a Coulter counter. (From [143])

In invertebrate cells amino acids and other nonelectrolytes play an important role in the RVI response (for review see [265, 94]. In vertebrate cells the RVI response is mainly accounted for by electrolytes, although nonelectrolytes may play a minor role [96, 143, 298].

Figure 34 gives a summary of the membrane transport mechanisms proposed to be involved in RVI in different cell types. The evidence for the suggested mechanisms is given below.

Volume-Activated Cotransport

Electroneutral $Na^+,K^+,2Cl^-$ cotransport systems were first described in avian erythrocytes [168, 169, 222, 224, 247], in Ehrlich ascites tumor cells [90], and in human red blood cells [42, 60]. Recently, a highly active cotransport system has been demonstrated in ferret red cells [75–77]. The cotransport systems in avian erythrocytes and in Ehrlich cells are discussed below. For human red blood cells see [29, 39, 57, 155, 324]. For recent reviews on the $Na^+,K^+,2Cl^-$ cotransport system, see [43, 88, 198, 243].

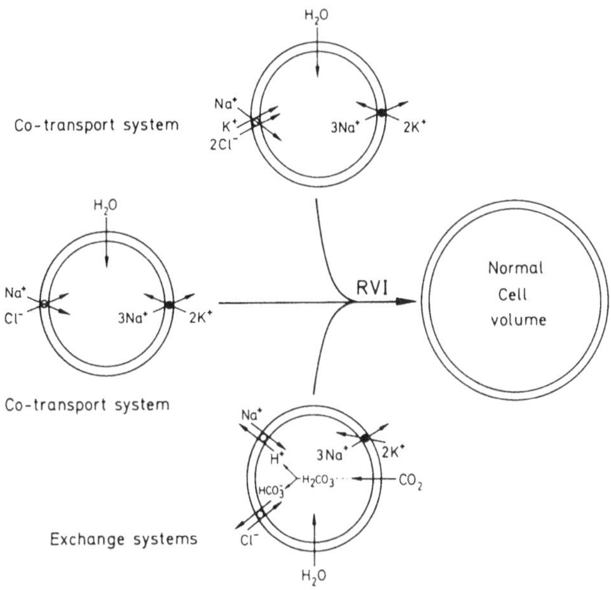

Fig. 34. Principal types of ion fluxes activated in various cell types during regulatory volume increase (*RVI*) (a), $Na^+,K^+,2Cl^-$ cotransport; (b) Na^+,Cl^- cotransport; and (c) electroneutral Na^+/H^+ exchange functionally coupled to Cl^-/HCO_3^- exchange, giving NaCl uptake without change in pH_i. Note that in all the RVI mechanisms the cell takes up NaCl with subsequent replacement of Na^+ by K^+ via the ATP-driven Na^+/K^+ pump, which is the only primary active transport involved. (From [143])

In epithelial cells, Na^+,Cl^- cotransport systems, as originally proposed by Nellans et al. [239] for the mucosal membrane of intestinal epithelial cells, have been widely demonstrated [91] and recognized as playing an important role in epithelial transport (see review by Frizzell et al. [81]). In several epithelia, however, the Cl^--dependent cation cotransport can be attributed to a $Na^+,K^+,2Cl^-$ cotransport system (see review by Warnock et al. [322], Greger [97], and O'Grady et al. [243]).

Cl^--dependent, furosemide- or bumetanide-sensitive cation cotransport systems have also been reported in several cultured cells including HeLa cells, BC_3H1 smooth muscle cells, primary chick heart cells [4], L cells [86], MDCK cells [225], 3T3 fibroblasts [7], and BALB/C 3T3 preadipose cells [242]. For recent reviews see Palfrey and Rao [248] and Saier and Boyden [277].

It should be noted that the demonstration of Cl^- dependence is not unequivocal evidence for the involvement of cotransport systems. In the case of the catecholamine-induced swelling of trout red cells in isotonic medium, for instance, the apparent Cl^- dependence of the Na^+ uptake, which suggested the involvement of Na^+,Cl^- cotransport, has recently been accounted for by the finding that in these cells substitution of nitrate for Cl^- strongly inhibits the rate of Na^+/H^+ exchange [24].

Avian Red Blood Cells

The involvement of cotransport systems in RVI was first demonstrated in avian red blood cells. When bird red cells (e.g., pigeon red cells [330] and duck red cells [165]) are suspended in hypertonic media, the cells at first shrink osmotically, but then swell back towards their initial volume, provided that the external K^+ concentration is elevated (above 5 mM). The RVI response seems to involve influx of Na^+ and K^+, accompanied by Cl^- [165]. Careful kinetic studies confirmed the coupled influx of Na^+ and K^+ after cell shrinkage [284], and net flux measurements under a variety of conditions indicated a Na^+:K^+ stoichiometry of 1:1 [285]. Norepinephrine and other catecholamines mimick the RVI response and produce a cell swelling in isotonic medium when the external K^+ concentration is elevated [122, 123, 166, 268, 285, 286]. The activation by norepinephrine seems to be mediated by changes in intracellular cAMP, and any agent that raises cAMP activates the cotransport system [14, 168, 247, 248, 268, 277]. The activation may involve an increase in the phosphorylation of specific sites on a 230-kDa protein [5, 6]. cAMP is, however, not involved in the volume-induced RVI [171]. Experiments with ATP-depleted cells suggest that both volume-dependent and catecholamine-stimulated fluxes are dependent on cell metabolism [248]. In both cases, the uptake of Na^+ and K^+ is insensitive to ouabain [165, 166] but is inhibited by "loop" diuretics such as furosemide [284, 285] and bumetanide [246]. The concentrations required for half-maximal inhibition of transport are $2\text{–}4 \times 10^{-7}$ M for bumetanide and $1\text{–}2 \times 10^{-5}$ M for furosemide [246a].

The involvement of Cl^- in Na^+ plus K^+ cotransport was first reported by Kregenow and Caryk [169] and Bakker-Grunwald [14]. Cation-coupled Cl^- fluxes could be demonstrated when SITS was present to inhibit the anion exchange system [169]. Definite evidence for electroneutral cotransport was provided by thermodynamic considerations by Haas et al. [123]. Using valinomycin to change the membrane potential, the Na^+ flux was found to be unaffected by the membrane potential, but dependent on the chemical Cl^- gradient. Moreover, it was demonstrated that a Cl^- gradient can drive Na^+ uphill [123]. These data strongly suggest that cell shrinkage or catecholamines activate a Na^+,K^+,$2Cl^-$ cotransport mechanism. It may be noted, as discussed by Cala [34], that under conditions where Cl^- is at electrochemical equilibrium, the equation for the driving force assuming electrogenic Na^+ plus K^+ cotransport [286] is exactly the same as the equation obtained on the basis of electroneutral Na^+ plus K^+ plus $2Cl^-$ cotransport [123]. It is suggested from kinetic studies that bumetanide binds to a chloride site which is affected by cations. The K_i for bumetanide shows a minimum at optimal concentrations of Na^+ and K^+ [121, 246]. Cl^- at very low levels is a prerequisite for bumetanide binding, but Cl^- at higher concentrations competitively inhibits the binding [121]. The movement of ions through the transport pathway depends on the sum of chemical potential gradients of Na^+, K^+, and Cl^-. If the cells are suspended in a medium with plasmalike concentrations of the transported ions, the driving force for the transport is zero, and no net movement occurs neither after hormonal stimulation nor after cell shrinkage. Only after reduction of intracellular ion content or elevation of extracellular K^+ can net inward transport be initiated [123].

Ehrlich Ascites Tumor Cells

Evidence for electroneutral cotransport of anions and cations in single mammalian cells was first presented by Geck et al. [90] who reported an electrically silent, furosemide-sensitive influx of Na^+, K^+ and Cl^- with a stoichiometry of 1:1:2 into Ehrlich cells which had previously been depleted of K^+ and loaded with Na^+, and in which the Na^+, K^+ pump was inhibited by ouabain. The flux was linearly related to the combined chemical potential gradients for the ions, but is nonzero when the calculated combined chemical potential is zero, suggesting possible coupling to metabolic energy (or inaccuracies in estimating the chemical potential gradient). No evidence of a primary active transport process was found, however, since the ATP consumption in respiration-blocked cells as well as the lactate production was found to be unaffected by furosemide inhibition of the cotransport system.

In Ehrlich cells at steady-state, a K^+,Cl^- cotransport with a stoichiometry near 1:1 has been reported [11, 13], at variance with a Na^+,K^+,$2Cl^-$ cotransport. Levinson [206] has examined the role of external Na^+ in cotransport under steady-state conditions, and proposed that the Cl^--dependent cation cotransport system, when operating at steady-state, mediates an exchange of KCl for KCl or NaCl for NaCl, with the relative proportions of each being determined by the extracellular $[Na^+]$, and with the sum (KCl + NaCl) being maintained relatively constant. At low extracellular Na^+ concentrations the cotransport system mediates the flux of KCl, but with increasing Na^+ concentration (above 60 m*M*) progressively less KCl is transported.

Geck and Pfeiffer [89] have reported the cotransport rate at steady-state to increase with external Na^+ concentration up to an optimal concentration, above which the cotransport rate decreases again. The results are explained in terms of a feed-back control system, the cotransport system being switched off when cell volume exceeds a critical value. This is consistent with previous findings of Hoffmann et al. [147] who reported the cotransport system to be activated in shrunken cells, but silent in control cells at steady-state, at a cell volume slightly above the threshold volume reported by Geck and Pfeiffer [89]. On the other hand, Levinson [207] has presented evidence that the cell volume is not a major factor regulating the activity of the cotransport system in Ehrlich cells.

Activation of Cotransport during RVI. When Ehrlich cells are preincubated in hypotonic medium in order to reduce their intracellular ion content (Fig. 35) and then resuspended in isotonic medium, the initial osmotic cell shrinkage is followed by a rapid volume increase back to the original cell volume [147]. This

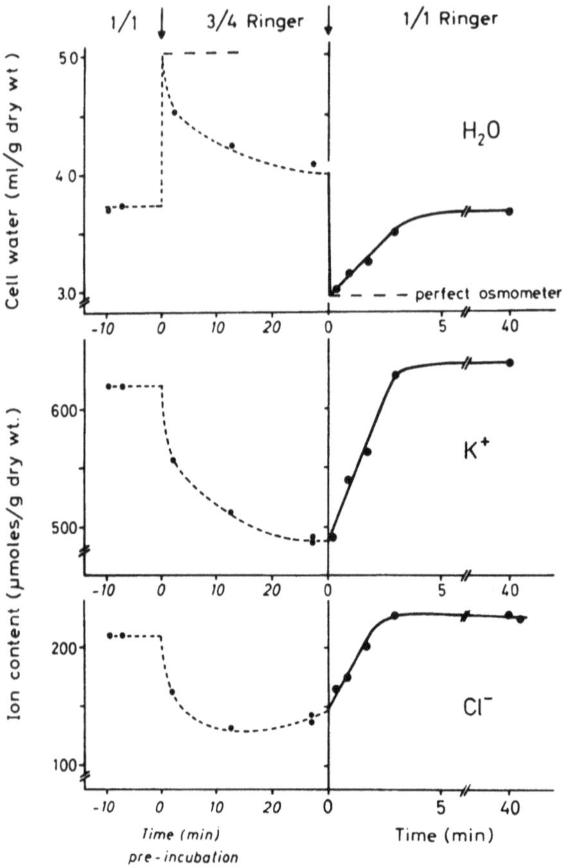

Fig. 35. Uptake of KCl during regulatory volume increase in Ehrlich ascites tumor cells after increase in external osmolarity. The cells were pretreated by exposure to low external osmolarity (225 mOsm) for 20–40 min. At time zero a tonicity of 300 mOsm was restored by addition of one-fourth volume of a double-strength saline solution, and cell water, K^+ and Cl^- content were followed with time. *Broken line*, the cell water content calculated for a perfect osmometer. The original cell water content in 300 mOsm saline solution before the hypotonic pretreatment is indicated in the figure for reference. (Redrawn from [147])

involves a net KCl uptake followed by a concomitant water uptake (Fig. 35). It was proposed that the primary process is an activation of an otherwise quiescent bumetanide-sensitive Na^+,Cl^- and/or Na^+,K^+,2Cl cotransport system, with subsequent replacement of Na^+ by K^+ via the Na^+,K^+ pump, stimulated by the Na^+ influx [147]. The evidence for this hypothesis can be summarized as follows:

1. The net K^+ and water uptake was found to be Cl^- dependent as well as Na^+ dependent [147]. The RVI is inhibited by furosemide or bumetanide, but is not inhibited by DIDS or amiloride (arguing against the involvement of functionally coupled Na^+/H^+ and Cl^-/HCO_3^- exchangers).

2. The observed net Cl^- flux during RVI was more than ten-fold larger than the calculated conductive Cl^- flux [147]. Under steady-state conditions the bumetanide-sensitive component of the unidirectional ^{36}Cl flux was negligible, and the unidirectional ^{36}Cl flux was unaffected by substitution of choline for external Na^+. In contrast, during RVI a large bumetanide-sensitive ^{36}Cl influx was observed which amounted to about half the total unidirectional ^{36}Cl flux [147].
3. Under conditions where the Na^+, K^+ pump was inhibited with ouabain, an uptake of Na^+ and Cl^- in about equimolar amounts could be demonstrated, whereas the uptake of K^+ was negligible. The uptake of Na^+ and Cl^- was inhibited by bumetanide. These findings provide direct experimental evidence for the involvement of the Na^+,K^+ pump, and supported the conclusion that a Na^+,Cl^- contransport is involved rather than a Na^+,K^+,$2Cl^-$ cotransport system [147]. It is unclear how the Na^+,K^+ pump is activated. Only a small increase in the Na^+ concentration in cell water was observed about 30 s after restoration of tonicity, and the further increase in cell Na^+ during RVI was insignificant.
4. The flux ratio for the bumetanide-sensitive component of unidirectional Cl^- influx and efflux during RVI was 1.9 [147]. Flux ratio analysis of the bumetanide-sensitive Cl^- flux under these conditions indicated a Na^+,Cl^- cotransport, since the ratio of the driving forces for Na^+,K^+,$2Cl^-$ cotransport would hardly be sufficient to account for the observed magnitude of the ratio of influx/efflux [147]. Moreover, a net uptake of KCl could be demonstrated even at low external K^+, under conditions where the calculated driving force for a Na^+,K^+,$2Cl^-$ cotransport system is in the outward direction. However, in this calculation it was assumed that the activity coefficients for the ions in the medium and in the cytosol are equal, which is unlikely. Whether the differences in activity coefficients could be big enough to make the energy deficit vanish, as suggested by Geck [87] is still an open question.

We have recently shown that Na^+,K^+,$2Cl^-$ cotransport is responsible for RVI in Ehrlich cells when the intracellular concentrations are sufficiently low [153]. The evidence is that Cl^- influx during RVI required K^+. In addition, there was a bumetanide-inhibitable K^+ influx during RVI which was dependent on Na^+ and Cl^-. The two transport systems are distinguishable, but may represent alternate modes of operation of the same transporter. This question is unsolved for Ehrlich cells as well as for avian red cells (see above section) and in the medullary loop of Henle [70]. Recently Sun et al. [302] reported, however, that the peptide hormone, arginine vasopressin, changes the ions transported by the loop diuretic-sensitive cotransporter in the mouse medullary thick ascending limb from an Na^+, Cl^- to an Na^+, K^+, $2Cl^-$ mode.

In *Amphiuma* red blood cells, lymphocytes and dog red blood cells the RVI response involves Na^+/H^+ exchange functionally coupled to Cl^-/HCO_3^- exchange (see "Na^+/H^+ Exchange"). An amiloride-sensitive Na^+/H^+ exchange can also be demonstrated in Ehrlich cells [55] but does not seem to play any significant role in the RVI response.

In Ehrlich cells stimulation of the Na^+/H^+ antiport can be accomplished by lowering the cytoplasmic pH. We have used the sodium salt of the weak organic acid, propionic acid, to acid-load Ehrlich cells. Intracellular dissociation of the free, lipid-soluble propionic acid brings about cytoplasmic acidification, which in turn activates the Na^+/H^+ exchanger (Fig. 36). When Ehrlich cells are suspended in sodium propionate medium, the operation of the antiport can be detected as a change in cellular volume, resulting from intracellular accumulation of sodium propionate. A typical experiment is illustrated in Fig. 36 (lower panel). As expected, the cell swelling is prevented by amiloride. For comparison, the upper frame shows that the cell swelling which occurs during RVI in Ehrlich cells is almost insensitive to amiloride. Thus, Na^+/H^+ exchange does not seem to play any significant role in the RVI response in Ehrlich cells.

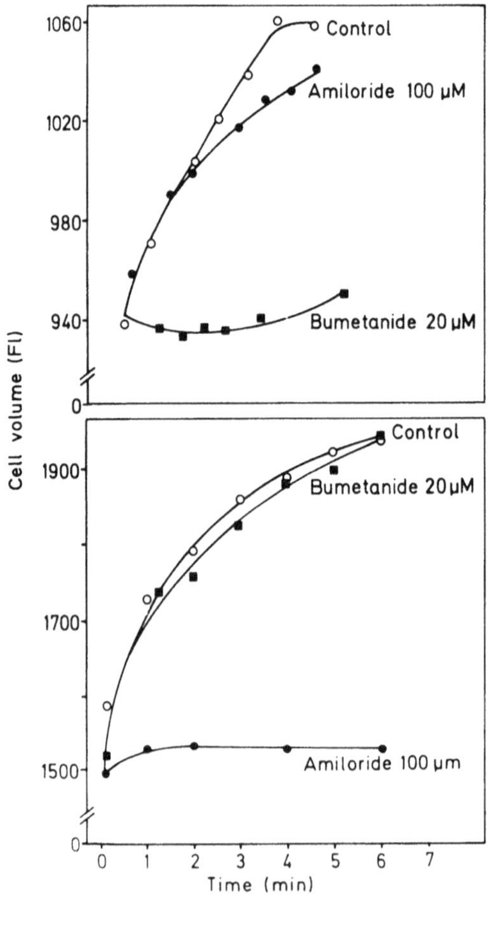

Fig. 36. Effect of bumetanide and amiloride upon regulatory volume increase (*upper panel*) and upon cell swelling in Na-propionate medium (*lower panel*) in Ehrlich ascites tumor cells. *Upper panel*, the cells were pretreated by exposure to low external osmolarity (225 mOsm) for 15 min. At time zero a sample of the cell suspension was diluted 1000-fold with isotonic (300 mOsm) medium, and the cell volume was followed with time using a Coulter counter. The regulatory volume increase (*control*) is blocked by bumetanide (20 μM) but unaffected by amiloride (100 μM). *Lower panel*, cell swelling under isotonic conditions was induced by incubation of the cells with the Na^+ salt of a weak organic acid following the technique introduced by Grinstein and coworkers [110]. At time zero a sample of cell suspension was diluted with isotonic Na-propionate medium at pH 6.6 and the cell volume followed with time. The cell swelling resulting from acid loading of the cells and subsequent activation of Na^+/H^+ exchange (*control*) is blocked by amiloride (100 μM) but unaffected by bumetanide (20 μM). (From [143])

Cotransport of $Na^+ + K^+ + 2Cl$ Responsible for High Cl^- Concentration in Principal Cells of the Frog Skin

As already mentioned, the frog skin epithelium looses KCl to the inside bathing solution when the latter is diluted. Such a loss of K^+ and Cl^- is also observed when the inside of the skin is exposed to isotonic solutions lacking potassium ions [308] or chloride [220, 208, 73, 95]. The K^+ and Cl^- lost is recovered when the inside of the skin is exposed to normal Ringer. It has been proposed [73] that this recovery is due to a coupled uptake of Na^+ and Cl^- from the inside bath. Ussing [309, 315] demonstrated that the recovery of KCl and volume by epithelia which had lost KCl was inhibited by furosemide and concluded that a cotransport process must be responsible for the process, but it was left undecided, whether the process was cotransport of $Na^+ + Cl^-$ or $Na^+ + K^+ + 2Cl^-$. However, the necessity of K^+ in the inside medium for maintenance of normal cell volume [308] suggested a role of the $Na^+ + K^+ + 2Cl^-$ cotransport.

These considerations led to a systematic study of the roles of K^+, Na^+ and Cl^- in the recovery of electrolytes and water after loss of K^+ and Cl^- during hypertonic shock [310–312]. The measurements of epithelial volume were carried out according to the method described by MacRobbie and Ussing [308] with later improvements [320]. In all the experiments discussed in the following, the outside medium was 1/20 ordinary frog Ringer or 1/20 nitrate Ringer (i.e., a Ringer in which Cl^- has been replaced by nitrate). The use of 1/20 Ringer has the advantage that the Na^+-concentrations in the outside medium and in the cells are nearly the same. This means that the sodium-diffusion potential across the apical membrane is close to zero, so that the transepithelial PD and that across the basolateral membranes of the principal cells are about the same. Also, with 1/20 Ringer outside, the anion shunt via the MR cells becomes negligible. Figure 37 illustrates the standard procedure of the volume measurements.

The recovery of K^+, Cl^-, and volume can be inhibited completely by furosemide [309] and bumetanide [310]. Since, with the technique applied, the epithelial volume is determined as the chloride free space plus the chloride space, it is of special interest to discuss how the chloride space of the epithelium depends on the chloride concentration in the inside bath. The results from a series of such experiments are shown in Fig. 38. The abscissa is the fraction in percent of chloride in the Ringer isotonic recovery solution, where the sum of the concentrations of chloride and thiocyanate was kept constant equal to the chloride concentration of ordinary Ringer. The ordinate gives the ratio between the chloride space and the total cellular volume. It is seen that there is an almost linear relationship between the two parameters. This is exactly what we might expect for a Na^+,K^+,2Cl cotransport. The equilibrium between cytoplasm and bath is determined by the equation

$$[Na_c] \times [K_c] \times [Cl_c]^2 = [Na_i] \times [K_i] \times [Cl_i]^2$$

where the symbols stand for concentrations (or rather activities) in the cells and in the inside bath, respectively. We shall make the reasonable assumption that

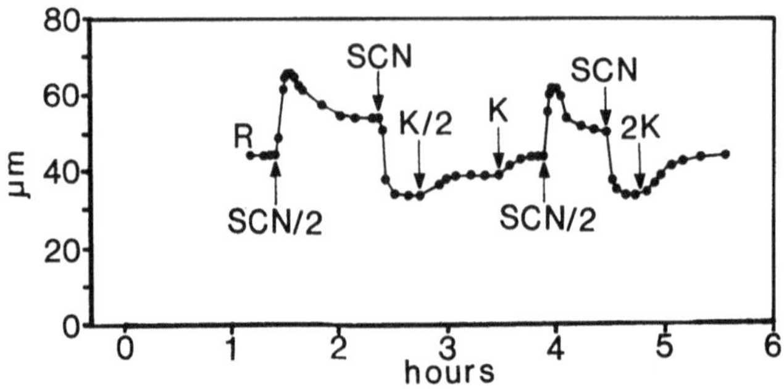

Fig. 37. Showing that the recovery of epithelial volume after chloride depletion depends on the potassium concentration of the inside bathing medium. *Ordinate*, thickness of the epithelium in micrometers. *Abscissa*, time in hours. Inside bath initially ordinary Ringer. At *first arrow* change of inside bath to half-strength thiocyanate Ringer (i.e., Ringer where chloride has been replaced by thiocyanate). At *second arrow* (*SCN*) the inside bath is made full thiocyanate). At *third arrow* (*K*/2) the inside bath is made ordinary Ringer with only half the usual K^+ concentration, i.e., 0.9 m*M*. At the *fourth arrow* (*K*) the inside bath is made ordinary Ringer with the usual K^+ concentration (1.9 m*M*). Thereafter, a new sequence with half thiocyanate, full thiocyanate, and recovery is started, but in this sequence the recovery medium (2*K*) is ordinary Ringer with double the usual K^+ concentration (i.e., 3.6 m*M*). (From [310])

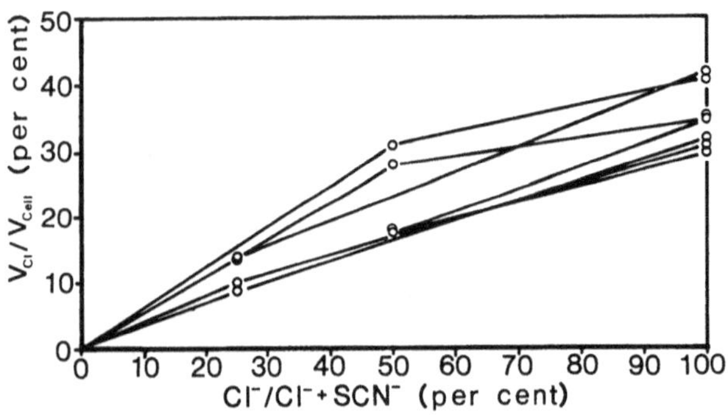

Fig. 38. Dependence of epithelial volume on the chloride concentration of the inside bath. Results from a series of experiments. Experimental procedure as in Fig. 37, except that here the concentrations of Na^+ and K^+ are kept constant, whereas the Cl^- concentration is varied by partial replacement of chloride by thiocyanate. *Abscissa*, the fraction (in percent) of chloride in the recovery medium. *Ordinate*, the final epithelial chloride space (in percent of the final total epithelial volume). (From [310])

the Na^+/K^+ pump tends to keep the cellular sodium and potassium concentrations constant. It is then possible to calculate the cellular chloride volume as a function of the concentrations of Na^+, K^+ and Cl^- in the bath. It was found that in the normal cells the product $Na^+ \times K^+ \times Cl^2$ is slightly lower than in the Ringer. This is compatible with the idea that under normal circumstances the cotransport is running very slowly from medium to cells, and that the very slow uptake of chloride is balanced by a slow leakage from cells to inside medium.

The problem of what activates the cotransport is still unsettled. It was suggested [309] that the system is activated when the chloride concentration in the cells drops. If the system is really a $Na^+,K^+,2Cl$ cotransport, all sites for the three ionic species might be saturated in the steady state with normal Ringer in the bath whereas a drop in any one of the three ion concentrations might create free carrier molecules. However, in analogy with red cells there may also be activation of cotransport by cAMP or other local messengers (see "Activation Mechanisms").

The cotransport mechanism is highly specific with respect to ionic species. Among the anions so far tested, only bromide can replace chloride. Potassium can be replaced by rubidium. Cesium, on the other hand, does not participate in the cotransport, although it does not inhibit the cotransport if potassium is also present.

Li^+ can replace sodium in the cotransport system, but the analysis of the effects of Li^+ in the recovery medium is complicated by the fact that Li^+ can also pass from the inside bathing medium into the cells via the Na^+/H^+ exchanger [314]. Apparently, Li^+ resembles the hydrogen ion both with respect to being able to participate in Na^+/H^+ exchange and as an activator of the system which is normally almost quiescent at the cell pH of our standard experiment procedure. It is via this system that Li^+, added to the outside medium, is transported from the cells into the inside medium, despite the fact that the Na^+/K^+ pump transports Li^+ only slowly or not at all. In order to avoid interference via the Na^+/H^+ (Li^+) exchanger, experiments were performed with 10^{-4} *M* amiloride added to the inside medium. This will inhibit the exchanger but leaves the cotransporter intact. Figure 39 shows one such experiment.

Initially the inside solution is ordinary Ringer. At the first arrow the inside medium is made half thiocyanate Ringer, and at the second arrow it is made full thiocyanate Ringer. Now, at the third arrow, marked $Li^+ + Am$, the inside bath is changed to a Ringer where all Na^+ has been replaced by Li^+ and amiloride has been added to make the solution 10^{-4} *M* with respect to this drug. A rather fast swelling ensues, a swelling which can be inhibited with bumetanide. However, only about half the volume lost during the foregoing sequence of events is recovered, but after change of the inside solution to normal Na^+-containing Ringer (fourth arrow), the tissue swells to its original value.

The cotransport evidently rapidly establishes a steady-state volume of the epithelium which is significantly smaller than that obtained with Na^+ as the participant. The decrease in steady-state volume has nothing to do with the presence of amiloride, as it is seen also in its absence. An obvious explanation is that because the Na^+/K^+ pump does not transport Li^+ [314], this ion will accumulate in the cells. The equation

$$[Li_c] \times [K_c] \times [Cl_c]^2 = [Li_i] \times [K_i] \times [Cl_i]^2$$

therefore predicts a lower cellular chloride concentration and hence a lower

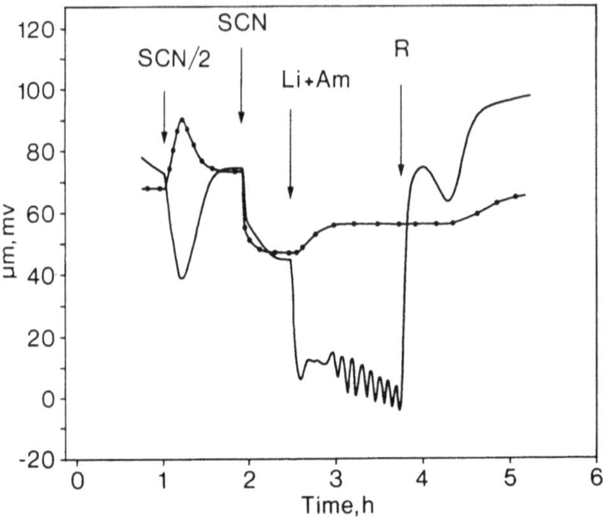

Fig. 39. Lithium replacement of sodium in the cotransport process. Outside medium 1/20 nitrate Ringer throughout. *Ordinate*, Transepithelial PD (in millivolts; *smooth curve*) and epithelial thickness (in micrometers; *dotted curve*). *Abscissa*, time (in hours). Inside bath initially ordinary Ringer. At *first arrow* change to half-strength thiocyanate Ringer. At *second arrow* change to full-strength thiocyanate Ringer. At *third arrow* change to lithium Ringer (all Na^+ replaced by lithium). The solution is made 2 m*M* with respect to amiloride in order to inhibit the basolateral Na^+/H^+ exchanger. At the *fourth arrow* the inside medium is again made full-strength ordinary Ringer

chloride volume when Li^+ is the dominating cation than if the major cation had been Na^+.

Other Cell Types

A role for Na^+,Cl^- or $Na^+,K^+,2Cl^-$ cotransport in cell volume regulation has, in addition to avian red blood cells, Ehrlich cells, and frog skin epithelial cells (see above), also been established in red blood cells from rat [56] and ferret [226] and in L cells [86], simian virus-transformed 3T3 cells [15], chick cardiac cells [80], astrocytes from rat cerebral cortex [156], and the apical membrane of rabbit kidney medullary thick ascending limb of Henle's loop (see review by Eveloff and Warnock [70]). (It may be noted that the $Na^+,K^+,2Cl^-$ cotransport system in red blood cells from other species, e.g., humans [57], is volume insensitive.) In the medullary loop of Henle (MTAL) both Na^+,Cl^- and $Na^+,K^+,2Cl^-$ cotransport systems have been identified, with the $Na^+,K^+,2Cl^-$ system being proposed to predominate during the RVI response [70]. As is suggested for Ehrlich cells (see above), MTAL cells can apparently shift between K^+-independent and K^+-dependent transport modes [302]. In tracheal epithelial cells a K^+-independent Na^+,Cl^- cotransport is activated by osmotic shrinkage [233].

Structure of the Na^+,K^+,Cl^- Cotransporter

Several studies on different tissues have attempted to identify proteins comprising the Na^+,K^+,Cl^- cotransport system by photoaffinity labeling using radiolabeled loop diuretics. Haas and Forbush ([119, 120]; see also [118]) have used [^{3}H]4-benzoyl-5-sulfamoyl-3-(3-thenyloxy) benzoic acid ([^{3}H]BSTBA) to photolabel membranes from dog kidney, duck red cells, and shark rectal gland. BSTBA is a bumetanide analog, which is a potent loop diuretic and which contains a photoreactive benzophenone group. BSTBA inhibits Na^+,K^+,Cl^- cotransport in duck red cells and competes with [^{3}H]bumetanide for binding sites on dog kidney membranes and duck red cells with an affinity similar to that of bumetanide itself. They find a single, broad peak of specific [^{3}H]BSTBA photoincorporation which is centered at about 150 kDa for dog kidney and duck red cell membranes and at about 200 kDa for shark rectal gland membranes [118].

To isolate proteins of the Ehrlich cell cotransporter, we employed bumetanide, a sulfamoylbenzoic acid diuretic which inhibits Na^+,K^+,Cl^- cotransport with high affinity [246]. We have shown that the bumetanide binding to the Na^+,K^+,Cl^- cotransporter in Ehrlich ascites cells is highly specific, the

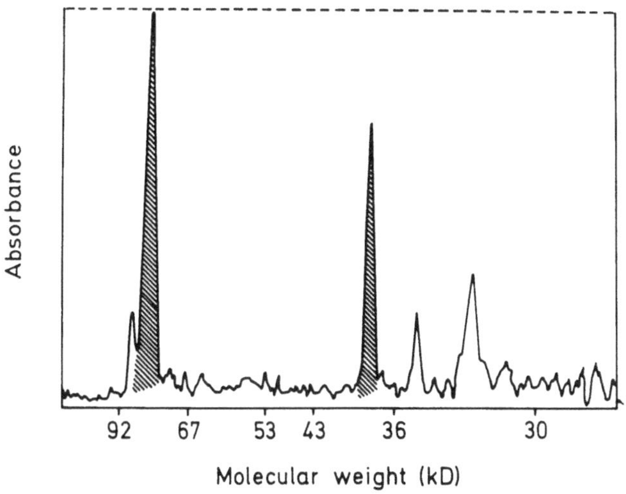

Fig. 40. SDS-PAGE electrophoresis of membrane proteins from Ehrlich ascites tumor cells, retained by affinity chromatography using bumetanide as ligand. The cells were disintegrated by high-pressure decompression and the membranes purified by differential centrifugation. The cell membranes were solubilized in 25 m*M* imidazole/2% *n*-octyl glucoside/1 m*M* EDTA, and the membrane polypeptides separated on an affinity chromatography column using bumetanide as ligand. The bumetanide affinity gel was prepared by reacting azido-bumetanide with activated thiol-Sepharose 4B gel. The membrane polypeptides retained on the affinity column were eluted with 1 m*M* bumetanide (at 50 m*M* NaCl, 5 m*M* KCl) and analyzed by SDS-PAGE electrophoresis. The major membrane polypeptides retained on the affinity column are indicated by hatching. (From [72])

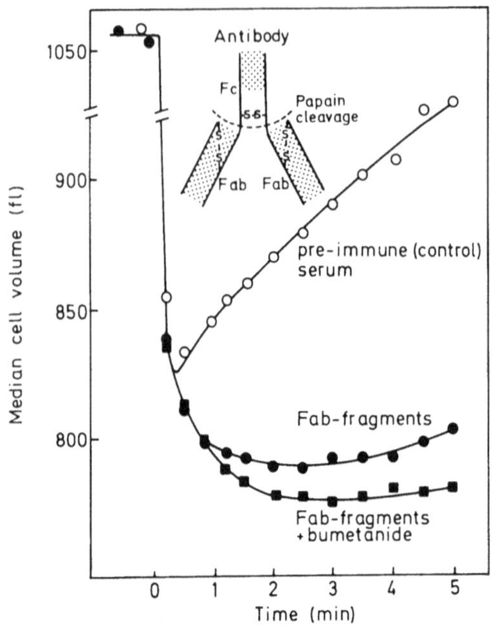

Fig. 41. Effect on RVI of F_{ab} fragments of the antibodies against bumetanide-binding proteins from Ehrlich cells. F_{ab} fragments were prepared either from the antiserum directly or from immunoglobulins purified from the antiserum, both giving similar results. Cells were preincubated 20 min in hypotonic media containing nonimmune serum or a F_{ab} fragment preparation. The nonimmune serum was diluted fivefold. The F_{ab} fragment preparations were at concentrations equivalent to five fold dilutions of the antiserum. The median cell volumes in hypotonic media were 1423 fl, cells in nonimmune serum, and 1291 fl, cells in the F_{ab} fragment preparation. At time zero the hypertonic challenge was administered. One aliquot received bumetanide (25 μM) at the same time. Similar results were obtained in three other experiments, including one with a different F_{ab} fragment preparation. F_{ab} fragments prepared from immunoglobulins purified from nonimmune serum had no effect on RVI. (Redrawn from [59])

evidence being a 1:1 relationship between [^{3}H]bumetanide binding and inhibition of transport [145a]. To purify candidate proteins of the cotransporter, we used a bumetanide-Sepharose affinity column [72a]. Isolated plasma membranes from Ehrlich cells were solubilized and applied to the column. Two main proteins were retarded by the affinity column, and subsequently eluted by bumetanide in solution [72a] (Fig. 40). These proteins were proposed to be constituents of the Na^+,K^+,Cl^- cotransporter. However, the specificity of binding of bumetanide to the cotransporter provided insufficient evidence for this proposal. We now have evidence that the purified protein contains components of the cotransporter. Antiserum raised against the purified proteins strongly inhibits Na^+,K^+,Cl^- cotransport measured by two independent methods [59]: cotransport was induced by hypertonic challenge and was measured as regulatory cell volume increase and as the bumetanide-sensitive component of the unidirectional Cl^- influx. In both assays, cotransport was strongly inhibited by the antiserum. Moreover, F_{ab} fragments of the purified IgG antibodies inhibited cotransport to a similar extent (Fig. 41).

Na^+/H^+ Exchange Functionally Coupled to Cl^-/HCO_3^- Exchange

Amphiuma Red Blood Cells

A model of the RVI response of red blood cells from the salamander *Amphiuma means* was given in Fig. 34. The principal mechanism activated during RVI in these cells is an amiloride-sensitive Na^+/H^+ exchange [32–35, 37, 168, 170, 293]. The H^+ efflux, driven by the Na^+ gradient, produces an increase of intracellular pH and cytoplasmic HCO_3^- concentration. The HCO_3^- gradient in turn causes the Cl^-/HCO_3^- exchange to cycle, resulting in a net Cl^- influx. Moreover, as in the RVI response in duck red blood cells and Ehrlich ascites tumor cells, secondary movements of Na^+ and K^+ occur through the Na^+/K^+ pump. The entering of Na^+ and Cl^- in exchange for H^+ and HCO_3^-, respectively, results in an increase in cellular Na^+ and Cl^- content and in cell volume [32, 293]. Importantly, the increase in cell volume results from the exchange of osmotically active Na^+ for H^+, almost all of which comes from intracellular buffers, such as hemoglobin, and therefore is osmotically inactive, and from exchange of Cl^- for HCO_3^-, which is derived from CO_2. As the cytoplasmic Na^+ concentration increases, resulting in a stimulation of the Na^+ pump; hence part of the Na^+ entering the cell is exchanged for K^+, and net increases in both Na^+ and K^+ are observed during RVI [293]. The net K^+ uptake is entirely blocked by ouabain [293]. The Na^+/H^+ entry step is completely blocked by 1 m*M* amiloride [32, 168, 170, 293] and the RVI response is, in contrast to the RVI response in duck red blood cells and in Ehrlich cells, insensitive to furosemide and bumetanide [292, 293].

As mentioned, the exchange of Na^+ for H^+ results in a disequilibrium of H^+ and HCO_3^-, despite the effect of intracellular buffers to minimize pH changes. The anion exchanger reequilibrates anions so that

$$\frac{[Cl^-]_i}{[Cl^-]_o} = \frac{[HCO_3^-]_i}{[HCO_3^-]_o} = \frac{[H^+]_o}{[H^+]_i} \tag{1}$$

where subscripts i and o refer to intracellular and extracellular concentrations, respectively. Thus, in the *Amphiuma* red blood cell, Cl^- moves in through the anion exchanger. The net Cl^- uptake is blocked by inhibition of the anion exchanger by SITS or DIDS [32, 168]. In addition, by blocking reequilibration of pH, SITS also greatly increases the alkalinization of the cell [168, 170]. Findings consistent with the model shown in Fig. 34 are as follows. (a) In cells where the Na^+/K^+ pump is blocked with ouabain, and the anion exchanger is simultaneously blocked with SITS or DIDS, the amount of H^+ released to the medium during volume regulation is equal to the amount of Na^+ entering the cell [32, 168, 170, 292]. (b) The SITS-treated cells gain less water for a given Na^+ uptake than untreated cells, because the SITS-treated cells gain no Cl^- [168, 170]. (c) Volume recovery is blocked by 1 m*M* amiloride or by replacement of external Na^+ with choline [293]. (d) Cells in hypertonic Na^+-free choline medium lose Na^+ and alkalinize the medium, implying that the Na^+/H^+ exchange process is reversible [170, 291]. (e) There is no major change in the membrane potential measured with microelectrodes [32] or estimated from chloride concentration ratios [293]. (f) The net Na^+ flux proceeds according to ($\mu Na^+ - \mu H^+$) where μNa^+ and μH^+ are the difference in chemical potentials for Na^+ and H^+, respectively [35, 36]. (g) Substitution of Br^-, I^-, SCN^-, or NO_3^- for medium Cl^- does not substantially affect volume-regulatory Na^+ uptake [170]. Some of these anions are, however, for reasons that are at present unknown, partially inhibitory [33, 170], even when the anion exchanger is blocked by inhibitors [33, 34].

In the *Amphiuma* red cell RVI response, (a) Na^+ uptake is a saturable function of $[Na^+]_o$ ($K_m = 39$ m*M*), (b) Li^+ substitutes for Na^+, whereas K^+ and choline do not, and (c) amiloride acts as a reversible competitive inhibitor of Na^+ uptake, with a K_i of about 2×10^{-6} *M* [292, 293]. The Na^+/H^+ exchanger of the *Amphiuma* red cell RVI response is qualitatively similar to that found in several systems. For a review, see Cala [37]. It has been proposed [37] that the same transporter may be able to effect the Na^+/H^+ exchange activated during RVI and the K^+/H^+ exchange activated during RVD in *Amphiuma* red cells (see "K^+/H^+ Exchange").

Lymphocytes

Lymphocytes do not regulate their volume when osmotically shrunken in hypertonic media [101, 114, 133, 271], unless the "RVI-after-RVD protocol" is used [101, 114], similar to the findings in the case of Ehrlich ascites tumor cells (Fig. 14). The reason for this behavior is unknown. As in Ehrlich cells, $[Cl^-]_i$ may play an important role in the activation of the RVI response in lymphocytes [101, 114].

The RVI response of lymphocytes appears to involve an amiloride-sensitive Na^+/H^+ exchange [101, 114], similar to that reported for *Amphiuma* red blood cells (see previous section) and dog red cells (see following section). The response is ouabain insensitive, Na^+ dependent, and electroneutral. In isotonic cells, stimulation of the Na^+/H^+ antiport can be accomplished by lowering the cytoplasmic pH or by addition of phorbol diesters (see "Activation Mechanisms"). These procedures are also found to induce cellular swelling in lymphocytes, thus resembling RVI [103].

Analysis of the cellular ion content following RVI indicates a net gain of K^+ with only a moderate increase in Na^+. In the presence of ouabain, however, osmotically shrunken cells still volume regulate, but associated with an uptake of Na^+ [101]. It was therefore concluded that net Na^+ uptake is the primary event in RVI, and that the increase in K^+ content occurs secondarily, via acceleration of the Na^+/K^+ pump by the elevated intracellular Na^+ concentration [101]. Amiloride substantially inhibited both RVI and the associated Na^+ uptake [101].

To test the involvement of Na^+/H^+ exchange in RVI more directly, Grinstein and coworkers undertook measurements of transmembrane H^+ transport during RVI, where the H^+ fluxes were estimated from changes either in the cytoplasmic pH (pH_i) or of the extracellular pH (pH_o) measured in lightly buffered media. Hypertonic shrinking of thymic lymphocytes results in a marked cytoplasmic alkalinization with commensurate acidification of the external medium, demonstrating that the internal alkalinization is due to a transmembrane displacement of H^+ equivalents. The process is strictly dependent on the presence of external Na^+ and is completely prevented by amiloride [109, 110]. Thus, the available data support an activation of an electroneutral Na^+/H^+ exchange as the primary event underlying RVI.

The mechanism of activation of Na^+/H^+ exchange during RVI was analyzed by comparing the kinetic parameters of transport in resting (isotonic) and hyperosmotically stressed cells. It is concluded that changes in the extracellular cation affinity of the antiport are not involved in the osmotically induced activation [109, 110]. Cellular pH is a crucial determinant of the rate of Na^+/H^+ exchange in isotonic cells [105, 111], and the pH sensitivity is therefore a likely target for regulation. For this reason Grinstein and coworkers have compared the pH_i dependence of the rate of H^+ extrusion in normal and osmotically shrunken cells. Osmotic shrinking produces a 0.2–0.3 pH unit alkaline shift in the pH dependence curve which activates the quiescent transporter at physiological pH_i (7.0–7.1), with the activation persisting until pH_i attains the new "set point" [109, 110]. The shift in the pH_i dependence of the Na^+/H^+ exchanger presumably reflects an altered pH_i sensitivity of an allosteric "modifier" site different from the transport site [9]. The regulation of the Na^+/H^+ exchanger is reviewed by Grinstein and Rothstein [113].

As discussed in the case of the *Amphiuma* red cell (see above) reswelling of the shrunken cells following activation of Na^+/H^+ exchange occurs because the cytoplasmic buffering power replenishes most of the extruded H^+, and the elevated cytoplasmic pH in turn increases the HCO_3^- content of the cells, driving Cl^- into the cell through the Cl^-/HCO_3^- exchange system (see [33, 114] for detailed discussion). It may be noted that a substantial Cl^- uptake will results only if the Cl^- conductance is low relative to the rate of Cl^-/HCO_3^- exchange, which has been previously demonstrated to be the case in human lymphocytes [114].

Dog Red Blood Cells

Dog red blood cells, except for a few rare breeds, and also cat red cells, lack a Na^+,K^+ pump in their cell membrane, resulting in cytoplasmic Na^+ and K^+ concentrations close to that of plasma [250].

The dog red cell RVI response has long been known to involve passive movements of sodium [262] which now also appears to be an amiloride-sensitive Na^+/H^+ exchange process [255, 257, 259]. Shrinkage-stimulated Na^+ fluxes are almost completely blocked by amiloride [255], and evidence consistent with direct coupling between Na^+ and H^+ movements has been presented [260]. Given appropriate ion gradients, Na^+-driven H^+ movements and H^+-driven Na^+ movements can be demonstrated [260].

A finding in the dog red cell RVI response that remains to be explained is that replacement of external Cl^- with SCN^- is completely inhibitory, and with NO_3^- partially inhibitory, although both of these anions are transported by the anion exchanger [251, 255]. Experiments in which glutaraldehyde was used to fix the transporter in a functional or nonfunctional state suggested [256, 257] that the chloride requirement of the Na^+/H^+ exchanger involves the activation or triggering of the exchange mechanism rather than the transport function per se.

External Li^+ inhibits H^+ influx and Na^+ efflux. At high external Li^+, H^+ efflux in exchange for Li^+ influx can be demonstrated [255]. External Li^+, however, has only little inhibiting effect on Na^+ influx [258].

A model for the Na^+/H^+ exchanger in dog red blood cells and its interaction with chloride and lithium ions has been presented (see review by Parker [257]). The model postulates that protons, Li^+ ions, and chloride have "allosteric" effects on the activation of the exchanger that are separate from the role of these ions as transported substrates. Chloride is necessary for the triggering of the Na^+/H^+ exchanger by cell shrinkage. Lithium is regarded by this transporter as being more like a proton than like Na^+. Raising the concentration of internal Li^+ activates the exchanger, in analogy with raising the concentration of internal H^+.

The Involvement of the Na^+/H^+ Exchanger in "Regulatory Volume Increase" in Frog Skin

In several cell types the Na^+/H^+ exchanger (sometimes in collaboration with a Cl^-/HCO_3^- exchanger) plays a role in volume regulation. Under most experimental conditions such effects are not important in the frog skin epithelium, but they can be made visible if the normal mechanisms are inhibited. As an example

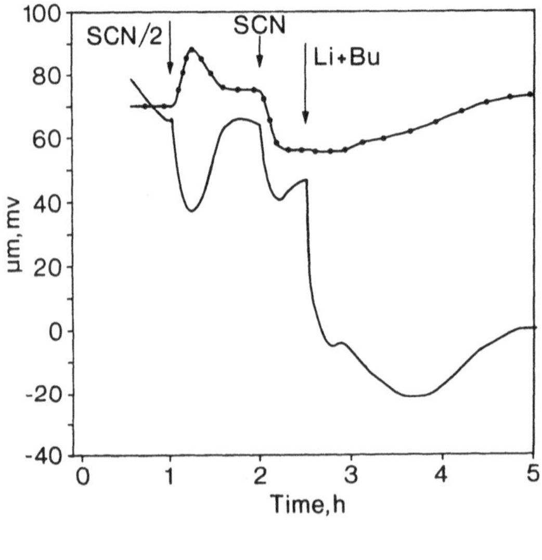

Fig. 42. Showing that a chloride-depleted epithelium is slowly able to recover volume with Li^+ Ringer as inside bath, even when the cotransport of Li^+,K^+,2Cl is inihibited by amiloride. Ordinates and abscissa as in Fig. 39. Outside medium 1/20 nitrate Ringer throughout. Inside bath initially ordinary Ringer. At the *first arrow* switch to half thiocyanate Ringer and at the *second arrow* full thiocyanate Ringer. At the *third arrow* finally Li^- Ringer, made 50 μ*M* with respect to bumetanide

let us consider the case where, after depletion of the epithelium of chloride by the usual sequence (half thiocyanate Ringer and full thiocyanate Ringer), we use lithium Ringer with 50 μ*M* bumetanide as recovery solution at the inside of the skin. In this case the cotransport of Na^+,K^+,2Cl is inhibited (Fig. 42). It can be seen that there is a very slow recovery of voloume, but in contrast to the experiment shown in Fig. 39 where the cotransport is active, the swelling does not stop at a level lower than the original volume of the epithelium. In this case the swelling is driven by the $Na^+/H^+(Li^+)$ exchanger. Li^+ inhibits the potassium channels of the basolateral membrane [314]. Thus, K^+ can no longer leave the cells. However, Na^+ which can enter the cells from the outside medium has two possibilities: it can exchange with Li^+ from the inside medium, or it can be pumped onward via the Na^+/K^+ pump, in which case, for every three Na^+ pumped, two K^+ ions are brought into the cell, from where they cannot escape. Thus Li^+ and K^+ accumulate in the cell. Electrically they must be balanced by an uptake of anions from the inside solution, but the violent depolarization leads to opening of the anion channels (see above). There is no upper limit to the swelling, in contrast to the cotransport case where the ionic product of the participating ions defines the equilibrium.

A situation rather similar to that created by Li^+ added to the inside solution can be brought about by increasing the CO_2 tension of the bathing solutions. It is well established that acidification of the cytoplasm of the frog skin epithelium closes the K^+ channels [127] and at the same time activates the Na^+/H^+ exchanger. Figure 43 shows an experiment where during the "recovery phase" the bathing solutions on both sides of the skin are saturated with a mixture of 90% oxygen and 10% CO_2. This leads to a very dramatic swelling of the

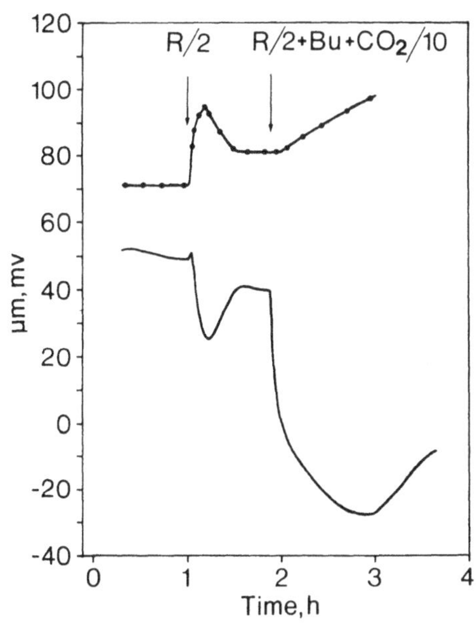

Fig. 43. Showing that acidification of the cytoplasm of the epithelium cells will induce violent swelling with half Ringer inside bath. Abscissa and ordinates as in Figs. 39 and 42. Outside medium 1/20 nitrate Ringer throughout. Inside bath initially Ringer. At the *first arrow* change to half Ringer. At the *second arrow* both inside and outside media are equilibrated with a mixture of 90% oxygen and 10% CO_2. Bumetanide is added in order to inhibit the Na^+,K^+,$2Cl^-$ transport system

epithelium. In many cases the swelling goes far beyond the original volume. Thus it is not a RVI in the true sense of the word. The situation resembles that seen with Li^+ Ringer: K^+ cannot escape from the cells. Na^+ from the outside medium as well as Na^+ entering the cell from the inside medium in exchange for H^+, formed during the acidification of the cell, are constantly pumped from the cell to the inside medium, and the return freight of K^+ cannot escape.

Other Cell Types

Activation of Na^+/H^+ exchange by osmotic shrinkage has also been demonstrated in trout erythrocytes [25, 27] in frog skeletal muscle, human epidermal carcinoma (A431) cells, and rabbit proximal tubule cells (for references, see Grinstein and Rothstein [113] and Moolenaar [230]), and recently in rabbit red blood cells [152]. Activation of coupled Na^+/H^+ and Cl^-/HCO_3^- exchange by osmotic shrinkage has been demonstrated in the apical membrane of *Necturus* gallbladder epithelial cells [190], and in the basolateral membrane of mouse kidney medullary loop of Henle (see review by Eveloff and Warnock [70]).

Activation Mechanisms

The actual mechanism of activation of the RVI response in osmotically shrunken cells is in most cases unknown. Ca^{2+}, calmodulin and the cytoskeleton do not seem to be involved in RVI, in Ehrlich cells [89] or red blodd cells [34] nor in gallbladder epithelial cells [78]. A decrease of cytosolic free Ca^{2+} has been reported during RVI in*Amphiuma* red cells [38] using the Ca^{2+}-sensitive dye arsenazo III and a null-point method. It is suggested that the internal Cl^- concentration may be involved in the activation of the RVI response in both Ehrlich cells [147] and lymphocytes [102]. Cl^- is, however, thought to play only a permissive role in these cells, modulating the response to cell shrinkage, as first suggested for frog skin epithelial cells [309]. In avian red cells cAMP may influence the operation of the Na^+, K^+, $2Cl^-$ cotransport system by inducing a conformational change similar to that induced by cell shrinkage [248, 291]. The activation of the cotransport system by catecholamines is mediated via cAMP and may involve phosphorylation of a 230-kDa protien (see Section F.I). Cell shrinkage does not, however, increase cAMP levels in avian red cells [171], nor does it increase phosphorylation in these cells of the 23-kDa protein, which is phosphorylated during catecholamine-induced activation of the cotransport systems [5]. Neither the cAMP level, nor cytosolic Ca^{2+} seem to play any role in the volume-sensitive cotransport in Ehrlich cells [89], and both have been reported to even inhibit cotransport in human red blood cells [85].

Theophylline and other alkylxanthines inhibit the furosemide-sensitive cotransport in Ehrlich cells, presumably by reducing the critical cell volume to a level above which the cotransport system becomes quiescent. The cotransport inhibition by these drugs is not mediated by cAMP or cGMP [89].

An interesting finding is that the furosemide-sensitive cotransport system of mammalian erythrocytes is volume insensitive in most species (including humans) and can only be activated by cell shrinkage in red blood cells from rats ([56]; see also [89]) and ferrets (226).

The Na^+/H^+ exchange in lymphocytes is activated by cell shrinkage in hypertonic medium and can also be activated by phorbol diester or DAG-induced activation of protein kinase C [103]. In both cases the activation is attributed to an alkaline shift in the pH sensitivity of the modifier site (see Section F.3).

The responses to cell shrinking and to phorbol esters are not additive. The responses are remarkably similar with respect to extent and time course of onset and reversal. Moreover, both responses are ATP-dependent and inhibited by similar concentrations of trifluoperazine and NEM [104, 106, 112]. At least one membrane polypeptide was demonstrated to be phosphorylated by both activation procedures, and in both cases the phosphorylation was prevented by trifluoperazine. These findings suggested that the activation by cell shrinkage was mediated via protein kinase C.

The possibility that protein kinase C is activated during cell shrinkage by activation of phospholipase C (phosphoinositidase) with resulting release of DAG was tested by Grinstein et al. [104, 112]. The level of phosphatidylinositol 4,5-bisphosphate was found to be unaffected by cell shrinkage. Moreover, the cytoplasmic concentration of inositol 1,4-diphosphate and IP_3 was also unaffected by osmotic activation. Based on these findings it was concluded that activation of phospholipase C is not involved in the RVI response. An intracellular redistribution of protein kinase C from soluble to membrane-bound form was detected following activation by phorbol esters but not following cell shrinkage. These results suggest that cell shrinkage involves activation of a protein kinase that is similar, but not identical to protein kinase C. The kinase phosphorylates the Na^+/H^+ exchanger or an auxillary protein, which produces an alkaline shift in the pH_i dependence of the exchanger [104, 112]. In the *Amphiuma* red blood cell both Na^+/H^+ and K^+/H^+ exchange can be activated by phorbol diester, suggesting a role for protein kinase C in the activation mechanism. Since, however, cell shrinkage and swelling activate Na^+/H^+ and K^+/H^+ exchange, respectively, it appears that additional reactions must be involved in the volume-induced activations [37]. In mouse meduallary thick ascending limbs cAMP activates Na^+/H^+ and Cl^-/HCO_3^- pathways which mediate RVI [129, 130].

Concluding Remarks

Volume regulatory mechanisms have been conserved from the earliest forms of bacterial life, and are general characteristics of vertebrate cells today, although the majority of body cells in a terrestrial vertebrate are never exposed to more than negligible variations in the osmolality of body fluids.

Cell volume regulation conceivably plays an essential role in cell growth and proliferation. Moreover, cell volume regulatory processes appear to be an integrated part of the function of secreting and absorbing epithelia. Furthermore, it is demonstrated that many hormones and growth factors also affect the ion transport mechanisms included in volume regulation and via the same "second messengers." Studying the molecular structure and function and the biochemical regulation of the passive membrane transport systems involved in cell volume regulation can thus provide useful knowledge with important implications.

Acknowledgements. The authors wish to thank Dr. L. O. Simonsen for helpful discussions and for critical reading of the manuscript.

References

1. Aabin B, Hoffmann EK (1986) Inhibition of the volume- and Ca^{2+}-activated Cl^- transport pathway in Ehrlich ascites tumor cells. Acta Physiol Scand 128:42A
2. Aabin B, Kristensen BI (1987) Identification of calmodulin in Ehrlich ascites tumor cells and its Ca^{2+}-dependent binding to the cell membrane. Acta Physiol Scand 129:18A
3. Adorante JS, Cala PM (1987) Activation of electroneutral K flux in *Amphiuma* red blood cells by N-ethylmaleimide. Distinction between K/H exchange and KCl cotransport. J Gen Physiol 90:209–227
4. Aiton JE, Chipperfield AR, Lamb JF, Ogden P, Simmons NL (1981) Occurrence of passive furosemide-sensitive transmembrane potassium transport in cultured cells. Biochim Biophys Acta 646:389–398
5. Alper SL, Beam KG, Greengard P (1980) Hormonal control of Na^+–K^+ co-transport in turkey erythrocytes: multiple site phosphorylation of goblin, a high molecular weight protein of the plasma membrane. J Biol Chem 255:4864–4871
6. Alper SL, Palfrey HC, DeRiemer SA, Greengard P (1980) Hormonal control of protein phosphorylation in turkey erythrocytes: phosphorylation by cAMP-dependent and Ca^{2+}-dependent protein kinases of distinct sites in goblin, a high molecular weight protein of the plasma membrane. J Biol Chem 255:11 029–11 039
7. Amsler K, Donahue JJ, Slayman CW, Adelberg EA (1985) Stimulation of bumetanide-sensitive K^+ transport in Swiss 3T3 fibroblasts by serum and mitogenic hormones. J Cell Physiol 123:257–263
8. Amsler K, Kinne R (1986) Photoinactivation of sodium-potassium-chloride cotransport in LLC-PK_1/Cl_4 cells by bumetanide. Am J Physiol 250:C799–C806
9. Aronson PS, Nee J, Suhm MA (1982) Modifier role of internal Na^+ in activating the Na^+/H^+ exchanger in renal microvillus membrane vesicles. Nature 299:161–163
10. Arslan P, Virgilio FD, Beltrame M, Tsien RY, Pozzan T (1985) Cytosolic Ca^{2+} homeostasis in Ehrlich and Yoshida carcinomas. A new, membrane-permeant chelator of heavy metals reveals that these ascites tumor cell lines have normal cytosolic free Ca^{2+}. J Biol Chem 260:2719–2727
11. Aull F (1981) Potassium chloride cotransport in steady-state ascites tumor cells. Does bumetanide inhibit? Biochim Biophys Acta 643:339–345
12. Bagnasco S, Balaban R, Fales HM, Yang Y-M, Burg M (1986) Predominant osmotically active organic solutes in rat and rabbit renal medullas. J Biol Chem 261:5872–5877
13. Bakker-Grunwald T (1978) Effect of anions on potassium self-exchange in ascites tumor cells Biochim Biophys Acta 513:292–295
14. Bakker-Grunwald T (1981) Hormone-induced diuretic-sensitive potassium transport in turkey erythrocytes in anion dependent. Biochim Biophys Acta 641:427–431

15. Bakker-Grunwald T, Ogden P, Lamb JF (1982) Effects of ouabain and osmolarity on bumetanide-sensitive potassium transport in Simian virus-transformed 3T3 cells. Biochim Biophys Acta 687:333–336
16. Ballantyne JS, Moyes CD, Moon TW (1987) Compatible and conteracting solutes and the evolution of ion and osmoregulation in fishes. Can J Zool 65:1883–1888
17. Barish ME (1983) A transient calcium-dependent chloride current in the immature *Xenopus* oocyte. J Physiol (Lond) 342:309–325
18. Bauer J Lauf PK (1983) Thiol-dependent passive K/Cl transport in sheep red cells III. Differential reactivity of membrane SH groups with N-ethylmaleimide and iodoacetamide. J Membr Biol 73:257–261
19. Begenisich T, Smith C (1984) Multi-ion nature of potassium channels in squid axons. Curr Top Membr Transp 22:353–369
20. Bennekou P, Stampe P (1988) The effect of ATP, intracellular calcium and the anion exchange inhibitor DIDS on conductive anion fluxes across the human red cell membrane. Biochim Biophys Acta 942:179–185
21. Ben-Sasson S, Shaviv R, Bentwich Z, Slavin S, Doljanski F (1975) Osmotic behavior of normal and leukemic lymphocytes. Blood 46:891–899
22. Berridge MJ (1984) Inositol trisphosphate and diacylglycerol as second messengers. Biochem J 220:345–360
23. Berridge MJ (1987) Inositol trisphosphate and diacylglycerol: two interacting second messengers. Annu Rev Biochem 56:159–193
24. Borgese F, Garcia-Romeu F, Motais R (1986) Catecholamine-induced transport systems in trout erythrocyte. Na^+/H^+ countertransport or NaCl cotransport? J Gen Physiol 87:551–566
25. Borgese F, Garcia-Romeu F, Motais R (1987) Control of cell volume and ion transport by β-adrenergic catecholamines in erythrocytes of rainbow trout, *Salmo gairdneri*. J Physiol (Lond) 382:123–144
26. Boucher RC, Larsen EH (1988) Comparison of ion transport by cultured secretory and absorptiive canine airway epithelia. Am J Physiol 254:C535–C547
27. Bourne PK, Cossins AR (1984) Sodium and potassium transport in trout (*Salmo gairdneri*) erythrocytes. J Physiol (Lond) 347:361–375
28. Breuer WV, Mack E, Klip A (1988) Calcium activated chloride pathways in L_6H_9 muscle cells. Arch Biochem Biophys (in press)
28a. Breuer WV, Mack E, Rothstein A (1988) Activation of K^+ and Cl^- channels by Ca^{2+} and cyclic AMP in dissociatile kidney epithelical (MDCK) cells. Pflugers Arch, 411(4) pp 450–455
29. Brugnara C, Canessa M, Cusi D, Tosteson DC (1986) Furosemide-sensitive Na and K fluxes in human red cells. J Gen Physiol 87:91–112
30. Burg MB (1988) Role of aldose reductase and sorbitol in maintaining the medullary intracellular milieu. Kidney Int 33:635–641
31. Busa WB (1986) Mechanisms and consequences of pH-mediated cell regulation. Annu Rev Physiol 48:389–402
32. Cala PM (1980) volume regulation by *Amphiuma* red blood cells. The membrane potential and its implications regarding the nature of the ion-flux pathways. J Gen Physiol 76:683–708
33. Cala PM (1983) Volume regulation by red blood cells: mechanisms of ion transport. Mol Physiol 4:33–52
34. Cala PM (1983) Cell volume regulation by *Amphiuma* red blood cells. The role of Ca^{2+} as a modulator of alkali metal/H^+ exchange. J Gen Physiol 82:761–784
35. Cala PM (1985a) Volume regulation by *Amphiuma* red blood cells: characteristics of volume-sensitive K/H and Na/H exchange. Mol Physiol 8: 199–214
36. Cala PM (1985b) Volume regulation by *Amphiuma* red blood cells: strategies for identifying alkali metal/H^+ transport. Fed Proc 44:2500–2507
37. Cala PM (1986) Volume-sensitive alkali metal-H transport in *Amphiuma* red blood cells. Curr Top Membr Transp 26:79–99
38. Cala PM, Mandel LJ, Murphy E (1986) Volume regulation by *Amphiuma* red blood cells: cytosolic free Ca and alkali metal-H exchange. Am J Physiol 250:C423–C429
39. Canessa M, Brugnara C, Cusi D, Tosteson DC (1986) Modes of operation and variable stoichiometry of the furosemide-sensitive Na^+ and K^+ fluxes in human red cells. J Gen Physiol 87:113–142
40. Chamberlin ME, Strange K (1989) Anisosmotic cell volume regulation: a comparative view Am J Physiol 257 (Cell Physiol, 26):C159–C173

41. Cheung RK, Grinstein S, Dosch H-M, Gelfand EW (1982) Volume regulation by human lymphocytes: characterization of the ionic basis for regulatory volume decrease. J Cell Physiol 112:189–196
42. Chipperfield AR (1980) An effect of chloride on (Na + K) co-transport in human red blood cells. Nature 286:281–282
43. Chipperfield AR (1986) The ($Na^+ - K^+ - Cl^-$) co-transport system. Clin Sci 71:465–476
44. Christensen O (1987) Mediation of cell volume regulation by Ca^{2+} influx through stretch-activated channels. Nature 330:66–68
45. Christensen O, Hoffmann EK (1991) Cell swelling activates K^+ and Cl^- channels as well as nonselective stretch-activated cation channels in Ehrlich ascites tumor cells. J Memb. Biol (in press)
46. Christensen S, Hoffmann EK, Saermark T, Simonsen LO (1988) Inositol trisphosphate may be a second messenger in regulatory volume decrease in Ehrlich mouse ascites-tumour cells. J Physiol (Lond) 403:109P
47. Cittadini A, Dani AM, Wolf F, Bossi D, Calviello G (1982) Calcium permeability of Ehrlich ascites tumour cell plasma membrane in vivo. Biochim Biophys Acta 686:27–35
48. Cornet M, Delpire E, Gilles R (1987) Study of microfilaments network during volume regulation process of cultured PC12 cells. Pflugers Arch 410:223–225
49. Costantin J, Alcalen S, Otero A, Dubinsky WP, Schultz S (1991) Reconstitution of an inwardly rectifying potassium channel from the basolateral membranes of Necturus enterocytes into planar lipid bilayers. Proc Natl Acad Sci USA 86(13) pp 5212–5216
50. Davis CW, Finn AL (1982) Sodium transport inhibition by amiloride reduces basolateral membrane potassium conductance in tight epithelia. Science (NY) 216:525–527
51. Davis CW, Finn AL (1985) Cell volume regulation in frog urinary bladder. Fed Proc 44:2520–2525
52. Delpire E, Gilles R, Duchéne C, Goessens G (1985) Effects of osmotic shocks on the ion content and ultrastructure of rat pheochromocytoma cells of line PC 12. Mol Physiol 8: 293–307
53. Dissing S, Nauntofte B (1991) Na^+ transport properties of isolated rat parotid acini. Am J Physiol 22:G1044–G1055
54. DiStefano A, Wittner M, Schlatter E, Lang HJ, Englert H, Greger R (1985) Diphenylamine-2-carboxylate, a blocker of the Cl^--conductive pathway in Cl^--transporting epithelia. Pflugers Arch 405 [Suppl 1]:S95–S100
55. Doppler W, Maly K, Grunicke H (1986) Role of the Na^+/H^+ antiport in the regulation of the internal PH of Ehrlich ascites tumor cells in culture. J Membr Biol 91:147–155
56. Duhm J, Göbel BO (1984) Na^+–K^+ transport and volume of rat erythrocytes under dietary K^+ deficiency. Am J Physiol 246:C20–C29
57. Duhm J, Göbel BO (1984) Role of the furosemide-sensitive Na^+/K^+ transport system in determining the steady-state Na^+ and K^+ content and volume of human erythrocytes in vitro and in vivo. J Membr Biol 77:243–254
58. Dunham PB, Ellory JC (1981) Passive potassium transport in low potassium sheep red cells: dependence upon cell volume and chloride. J Physiol (Lond) 318:511–530
59. Dunham PB, Jessen F, Hoffmann EK (1990) Inhibition of Na – K – Cl cotransport in Ehrlich ascites cells by antiserum against purified proteins of the cotransporter. Proc Natl Acad Sci USA 87:6828–6832
60. Dunham PB, Stewart GW, Ellory JC (1980) Chloride-activated passive potassium transport in human erythrocytes. Proc Natl Acad Sci USA 77:1711–1715
61. Dürr JE, Hviid Larsen E (1986) Indacrinone (MK-196)—a specific inhibitor of the voltage-dependent Cl^- permeability in toad skin. Acta Physiol Scand 127:145–153
62. Ehrenfeld J, Harvey BJ (1986) Evidence for basolateral sodium/hydrogen exchange in frog skin epithelium and its role in sodium transport and intracellular pH regulation. Proc Physiol Soc Sept 1986, p 144P
63. Ellory JC, Dunham PB, Logue PJ, Stewart GW (1982) Anion-dependent cation transport in erythrocytes. Philos Trans R Soc Lond [Biol] 229:483–495
64. Ellory JC, Hall AC, Stewart GW (1985a) Volume-sensitive passive potassium fluxes in red cells. In: Gilles R, Gilles-Baillien M (eds) Transport processes, Iono- and osmoregulation. Current comparative approaches. Springer, Berlin Heidelberg New York, pp 401–410
65. Ellory JC, Hall AC, Stewart GW (1985b) Volume-sensitive cation fluxes in mammalian red cells. Mol Physiol 8:235–246

66. Escobales N, Canessa M (1986) Amiloride-sensitive Na^+ transport in human red cells: evidence for a Na/H exchange system. J Membr Biol 90:21–28
67. Eskesen K, Ussing HH (1985) Determination of the electromotive force of active sodium transport in frog skin epithelium (Rana temporaria) from presteady state flux ratio experiments. J Membr Biol 86:105–111
68. Eskesen K, Ussing HH (1986) Single-File diffusion through K^+ channels in frog skin epithelium. J Membr Biol 91:245–250
69. Eskesen K, Ussing HH (1989) Transport pathways for Na^+ and Br^- (Cl^-) in noradrenaline-stimulated frog skin (*Rana temporaria*) Acta Physiol Scand 136:535–546
70. Eveloff JL, Warnock DG (1987) Activation of ion transport systems during cell volume regulation. Am J Physiol 252: F1–F10
71. Falke LC, Misler S (1989) Activity of ion channels during volume regulation by clonal N1E115 neuroblastoma cell. Proc Natl Acad Sci USA 86:3919–3923
72. Feinstein MB, Sháafi RI (1983) Role of calcium in arachidonic acid metabolism and in the actions of arachidonic acid-derived metabolites. In: Cheung WY (ed) Calcium and cell function, Vol IV. Academic, New York, pp 337–376
72a. Feit PW, Hoffmann EK, Schiødt M, Kristensen P, Jessen F, Dunham PB (1988) Purification of proteins of the Na/Cl cotransporter from membranes of Ehrlich ascites cells using a bumetanide-sepharose affinity column. J Membr Biol 103:135–147
73. Ferreira KTG, Ferreira HG (1981) The regulation of volume and ion composition in frog skin. Biochim Biophys Acta 646:193–202
74. Fincham DA, Wolowyk MW, Young JD (1987) volume-sensitive taurine transport in fish erythrocytes. J Membr Biol 96:45–56
75. Flatman PW (1983) Sodium and potassium transport in ferret red cells. J Physiol (Lond) 341:545–557
76. Flatman PW (1987) The effects of calcium on potassium transport in ferret red cells. J Physiol (Lond) 386:407–423
77. Flatman PW (1988) The effects of magnesium on potassium transport in ferret red cells. J Physiol (Lond) 397:471–487
78. Foskett JK, Spring KR (1985) Involvement of calcium and cytoskeleton in gallbladder epithelial cell volume regulation. Am J Physiol 248:C27–C36
79. Freeman CJ, Bookchin RM, Ortiz OE, Lew VL (1987) K-permeabilized human red cells lose an alkaline, hypertonic fluid containing excess K^+ over diffusible anions. J Membr Biol 96:235–241
80. Frega NS, Oiboma DR, Geuvfler B, Leaf A (1976) Ischemic renal injury. Kidney Int 10:S17–S25
80a. Frelin C, Chassande O, Lazdunski M (1986) Biochemical characterization of the $Na^+/K^+/Cl^-$ co-transport in chick cardiac cells. Biochem Biophys Res Commun 134:326–331
81. Frizzell RA, Field M, Schultz SG (1979) Sodium-coupled chloride transport by epithelial tissues. Am J Physiol 236:F1–F8
81a. Frizzell RA, Rechkemmer G, Shoemaker RL (1986) Altered regulation of airway epithelial cell chloride channels in cystic fibrosis. Science (NY) 233:558–560
82. Fugelli K (1980) Amino acid transport coupled to cell volume regulation in anisosmotic media. In: Gilles R (ed) *Animals and environmental fitness. Physiological and biochemical aspects of adaptation and ecology*, Vol 1. Pergamon, Oxford, pp 27–41
83. Fugelli K, Rohrs H (1980) The effect of Na^+ and osmolality of the influx and steady state distribution of taurine and gamma-aminobutyric acid in flounder (Platichthys flesus) erythrocytes. Comp Biochem Physiol 67A:545–551
84. Fugelli K, Thoroed SM (1986) Taurine transport associated with cell volume regulation in flounder erythrocytes under anisosmotic conditions. J Physiol (Lond) 374:245–261
85. Garay RP (1982) Inhibition of the Na^+/K^+ cotransport system by cyclic AMP and intracellular Ca^{2+} in human red cells. Biochim Biophys Acta 688:786–792
86. Gargus JJ, Slayman CW (1980) Mechanism and role of furosemide-sensitive K^+ transport in L cells: a genetic approach. J Membr Biol 52:245–256
86a. Gazzola GC, Dall'Asta V, Nucci FA, Rossi PA, Bussolati O, Hoffmann EK, Guidoffi GG (1991) Role of amino acid transport system A in the control of cell volume in cultured human fibroblasts. Cell Physiol Biochem

87. Geck P (1990) Volume regulation in Ehrlich cells. In: Beyenbach KW (ed) Cell volume regulation Karger, Basel, pp 26–59
88. Geck P, Heinz E (1986) The Na – K – 2Cl contransport system. J Membr Biol 91: 91–105
89. Geck P, Pfeiffer B (1985) $Na^+ + K^+ 2Cl^-$ cotransport in animal cells—its role in volume regulation. Ann NY Acad Sci 456:166–182
90. Geck P, Pietrzyk C, Burckhardt B-C, Pfeiffer B, Heinz E (1980) Electrically silent cotransport of Na^+, K^+ and Cl^- in Ehrlich cells. Biochim Biophys Acta 600:432–447
91. Gerencser GA, Lee S-H (1983) Cl^--stimulated adenosine triphosphatase: existence, location and function. J Exp Biol 106:143–161
92. Gilles R (1979) Intracellular organic osmotic effectors. In: Gilles R (ed) *Mechanisms of osmoregulation in animals. Maintenance of cell volume.* Wiley, New York, pp 111–154
93. Gilles R (1983) Volume maintenance and regulation in animal cells: some features and trends. Mol Physiol 4:3–16
94. Gilles R (1987) Volume regulation in cells of euryhaline invertebrates. Curr Top Membr Transp 30:205–247
95. Giraldez F, Ferreira KTG (1984) Intracellular chloride activity and membrane potential in stripped frog skin (*Rana temporaria*). Biochim Biophys Acta 769:625–628
96. Goldstein L, Kleinzeller A (1987) Cell volume regulation in lower vertebrates. Curr Top Membr Transp 30:181–204
97. Greger R (1985) Ion transport mechanisms in the thick ascending limb of Henle's loop of mammalian nephron. Physiol Rev 65:760–797
98. Greger R, Gögelein H, Schlatter E (1988) Stimulation of NaCl secretion in the rectal gland of the dogfish *Squalus acanthias.* Comp Biochem Physiol 90A:733–739
99. Grinstein S, Clarke CA, DuPre A, Rothstein A (1982a) Volume-induced increase of anion permeability in human lymphocytes. J Gen Physiol 80:801–823
100. Grinstein S, Clarke CA, Rothstein A (1982b) Increased anion permeability during volume regulation in human lymphocytes. Philos Trans R Soc Lond [Biol] 299:509–518
101. Grinstein S, Clarke CA, Rothstein A (1983) Activation of Na^+/H^+ exchange in lymphocytes by osmotically induced volume changes and by cytoplasmic acidification. J Gen Physiol 82:619–638
102. Grinstein S, Clarke CA, Rothstein A, Gelfand EW (1983) Volume induced anion conductance in human B lymphocytes is cation independent. Am J Physiol 245:C160–C163
103. Grinstein S, Cohen S, Goetz JD, Rothstein A (1985) Osmotic and phorbol ester-induced activation of Na^+/H^+ exchange: possible role of protein phosphorylation in lymphocyte volume regulation. J Cell Biol 101:269–276
104. Grinstein S, Cohen S, Goetz JD, Rothstein A, Mellor SA, Gelfand EW (1986) Activation of the Na^+–H^+ antiport by changes in cell volume and by phorbol esters; possible role of protein kinase. Curr top Membr transp 26:115–134
105. Grinstein S, Cohen S, Rothstein A (1984) Cytoplasmic pH regulation in thymic lymphocytes by an amiloride-sensitive Na^+/H^+ antiport. J Gen Physiol 83:341–369
106. Grinstein S, Cohen S, Rothstein A (1985) Chemical modification of the Na^+/H^+ exchanger of thymic lymphocytes. Inhibition by N-ethylmaleimide. Biochim Biophys Acta 812:213–222
107. Grinstein S, Cohen S, Sarkadi B, Rothstein A (1983) Induction of ^{86}Rb fluxes by Ca^{2+} and volume changes in thymocytes and their isolated membranes. J Cell Physiol 116:352–362
108. Grinstein S, DuPre A, Rothstein A (1982) Volume regulation by human lymphocytes. Role of calcium. J Gen Physiol 79:849–868
109. Grinstein S, Goetz JD, Cohen S, Furuya W, Rothstein A, Gelfand EW (1985) Mechanism of regulatory volume increase in osmotically shrunken lymphocytes. Mol Physiol 8:185–198
110. Grinstein S, Goetz JD, Cohen S, Rothstein A, Gelfand EW (1985) Regulation of Na^+/H^+ exchange in lymphocytes. Ann NY Acad Sci 456:207–219
111. Grinstein S, Goetz JD, Rothstein A (1984) $^{22}Na^+$ fluxes in thymic lymphocytes. II. Amiloride-sensitive Na^+/H^+ exchange pathway; reversibility of transport and asymmetry of the modifier site. J Gen Physiol 84:585–600
112. Grinstein S, Goetz-Smith JD, Stewart D, Beresford BJ, Mellors A (1986) Protein phosphorylation during activation of Na^+/H^+ exchange by phorbol esters and by osmotic shrinking. Possible relation to cell pH and volume regulation. J Biol Chem 261:8009–8016
113. Grinstein S, Rothstein A (1986) Mechanisms of regulation of the Na^+/H^+ exchanger. J Membr Biol 90:1–12
114. Grinstein S, Rothstein A, Sarkadi B, Gelfand EW (1984) Responses of lymphocytes to anisotonic media: volume-regulating behavior. Am J Physiol 246:C204–C215

115. Grygorczyk R, Schwarz W, Passow H (1984) Ca^{2+}-activated K^+ channels in human red cells. Comparison of single-channel currents with ion fluxes. Biophys J 45:693–698
116. Gstrein E, Paulmichl M, Lang F (1987) Electrical properties of Ehrlich ascites tumor cells. Pflugers Arch 408:432–437
117. Guharay F, Sachs F (1984) Stretch-activated single ion channel currents in tissue-cultured embryonic chick skeletal muscle. J Physiol (Lond) 352:685–701
118. Haas M (1989) Properties and diversity of (Na–K–Cl) cotransporters. Annu Rev Physiol 51:443–457
119. Haas M, Forbush B III (1987) Photolabeling of a 150-kDa (Na + K + Cl) cotransport protein from dog kidney with a bumetanide analogue. Am J Physiol 253:C243–C250
120. Haas M, Forbush B III (1988) Photoaffinity labelling of a 150-kDa (Na + K + Cl) cotransport protein from duck red cells with an analog of bumetanide. Biochim Biophys Acta 939:131–144
121. Haas M, McManus TJ (1983) Bumetanide inhibits (Na + K + 2Cl) cotransport at a chloride site. Am J Physiol 245:C235–C240
122. Haas M, McManus TJ (1985) Effect of norepinephrine on swelling-induced potassium transport in duck red cells. Evidence against a volume-regulatory decrease under physiological conditions. J Gen Physiol 85:649–667
123. Haas M, Schmidt WF III, McManus TJ (1982) Catecholamine-stimulated ion transport in duck red cells. Gradient effects in electrically neutral (Na + K + 2Cl) cotransport. J Gen Physiol 80:125–147
124. Hall AC, Ellory JC (986) Effects of high hydrostatic pressure on 'passive' monovalent cation transport in human red cells. J Membr Biol 94:1–17
125. Hamill OP (1983) Potassium and chloride channels in red blood cells. In: Sakmann B, Neher E (eds) *Single-channel recording.* Plenum, New York, pp 451–471
126. Hannafin J, Kinne-Saffran E, Friedman D, Kinne R (1983) Presence of a sodium-potassium chloride contransport system in the rectal gland of *Squalus acanthias.* J Membr Biol 75:73–83
127. Harvey J, Ehrenfeld J (1988) Epithelial pH and ion transport regulation by proton pumps and exchangers. In: Proton passage across cell membrances. Ciba Found Symp 139:139–164
128. Hazama A, Okada Y (1988) Ca^+ sensitivity of volume-regulatory K^+ and Cl^- channels in cultured human epithelial cells. J Physiol (Lond) 402:687–702
129. Hebert SC (1986) Hypertonic cell volume regulation in mouse thick limbs. I. ADH dependency and nephron heterogeneity. Am J Physiol 250:C907–C919
130. Hebert SC (1986) Hypertonic cell volume regulation in mouse thick limbs. II. Na^+–H^+ and Cl^-–HCO_3^- exchange in basolateral membranes. Am J Physiol 250:C920–C931
131. Heinz E, Geck P, Pietrzyk C (1975) Driving forces of amino acid transport in animal cells. Ann NY Acad Sci 264: 428–441
132. Hellman SI, Nagel W, Fisher RS (1979) Ouabain on active transepithelial Na transport in frog skin: studies with microelectrodes. J Gen Physiol 74:105–127
133. Hempling HG, Thompson S, Dupre A (1977) Osmotic properties of human lymphocyte. J Cell Physiol 93:293–302
134. Hendil KB, Hoffmann EK, (1974) Cell volume regulation in Ehrlich ascites tumor cells. J Cell Physiol 84:115–125
134a. Hladky SB, Rink TJ (1977) pH equilibrium across the red cell membrane. In: Ellory JC, Lev VL (eds) Membrane transport in red cells. Academic, London, pp 115–135
135. Hodgkin AL, Keynes RD (1955) The potassium permeability of a giant nerve fibre. J Physiol (Lond) 128:61–88
136. Hoffmann EK (1978) Regulation of cell volume by selective changes in the leak permeabilities of Ehrlich ascites tumor cells. In: Jørgensen CB, Skadhauge E (eds) *Alfred Benzon Symposium XI. Osmotic and volume regulation.* Munksgaard, Coppenhagen, pp 397–417
137. Hoffmann EK (1983) Volume regulation by animal cells. In: Cossins AR, Sheterline P (eds) *Cellular acclimatisation to environmental change.* Cambridge University Press, Cambridge, (Soc Exp Biol Semin Ser 17) pp 55–80
138. Hoffmann EK (1985) Role of separate K^+ and Cl^- channels and of Na^+/Cl^- cotransport in volume regulation in Ehrlich cells. Fed. Proc. 44:2513–2519
139. Hoffmann EK, Hendil KB (1976) The role of amino acids and taurine in isosmotic intracellular regulation in Ehrlich ascites mouse tumour cells. J Comp Physiol 108:279–286
140. Hoffmann EK, Kolb A (1991) The mechanisms of activation of regulatory volume responses after cell swelling. In: Gilles R, Bolis L, Hoffmann E (eds) Volume and Osmolality Control in animal cells. ACEP series Vol. 9, Springer, Berlin Heidelberg New York, pp 140–177

141. Hoffmann EK, Lambert IH, (1983) Amino acid transport and cell volume regulation in Ehrlich ascites tumour cells. J Physiol (Lond) 338:613–625
142. Hoffmann EK, Lambert IH, Simonsen LO (1986) Separate, Ca^{2+}-activated K^+ and Cl^- transport pathways in Ehrlich ascites tumor cells. J Membr Biol 91:227–224
143. Hoffmann EK, Simonsen LO (1989) Membrane mechanisms in volume and pH regulation in vertebrate cells. Physiol Rev 69:315–382
144. Hoffmann EK, Simonsen LO, Lambert IH (1984) Volume-induced increase of K^+ and Cl^- permeabilities in Ehrlich ascites tumor cells. Role of internal Ca^{2+}. J Membr Biol 78:211–222
144a. Hoffmann EK, Schiødt M, Dunham PB (1986) The number of chloride-cation cotransport sites on Ehrlich ascites cells measured with ^3H bumetanide. Am J Physiol 250 (Cell Physiol 19):C688–C693
145. Hoffmann EK, Lambert IH, Simonsen LO (1988) Mechanisms in volume regulation in Ehrlich ascites tumor cells. Renal Physiol Biochem 11:221–247
146. Hoffmann EK, Simonsen LO, Sjøholm C (1979) Membrane potential, chloride exchange, and chloride conductance in Ehrlich mouse ascites tumour cells. J Physiol (Lond) 296:61–84
147. Hoffmann EK, Sjøholm C, Simonsen LO, (1983) Na^+,Cl^- cotransport in Ehrlich ascites tumor cells activated during volume regulation (regulatory volume increase). J Membr Biol 76:269–280
148. Horisberger J-O, Giebisch G (1988) Voltage dependence of the basolateral membrane conductance in the *Amphiuma* collecting tubule. J Membr Biol 105:257–263
149. Horowicz P, Gage PW, Eisenberg RS (1968) The role of the electrochemical gradient in determining potassium fluxes in frog striated muscle. J Gen Physiol 51 [Suppl 5(2)]:193s–203s
150. Hudson RL, Schultz SG (1988) Sodium-coupled glycine uptake by Ehrlich ascites tumor cells results in an increase in cell volume and plasma membrane channel activities. Proc Natl Acad Sci USA 85:279–283
151. Jacobs MH, Stewart DR (1942) The rôle of carbonic anhydrase in certain ionic exchanges involving the erythrocyte. J Gen Physiol 25:539–552
152. Jennings ML, Douglas SM, McAndrew PE (1986) Amiloride-sensitive sodium-hydrogen exchange in osmotically shrunken rabbit red blood cells. Am J Physiol 251:C32–C40
153. Jensen BS, Hoffmann EK (1990) Na^+, Cl^- or Na^+, K^+, $2Cl^-$ on the K^+-dependence of the cation-anion cotransport system in Ehrlich ascides human cells. Acta Physiol Scand 140(1):34A
154. Jessen F, Hoffmann EK, Aabin B (1989) Functionally reconstitution of an isolated K^+ channel from Ehrlich ascites tumor cells. Acta Phsiol Scand 136:10A
155. Kaji D, Kahn T (1985) Kinetics of Cl-dependent K influx in human erythrocytes with and without external Na: effect of NEM. Am J Physiol 249:C490–C496
156. Kawahara K, Hunter M, Giebisch G (1987) Potassium channels in Necturus proximal tubule. Am J Physiol 253:F488–F494
156a. Kimelberg HK, Frangakis MV (1986) Volume regulation in primary astrocyte cultures. Adv Biosci 61:177–186
157. King PA, Goldstein L (1983) Organic osmolytes and cell volume regulation in fish. Mol Physiol 4:53–66
158. Kinne R (1988) Sodium cotransport systems in epithelial secretion. Comp Biochem Physiol 90A:721–727
159. Kirk KL, DiBona DR, Schafer JA (1987a) Regulatory volume decrease in perfused proximal nephron: evidence for a dumping of cell K^+. Am J Physiol 252:F933–F942
160. Kirk KL, Schafer JA, DiBona DR (1987b) Cell volume regulation in rabbit proximal straight tubule perfused in vitro. Am J Physiol 252:F922–F932
161. Klaven NB, Pershadsingh HA, Henius GV, Laris PC, Long JW Jr, McDonald JM (1983) A high-affinity, calmodulin-sensitive (Ca^{2+} + Mg^{2+})-ATPase and associated calcium-transport pump in the Ehrlich ascites tumor cell plasma membrane. Arch Biochem Biophys 226:618–628
162. Knepper MA (1982) Measurement of osmolality in kidney slices using vapor pressure osmometry. Kidney Int 21:653–655
163. Kramhøft B, Lambert IH, Hoffmann EK, Jørgensen F (1986) Activation of Cl-dependent K transport in Ehrlich ascites tumor cells. Am J Physiol 251:C369–C379
164. Kregenow FM (1971) The response of duck erythrocytes to nonhemolytic hypotonic media. Evidence for a volume-controlling mechanism. J Gen Physiol 58:372–395
165. Kregenow FM (1971) The response of duck erythrocytes to hypertonic media. Further evidence for a volume-controlling mechanism. J Gen Physiol 58:396–412

166. Kregenow FM (1973) The response of duck erythrocytes to norepinephrine and an elevated extracellular potassium: volume regulation in isotonic media. J Gen Physiol 61:509–527
167. Kregenow FM (1974) Functional separation of the Na-K exchange pump from the volume controlling mechanism in enlarged duck red cells. J Gen Physiol 64:393–412
168. Kregenow FM (1981) Osmoregulatory salt transporting mechanisms: control of cell volume in anisotonic media. Annu Rev Physiol 43:493–505
169. Kregenow FM, Caryk T (1979) Co-transport of cations and Cl during the volume regulatory responses of duck erythrocytes. Physiologist 22:73
170. Kregenow FM, Caryk T, Siebens AW (1985) Further studies of the volume-regulatory response of *Amphiuma* red cells in hypertonic media. Evidence for amiloride-sensitive Na/H exchange. J Gen Physiol 86: 565–584
171. Kregenow FM, Robbie DE, Orloff J (1976) Effect of norepinephrine and hypertonicity on K influx and cyclic AMP in duck erythrocytes. Am J Physiol 231:306–312
172. Kristensen LØ (1986) Associations between transports of alanine and cations across cell membrane in rat hepatocytes. Am J Physiol 251:G575–G584
173. Kristensen LØ, Folke M (1984) Volume-regulatory K^+ efflux during concentrative uptake of alanine in isolated rat hepatocytes. Biochem J 221:265–268
174. Kristensen P (1983) Exchange diffusion, electrodiffusion and rectification in the chloride transport pathway of frog skin. J Membr Biol 72:141–151
175. Kristensen P, Larsen EH (1978) Relation between chloride exchange diffusion and a conductive chloride transport pathway across the isolated skin of the toad (Bufo bufo). Acta Physiol Scand 102:22–34
176. Kristensen P, Ussing HH (1985) Epithelial organization. In: Seldin DW, Giebisch G (eds) *The kidney: physiology and pathophysiology.* Raven, New York, pp 173–188
177. Krogh A (1946) The active and passive exchanges of inorganic ions through the surfaces of living cells and through living membranes generally. Proc R Soc Lond [Biol] 133:140–200
178. Lambert IH (1984) Na^+-dependent taurine uptake in Ehrlich ascites tumour cells. Mol Physiol 6:233–246
179. Lambert IH (1985) Taurine transport in Ehrlich ascites tumour cells. Specificity and chloride dependence. Mol Physiol 7:323–332
180. Lambert IH (1987) Effect of arachidonic acid, fatty acids, prostaglandins, and leukotrienes on volume regulation in Ehrlich ascites tumor cells. J Membr Biol 98:207–221
181. Lambert IH (1987) Activation of the volume- or ionophore A23187 plus Ca^{2+}-induced Cl^- transport pathway in Ehrlich mouse ascites tumour cells is mediated via leucotriene-D_4. J Physiol (Lond) 394:85P
181a. Lambert IH, Hoffmann EK (1990) Taurine transport and cell volume regulation in a mammalian cell. In: Posantes-Morales H. Martin DL, Shain W, Martin del Rio R. (eds) Taurine; functional neurochemistry, physiology and cardiology. Wiley-Liss, New York pp 267–276
182. Lambert IH, Hoffmann EK, Christensen P (1987) Role of prostaglandins and leukotrienes in volume regulation by Ehrlich ascites tumor cells. J Membr Biol 98:247–256
183. Lambert IH, Hoffmann EK, Jørgensen F (1987) Membrane potential, anion and cation conductances in Ehrlich asites tumor cells. J Membr Biol 111:113–132
184. Lang F, Messner G, Rehwald W (1986) Electrophysiology of sodium-coupled transport in proximal renal tubules. Am J Physiol 250:F953–F962
185. Lang F, Paulmichl M, Voelkl H, Gstrein E, Friedrich F (1987) Electrophysiology of cell volume regulation. In: Kovacevic Z, Guder WG (eds) *Molecular nephrology. Biochemical aspects of kidney function.* De Gruyter, Berlin, pp 133–139
186. Lang F, Völke H, Häussinger D (1990) General principles in cell volume regulation. In: Beyenbach KW (ed) *Cell volume regulation.* Karger, Basel, pp 1–26
187. Lansman JB, Hallam TJ, Rink TJ (1987) Single stretch-activated ion channels in vascular endothelial cells as mechanotransducers? Nature 325:811–813
188. Larsen EH (1973) Effect of amiloride, cyanide and ouabain on the active transport pathway in toad skin. In: Ussing HH, Torn NA (eds) Transport mechanisms in epithelia. Munksgaard, Copenhagen, pp 131–147 (Proceedings of the Alfred Benton symposium, vol 5)
189. Larson M, Spring KR (1984) Volume regulation by *Necturus* gallbladder: basolateral KCl exit. J Membr Biol 81:219–232
190. Larson M, Spring KR (1987) Volume regulation in epithelia. Curr Top Membr Transp 30:105–123

191. Lau KR, Hudson RL, Schultz SG (1984) Cell swelling increases a barium-inhibitable potassium conductance in the basolateral membrane of *Necturus* small intestine. Proc Natl Acad Sci USA 81:3591–3594
192. Lauf PK (1982) Evidence for chloride dependent potassium and water transport induced by hyposmotic stress in erythrocytes of the marine teleost, *Opsanus tau*. J. Comp Physiol 146: 9–16
193. Lauf PK (1983) Thiol-dependent passive K/Cl transport in sheep red cells: II. Loss of Cl^- and N-ethylmaleimide sensitivity in maturing high K^+ cells. J Membr Biol 73:247–256
194. Lauf PK (1985a) On the relationship between volume- and thiol-stimulated K^+Cl^- fluxes in red cell membranes. Mol Physiol 8:215–234
195. Lauf PK (1985b) K^+: Cl^- cotransport: sulphydryls, divalent cations, and the mechanism of volume activation in a red cell. J Membr Biol 88:1–13
196. Lauf PK, Adragna NC, Garay RP (1984) Activation by N-ethylmaleimide of a latent K^+–Cl^- flux in human red blood cells. Am J Physiol 246:C385–C390
197. Lauf PK, Mangor-Jensen A (1984) Effects of A23187 and Ca^{2+} on volume- and thiol-stimulated, ouabain-resistant K^+Cl^- fluxes in low K^+ sheep erythrocytes. Biochem Biophys Res Commun 125:790–796
198. Lauf PK, McManus TJ, Haas M, Forbush B III, Duhm J, Flatman PW, Saier MH Jr, Russel JM (1987) Physiology and biophysics of chloride and cation cotransport across cell membranes. Fed Proc 46:2377–2394
199. Lauf PK, Theg BE (1980) A chloride dependent K^+ flux induced by N-ethylmaleimide in genetically low K^+ sheep and goat erythrocytes. Biochem Biophys Res Commun 92:1422–1428
200. Law R (1985) Volume regulation by mammalian renal cells exposed to anisosmotic media. Mol Physiol 8:143–160
201. Law RO, Turner DPJ (1987) Are ninhydrin-positive substances volume-regulatory osmolytes in rat renal papillary cells? J Physiol (Lond) 386:45–61
202. Leaf A (1959) Maintenance of concentration gradients and regulation of cell volume. Ann NY Acad Sci 72:396–404
203. Leite MV, Goldstein L. (1987) Ca^{2+} ionophore and phorbol ester stimulate taurine efflux from skate erythrocytes. J Exp Zool 242:95–97
204. Lew VL, Bookchin RM (1986) Volume, pH, and ion-content regulation in human red cells: analysis of transient behavior with an integrated model. J Membr Biol 92:57–74
205. Lewis SA, Donaldsen P (1990) Ion channels and cell volume regulation: chaos in an organized system. News Physiol Sci 5:112–119
206. Levinson C (1985) Sodium-dependent ion cotransport in steady-state Ehrlich ascites tumor cells. J Membr Biol 87:121–130
207. Levinson C (1987) Volume regulatory activity of the Ehrlich ascites tumor cell and its relationship to ion transport. J Membr Biol 100:183–191
208. Lewis SA, de Moura JLC (1982) Incorporation of cytoplasmic vesicles into apical membrane of mammalian urinary bladder epithelium. Nature 297:685–688
209. Linshaw MA (1980) Effect of metabolic inhibitors on renal tubule cell volume. Am J Physiol 239:F571–F577
210. Linshaw MA, Stapleton FB (1978) Effect of ouabain and colloid osmotic pressure on renal tubule cell volume. Am J Physiol 235:F480–F491
211. Lise B, de Rouffinac C (1985) Urinary concentrating ability: insights from comparative anatomy. Am J Physiol 249:R643–R666
212. Livne A, Grinstein S, Rothstein A (1987) Volume-regulating behavior of human platelets. J Cell Physiol 131:354–363
213. Livne A, Hoffmann EK (1990) Cytoplasmic acidification and activation of Na^+/H^+ exchange during regulatory volume decrease in Ehrlich ascites tumour cells. J Membr Biol 114:153–157
214. Logue P, Anderson C, Kanik C, Farquharson B, Dunham P (1983) Passive potassium transport in LK sheep red cells: modification by N-ethyl maleimide. J Gen Physiol 81:861–885
215. Lohr JW, Grantham JJ (1986) Isovolumetric regulation of isolated S_2 proximal tubules in anisotonic media. J Clin Invest 78:1165–1172
216. Lytle C, McManus TJ (1987) Effect of loop diuretics and stilbene derivatives on swelling-induced K–Cl cotransport. J Gen Physiol 90:28a–29a
217. MacKnight ADC (1983) Volume regulation in mammalian kidney cells. Mol Physiol 4:17–31
218. MacKnight ADC (1984) Cellular response to injury. In: Staub NC, Taylor AE (eds) Edema. Raven, New York, pp 489–520

219. MacKnight ADC, Leaf A (1985) Cellular responses to extracellular osmolality. In: Seldin DW, Giebisch G (eds) *The kidney: physiology and pathophysiology.* Raven, New York, pp 117–132
220. MacRobbie EAC, Ussing HH (1961) Osmotic behaviour of the epithelial cells of frog skin. Acta Physiol Scand 53:348–365
221. McConnell FM, Goldstein L (1987) Hypotonic stress stimulates the inositol phosphate cycle in skate erythrocytes. J Gen Physiol 90:29a
222. McManus TJ (1982) Catecholamine-stimulated and volume-sensitive ion movements in avian red cells: alternate modes of chloride-dependent cation transport. Jacobs-Parpart-Ponder Memorial Lecture. Red Cell Club meeting. April 19, 1982
223. McManus TJ, Haas M, Starke LC, Lytle CY (1985) The duck red cell model of volume-sensitive chloride-dependent cation transport. Ann NY Acad Sci 456:183–186
224. McManus TJ, Schmidt WF III (1978) Ion and co-ion transport in avian red cells. In: Hoffman JF (ed) *Membrane transport processes,* Vol 1. Raven, New York, pp 79–106
225. McRoberts JA, Erlinger, S. Rindler, MJ, Saier MH Jr (1982) Furosemide-sensitive salt transport in the Madin-Darby canine kidney cell line: evidence for the cotransport of Na^+, K^+, and Cl^-. J Biol Chem 257:2260–2266
226. Mercer RW, Hoffman JF (1985) Bumetanide-sensitive Na/K cotransport in ferret red blood cells. Biophys J 47:157a
227. Mills JW (1987) The cell cytoskeleton: possible role in volume control. Curr Top Membr Transp 30:75–101
228. Mills JW, Lubin M (1986) Effect of adenosine 3′,5′-cyclic monophosphate on volume and cytoskeleton of MDCK cells. Am J Physiol 250:C319–C324
229. Mills JW, Skiest DJ (1985) Role of cyclic AMP and the cytoskeleton in volume control in MDCK cells. Mol Physiol 8:247–262
230. Moolenaar WH (1986) Effects of growth factors on intracellular pH regulation. Annu Rev Physiol 48:363–376
231. Moolenaar WH, Defize LHK, van der Saag PT, de Laat SW (1986) The generation of ionic signals by growth factors. Curr Top Membr Transp 26:137–156
232. Morris CE (1990) Mechanosensitive ion channels. J Membr Biol 113:93–107
233. Musch MW, Field M (1989) K-independent Na–Cl cotransport in bavine tracheal epithelial cells. Am J Physiol 256:C658–C665
234. Nagel W (1976) The intracellular electrical potential profile of frog skin epithelium. Pflugers Arch 365:135–143
235. Nagel W (1979) Inhibition of potassium conductance by barium in frog skin epithelium. Biochim Biophys Acta 552:346–357
236. Nakanishi T, Balaban RS, Burg MB (1988) Survey of osmolytes in renal cell lines. Am J Physiol 255:C181–C191
237. Nauntofte B, Poulsen JH (1984) Chloride transport in rat parotid acini: furosemide-sensitive uptake and calcium-dependent release. J Physiol (Lond) 357:61 P
238. Nauntofte B, Poulsen JH (1986) Effects of Ca^{2+} and furosemide on Cl^- transport and O_2 uptake in rat parotid acini. Am J Physiol 251:C175–C185
239. Nellans HN, Frizzell RA, Schultz SG (1973) Coupled sodium-chloride influx across the brush border of rabbit ileum. Am J Physiol 225:467–475
240 Nielsen R (1985) Ba^{2+}-induced changes in the Na^+ and K^+ permeability of the isolated frog skin. Acta Physiol Scand 124:61–70
241. Nielsen R (1990) Prostaglandin E_2 enhances the sodium conductance of exocrine glands in isolated frog skin (Rana Esculenta). J Membr Biol 113:31–38
242. O'Brien TG, Krzeminski K (1983) Phorbol ester inhibits furosemide-sensitive potassium transport in BALB/c 3T3 preadipose cells. Proc Natl Acad Sci USA 80:4334–4338
243. O'Grady SM, Palfrey HC, Field M (1987) Characteristics and functions of Na–K–Cl cotransport in epithelial tissues. Am J Physiol 253:C177–C192
244. Okada Y, Yada T, Ohno-Shosaku T, Oiki S (1987) Evidence for the involvement of calmodulin in the operation of Ca-activated K channels in mouse fibroblasts. J Membr Biol 96:121–128
245. Owen DG, Segal M, Barker JL (1984) A Ca-dependent Cl^- conductance in cultured mouse spinal neurones. Nature 311:567–570
246. Palfrey HC, Feit PW, Greengard P (1980) cAMP-stimulated cation cotransport in avian erythrocytes: inhibition by "loop" diuretics. Am J Physiol 238:C139–C148
246a. Palfrey HC, Greengard P, Feit PW (1980) Specific inhibition by "loop" diuretics of an anion-dependent $Na^+ + K^+$ cotransport system in avian erythrocytes. Ann NY Acad Sci 341: 134–138

247. Palfrey HC, Greengard P (1981) Hormone-sensitive ion transport systems in erythrocytes as models for epithelial ion pathways. Ann NY Acad Sci 372:291–308
248. Palfrey HC, Rao MC (1983) Na/K/Cl co-transport and its regulation. J Exp Biol 106:43–54
249. Parker JC (1973) Dog red blood cells. Adjustment of salt and water content in vitro. J Gen Physiol 62:147–156
250. Parker JC (1977) Solute and water transport in dog and cat red blood cells. In: Ellory JC, Lew VL (ed) *Membrane transport in red cells.* Academic, London, pp 427–465
251. Parker JC (1981) Heterogeneity among dog red blood cells. J Gen Physiol 78:141–150
252. Parker JC (1983a) Volume-responsive sodium movements in dog red cells. Am J Physiol 244:C324–C330
253. Parker JC (1983b) Hemolytic action of potassium salts on dog red blood cells. Am J Physiol 244:C313–C317
254. Parker JC (1983c) Passive calcium movments in dog red blood cells: anion effects. Am J Physiol 244:C318–C323
255. Parker JC (1983d) Volume-responsive sodium movements in dog red blood cells. Am J Physiol 244:C324–C330
256. Parker JC (1984) Glutaraldehyde fixation of sodium transport in dog red blood cells. J Gen Physiol 84:789–803
257. Parker JC (1986) Na-proton exchange in dog red blood cells. Curr Top Membr Transp 26:101–114
258. Parker JC (1986) Interactions of lithium and protons with the sodium-proton exchanger of dog red blood cells. J Gen Physiol 87:189–200
259. Parker JC (1988) Volume-activated transport systems in dog red blood cells. Comp Biochem Physiol [A] 90A:539–542
260. Parker JC, Castranova V (1984) Volume-responsive sodium and proton movements in dog red blood cells. J Gen Physiol 84:379–401
261. Parker JC, Gitelman HJ, Glosson PS, Leonard DL (1975) Role of calcium in volume regulation by dog red blood cells. J Gen Physiol 65:84–96
262. Parker JC, Hoffman JF (1965) Interdependence of cation permeability, cell volme, and metabolism in dog red blood cells. Fed Proc 24:589
263. Petersen OH (1988) The control of ion channels and pumps in exocrine acinar cells. Comp Biochem Physiol [A] 90A:717–721
264. Petersen OH, Gallacher DV (1988) Electrophysiology of pancreatic and salivary acinar cells. Annu Rev Physiol 50:65–80
265. Pierce SK (1982) Invertebrate cell volume control mechanism: coordinated use of intracellular amino acids and inorganic ions as osmotic solute. Biol Bull 163:405–419
266. Pierce SK, Politis AD (1990) Ca^{2+}-activated cell volume recovery mechanisms. Annu Rev Physiol 52:27–42
267. Pine MB, Rhodes D, Thorp K, Tsai Y (1979) Anion exchange and volume regulation during metabolic blockade of renal cortical slices. J Physiol (Lond) 297:387–403
268. Riddick DH, Kregenow FM, Orloff J (1971) The effect of norepinephrine and dibutyryl cyclic adenosine monophosphate on cation transport in duck erythrocytes. J Gen Physiol 57:752–766
269. Rink TJ, Sanchez A, Grinstein S, Rothstein A (1983) Volume restoration in osmotically swollen lymphocytes does not involve changes in free Ca^{2+} concentration. Biochim Biophys Acta 762:593–596
270. Rossi JPFC, Schatzmann HJ (1982) Is the red cell calcium pump electrogenic? J Physiol (Lond) 327:1–15
271. Roti Roti LW, Rothstein A (1973) Adaptation of mouse leukemic cells (L5178Y) to anisotonic media. I. Cell volume regulation. Exp Cell Res 79:295–310
272. Roy G, Sauvé R (1987) Effect of anisotonic media on volume, ion and amino-acid content and membrane potential of kidney cells (MDCK) in culture. J Membr Biol 100:83–96
272a. Rugolo M, Mastocola T, Flamigni A, Lenaz G (1989) Chloride transport in human fibroblasts is activated by hypotonic shock. Biochem Biophys Res Commun 160(3):1330–1338
273. Sachs F (1987) Baroreceptor mechanisms at the cellular level. Fed Proc 46:12–16
274. Sackin H (1987) Stretch-activated potassium channels in renal proximal tubule. Am J Physiol 253:F1253–F1262
275. Sackin H (1989) A stretch-activated K^+ channel sensitive to cell volume. Proc Natl Acad Sci USA 86:1731–1735

276. Sackin H, Palmer LG (1987) Basolateral potassium channels in renal proximal tubule. Am J Physiol 253:F476–F487
277. Saier MH Jr, Boyden DA (1984) Mechanism, regulation and physiological significance of the loop diuretic-sensitive NaCl/KCl symport system in animal cells. Mol Cell Biochem 59:11–32
278. Saltin B, Sjøgaard G, Strange S, Juel C (1987) Redistribution of K^+ in the human body during muscular exercise; its role to maintain whole body homeostasis. In: Shirakik, Yousef MK (eds) *Man in stressful environments. Thermal and work physiology.* Thomas, Springfield, pp 247–267
279. Sarkadi B, Cheung R, Mack E, Grinstein S, Gelfand EW, Rothstein A (1985) Cation and anion transport pathways in volume regulatory response of human lymphocytes to hyposmotic media. Am J Physiol 248:C480–C487
280. Sarkadi B, Mack E, Rothstein A (1984a) Ionic events during the volume response of human peripheral blood lymphocytes to hypotonic media. I. Distinctions between volume-activated Cl^- and K^+ conductance pathways. J Gen Physiol 83:497–512
281. Sarkadi B, Mack E, Rothstein A (1984b) Ionic events during the volume response of human peripheral blood lymphocytes to hypotonic media. II. Volume- and time-dependent activation and inactivation of ion transport pathways. J Gen Physiol 83:513–527
282. Sauvé R, Simoneau C, Monette R, Roy G (1986) Single-channel analysis of the potassium permeability in HeLa cancer cells: evidence for a calcium-activated potassium channel of small unitary conductance. J Membr Biol 92:269–282
283. Scharff O, Foder B, Skibsted U (1983) Hysteretic activation of the Ca^{2+} pump revealed by calcium transients in human red cells. Biochim Biophys Acta 730:295–305
284. Schmidt WF III, McManus TJ (1977) Ouabain-insensitive salt and water movements in duck red cells. I. Kinetics of cation transport under hypertonic conditions. J Gen Physiol 70:59–79
285. Schmidt WF III, McManus TJ (1977) Ouabain-insensitive salt and water movements in duck red cells. II. Norepinephrine stimulation of sodium plus potassium cotransport. J Gen Physiol 70:81–97
286. Schmidt WF III, McManus TJ (1977) Ouabain-insensitive salt and water movements in duck red cells. III. The role of chloride in the volume response. J Gen Physiol 70:99–121
287. Schultz SG (1989) Intracellular sodium activities and basolateral membrane potassium conductances of sodium-absorbing epithelial cells. Curr Top Membr Transp 34: 21–44
288. Schultz SG (1989) Volume preservation. Then and now . News Physiol Sci 4:169–172
289. Schultz SG, Hudson RL, Lapointe J-Y (1985) Electrophysiological studies of sodium cotransport in epithelia; towards a cellular model. Ann NY Acad Sci 456:127–135
290. Sekar MC, Hokin LE (1986) The role of phosphoinositides in signal transduction. J Membr Biol 89:193–210
291. Siebens AW (1985) Cellular volume control. In: Seldin DW, Giebisch G (eds) *The kidney: physiology and pathophysiology.* Raven, New York, pp 91–115
292. Siebens AW, Kregenow FM (1980) Analysis of amiloride-sensitive volume regulation in *Amphiuma* red cells. Fed Proc 39:379
293. Siebens AW, Kregenow FM (1985) Volume-regulatory responses of *Amphiuma* red cells in anisotonic media. The effect of amiloride. J Gen Physiol 86:527–564
294. Simonsen LO, Brown AM, Christensen S, Harbak H, Svane PC, Hoffmann EK (1990) Thrombin and bradykinin mimic the volume response induced by cell swelling in Ehrlich mouse ascites tumour cells. Renal Physiol Biochem 13:176
295. Simonsen LO, Hoffmann EK, Sjøholm C (1976) Chloride conductance of the Ehrlich ascites tumor cell membrane. FEBS symposium on the biochemistry of membrane transport. Zürich, 1976, p 225
296. Simmons NL (1984) Epithelial cell volume regulation in hypotonic fluids: studies using a model tissue culture renal epithelial cell system. Q J Exp Physiol 69:83–95
297. Smalley CE, Tucker EM, Dunham PB, Ellory JC (1982) Interaction of L antibody with low potassium-type sheep red cells: resolution of two separate functional antibodies. J Membr Biol 64:167–174
298. Spring KR, Siebens AW (1988) Solute transport and epithelial cell volume regulation. Comp Biochem Physiol [A] 90A:557–560
299. Sten-Knudsen O (1978) Passive transport processes. In: Giebisch G, Tosteson DC, Ussing HH (eds) *Membrane transport in biology. Concepts and models,* Vol 1. Springer, Berlin Heidelberg New York, pp 5–113
300. Stockbridge LL, French AS (1988) Stretch-activated cation channels in human fibroblasts. Biophys J 54:187–190

301. Stoddard JS, Jacobsen E, Helman SI (1985) Basolateral membrane chloride transport in isolated epithelia of frog skin. Am J Physiol 249:C318–C319
302. Sun A, Grossman EB, Lombard M, Hebert SC (1991) Vasopressin alters the mechanism of apical Cl^- entry from Na^+, Cl^- to Na^+, K^+, $2Cl^-$ cotransport in mouse medullary thick ascending limb. J Membr Biol 120:83–94
303. Thurston JH, Hauhart RE, Naccarato EF (1981) Taurine: possible role in osmotic regulation of mammalian heart. Science (NY) 214:1373–1374
304. Tosteson DC, Hoffman JF (1960) Regulation of cell volume by active cation transport in high and low potassium sheep red cells. J Gen Physiol 44:169–194
305. Ubl J, Murer H, Kolb H-A (1988) Ion channels activated by osmotic and mechanical stress in membranes of opossum kidney cells. J Membr Biol 104:223–232
306. Uchida S, Green N, Conn H, Triche T, Mims S, Burg MB (1987) High NaCl induces stable changes in phenotype and karyotype of renal cells in culture. Am J Physiol 253:C230–C242
307. Ussing HH (1960) Active and passive transport of the alkali metal ions. In: Ussing HH, Kruhøffer P, Hess Thaysen J, Thorn NA (eds) *The alkali metal ions in biology*. Berlin: Springer, Berlin Göttingen Heidelberg, pp 45–143 (p 67)
308. Ussing HH (1965) Transport of electrolytes and water across the isolated amphibian skin. The Harvey Lectures, Vol 59. Academic, New York, pp 1–30
309. Ussing HH (1982) Volume regulation of frog skin epithelium. Acta Physiol Scand 114:363–369
310. Ussing HH (1985) Volume regulation and basolateral co-transport of sodium, potassium, and chloride ions in frog skin epithelium. Pflugers Arch 405 [Suppl 1]:S2–S7
311. Ussing HH (1986) Epithelial cell volume regulation illustrated by experiments in frog skin. Renal Physiol 9: 38–46
312. Ussing HH (1987) Role of potassium in epithelial transport illustrated by experiments on frog skin epithelium. Curr Top Membr Transp 28:3–17
313. Ussing HH (1988a) Anion transport pathways in frog skin epithelium. In: Thorn A, Treiman M, Petersen OH (eds) Molecular mechanisms in secretion. Munksgaard, Copenhagen, pp 17–27
314. Ussing HH (1988b) Odd behaviour of Li^+ in frog skin transport. Comb Biochim Physiol [A] 90A:525–529
315. Ussing HH (182) Pathways for transport in epithelia. In: Corradino RA (ed) Functional Regulation at the cellular and molecular levels Elsevier/North Holland, Amsterdam, pp 285–297
316. Ussing HH (1990) Volume regulation of frog skin epithelium. In: Beyenbach KW (ed) Cell volume regulation. Comp Physiol, Vol 4. Basel, Karger, pp 87–113
317. Ussing HH, Eskesen K (1989) Mechanism of isotonic water transport in glands. Acta Physiol Scand 136: 443–454
318. Van den Bosch H (1980) Intracellular phospholipases A. Biochim Biophys Acta 604:191–246
319. Van Rossum GDV, Russo MA (1981) Ouabain-resistant mechanism of volume control and the ultra structural organization of liver slices recovering from swelling in vitro. J Membr Biol 59:191–209
319a. Vestergaard-Bogind B, Stampe P, Christophersen P (1985) Single-file diffusion through the Ca^{2+}-activated K^+ channel of human red cells. J Membr Biol 88:67–75
320. Vislie T (1983) Cell volume regulation in fish heart ventricles with special reference to taurine. Comp Biochem Physiol [A] 76A:507–514
320a. Voute C, Ussing HH (1970) Quantitative relation between hydrostatic pressure gradient, extracellular volume and active sodium transport in the epithelium of the frag skin (*Rana temporaria*) Exp Cell Res 62:375–383
321. Wade JV, Olson JP, Samson FE, Nelson SR, Pazdernik TL (1988) A possible role for taurine in osmoregulation within the brain. J Neurochem 51:740–745
322. Warnock DG, Greger R, Dunham PB, Benjamin MA, Frizzell RA, Field M, Spring KR, Ives HE, Aronson PS, Seifter I (1984) Ion transport processes in apical membranes of epithelia. Fed Proc 43:2473–2487
323. Welling PA, Linshaw MA, Sullivan LP (1985) Effect of barium on cell volume regulation in rabbit proximal straight tubules. Am J Physiol 249:F20–F27
324. Wiater LA, Dunham PB (1983) Passive transport of K^+ and Na^+ in human red blood cells: sulfhydryl binding agents and furosemide. Am J Physiol 245:C348–C356
325. Willumsen NJ, Larsen EH (1986) Membrane potentials and intracellular Cl^- activity of toad skin epithelium in relation to activation and deactivation of the transepithelial Cl^- conductance. J Membr Biol 94:173–190

326. Wong SME, Chase HS Jr (1986) Role of intracellular calcium in cellular volume regulation. Am J Physiol 250:C841–C852
327. Wong PY-K, Cheung WY (1979) Calmodulin stimulates human platelet phospholipase A_2. Biochem Biophys Res Commun 90:473–480
328. Yancey PH (1988) Osmotic effectors in kidneys of xeric and mesic rodents; corticomedullary distributions and changes with water availability. J Comp Physiol 158:369–380
329. Yancey PH, Clark ME, Hand SC, Bowlus RD and Somero GN (1982) Living with water stress: evolution of osmolyte systems. Science 217:1214–1222
330. Ørskov SL (1954) The potassium absorption by pigeon blood cells. A considerable potassium absorption by pigeon- and hen blood cells is observed when a hypertonic sodium chloride solution is added. Acta Physiol Scand 31:221–229

Subject Index

GPSR Compliance
The European Union's (EU) General Product Safety Regulation (GPSR) is a set of rules that requires consumer products to be safe and our obligations to ensure this.

If you have any concerns about our products, you can contact us on

ProductSafety@springernature.com

In case Publisher is established outside the EU, the EU authorized representative is:

Springer Nature Customer Service Center GmbH
Europaplatz 3
69115 Heidelberg, Germany

www.ingramcontent.com/pod-product-compliance
Lightning Source LLC
Chambersburg PA
CBHW072312240926
51846CB00011B/57
* 9 7 8 3 6 4 2 7 6 9 8 5 6 *